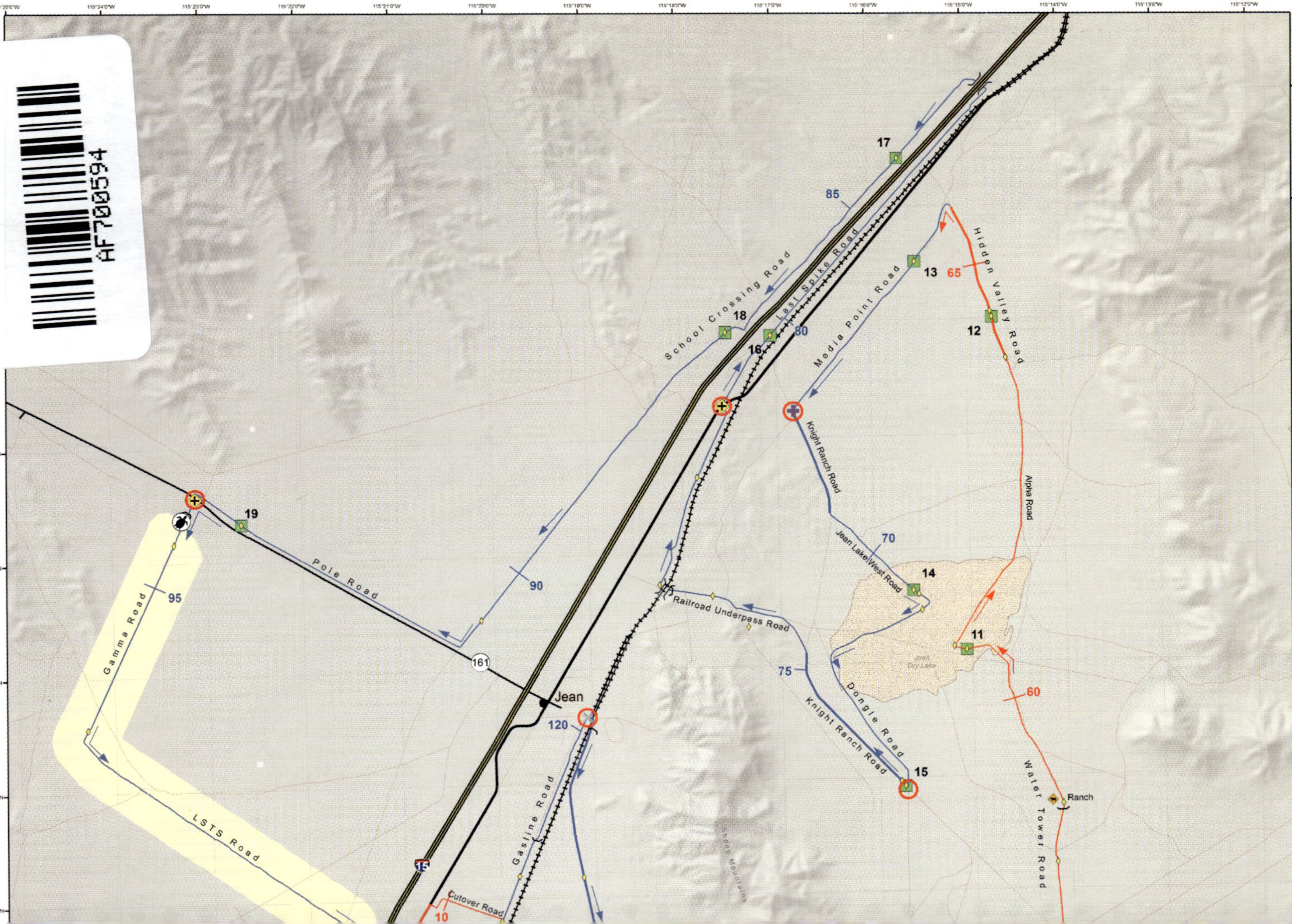
School Crossing Road
Last Spike Road
Media Point Road
Hidden Valley Road
Alpha Road
Knight Ranch Road
Jean Lake West Road
Railroad Underpass Road
Dongle Road
Knight Ranch Road
Pole Road
Gamma Road
LSTS Road
Gasline Road
Cutover Road
Water Tower Road
Ranch
Jean
161
15
10
60
65
70
75
80
85
90
95
120
11
12
13
14
15
16
17
18
19

Rory MacKean

Springer Tracts in Advanced Robotics

Volume 36

Editors: Bruno Siciliano · Oussama Khatib · Frans Groen

Martin Buehler, Karl Iagnemma and
Sanjiv Singh (Eds.)

The 2005 DARPA Grand Challenge

The Great Robot Race

Springer

Professor Bruno Siciliano, Dipartimento di Informatica e Sistemistica, Università di Napoli Federico II, Via Claudio 21, 80125 Napoli, Italy, E-mail: siciliano@unina.it

Professor Oussama Khatib, Robotics Laboratory, Department of Computer Science, Stanford University, Stanford, CA 94305-9010, USA, E-mail: khatib@cs.stanford.edu

Professor Frans Groen, Department of Computer Science, Universiteit van Amsterdam, Kruislaan 403, 1098 SJ Amsterdam, The Netherlands, E-mail: groen@science.uva.nl

Editors

Martin Buehler
Boston Dynamics
515 Massachusetts Avenue
Cambridge, MA 02139
USA
E-mail: buehler@bostondynamics.com

Karl Iagnemma
Department of Mechanical Engineering
Massachusetts Institute of Technology
77 Massachusetts Avenue
Cambridge, MA 02139
USA
E-mail: kdi@mit.edu

Sanjiv Singh
Carnegie Mellon University
Robotics Institute
5000 Forbes Avenue
Pittsburgh, PA 15213
USA
E-mail: ssingh@ri.cmu.edu

Library of Congress Control Number: 2007930554

ISSN print edition: 1610-7438
ISSN electronic edition: 1610-742X

Papers in this book first appeared in

- Special Issue on DARPA Grand Challenge, Part 1, Eds. Martin Buehler and Karl Iagnemma, Journal of Field Robotics, Volume 23, Issue 8, Pages 461–652(August 2006), Copyright 2006, John Wiley & Sons
- Special Issue on DARPA Grand Challenge, Part 2, Eds. Martin Buehler and Karl Iagnemma, Journal of Field Robotics, Volume 23, Issue 9, Pages 655–835(September 2006), Copyright 2006, John Wiley & Sons

Reprinted with permission of John Wiley & Sons, Inc.

ISBN-10 3-540-73428-7 Springer Berlin Heidelberg New York
ISBN-13 978-3-540-73428-4 Springer Berlin Heidelberg New York

Springer is a part of Springer Science+Business Media
springer.com

Printed in Germany

Typesetting: Digital data supplied by editor.
Data-conversion and production: SPS, Chennai, India
Cover-Design: Eric Kichner, Heidelberg
Printed on acid-free paper SPIN: 12082346 89/SPS 5 4 3 2 1 0

Foreword

At the dawn of the new millennium, robotics is undergoing a major transformation in scope and dimension. From a largely dominant industrial focus, robotics is rapidly expanding into the challenges of unstructured environments. Interacting with, assisting, serving, and exploring with humans, the emerging robots will increasingly touch people and their lives.

The goal of the new series of *Springer Tracts in Advanced Robotics (STAR)* is to bring, in a timely fashion, the latest advances and developments in robotics on the basis of their significance and quality. It is our hope that the wider dissemination of research developments will stimulate more exchanges and collaborations among the research community and contribute to further advancement of this rapidly growing field.

The volume edited by Martin Buehler, Karl Iagnemma and Sanjiv Singh presents a unique and extensive collection of the scientific results by the teams which took part into the DARPA Grand Challenge in October 2005 in the Nevada desert. This event reached an incredible peak of popularity in the media, the race of the century like someone called it! The Grand Challenge demonstrated the fast growing progress toward the development of robotics technology, as it showed the feasibility of using mobile robots operating autonomously in real world scenarios. As such, not only has it revealed the great potential for a number of field robotics applications in hostile or hazardous environments, but also it has set an unprecedented milestone for competitions and benchmarking which is likely to become a format adopted for future events of the same kind in other areas of robotics research.

The original papers were earlier published in two special issues of the Journal of Field Robotics. Our series is very proud to reprise them and offer archival publication as a special STAR volume!

Naples, Italy
June 2007

Bruno Siciliano
STAR Editor

Foreword

Foreword

Some called it the race of the century.

The Defense Advanced Research Projects Agency (DARPA) has a rich history of pursuing innovative technical ideas that lead to capabilities never before dreamed possible. DARPA's past projects resulted in the ARPANET (which led to the Internet), stealth, and the Predator and Global Hawk unmanned aerial vehicles.

In 2003, Dr. Tony Tether, DARPA Director, believed the time was right to push autonomous ground vehicle technology. The thinking was that if such vehicles were possible, they could be used in dangerous military missions and save the lives of young American men and women in uniform.

At that time, many experts believed fully autonomous ground vehicles that could travel great distances at speeds important to military operations were not possible within any near-term timeline. Hence, Dr. Tether created the DARPA Grand Challenge to accelerate autonomous ground vehicle technology as an experiment. The expectation was that only a few people would respond, but the Grand Challenge awakened a new surge of creativity among thousands from the United States and other countries

The first Grand Challenge was held in March 2004 and, from an initial field of 106 applicants, 15 teams competed in the final event. The best vehicle managed to travel 5 percent of the route before failing. It was clear then that the challenge was indeed "grand."

At the 2004 closing ceremony, DARPA announced another competition and doubled the prize to $2 million. October 2005 was chosen for the second competition—a mere 18 months from the first event. America's innovators again rolled up their sleeves and got to work. Eighteen months is not a long time to create technology that may change the world.

DARPA received 197 applications to compete in the 2005 event, and each of the 23 teams that made it to the final event had a vehicle that proved, through a series of difficult qualifying trials, to be better than the best vehicle in the 2004 race.

The 2005 course was tough: 132 miles of difficult desert roads across Nevada that contained a mixture of featureless terrain, dust, global positioning system drop-outs, sharp turns, narrow openings, bridges, railroad overpasses, long tunnels, obstacles, and a narrow winding mountain road with a 200-foot drop-off. The actual route was kept secret until 2 hours before the start.

Remarkably, five teams finished the course, four of them under the 10-hour limit and within 37 minutes of the winning time of Stanford's "Stanley" at 6 hours and 53 minutes and an average speed of 19.2 miles per hour. In another first for autonomous vehicle operations, "Terramax" finished the course on the second day after remaining parked overnight in autonomous mode. The vehicles that did not finish the course suffered mechanical, system, or software problems. In the end, 22 of the 23 teams traveled farther than the best vehicle did in 2004. By all accounts, Grand Challenge 2005 was a spectacular success.

The primary reason for the success of the Grand Challenge lies with the teams—the students, engineers, scientists, and backyard mechanics—all inventors who brought fresh ideas to solve a very difficult technical problem. They were individuals, but learned to work together in teams. Unlike other prize events, the DARPA Grand Challenge set a specific date for the competition, and all the teams spent countless long hours overcoming setbacks to be there for the final event.

The teams accelerated autonomous ground vehicle technology beyond expectation. Their vehicles have served to prove the technical feasibility and promote acceptance of unmanned autonomous ground vehicles within the Defense community, much as unmanned air vehicles have come to be accepted as essential partners in the air.

DARPA thanks all the Grand Challenge participants for their hard work and willingness to tackle a problem important to the Department of Defense. They proved that a challenge can fuel creativity and obtain accomplishments that prove conventional wisdom wrong. DARPA also thanks the staff and the many organizations that helped make the event a success. The excitement in their eyes often matched that of the competitors.

Ron Kurjanowicz
2005 DARPA Grand Challenge Program Manager

Introduction

On July 30, 2002 the Defense Advanced Research Projects Agency (DARPA) announced a "Grand Challenge" to robotics researchers, engineers, inventors, and hobbyists across the country: develop a mobile robot that can autonomously traverse a desert route of 140 miles from Barstow, California to Primm, Nevada on Saturday, March 13, 2004. The course would run over dirt roads, trails, lakebeds, rocky terrain, and gullies — and the team whose robot completed the course the fastest (within a 10 hour time limit) would receive 1 million dollars.

The fact that none of the 15 finalists completed more than 7.4 miles of the course is testimony to the difficulty of developing autonomous robots that are robust, perceptive, and intelligent enough to travel long distances in unstructured terrain. The competition also served as a wake-up call to robotics researchers, as it demonstrated that algorithms and robot systems that function perfectly in simulations or in the laboratory are often less effective in the real world.

The second DARPA Grand Challenge drew a larger pool of entries. Of the original 195 applicants, 43 teams were selected to participate in the National Qualification Event (NQE) held at California Speedway in Fontana from September 27 through October 5, 2004. Robot performances at the NQE were judged by (1) the elapsed time required to complete a short test course; (2) the number of obstacles successfully passed without contact; (3) the number of gates successfully passed. The results of the NQE were used to whittle the 43 teams down to 23 finalists to race in the second DARPA Grand Challenge.

On race day, October 8, 2005, the finalists competed on a twisting, unpaved course over 132 miles in the high desert near Primm, Nevada. Although no obstacles were placed in the vehicles' path, the route was often rugged and bumpy, and passed through tight tunnels and a narrow, twisting mountain pass. Remarkably, 22 of the 23 finalists traveled further than the 7.4 miles traversed by the most successful entry from 2004. Even more remarkably, five vehicles successfully completed the course. The Stanford Racing Team was awarded the 2 million dollar prize with a winning time of 6 hours, 53 minutes. Their achievement — and the achievement of all of the finalists — marks a significant milestone in robotics technology, and promises a bright future for the increased use of mobile robots in real world scenarios.

This book presents 15 technical papers describing 16 of the 23 vehicles that competed as finalists in the 2005 DARPA Grand Challenge. These papers originally appeared in two special issues of *The Journal of Field Robotics,* in September and October, 2006. They document the mechanical, algorithmic, and sensory solutions developed by the various teams. Also included is a new picture gallery of the participating robots, with a description of each team's race results and (where appropriate) failure descriptions, and a foreword by Ron Kurjanowitz, the DARPA program manager who oversaw the Grand Challenge contest.

Stanley, the winning robot, is described in the paper by Thrun et al. The Stanford team focused on software, applying new and existing machine learning and probabilistic techniques for long-range terrain perception, real-time collision avoidance, and stable vehicle control on slippery and rugged terrain. An important part of their successful strategy was to minimize custom hardware development to achieve robustness: their system was based on off-the-shelf sensors, standard Pentium computers, and a Touareg R5 Turbodiesel with support by a dedicated Volkswagen team.

The next paper, by Urmson et al., is a description of the two Carnegie Mellon robots that earned second and third place, with finishing times of 7 hours, 5 minutes, and 7 hours, 14 minutes — at the heels of Stanley. It describes their hardware, software and sensing systems, strategic approaches, vehicles, and testing. A detailed performance description and insightful failure analysis describes what went wrong, and more importantly, how the systems performed well despite several problems that occurred during the race.

Team Gray's entry is described in the paper by Trepagnier et al. This robot was the fourth to successfully complete the course, in a time of 7 hours and 30 minutes—less than 40 minutes after the winner. The paper describes practical issues related to Team Gray's vehicle, sensor, and actuator selection, sensor processing algorithms, software design, and low-level vehicle control. Analysis of Team Gray's field test results is also presented.

With its 30,000 pound weight and twelve cylinder, 425 hp engine, TerraMax was the heavy-weight entry in the DGC. Based on a 27 ft long commercial Oshkosh truck and barely fitting through a 9 x 9 ft tunnel, TerraMax was designed for robustness rather than speed. The vehicle was one of five to finish the course, after an all-night pause. The paper by Braid, Broggi and Schmiedel describes the TerraMax vehicle, its computers and sensors, and a rugged implementation of Collin Rockwell's intelligent Vehicle Management System (iVMS). In addition to multiple laser scanners, a forward-looking trinocular vision system provided sensing for obstacle and path detection. It relied on image disparity to estimate the average terrain slope and compute the free space in front of the vehicle.

The 'twin contenders', Cliff and Rocky, are described in a paper by a team of Virginia Tech engineering students. Both bots were halted by hardware problems after traversing approximately 40 miles, to place them at rank 8 and 9. Each implemented a different navigation strategy – a reactive one ("Dynamic Expanding Zones") for Cliff, and a more deliberative one ("Non-Uniform Terrain Search") for Rocky – and the pros and cons of each strategy for the DGC are evaluated.

Team Desert Buckeye's Intelligent Off-road Navigator (ION) came in 10th. Chen and Özgüner's paper details their integration of sensor fusion, navigation, vehicle control, signal processing, drive-by-wire technology, reliable software and mapping. ION's overall behavior was driven by a finite state machine that aimed to accommodate all possible situations and state transitions. The navigation module's main responsibility was to generate smooth and obstacle-free local paths with appropriate speed set points. The obstacle avoidance algorithm employed fuzzy controllers for both steering and speed control.

The paper by Mason et al. describes the Golem Group / UCLA's entry into the DGC. This team's entry was based on a robot from the 2004 Grand Challenge, and the paper describes some aspects of the design evolution to the 2005 system. Simple yet effective approaches to obstacle detection and path planning are presented, and a state estimation approach is described in detail. An approach to path detection in off-road environments is also presented.

Another second-time contestant, CajunBot, was built on an all-terrain, six-wheeled ATV vehicle base. The team's paper by Lakhotia et al. describes their complete hardware and software systems, including their transformation of an ATV into an AGV. The paper also presents an innovative obstacle detection system that took advantage of shocks and bumps to improve visibility, and their efficient software implementation that permitted all code to run on a single Pentium processor.

SciAutonics / Auburn University's entry is described in the paper by Travis et al. Areas discussed in detail include the team makeup and strategy, vehicle selection, software architecture, vehicle control, navigation, path planning, and obstacle detection. The team's results at the National Qualifying Event and Grand Challenge are also described.

Team CIMAR finished eighth in the 2004 Grand Challenge and came back to race in 2005 with a completely new NaviGator vehicle. Their paper describes in detail the entire system, including the completely custom-built vehicle, a JAUS-based system architecture, their modular "Smart Sensor" concept, terrain mapping based on a traversability grid, as well as planning, control, perception, and localization methods.

Princeton University's entry into the DGC is described in the paper by Atreya et al. The paper describes their robot, Prospect Eleven, which completed 9.3 miles of the course on race day and extensive portions of the 2004 and 2005 courses in later tests. Simple approaches to obstacle detection, path planning, and control are described. Interestingly, Prospect Eleven relied solely on stereo vision for perception, a unique approach among DGC finalists.

The paper by Miller et al. describing Cornell University's system addresses issues related to sensor fusion, gimbal pointing, and presents a cubic spline-based path planner. The paper also briefly describes their approach to terrain mapping, which is further detailed in the next paper (authored by Miller and Campbell). There, a real-time terrain mapping and estimation algorithm based on Gaussian mixture models is presented, along with experimental results of a simple obstacle detection task.

Alice is the successor to Bob, Caltech team's DGC 2004 race vehicle. In their paper, Cremean et al. describe Alice's complete system, developed by a team of over 50 undergraduate students. They fused sensor data into speed maps, and employed a receding horizon optimization-based trajectory planner. Contingency management was emphasized, and they employed a hybrid architecture with a state table and a series of rule-based filtering layers. The paper concludes with a description of the system's performance in the qualifying event and the race, a failure analysis and lessons learned.

MITRE's entry into the 2005 DGC is described in the paper by Grabowski et al. The paper presents a high-level description of the vehicle, software architecture, navigation and planning system, and sensing system. Descriptions of vehicle field testing and performance at the National Qualification Event are presented, as is an analysis of the system's failure at the DGC.

Together, the achievements that made this race so successful and exciting demonstrate how far the academic and industrial robotics communities have come towards building robust driverless vehicles. Successful teams integrated many sophisticated components, from sensing and sensor fusion algorithms, to localization methods, planning algorithms, failure detection and recovery methods, and vehicle automation techniques. Just as important as solving the technical challenges were good team management, plenty of testing, sufficient funding, and a healthy dose of good luck. The 2005 Grand Challenge was made manageable by removing obstacles from the vehicles' path, and providing dense and precise GPS data of the race route to the finalists. These and other issues will have to be dealt with head-on in the upcoming DARPA Urban Challenge, where vehicles will be required to drive autonomously in traffic.

We hope that readers will find this book interesting, useful, and inspiring. As autonomous ground vehicles become more advanced and their use more widespread, we feel that their study can only grow in importance.

Finally, we would like to express our gratitude to the many individuals, listed at the end of the book, who reviewed these papers for *The Journal of Field Robotics*. These individuals offered expert commentary, often through several drafts, and contributed to the papers' uniformly high quality.

Martin Buehler
Karl Iagnemma
Sanjiv Singh

Picture Gallery

Vehicle Name	Team Name	Miles Driven	Result
Stanley	Stanford Racing Team	132	Finished first in 6 h 53 min.

Source: DARPA

Vehicle Name	**Team Name**	**Miles Driven**	**Result**
Sandstorm	Red Team	132	Finished second in 7 h 5 min.

Source: DARPA

Vehicle Name	Team Name	Miles Driven	Result
H1ghlander	Red Team Too	132	Finished third in 7 h 14 min.

Source: DARPA

Vehicle Name	Team Name	Miles Driven	Result
KAT-5	Team Gray	132	Finished fourth in 7 h 30 min.

Source: DARPA

Vehicle Name	Team Name	Miles Driven	Result
TerraMax	Team TerraMax	132	Finished fifth in 12 h 51 min.

Source: DARPA

Vehicle Name	Team Name	Miles Driven	Result
DEXTER	Team ENSCO	81	A tire blew out after going off-course due to a network communication failure.

Source: DARPA

Vehicle Name	Team Name	Miles Driven	Result
Spirit	Axion Racing	66	Spirit got hung up on a rock after a mechanical failure.

Source: DARPA

Vehicle Name	Team Name	Miles Driven	Result
Cliff	Virginia Tech Grand Challenge Team	44	Cliff's drive engine stalled when it briefly slowed to an idle.

Source: DARPA

Vehicle Name	**Team Name**	**Miles Driven**	**Result**
Rocky	Virginia Tech Team Rocky	39	The on-board generator shut down due to a suspected false low-oil reading.

Source: DARPA

Vehicle Name	Team Name	Miles Driven	Result
ION	Desert Buckeyes	29	The team suspects that slowness in gear shifting gave the impression that ION had come to a halt and was terminated. ION was in the middle of the road and still drivable.

Source: DARPA

Vehicle Name	Team Name	Miles Driven	Result
'DAD, Are We There Yet?'	Team DAD (Digital Auto Drive)	26	The mounting for custom LIDAR scanner failed, resulting in a disconnected power supply to the vehicle navigation system.

Source: DARPA

Vehicle Name	Team Name	Miles Driven	Result
Desert Rat	Insight Racing	26	Desert Rat lost a heading sensor early in the race which slowed it down significantly. It continued till late in the day when its sensors got blinded by the low sun and it got hung up on a berm.

Source: DARPA

Vehicle Name	Team Name	Miles Driven	Result
Xboxx	Mojavaton	23	After the throttle motor failed, Xboxx dropped to 3 mph idle speed, and stalled at an uphill grade.

Source: DARPA

Vehicle Name	Team Name	Miles Driven	Result
Golem 2	The Golem Group / UCLA	22	The computer crashed due to faulty memory management (memory over-allocation). This problem did not manifest itself during earlier shorter desert runs, and longer, less sensor-rich endurance runs on a track. The uncontrolled vehicle departed from the course boundaries at high speed, crashing through vegetation. The DARPA "pause" button was no longer functional, since no software was running, and the DARPA observers did not press the "disable" button in case the vehicle might recover. Golem 2 hurtled more than half a mile off the course before pounding from the rough terrain finally shook connectors free from its fuse box, killing the engine.

Source: Sanjiv Singh

Vehicle Name	Team Name	Miles Driven	Result
CajunBot	Team CajunBot	17	The brake motor burned out when CajunBot was paused for about fifty minutes.

Source: DARPA

Vehicle Name	Team Name	Miles Driven	Result
RASCAL	SciAutonics / Auburn Engineering	16	The team suspects that the USB hubs lost power or overheated, severing the computer's connection to the LIDAR and other sensors.

Source: DARPA

Vehicle Name	Team Name	Miles Driven	Result
Desert Tortoise	Intelligent Vehicle Safety Technologies	14	Problems appeared at an intersection just beyond mile 12, from a wide (~25 ft) well-defined road onto a narrow (~10 ft) visually-obscured route segment. Road following behavior was inadvertently left on, causing the left turn to be interrupted for approximately 200 ms. The vehicle also responded to a false positive from the obstacle detection system that blocked the nominal path, causing the vehicle to leave the road. It drove over extremely rugged terrain at 10 to 15 mph and, after several attempts, returned to the course. However, at this point a navigation error caused the vehicle to behave erratically, leading to its termination, prior to contact with any manmade hazards or obstacles. Post race inspections found a lose connector as a possible cause for the navigation error, and a misaligned main sensor rack.

Source: Sanjiv Singh

Vehicle Name	Team Name	Miles Driven	Result
NaviGATOR	CIMAR	14	A 20 ft (6 m) position error made path planning based on sensing the road impossible and NaviGATOR drove off course into bushes.

Source: DARPA

Vehicle Name	Team Name	Miles Driven	Result
Prospect Eleven	Princeton University	9	Due to a software bug in the obstacle tracking code, old obstacles weren't entirely cleared from the list of tracked obstacles. This overwhelmed the processor, eventually slowing the control loop to run at 0.3 Hz instead the nominal 16-20 Hz. As a result, steering control became unstable, and Prospect Eleven was disabled.

Source: DARPA

Vehicle Name	**Team Name**	**Miles Driven**	**Result**
Spider	Team Cornell	9	State estimation errors (mostly pitch) likely caused ground to look like obstacles to the LIDAR. Trying to avoid this phantom obstacle caused Spider to veer into a concrete wall. Backing up was not implemented and Spider failed to turn free from the wall.

Source: DARPA

Vehicle Name	Team Name	Miles Driven	Result
Alice	Team Caltech	8	Proximity to power lines caused a temporary GPS outage. Large localization error after GPS re-acquisition caused Alice to veer towards concrete barriers. The mid-range sensors to detect these entered into error mode early on in the race and Alice crashed into the barriers.

Source: DARPA

Vehicle Name	Team Name	Miles Driven	Result
JackBot	MonsterMoto	7	JackBot was running flawlessly until the steering feedback potentiometer came loose from the drive mechanism.

Source: DARPA

Vehicle Name	Team Name	Miles Driven	Result
The Meteor	The MITRE Meteorites	1	Interpreting dust clouds as obstacles, the Meteor drove off the road into bushes and tall weeds, from where it couldn't recover.

Source: DARPA

Contents

1

Stanley: The Robot That Won the DARPA Grand Challenge

Sebastian Thrun[1], Mike Montemerlo[1], Hendrik Dahlkamp[1], David Stavens[1], Andrei Aron[1], James Diebel[1], Philip Fong[1], John Gale[1], Morgan Halpenny[1], Gabriel Hoffmann[1], Kenny Lau[1], Celia Oakley[1], Mark Palatucci[1], Vaughan Pratt[1], Pascal Stang[1], Sven Strohband[2], Cedric Dupont[2], Lars-Erik Jendrossek[2], Christian Koelen[2], Charles Markey[2], Carlo Rummel[2], Joe van Niekerk[2], Eric Jensen[2], Philippe Alessandrini[2], Gary Bradski[3], Bob Davies[3], Scott Ettinger[3], Adrian Kaehler[3], Ara Nefian[3], and Pamela Mahoney[4]

[1] Stanford Artificial Intelligence Laboratory
Stanford University
Stanford, CA 94305

[2] Volkswagen of America, Inc.
Electronics Research Laboratory
4009 Miranda Avenue, Suite 100
Palo Alto, CA 94304

[3] Intel Research
2200 Mission College Bvld.
Santa Clara, CA 95052

[4] Mohr Davidow Ventures
3000 Sand Hill Road, Bldg. 3, Suite 290
Menlo Park, CA 94025

Summary. This article describes the robot Stanley, which won the 2005 DARPA Grand Challenge. Stanley was developed for high-speed desert driving without human intervention. The robot's software system relied predominately on state-of-the-art AI technologies, such as machine learning and probabilistic reasoning. This article describes the major components of this architecture, and discusses the results of the Grand Challenge race.

1.1 Introduction

The Grand Challenge was launched by the Defense Advanced Research Projects Agency (DARPA) in 2003 to spur innovation in unmanned ground vehicle navigation. The goal of the Challenge was the development of an autonomous robot capable of traversing unrehearsed, off-road terrain. The first competition, which carried a prize of \$1M, took place on March 13, 2004. It required robots to navigate a 142-mile long course through the Mojave desert in no more than 10 hours. 107 teams registered and 15 raced, yet none of the participating robots navigated more than 5% of the entire course. The challenge was repeated on October 8, 2005, with an increased prize of \$2M. This time, 195 teams registered

M. Buehler, K. Iagnemma, and S. Singh (Eds.): DARPA'05, STAR 36, pp. 1–43, 2007.
springerlink.com

(a) (b)

Fig. 1.1. (a) At approximately 1:40pm on Oct 8, 2005, Stanley is the first robot to complete the DARPA Grand Challenge. (b) The robot is being honored by DARPA Director Dr. Tony Tether.

and 23 raced. Of those, five teams finished. Stanford's robot "Stanley" finished the course ahead of all other vehicles in 6 hours 53 minutes and 58 seconds and was declared the winner of the DARPA Grand Challenge; see Fig. 1.1.

This article describes the robot Stanley, and its software system in particular. Stanley was developed by a team of researchers to advance the state-of-the-art in autonomous driving. Stanley's success is the result of an intense development effort led by Stanford University, and involving experts from Volkswagen of America, Mohr Davidow Ventures, Intel Research, and a number of other entities. Stanley is based on a 2004 Volkswagen Touareg R5 TDI, outfitted with a 6 processor computing platform provided by Intel, and a suite of sensors and actuators for autonomous driving. Fig. 1.2 shows images of Stanley during the race.

The main technological challenge in the development of Stanley was to build a highly reliable system capable of driving at relatively high speeds through diverse and unstructured off-road environments, and to do all this with high precision. These requirements led to a number of advances in the field of autonomous navigation, as surveyed in this article. New methods were developed, and existing methods extended, in the areas of long-range terrain perception, real-time collision avoidance, and stable vehicle control on slippery and rugged terrain. Many of these developments were driven by the speed requirement, which rendered many classical techniques in the off-road driving field unsuitable. In pursuing these developments, the research team brought to bear algorithms from diverse areas including distributed systems, machine learning, and probabilistic robotics.

1.1.1 Race Rules

The rules [DARPA, 2004] of the DARPA Grand Challenge were simple. Contestants were required to build autonomous ground vehicles capable of traversing

Fig. 1.2. Images from the race

a desert course up to 175 miles long in less than 10 hours. The first robot to complete the course in under 10 hours would win the challenge and the $2M prize. Absolutely no manual intervention was allowed. The robots were started by DARPA personnel and from that point on had to drive themselves. Teams only saw their robots at the starting line and, with luck, at the finish line.

Both the 2004 and 2005 races were held in the Mojave desert in the southwest United States. Course terrain varied from high quality, graded dirt roads to winding, rocky, mountain passes; see Fig. 1.2. A small fraction of each course traveled along paved roads. The 2004 course started in Barstow, CA, approximately 100 miles northeast of Los Angeles, and finished in Primm, NV, approximately 30 miles southwest of Las Vegas. The 2005 course both started and finished in Primm, NV.

The specific race course was kept secret from all teams until two hours before the race. At this time, each team was given a description of the course on CD-ROM in a DARPA-defined Route Definition Data Format (RDDF). The RDDF is a list of longitudes, latitudes, and corridor widths that define the course boundary, and a list of associated speed limits; an example segment is shown

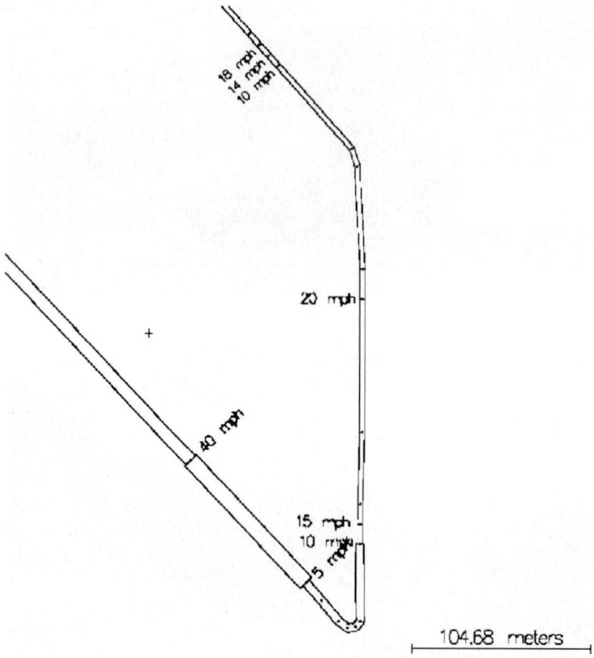

Fig. 1.3. A section of the RDDF file from the 2005 DARPA Grand Challenge. The corridor varies in width and maximum speed. Waypoints are more frequent in turns.

in Fig. 1.3. Robots that travel substantially beyond the course boundary risk disqualification. In the 2005 race, the RDDF contained 2,935 waypoints.

The width of the race corridor generally tracked the width of the road, varying between 3 and 30 meter in the 2005 race. Speed limits were used to protect important infrastructure and ecology along the course and to maintain the safety of DARPA chase drivers who followed behind each robot. The speed limits varied between 5 and 50 MPH. The RDDF defined the approximate route that robots would take, so no global path planning was required. As a result, the race was primarily a test of high-speed road finding and obstacle detection and avoidance in desert terrain.

The robots all competed on the same course, starting one after another at 5 minute intervals. When a faster robot overtook a slower one, the slower robot was paused by DARPA officials, allowing the second robot to pass the first as if it were a static obstacle. This eliminated the need for robots to handle the case of dynamic passing.

1.1.2 Team Composition

The Stanford Racing Team team was organized in four major groups. The *Vehicle Group* oversaw all modifications and component developments related to the core vehicle. This included the drive-by-wire systems, the sensor and computer mounts, and the computer systems. The group was led by researchers from Volkswagen of America's Electronics Research Lab. The *Software Group* developed all software, including the navigation software and the various health monitor and safety systems. The software group was led by researchers affiliated with Stanford University. The *Testing Group* was responsible for testing all system components and the system as a whole, according to a specified testing schedule. The members of this group were separate from any of the other groups. The testing group was led by researchers affiliated with Stanford University. The *Communications Group* managed all media relations and fund raising activities of the Stanford Racing Team. The communications group was led by employees of Mohr Davidow Ventures, with participation from all other sponsors. The operations oversight was provided by a steering board that included all major supporters.

1.2 Vehicle

Stanley is based on a diesel-powered Volkswagen Touareg R5. The Touareg has four wheel drive, variable-height air suspension, and automatic, electronic locking differentials. To protect the vehicle from environmental impact, Stanley has been outfitted with skid plates and a reinforced front bumper. A custom interface enables direct, electronic actuation of both throttle and brakes. A DC motor attached to the steering column provides electronic steering control. A linear actuator attached to the gear shifter shifts the vehicle between drive, reverse, and parking gears (Fig. 1.4c). Vehicle data, such as individual wheel speeds and steering angle, are sensed automatically and communicated to the computer system through a CAN bus interface.

The vehicle's custom-made roof rack is shown in Fig. 1.4a. It holds nearly all of Stanley's sensors. The roof provides the highest vantage point from the vehicle; from this point the visibility of the terrain is best, and the access to GPS signals is least obstructed. For environment perception, the roof rack houses five SICK laser range finders. The lasers are pointed forward along the driving direction of the vehicle, but with slightly different tilt angles. The lasers measure cross-sections of the approaching terrain at different ranges out to 25 meters in front of the vehicle. The roof rack also holds a color camera for long-range road perception, which is pointed forward and angled slightly downwards. For long-range detection of large obstacles, Stanley's roof rack also holds two 24 GHz RADAR sensors, supplied by Smart Microwave Sensors. Both RADAR sensors cover the frontal area up to 200 meter, with a coverage angle in azimuth of about 20 degrees. Two antennae of this system are mounted on both sides of the laser sensor array. The lasers, camera, and radar system comprise the *environment sensor group* of the system. That is, they inform Stanley of the terrain ahead, so that Stanley can decide where to drive, and at what speed.

Further back, the roof rack holds a number of additional antennae: one for Stanley's GPS positioning system and two for the GPS compass. The GPS positioning unit is a L1/L2/Omnistar HP receiver. Together with a trunk-mounted

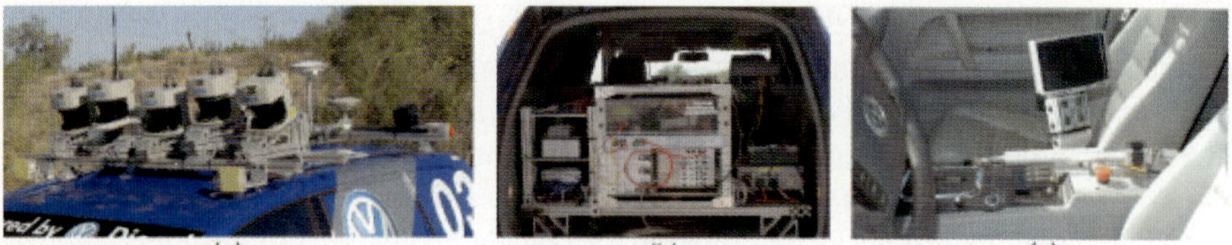

(a) (b) (c)

Fig. 1.4. (a) View of the vehicle's roof rack with sensors. (b) The computing system in the trunk of the vehicle. (c) The gear shifter, control screen, and manual override buttons.

inertial measurement unit (IMU), the GPS systems are the *proprioceptive sensor group*, whose primary function is to estimate the location and velocity of the vehicle relative to an external coordinate system.

Finally, a radio antenna and three additional GPS antennae from the DARPA E-Stop system are also located on the roof. The E-Stop system is a wireless link that allows a chase vehicle following Stanley to safely stop the vehicle in case of emergency. The roof rack also holds a signaling horn, a warning light, and two manual E-stop buttons.

Stanley's computing system is located in the vehicle's trunk, as shown in Fig. 1.4b. Special air ducts direct air flow from the vehicle's air conditioning system into the trunk for cooling. The trunk features a shock-mounted rack that carries an array of six Pentium M computers, a Gigabit Ethernet switch, and various devices that interface to the physical sensors and the Touareg's actuators. It also features a custom-made power system with backup batteries, and a switch box that enables Stanley to power-cycle individual system components through software. The DARPA-provided E-Stop is located on this rack on additional shock compensation. The trunk assembly also holds the custom interface to the Volkswagen Touareg's actuators: the brake, throttle, gear shifter, and steering controller. A six degree-of-freedom IMU is rigidly attached to the vehicle frame underneath the computing rack in the trunk.

The total power requirement of the added instrumentation is approximately 500 W, which is provided through the Touareg's stock alternator. Stanley's backup battery system supplies an additional buffer to accommodate long idling periods in desert heat.

The operating system run on all computers is Linux. Linux was chosen due to its excellent networking and time sharing capabilities. During the race, Stanley executed the race software on three of the six computers; a fourth was used to log the race data (and two computers were idle). One of the three race computers was entirely dedicated to video processing, whereas the other two executed all other software. The computers were able to poll the sensors at up to 100 Hz, and to control the steering, throttle and brake at frequencies up to 20 Hz.

An important aspect in Stanley's design was to retain street legality, so that a human driver could safely operate the robot as a conventional passenger car. Stanley's custom user interface enables a driver to engage and disengage the computer system at will, even while the vehicle is in motion. As a result, the driver can disable computer control at any time of the development, and regain manual control of the vehicle. To this end, Stanley is equipped with several manual override buttons located near the driver seat. Each of these switches controls one of the three major actuators (brakes, throttle, steering). An additional central emergency switch disengages all computer control and transforms the robot into a conventional vehicle. While this feature was of no relevance to the actual race (in which no person sat in the car), it proved greatly beneficial during software development. The interface made it possible to operate Stanley autonomously with people inside, as a dedicated safety driver could always catch computer glitches and assume full manual control at any time.

During the actual race, there was of course no driver in the vehicle, and all driving decisions were made by Stanley's computers. Stanley possessed an operational control interface realized through a touch-sensitive screen on the driver's console. This interface allowed Government personnel to shut down and restart the vehicle, if it became necessary.

1.3 Software Architecture

1.3.1 Design Principles

Before both the 2004 and 2005 Grand Challenges, DARPA revealed to the competitors that a stock 4WD pickup truck would be physically capable of traversing the entire course. These announcements suggested that the innovations necessary to successfully complete the challenge would be in designing intelligent driving software, not in designing exotic vehicles. This announcement and the performance of the top finishers in the 2004 race guided the design philosophy of the Stanford Racing Team: *treat autonomous navigation as a software problem.*

In relation to previous work on robotics architectures, Stanley's software architecture is probably best thought of as a version of the well-known *three layer architecture* [Gat, 1998], albeit without a long-term symbolic planning method. A number of guiding principles proved essential in the design of the software architecture:

Control and data pipeline. There is no centralized master-process in Stanley's software system. All modules are executed at their own pace, without interprocess synchronization mechanisms. Instead, all data is globally time-stamped, and time stamps are used when integrating multiple data sources. The approach reduces the risk of deadlocks and undesired processing delays. To maximize the configurability of the system, nearly all inter-process communication is implemented through publish-subscribe mechanisms. The information from sensors to actuators flows in a single direction; no information is received more than once by the same module. At any point in time, all modules in the pipeline are working simultaneously, thereby maximizing the information throughput and minimizing the latency of the software system.

State management. Even though the software is distributed, the *state* of the system is maintained by local authorities. There are a number of state variables in the system. The health state is locally managed in the health monitor; the parameter state in the parameter server; the global driving mode is maintained in a finite state automaton; and the vehicle state is estimated in the state estimator module. The environment state is broken down into multiple maps (laser, vision, and radar). Each of these maps are maintained in dedicated modules. As a result, all other modules will receive values that are mutually consistent. The exact state variables are discussed in later sections of this article. All state variables are broadcast to relevant modules of the software system through a publish-subscribe mechanism.

Reliability. The software places strong emphasis on the overall reliability of the robotic system. Special modules monitor the health of individual software and hardware components, and automatically restart or power-cycle such components when a failure is observed. In this way, the software is robust to certain occurrences, such as crashing or hanging of a software modules or stalled sensors.

Development support. Finally, the software is structured so as to aid development and debugging of the system. The developer can easily run just a sub-system of the software, and effortlessly migrate modules across different processors. To facilitate debugging during the development process, all data is logged. By using a special replay module, the software can be run on recorded data. A number of visualization tools were developed that make it possible to inspect data and internal variables while the vehicle is in motion, or while replaying previously logged data. The development process used a version control process with a strict set of rules for the release of race-quality software. Overall, we found that the flexibility of the software during development was essential in achieving the high level of reliability necessary for long-term autonomous operation.

1.3.2 Processing Pipeline

The race software consisted of approximately 30 modules executed in parallel (Fig. 1.5). The system is broken down into six layers which correspond to the following functions: sensor interface, perception, control, vehicle interface, user interface, and global services.

1. The **sensor interface layer** comprises a number of software modules concerned with receiving and time-stamping all sensor data. The layer receives data from each laser sensor at 75 Hz, from the camera at approximately 12 Hz, the GPS and GPS compass at 10 Hz, and the IMU and the Touareg CAN bus at 100 Hz. This layer also contains a database server with the course coordinates (RDDF file).
2. The **perception layer** maps sensor data into internal models. The primary module in this layer is the UKF vehicle state estimator, which determines the vehicle's coordinates, orientation, and velocities. Three different mapping modules build 2-D environment maps based on lasers, the camera, and the

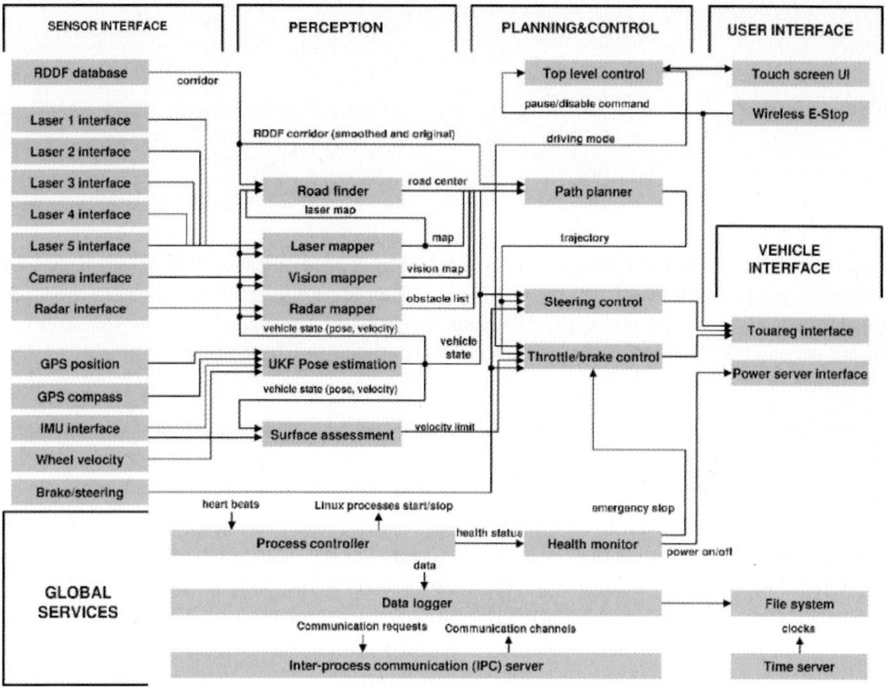

Fig. 1.5. Flowchart of Stanley Software System. The software is roughly divided into six main functional groups: sensor interface, perception, control, vehicle interface, and user interface. There are a number of cross-cutting services, such as the process controller and the logging modules.

radar system. A road finding module uses the laser-derived maps to find the boundary of a road, so that the vehicle can center itself laterally. Finally, a surface assessment module extracts parameters of the current road for the purpose of determining safe vehicle speeds.

3. The **control layer** is responsible for regulating the steering, throttle, and brake response of the vehicle. A key module is the path planner, which sets the trajectory of the vehicle in steering- and velocity-space. This trajectory is passed to two closed loop trajectory tracking controllers, one for the steering control and one for brake and throttle control. Both controllers send low-level commands to the actuators that faithfully execute the trajectory emitted by the planner. The control layer also features a top level control module, implemented as a simple finite state automaton. This level determines the general vehicle mode in response to user commands received through the in-vehicle touch screen or the wireless E-stop, and maintains gear state in case backwards motion is required.
4. The **vehicle interface layer** serves as the interface to the robot's drive-by-wire system. It contains all interfaces to the vehicle's brakes, throttle, and

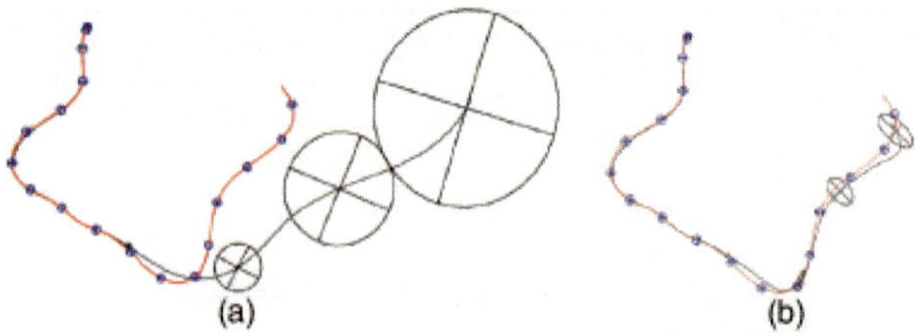

Fig. 1.6. UKF state estimation when GPS becomes unavailable. The area covered by the robot is approximately 100 by 100 meter. The large ellipses illlustrate the position uncertainty after losing GPS. (a) Without integrating the wheel motion the result is highly erroneous. (b) The wheel motion clearly improves the result.

steering wheel. It also features the interface to the vehicle's server, a circuit that regulates the physical power to many of the system components.

5. The **user interface layer** comprises the remote E-stop and a touch-screen module for starting up the software.
6. The **global services layer** provides a number of basic services for all software modules. Naming and communication services are provided through CMU's Inter-Process Communication (IPC) toolkit [Simmons and Apfelbaum, 1998]. A centralized parameter server maintains a database of all vehicle parameters and updates them in a consistent manner. The physical power of individual system components is regulated by the power server. Another module monitors the health of all systems components and restarts individual system components when necessary. Clock synchronization is achieved through a time server. Finally, a data logging server dumps sensor, control, and diagnostic data to disk for replay and analysis.

The following sections will describe Stanley's core software processes in greater detail. The paper will then conclude with a description of Stanley's performance in the Grand Challenge.

1.4 Vehicle State Estimation

Estimating vehicle state is a key prerequisite for precision driving. Inaccurate pose estimation can cause the vehicle to drive outside the corridor, or build terrain maps that do not reflect the state of the robot's environment, leading to poor driving decisions. In Stanley, the *vehicle state* comprises a total of 15 variables. The design of this parameter space follows standard methodology [Farrell and Barth, 1999, van der Merwe and Wan, 2004]:

# values	state variable
3	position (longitude, latitude, altitude)
3	velocity
3	orientation (Euler angles: roll, pitch, yaw)
3	accelerometer biases
3	gyro biases

An unscented Kalman filter (UKF) [Julier and Uhlmann, 1997] estimates these quantities at an update rate of 100Hz. The UKF incorporates observations from the GPS, the GPS compass, the IMU, and the wheel encoders. The GPS system provides both absolute position and velocity measurements, which are both incorporated into the UKF. From a mathematical point of view, the sigma point linearization in the UKF often yields a lower estimation error than the linearization based on Taylor expansion in the EKF [van der Merwe, 2004]. To many, the UKF is also preferable from an implementation standpoint because it does not require the explicit calculation of any Jacobians; although those can be useful for further analysis.

While GPS is available, the UKF uses only a "weak" model. This model corresponds to a moving mass that can move in any direction. Hence, in normal operating mode the UKF places no constraint on the direction of the velocity vector relative to the vehicle's orientation. Such a model is clearly inaccurate, but the vehicle-ground interactions in slippery desert terrain are generally difficult to model. The moving mass model allows for any slipping or skidding that may occur during off-road driving.

However, this model performs poorly during GPS outages, however, as the position of the vehicle relies strongly on the accuracy of the IMU's accelerometers. As a consequence, a more restrictive UKF motion model is used during GPS outages. This model constrains the vehicle to only move in the direction it is pointed. Integration of the IMU's gyroscopes for orientation, coupled with wheel velocities for computing the position, is able to maintain accurate pose of the vehicle during GPS outages of up to 2 minutes long; the accrued error is usually in the order of centimeters. Stanley's health monitor will decrease the maximum vehicle velocity during GPS outages to 10 mph in order to maximize the accuracy of the restricted vehicle model. Fig. 1.6a shows the result of position estimation during a GPS outage with the weak vehicle model; Fig. 1.6b the result with the strong vehicle model. This experiment illustrates the performance of this filter during a GPS outage. Clearly, accurate vehicle modeling during GPS outages is essential. In an experiment on a paved road, we found that even after 1.3 km of travel without GPS on a cyclic course, the accumulated vehicle error was only 1.7 meters.

1.5 Laser Terrain Mapping

1.5.1 Terrain Labeling

To safely avoid obstacles, Stanley must be capable of accurately detecting non-drivable terrain at a sufficient range to stop or take the appropriate evasive action. The faster the vehicle is moving, the farther away obstacles must be detected. Lasers are used as the basis for Stanley's short and medium range obstacle avoidance. Stanley is equipped with five single-scan laser range finders mounted on the roof, tilted downward to scan the road ahead. Fig. 1.7a illustrates the scanning process. Each laser scan generates a vector of 181 range measurements spaced 0.5 degrees apart. Projecting these scans into the global

Fig. 1.7. (a) Illustration of a laser sensor: The sensor is angled downward to scan the terrain in front of the vehicle as it moves. Stanley possesses five such sensors, mounted at five different angles. (b) Each laser acquires a 3-D point cloud over time. The point cloud is analyzed for drivable terrain and potential obstacles.

coordinate frame according to the estimated pose of the vehicle results in a 3-D point cloud for each laser. Fig. 1.7b shows an example of the point clouds acquired by the different sensors. The coordinates of such 3-D points are denoted $(X_k^i \ \ Y_k^i \ \ Z_k^i)$; here k is the time index at which the point was acquired, and i is the index of the laser beam.

Obstacle detection on laser point clouds can be formulated as a classification problem, assigning to each 2-D location in a surface grid one of three possible values: occupied, free, and unknown. A location is occupied by an obstacle if we can find two nearby points whose vertical distance $|Z_k^i - Z_m^j|$ exceeds a critical vertical distance δ. It is considered drivable (free of obstacles) if no such points can be found, but at least one of the readings falls into the corresponding grid cell. If no reading falls into the cell, the drivability of this cell is considered unknown. The search for nearby points is conveniently organized in a 2-D grid, the same grid used as the final drivability map that is provided to the vehicle's navigation engine. Fig. 1.8 shows the example grid map. As indicated in this figure, the map assigns terrain to one of three classes: drivable, occupied, or unknown.

Unfortunately, applying this classification scheme directly to the laser data yields results inappropriate for reliable robot navigation. Fig. 1.9 shows such an instance, in which a small error in the vehicle's roll/pitch estimation leads to a massive terrain classification error, forcing the vehicle off the road. Small pose errors are magnified into large errors in the projected positions of laser points because the lasers are aimed at the road up to 30 meters in front of the vehicle. In our reference dataset of labeled terrain, we found that 12.6% of known drivable area is classified as obstacle, for a height threshold parameter $\delta = 15$cm. Such situations occur even for roll/pitch errors smaller than 0.5 degrees. Pose errors of this magnitude can be avoided by pose estimation systems that cost hundreds of thousands of dollars, but such a choice was too costly for this project.

The key insight to solving this problem is illustrated in Fig. 1.10. This graph plots the perceived obstacle height $|Z_k^i - Z_m^j|$ along the vertical axis for a

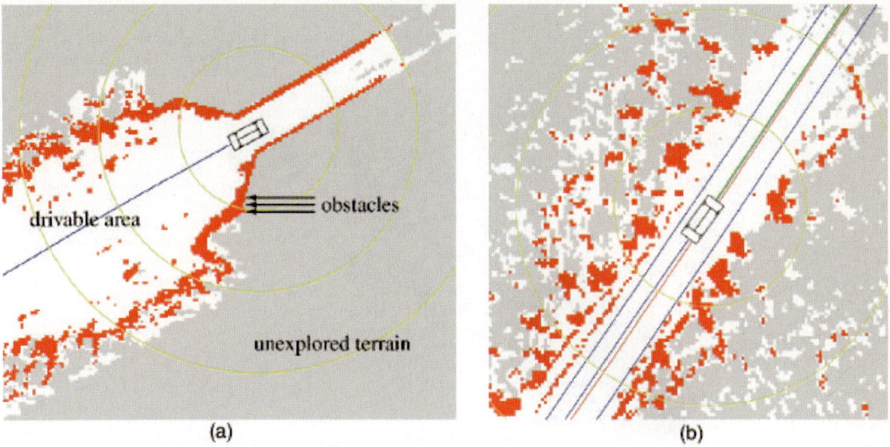

Fig. 1.8. Examples of occupancy maps: (a) an underpass, and (b) a road

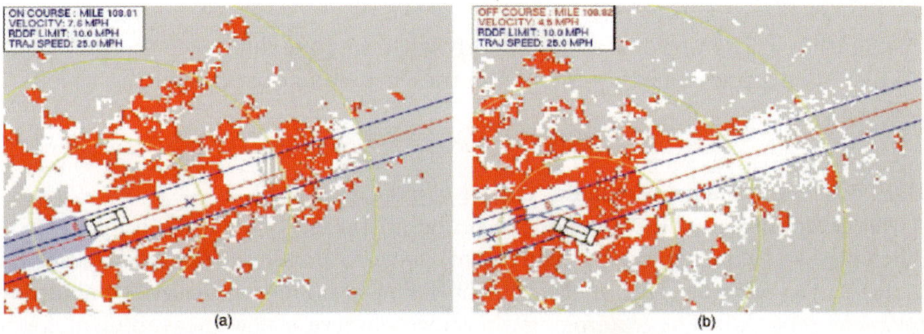

Fig. 1.9. Small errors in pose estimation (smaller than 0.5 degrees) induce massive terrain classification errors, which if ignored could force the robot off the road. These images show two consecutive snapshots of a map that forces Stanley off the road. Here obstacles are plotted in red, free space in white, and unknown territory in gray. The blue lines mark the corridor as defined by the RDDF.

collection of grid cells taken from flat terrain. Clearly, for some grid cells the perceived height is enormous—despite the fact that in reality, the surface is flat. However, this function is not random. The horizontal axis depicts the time difference $\Delta t\ |k - m|$ between the acquisition of those scans. Obviously, the error is strongly correlated with the elapsed time between the two scans.

To model this error, Stanley uses a first order Markov model, which models the drift of the pose estimation error over time. The test for the presence of an obstacle is therefore a *probabilistic* test. Given two points $(X_k^i \ \ Y_k^i \ \ Z_k^i)^T$ and $(X_m^j \ \ Y_m^j \ \ Z_m^j)^T$, the height difference is distributed according to a normal

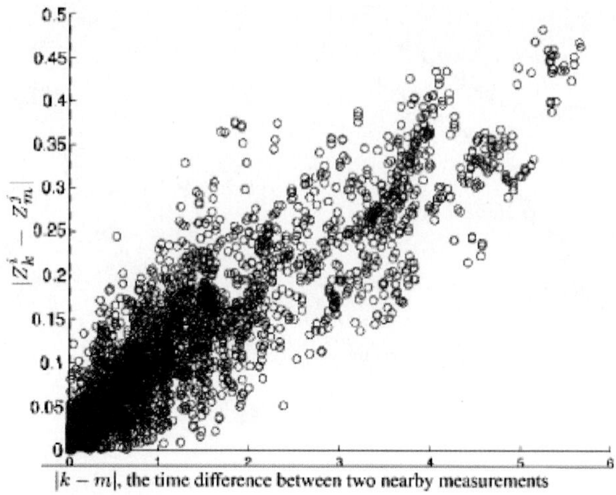

Fig. 1.10. Correlation of time and vertical measurement error in the laser data analysis

distribution whose variance scales linearly with the time difference $|k-m|$. Thus, Stanley uses a probabilistic test for the presence of an obstacle, of the type

$$p(|Z_k^i - Z_m^j| > \delta) > \alpha \tag{1.1}$$

Here α is a confidence threshold, e.g., $\alpha = 0.05$.

When applied over a 2-D grid, the probabilistic method can be implemented efficiently so that only two measurements have to be stored per grid cell. This is due to the fact that each measurement defines a bound on future Z-values for obstacle detection. For example, suppose we observe a new measurement for a cell which was previously observed. Then one or more of three cases will be true:

1. The new measurement might be a witness of an obstacle, according to the probabilistic test. In this case Stanley simply marks the cell as obstacle and no further testing takes place.
2. The new measurement does not trigger as a witness of an obstacle, but in future tests it establishes a tighter lower bound on the minimum Z-value than the previously stored measurement. In this case, our algorithm simply replaces the previous measurement with this new one. The rationale behind this is simple: If the new measurement is more restrictive than the previous one, there will not be a situation where a test against this point would fail while a test against the older one would succeed. Hence, the old point can safely be discarded.
3. The third case is equivalent to the second, but with a refinement of the upper value. Notice that a new measurement may refine simultaneously the lower and the upper bounds.

The fact that only two measurements per grid cell have to be stored renders this algorithm highly efficient in space and time.

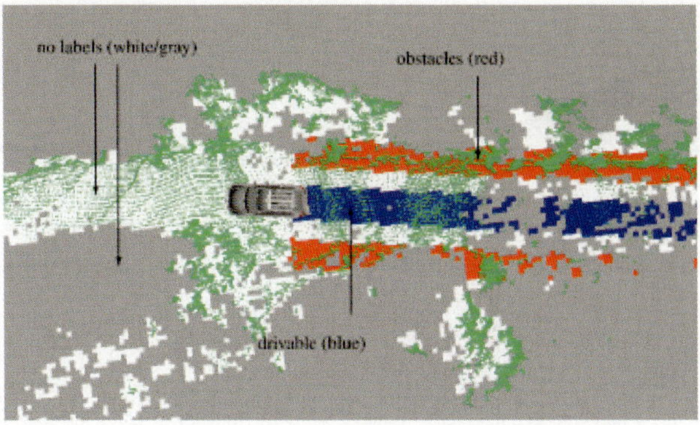

Fig. 1.11. Terrain labeling for parameter tuning: The area traversed by the vehicle is labeled as "drivable" (blue) and two stripes at a fixed distance to the left and the right are labeled as "obstacles" (red). While these labels are only approximate, they are extremely easy to obtain and significantly improve the accuracy of the resulting map when used for parameter tuning.

1.5.2 Data-Driven Parameter Tuning

A final step in developing this mapping algorithm addresses parameter tuning. Our approach, and the underlying probabilistic Markov model, possesses a number of unknown parameters. These parameters include the height threshold δ, the statistical acceptance probability threshold α, and various Markov chain error parameters (the noise covariances of the process noise and the measurement noise).

Stanley uses a discriminative learning algorithm for locally optimizing these parameters. This algorithm tunes the parameters in a way that maximizes the discriminative accuracy of the resulting terrain analysis on labeled training data.

The data is labeled through human driving, similar in spirit to [Pomerleau, 1993]. Fig. 1.11 illustrates the idea: A human driver is instructed to only drive over obstacle-free terrain. Grid cells traversed by the vehicle are then labeled as "drivable." This area corresponds to the blue stripe in Fig. 1.11. A stripe to the left and right of this corridor is assumed to be all obstacles, as indicated by the red stripes in Fig. 1.11. The distance between the "drivable" and "obstacle" is set by hand, based on the average road width for a segment of data. Clearly, not all of those cells labeled as obstacles are actually occupied by actual obstacles; however, even training against an approximate labeling is enough to improve overall performance of the mapper.

The learning algorithm is now implemented through coordinate ascent. In the outer loop, the algorithm performs coordinate ascent relative to a data-driven scoring function. Given an initial guess, the coordinate ascent algorithm modifies each parameter one-after-another by a fixed amount. It then determines if the new value constitutes an improvement over the previous value when evaluated over a logged data set, and retains it accordingly. If for a given interval size no improvement can be found, the search interval is cut in half and the search is

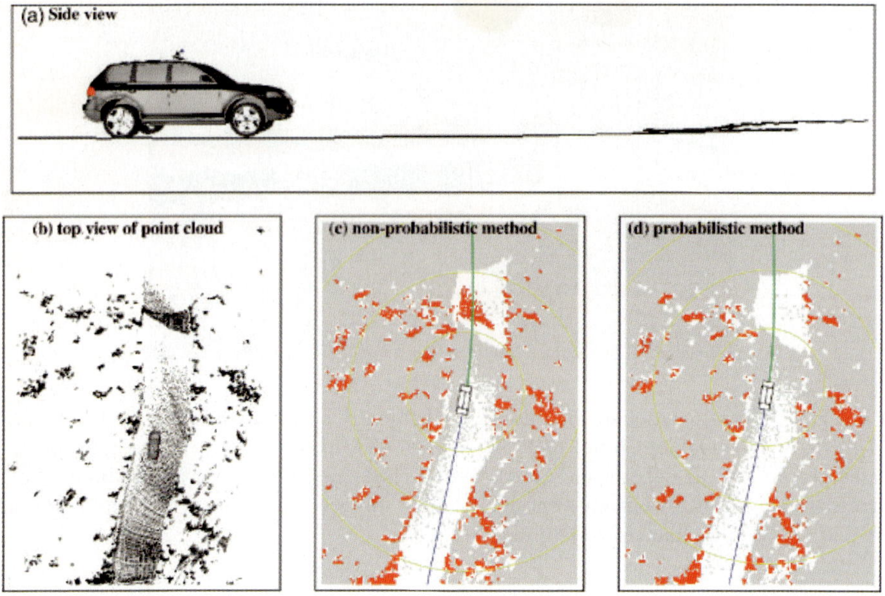

Fig. 1.12. Example of pitching combined with small pose estimation errors: (a) shows the reading of the center beam of one of the lasers, integrated over time. Some of the terrain is scanned twice. Panel (b) shows the 3-D point cloud; panel (c) the resulting map without probabilistic analysis, and (d) the map with probabilistic analysis. The map shown in Panel (c) possesses a phantom obstacle, large enough to force the vehicle off the road.

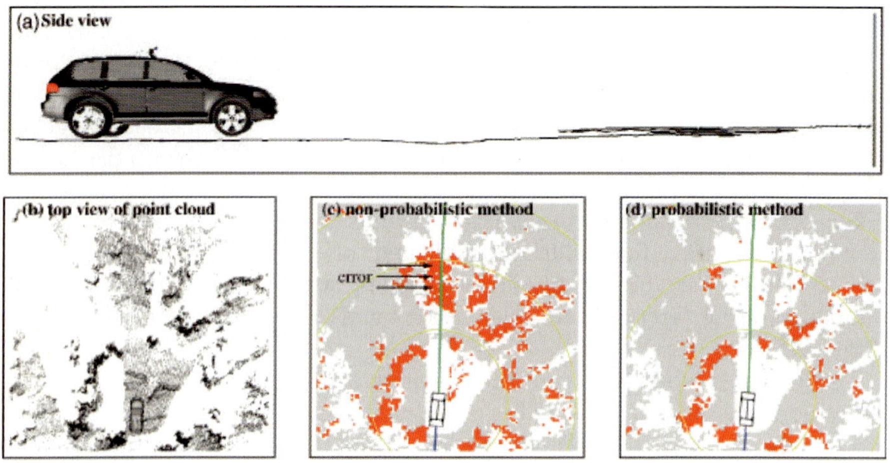

Fig. 1.13. A second example

continued, until the search interval becomes smaller than a pre-set minimum search interval (at which point the tuning is terminated).

The probabilistic analysis paired with the discriminative algorithm for parameter tuning has a significant effect on the accuracy of the terrain labels. Using an independent testing data set, we find that the false positive rate (the area labeled as drivable in Fig. 1.11) drops from 12.6% to 0.002%. At the same time, the rate at which the area off the road is labeled as obstacle remains approximately constant (from 22.6% to 22.0%). This rate is *not* 100% simply because most of the terrain there is still flat and drivable. Our approach for data acquisition *mislabels* the flat terrain as non-drivable. Such mislabeling however, does not interfere with the parameter tuning algorithm, and hence is preferable to the tedious process of labeling pixels manually.

Fig. 1.12 shows an example of the mapper in action. A snapshot of the vehicle from the side illustrates that part of the surface is scanned multiple times due to a change of pitch. As a result, the non-probabilistic method hallucinates a large occupied area in the center of the road, shown in Panel c of Fig. 1.12. Our probabilistic approach overcomes this error and generates a map that is good enough for driving. A second example is shown in Fig. 1.13.

1.6 Computer Vision Terrain Analysis

The effective maximum range at which obstacles can be detected with the laser mapper is approximately 22 meters. This range is sufficient for Stanley to reliably avoid obstacles at speeds up to 25 mph. Based on the 2004 race course, the development team estimated that Stanley would need to reach speeds of 35 mph in order to successfully complete the challenge. To extend the sensor range enough to allow safe driving at 35 mph, Stanley uses a color camera to find drivable surfaces at ranges exceeding that of the laser analysis. Fig. 1.14 compares laser and vision mapping side-by-side. The left diagram shows a laser map acquired during the race; here obstacles are detected at approximately 22 meter range. The vision map for the same situation is shown on the right side. This map extends beyond 70 meters (each yellow circle corresponds to 10 meters range).

Our work builds on a long history of research on road finding[Pomerleau, 1991, Crisman and Thorpe, 1993]; see also [Dickmanns, 2002]. To find the road, the vision module classifies images into drivable and non-drivable regions. This classification task is generally difficult, as the road appearance is affected by a number of factors that are not easily measured and change over time, such as the surface material of the road, lighting conditions, dust on the lens of the camera, and so on. This suggests that an adaptive approach is necessary, in which the image interpretation changes as the vehicle moves and conditions change.

The camera images are not the only source of information about upcoming terrain available to the vision mapper. Although we are interested in using vision to classify the drivability of terrain beyond the laser range, we already have such drivability information from the laser in the near range. All that is required from the vision routine is to extend the reach of the laser analysis. This is different

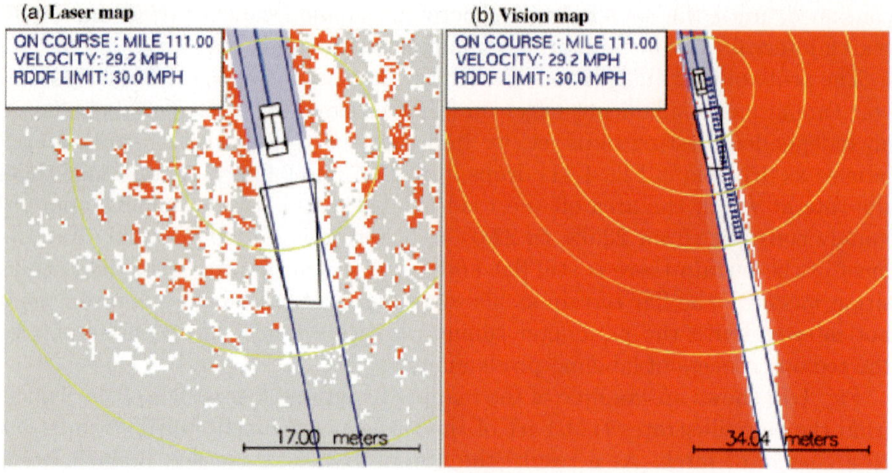

Fig. 1.14. Comparison of the laser-based (left) and the image-based (right) mapper. For scale, circles are spaced around the vehicle at 10 meter distance. This diagram illustrates that the reach of lasers is approximately 22 meters, whereas the vision module often looks 70 meters ahead.

Fig. 1.15. This figure illustrates the processing stages of the computer vision system: (a) a raw image; (b) the processed image with the laser quadrilateral and a pixel classification; (c) the pixel classification before thresholding; (d) horizon detection for sky removal

from the general-purpose image interpretation problem, in which no such data would be available.

Stanley finds drivable surfaces by projecting drivable area from the laser analysis into the camera image. More specifically, Stanley extracts a quadrilateral ahead of the robot in the laser map, so that all grid cells within this quadrilateral are drivable. The range of this quadrilateral is typically between 10 and 20 meters ahead of the robot. An example of such a quadrilateral is shown in Fig. 1.14a. Using straightforward geometric projection, this quadrilateral is then mapped into the camera image, as illustrated in Fig. 1.15a and b. An adaptive computer vision algorithm then uses the image pixels inside this quadrilateral as training examples for the concept of drivable surface.

The learning algorithm maintains a mixture of Gaussians that model the color of drivable terrain. Each such mixture is a Gaussian defined in the RGB color space of individual pixels; the total number of Gaussians is denoted n. The learning algorithm maintains for each mixture a mean RGB-color μ_i, a covariance Σ_i, and a number m_i that counts the total number of image pixels that were used to train this Gaussian.

When a new image is observed, the pixels in the drivable quadrilateral are mapped into a much smaller number of k "local" Gaussians using the EM algorithm[Duda and Hart, 1973], with $k < n$ (the covariance of these local Gaussians are inflated by a small value so as to avoid overfitting). These k local Gaussians are then merged into the memory of the learning algorithm, in a way that allows for slow and fast adaptation. The learning adapts to the image in two possible ways; by adjusting the previously found internal Gaussian to the actual image pixels, and by introducing new Gaussians and discarding older ones. Both adaptation steps are essential. The first enables Stanley to adapt to slowly changing lighting conditions; the second makes it possible to adapt rapidly to a new surface color (e.g., when Stanley moves from a paved to an unpaved road).

In detail, to update the memory, consider the j-th local Gaussian. The learning algorithm determines the closest Gaussian in the global memory, where closeness is determined through the Mahalanobis distance.

$$d(i,j) = (\mu_i - \mu_j)^T \, (\Sigma_i + \Sigma_j)^{-1} \, (\mu_i - \mu_j) \tag{1.2}$$

Let i be the index of the minimizing Gaussian in the memory. The learning algorithm then chooses one of two possible outcomes:

1. The distance $d(i,j) \leq \phi$, where ϕ is an acceptance threshold. The learning algorithm then assumes that the global Gaussian j is representative of the local Gaussian i, and adaptation proceeds slowly. The parameters of this global Gaussian are set to the weighted mean:

$$\mu_i \longleftarrow \frac{m_i \, \mu_i}{m_i + m_j} + \frac{m_j \, \mu_j}{m_i + m_j} \tag{1.3}$$

$$\Sigma_i \longleftarrow \frac{m_i \, \Sigma_i}{m_i + m_j} + \frac{m_j \, \Sigma_j}{m_i + m_j} \tag{1.4}$$

$$m_i \longleftarrow m_i + m_j \tag{1.5}$$

   Here m_j is the number of pixels in the image that correspond to the j-th Gaussian.
2. The distance $d(i,j) > \phi$ for any Gaussian i in the memory. This is the case when none of the Gaussian in memory are near the local Gaussian extracted from the image, where nearness is measured by the Mahalanobis distance. The algorithm then generates a new Gaussian in the global memory, with parameters μ_j, Σ_j, and m_j. If all n slots are already taken in the memory, the algorithm "forgets" the Gaussian with the smallest total pixel count m_i, and replaces it by the new local Gaussian.

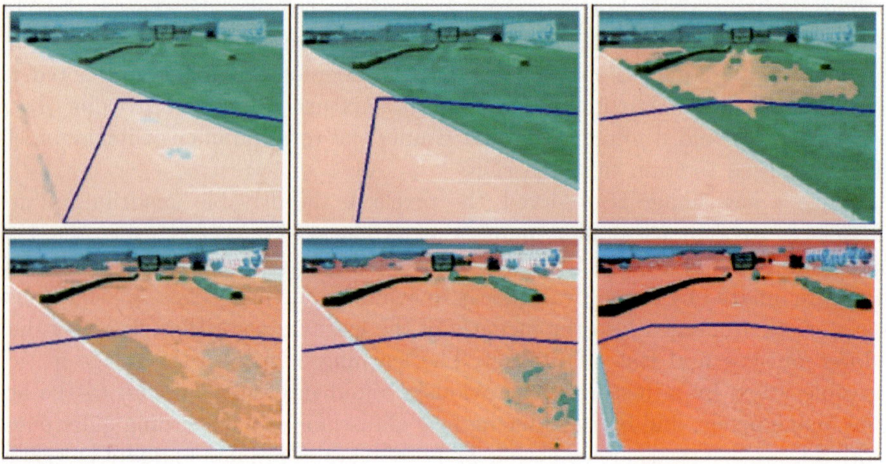

Fig. 1.16. These images illustrate the rapid adaptation of Stanley's computer vision routines. When the laser predominately screens the paved surface, the grass is not classified as drivable. As Stanley moves into the grass area, the classification changes. This sequence of images also illustrates why the vision result should not be used for steering decisions, in that the grass area is clearly drivable, yet Stanley is unable to detect this from a distance.

After this step, each counter m_i in the memory is discounted by a factor of $\gamma < 1$. This exponential decay term makes sure that the Gaussians in memory can be moved in new directions as the appearance of the drivable surface changes over time.

For finding drivable surface, the learned Gaussians are used to analyze the image. The image analysis uses an initial sky removal step defined in [Ettinger et al., 2003]. A subsequent flood-fill step then removes additional sky pixels not found by the algorithm in [Ettinger et al., 2003]. The remaining pixels are than classified using the learned mixture of Gaussian, in the straightforward way. Pixels whose RGB-value is near one or more of the learned Gaussians are classified as drivable; all other pixels are flagged as non-drivable. Finally, only regions connected to the laser quadrilateral are labeled as drivable.

Fig. 1.15 illustrates the key processing steps. Panel a in this figure shows a raw camera image, and Panel b shows the image after processing. Pixels classified as drivable are colored red, whereas non-drivable pixels are colored blue. The remaining two panels on Fig. 1.15 show intermediate processing steps: the classification response before thresholding (Panel c) and the result of the sky finder (Panel d).

Due to the ability to create new Gaussians on-the-fly, Stanley's vision routine can adapt to new terrain within seconds. Fig. 1.16 shows data acquired at the *National Qualification Event* of the DARPA Grand Challenge. Here the vehicle moves from a pavement to grass, both of which are drivable. The sequence in Fig. 1.16 illustrates the adaptation at work: the boxed areas towards the bottom

Fig. 1.17. Processed camera images in flat and mountainous terrain (Beer Bottle Pass)

of the image are the training region, and the red coloring in the image is the result of applying the learned classifier. As is easily seen in Fig. 1.16, the vision module successfully adapts from pavement to grass within less than a second while still correctly labeling the hay bales and other obstacles.

Under slowly changing lighting conditions, the system adapts more slowly to the road surface, making extensive use of past images in classification. This is illustrated in the bottom row of Fig. 1.17, which shows results for a sequence of images acquired at the Beer Bottle pass, the most difficult passage in the 2005 race. Here most of the terrain has similar visual appearance. The vision module, however, still competently segments the road. Such a result is only possible because the system balances the use of past images with its ability to adapt to new camera images.

Once a camera image has been classified, it is mapped into an overhead map, similar to the 2-D map generated by the laser. We already encountered such a map in Fig. 1.14b, which depicted the map of a straight road. Since certain color changes are natural even on flat terrain, the vision map is not used for steering control. Instead, it is used exclusively for velocity control. When no drivable corridor is detected within a range of 40 meters, the robot simply slows down to 25 mph, at which point the laser range is sufficient for safe navigation. In other words, the vision analysis serves as an early warning system for obstacles beyond the range of the laser sensors.

In developing the vision routines, the research team investigated a number of different learning algorithms. One of the primary alternatives to the generative mixture of Gaussian method was a discriminative method, which uses boosting and decision stumps for classification [Davies and Lienhart, 2006]. This method relies on examples of non-drivable terrain, which were extracted using an algorithm similar to the one for finding a drivable quadrilateral. A performance evaluation, carried out using independent test data gathered on the 2004 race

Table 1.1. Road detection rate for the two primary machine learning methods, broken down into different ranges. The comparison yields no conclusive winner.

	Flat Desert Roads		Mountain Roads	
	Discriminative training	Generative training	Discriminative training	Generative training
Drivable terrain detection rate, 10-20m	93.25%	90.46%	80.43%	88.32%
Drivable terrain detection rate, 20-35m	95.90%	91.18%	76.76%	86.65%
Drivable terrain detection rate, 35-50m	94.63%	87.97%	70.83%	80.11%
Drivable terrain detection rate, 50m+	87.13%	69.42%	52.68%	54.89%
False positives, all ranges	3.44%	3.70%	0.50%	2.60%

course, led to inconclusive results. Table 1.1 shows the classification accuracy for both methods, for flat desert roads and mountain roads. The generative mixture of Gaussian methods was finally chosen because it does not require training examples of non-drivable terrain, which can be difficult to obtain in flat open lake-beds.

1.7 Road Property Estimation

1.7.1 Road Boundary

One way to avoid obstacles is to detect them and drive around them. This is the primary function of the laser mapper. Another effective method is to drive in such a way that minimizes the a priori chances of encountering an obstacle. This is possible because obstacles are rarely uniformly distributed in the world. On desert roads, obstacles such as rocks, brush, and fence posts exist most often along the sides of the road. By simply driving down the middle of the road, most obstacles on desert roads can be avoided without ever detecting them!

One of the most beneficial components of Stanley's navigation routines, thus, is a method for staying near the center of the road. To find the road center, Stanley uses probabilistic low-pass filters to determine both road sides based using the laser map. The idea is simple; in expectation, the road sides are parallel to the RDDF. However, the exact lateral offset of the road boundary to the RDDF center is unknown and varies over time. Stanley's low-pass filters are implemented as one-dimensional Kalman filters. The state of each filter is the lateral distance between the road boundary and the center of the RDDF. The KFs search for possible obstacles along a discrete search pattern orthogonal to the RDDF, as shown in Fig. 1.18a. The largest free offset is the "observation" to the KF, in that it establishes the local measurement of the road boundary. So if multiple parallel roads exist in Stanley's field of view separated by a small berm, the filter will only trace the innermost drivable area.

By virtue of KF integration, the road boundaries change slowly. As a result, small obstacles or momentary situations without side obstacles affect the road boundary estimation only minimally; however, persistent obstacles that occur over extended period of time do have a strong effect.

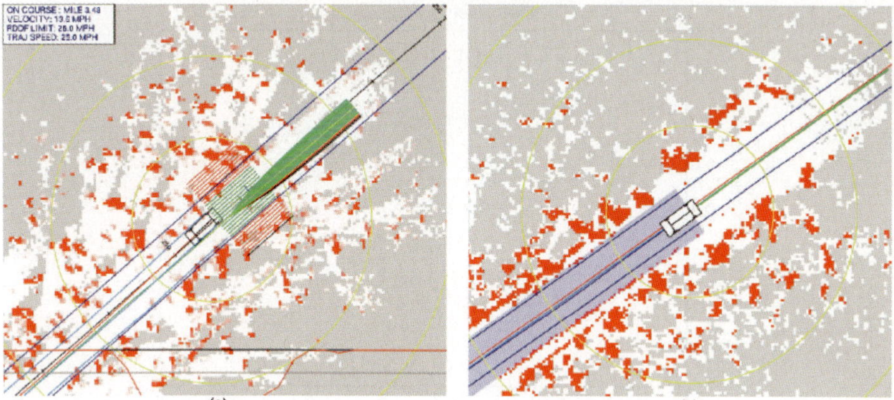

(a) (b)

Fig. 1.18. (a) Search regions for the road detection module: the occurrence of obstacles is determined along a sequence of lines parallel to the RDDF. (b) The result of the road estimator is shown in blue, behind the vehicle. Notice that the road is bounded by two small berms.

Based on the output of these filters, Stanley defines the road to be the center of the two boundaries. The road center's lateral offset is a component in scoring trajectories during path planning, as will be discussed further below. In the absence of other contingencies, Stanley slowly converges to the estimated road center. Empirically, we found that this driving technique stays clear of the vast majority of natural obstacles on desert roads. While road centering is clearly only a heuristic, we found it to be highly effective in extensive desert tests.

Fig. 1.18b shows an example result of the road estimator. The blue corridor shown there is Stanley's best estimate of the road. Notice that the corridor is confined by two small berms, which are both detected by the laser mapper. This module plays an important role in Stanley's ability to negotiate desert roads.

1.7.2 Terrain Ruggedness

In addition to avoiding obstacles and staying centered along the road, another important component of safe driving is choosing an appropriate velocity [Iagnemma et al., 2004]. Intuitively speaking, desert terrain varies from flat and smooth to steep and rugged. The type of the terrain plays an important role in determining the maximum safe velocity of the vehicle. On steep terrain, driving too fast may lead to fishtailing or sliding. On rugged terrain, excessive speeds may lead to extreme shocks that can damage or destroy the robot. Thus, sensing the terrain type is essential for the safety of the vehicle. In order to address these two situations, Stanley's velocity controller constantly estimates terrain slope and ruggedness and uses these values to set intelligent maximum speeds.

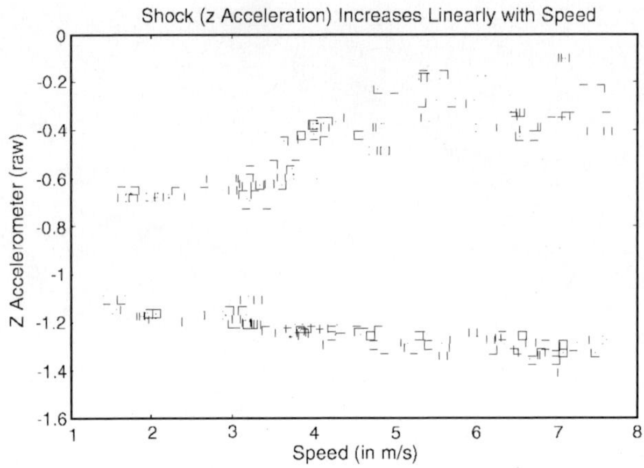

Fig. 1.19. The relationship between velocity and imparted acceleration from driving over a fixed sized obstacle at varying speeds. The plot shows two distinct reactions to the obstacle, one up and one down. While this relation is ultimately non-linear, it is well modeled by a linear function within the range relevant for desert driving.

The terrain slope is taken directly from the vehicle's pitch estimate, as computed by the UKF. Borrowing from [Brooks and Iagnemma, 2005], the terrain ruggedness is measured using the vehicle's z accelerometer. The vertical acceleration is band-pass filtered to remove the effect of gravity and vehicle vibration, while leaving the oscillations in the range of the vehicle's resonant frequency. The amplitude of the resulting signal is a measurement of the vertical shock experienced by the vehicle due to excitation by the terrain. Empirically, this filtered acceleration appears to vary linearly with velocity. (See Fig. 1.19.) In other words, doubling the maximum speed of the vehicle over a section of terrain will approximately double the maximum differential acceleration imparted on the vehicle. In Section 1.9.1, this relationship will be used to derive a simple rule for setting maximum velocity to approximately bound the maximum shock imparted on the vehicle.

1.8 Path Planning

As was previously noted, the RDDF file provided by DARPA largely eliminates the need for any global path planning. Thus, the role of Stanley's path planner is primarily local obstacle avoidance. Instead of planning in the global coordinate frame, Stanley's path planner was formulated in a unique coordinate system: perpendicular distance, or "lateral offset" to a fixed base trajectory. Varying lateral offset moves Stanley left and right with respect to the base trajectory, much like a car changes lanes on a highway. By changing lateral offset intelligently, Stanley can avoid obstacles at high speeds while making fast progress along the course.

The base trajectory that defines lateral offset is simply a smoothed version of the skeleton of the RDDF corridor. It is important to note that this base trajectory is not meant to be an optimal trajectory in any sense; it serves as a baseline coordinate system upon which obstacle avoidance maneuvers are continuously layered. The following two sections will describe the two parts to Stanley's path planning software: the path smoother that generates the base trajectory before the race, and the online path planner which is constantly adjusting Stanley's trajectory.

1.8.1 Path Smoothing

Any path can be used as a base trajectory for planning in lateral offset space. However, certain qualities of base trajectories will improve overall performance.

- **Smoothness.** The RDDF is a coarse description of the race corridor and contains many sharp turns. Blindly trying to follow the RDDF waypoints would result in both significant overshoot and high lateral accelerations, both of which could adversely affect vehicle safety. Using a base trajectory that is smoother than the original RDDF will allow Stanley to travel faster in turns and follow the intended course with higher accuracy.
- **Matched curvature.** While the RDDF corridor is parallel to the road in expectation, the curvature of the road is poorly predicted by the RDDF file in turns, again due to the finite number of waypoints. By default, Stanley will prefer to drive parallel to the base trajectory, so picking a trajectory that exhibits curvature that better matches the curvature of the underlying desert roads will result in fewer changes in lateral offset. This will also result in smoother, faster driving.

Stanley's base trajectory is computed before the race in a four-stage procedure.

1. First, points are added to the RDDF in proportion to the local curvature (see Fig. 1.20a).
2. The coordinates of all points in the upsampled trajectory are then adjusted through least squares optimization. Intuitively, this optimization adjusts

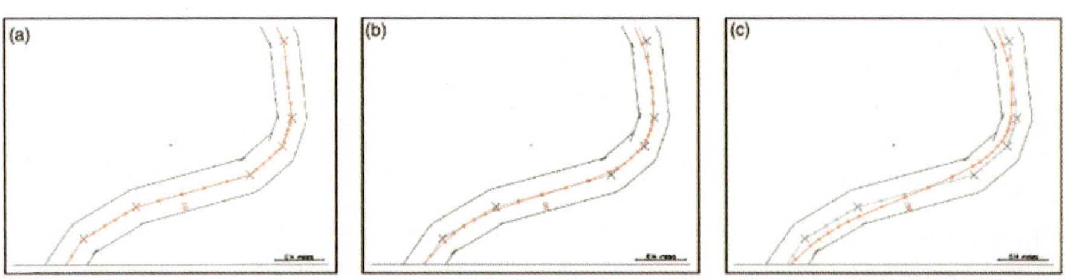

Fig. 1.20. Smoothing of the RDDF: (a) adding additional points; (b) the trajectory after smoothing (shown in red); (c) a smoothed trajectory with a more aggressive smoothing parameter. The smoothing process takes only 20 seconds for the entire 2005 course.

each waypoint so as to minimize the curvature of the path while staying as close as possible to the waypoints in the original RDDF. The resulting trajectory is still piecewise linear, but it is significantly smoother than the original RDDF.

Let $x_1, \ldots, x_N$ be the waypoints of the base trajectory to be optimized. For each of these points, we are given a corresponding point along the original RDDF, which shall be denoted y_i. The points $x_1, \ldots, x_N$ are obtained by minimizing the following additive function:

$$\operatorname*{argmin}_{x_1,\ldots,x_N} \sum_i |y_i - x_i|^2 \quad - \quad \beta \sum_n \frac{(x_{n+1} - x_n) \cdot (x_n - x_{n-1})}{|x_{n+1} - x_n| \; |x_n - x_{n-1}|} \quad + \quad \sum_n f_{\text{RDDF}}(x_n) \tag{1.6}$$

Here $|y_i - x_i|^2$ is the quadratic distance between the waypoint x_i and the corresponding RDDF anchor point y_i; the index variable i iterates over the set of points x_i. Minimizing this quadratic distance for all points i ensures that the base trajectory stays close to the original RDDF. The second expression in Eq. (1.6) is a curvature term; It minimizes the angle between two consecutive line segments in the base trajectory by minimizing the dot product of the segment vectors. Its function is to smooth the trajectory: the smaller the angle, the smoother the trajectory. The scalar β trades off these two objectives and is a parameter in Stanley's software. The function $f_{\text{RDDF}}(x_n)$ is a differentiable barrier function that goes to infinity as a point x_n approaches the RDDF boundary, but is near zero inside the corridor away from the boundary. As a result, the smoothed trajectory is always inside the valid RDDF corridor. The optimization is performed with a fast version of conjugate gradient descent, which moves RDDF points freely in 2-D space.

3. The next step of the path smoother involves cubic spline interpolation. The purpose of this step is to obtain a path that is differentiable. This path can then be resampled efficiently.
4. The final step of path smoothing pertains to the calculation of the speed limit attached to each waypoint of the smooth trajectory. Speed limits are the minimum of three quantities: (a) the speed limit from corresponding segment of the original RDDF, (b) a speed limit that arises from a bound on lateral acceleration, and (c) a speed limit that arises from a bounded deceleration constraint. The lateral acceleration constraint forces the vehicle to slow down appropriately in turns. When computing these limits, we bound the lateral acceleration of the vehicle to 0.75 m/sec^2, in order to give the vehicle enough maneuverability to safely avoid obstacles in curved segments of the course. The bounded deceleration constraint forces the vehicle to slow down in anticipation of turns and changes in DARPA speed limits.

Fig. 1.20 illustrates the effect of smoothing on a short segment of the RDDF. Panel a shows the RDDF and the upsampled base trajectory before smoothing. Panels b and c show the trajectory after smoothing (in red), for different values of the parameter β. The entire data pre-processing step is fully automated, and requires only approximately 20 seconds of compute time on a 1.4 GHz laptop, for the entire 2005 race course. This base trajectory is transferred onto Stanley, and the software is ready to go. No further information about the environment or the race is provided to the robot.

It is important to note that Stanley does not modify the original RDDF file. The base trajectory is only used as the coordinate system for obstacle avoidance. When evaluating whether particular trajectories stay within the designated race course, Stanley checks against the original RDDF file. In this way, the preprocessing step does not affect the interpretation of the corridor constraint imposed by the rules of the race.

1.8.2 Online Path Planning

Stanley's online planning and control system is similar to the one described in [Kelly and Stentz, 1998]. The online component of the path planner is responsible for determining the actual trajectory of the vehicle during the race. The goal of the planner is to complete the course as fast as possible while successfully avoiding obstacles and staying inside the RDDF corridor. In the absence of obstacles, the planner will maintain a constant lateral offset from the base trajectory. This results in driving a path parallel to the base trajectory, but possibly shifted left or right. If an obstacle is encountered, Stanley will plan a smooth change in lateral offset that avoids the obstacle and can be safely executed. Planning in lateral offset space also has the advantage that it gracefully handles GPS error. GPS error may systematically shift Stanley's position estimate. The path planner will simply adjust the lateral offset of the current trajectory to recenter the robot in the road.

The path planner is implemented as a search algorithm that minimizes a linear combination of continuous cost functions, subject to a fixed vehicle model. The vehicle model includes several kinematic and dynamic constraints including maximum lateral acceleration (to prevent fishtailing), maximum steering angle (a joint limit), maximum steering rate (maximum speed of the steering motor), and maximum deceleration (due to the stopping distance of the Touareg). The cost functions penalize running over obstacles, leaving the RDDF corridor, and the lateral offset from the current trajectory to the sensed center of the road surface. The soft constraints induce a ranking of admissible trajectories. Stanley chooses the best such trajectory. In calculating the total path costs, unknown territory is treated the same as drivable surface, so that the vehicle does not swerve around unmapped spots on the road, or specular surfaces such as puddles.

At every time step, the planner considers trajectories drawn from a two-dimensional space of maneuvers. The first dimension describes the amount of lateral offset to be added to the current trajectory. This parameter allows Stanley to move left and right, while still staying essentially parallel to the base trajectory. The second dimension describes the rate at which Stanley will attempt to change to this lateral offset. The lookahead distance is speed-dependent and ranges from 15 to 25 meters. All candidate paths are run through the vehicle model to ensure that obey the kinematic and dynamic vehicle constraints. Repeatedly layering these simple maneuvers on top of the base trajectory can result in quite sophisticated trajectories.

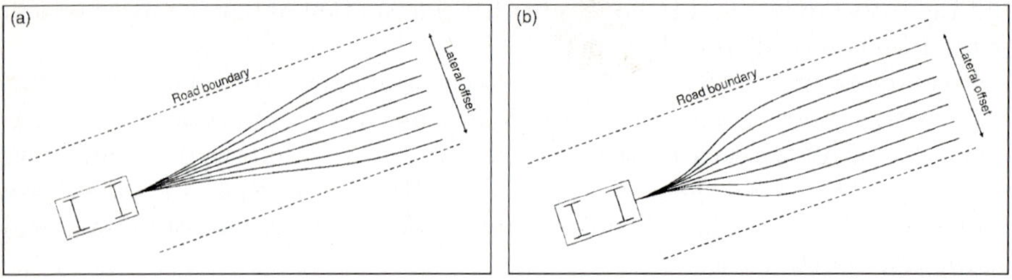

Fig. 1.21. Path planning in a 2-D search space: (a) shows paths that change lateral offsets with the minimum possible lateral acceleration (for a fixed plan horizon); (b) shows the same for the maximum lateral acceleration. The former are called "nudges," and the latter are called "swerves."

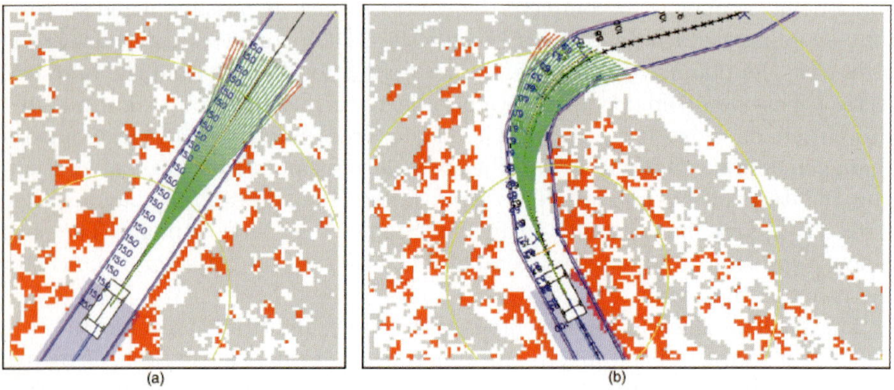

Fig. 1.22. Snapshots of the path planner as it processes the drivability map. Both snapshots show a map, the vehicle, and the various nudges considered by the planner. The first snapshot stems from a straight road (Mile 39.2 of the 2005 race course). Stanley is traveling 31.4 mph, hence can only slowly change lateral offsets due to the lateral acceleration constraint. The second example is taken from the most difficult part of the 2005 DARPA Grand Challenge, a mountainous area called Beer Bottle Pass. Both images show only nudges for clarity.

The second parameter in the path search allows the planner to control the urgency of obstacle avoidance. Discrete obstacles in the road, such as rocks or fence posts often require the fastest possible change in lateral offset. Paths that change lateral offset as fast as possible without violating the lateral acceleration constraint are called "swerves." Slow changes in the positions of road boundaries require slow, smooth adjustment to the lateral offset. Trajectories with the slowest possible change in lateral offset for a given planning horizon are called "nudges." Swerves and nudges span a spectrum of maneuvers appropriate for high speed obstacle avoidance: fast changes for avoiding head on obstacles, and slow changes for smoothly tracking the road center. Swerves and nudges are illustrated in Fig. 1.21. On a straight road, the resulting trajectories are similar to those of Ko and Simmons's lane curvature method [Ko and Simmons, 1998].

The path planner is executed at 10 Hz. The path planner is ignorant to actual deviations from the vehicle and the desired path, since those are handled by the low-level steering controller. The resulting trajectory is therefore always continuous. Fast changes in lateral offset (swerves) will also include braking in order to increase the amount of steering the vehicle can do without violating the maximum lateral acceleration constraint.

Fig. 1.22 shows an example situation for the path planner. Shown here is a situation taken from Beer Bottle Pass, the most difficult passage of the 2005 Grand Challenge. This image only illustrates one of the two search parameters: the lateral offset. It illustrates the process through which trajectories are generated by gradually changing the lateral offset relative to the base trajectory. By using the base trajectory as a reference, path planning can take place in a low-dimensional space, which we found to be necessary for real-time performance.

1.9 Real-Time Control

Once the intended path of the vehicle has been determined by the path planner, the appropriate throttle, brake, and steering commands necessary to achieve that path must be computed. This control problem will be described in two parts: the velocity controller and steering controller.

1.9.1 Velocity Control

Multiple software modules have input into Stanley's velocity, most notably the path planner, the health monitor, the velocity recommender, and the low-level velocity controller. The low-level velocity controller translates velocity commands from the first three modules into actual throttle and brake commands. The implemented velocity is always the minimum of the three recommended speeds. The path planner will set a vehicle velocity based on the base trajectory speed limits and any braking due to swerves. The vehicle health monitor will lower the maximum velocity due to certain preprogrammed conditions, such as GPS blackouts or critical system failures.

The velocity recommender module sets an appropriate maximum velocity based on estimated terrain slope and roughness. The terrain slope affects the maximum velocity if the pitch of the vehicle exceeds 5 degrees. Beyond 5 degrees of slope, the maximum velocity of the vehicle is reduced linearly to values that, in the extreme, restrict the vehicle's velocity to 5 mph. The terrain ruggedness is fed into a controller with hysteresis that controls the velocity setpoint to exploit the linear relationship between filtered vertical acceleration amplitude and velocity; see Sect. 1.7.2. If rough terrain causes a vibration that exceeds the maximum allowable threshold, the maximum velocity is reduced linearly such that continuing to encounter similar terrain would yield vibrations exactly meeting the shock limit. Barring any further shocks, the velocity limit is slowly increased linearly with distance traveled.

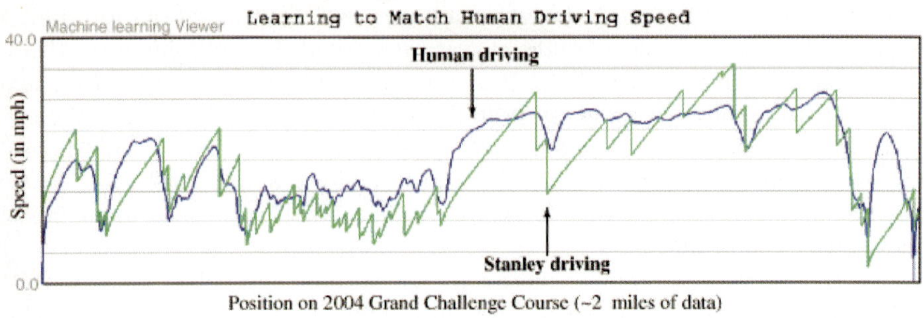

Fig. 1.23. Velocity profile of a human driver and of Stanley's velocity controller in rugged terrain. Stanley identifies controller parameters that match human driving. This plot compares human driving with Stanley's control output.

This rule may appear odd, but it has great practical importance; it reduces the Stanley's speed when the vehicle hits a rut. Obviously, the speed reduction occurs after the rut is hit, not before. By slowly recovering speed, Stanley will approach nearby ruts at a much lower speed. As a result, Stanley tends to drive slowly in areas with many ruts, and only returns to the base trajectory speed when no ruts have been encountered for a while. While this approach does not avoid isolated ruts, we found it to be highly effective in avoiding many shocks that would otherwise harm the vehicle. Driving over wavy terrain can be just as hard on the vehicle as driving on ruts. In bumpy terrain, slowing down also changes the frequency at which the bumps pass, reducing the effect of resonance.

The velocity recommender is characterized by two parameters: the maximum allowable shock, and the linear recovery rate. Both are learned from human driving. More specifically, by recording the velocity profile of a human in rugged terrain, Stanley identifies the parameters that most closely match the human driving profile. Fig. 1.23 shows the velocity profile of a human driver in a mountainous area of the 2004 Grand Challenge Course (the "Daggett Ridge"). It also shows the profile of Stanley's controller for the same data set. Both profiles tend to slow down in the same areas. Stanley's profile, however, is different in two ways: the robot deceleerates much faster than a person, and its recovery is linear whereas the person's recovery is nonlinear. The fast acceleration is by design, to protect the vehicle from further impact.

Once the planner, velocity recommender, and health monitor have all submitted velocities, the minimum of these speeds is implemented by the velocity controller. The velocity controller treats the brake cylinder pressure and throttle level as two opposing, single-acting actuators that exert a longitudinal force on the car. This is a very close approximation for the brake system, and was found to be an acceptable simplification of the throttle system. The controller computes a single error metric, equal to a weighted sum of the velocity error and the integral of the velocity error. The relative weighting determines the trade-off between disturbance rejection and overshoot. When the error metric is positive,

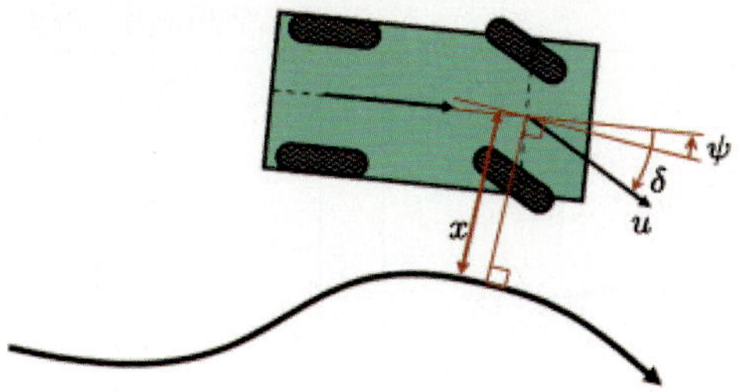

Fig. 1.24. Illustration of the steering controller. With zero cross-track error, the basic implementation of the steering controller steers the front wheels parallel to the path. When cross-track error is perturbed from zero, it is nulled by commanding the steering according to a non-linear feedback function.

the brake system commands a brake cylinder pressure proportional to the PI error metric, and when it is negative, the throttle level is set proportional to the negative of the PI error metric. By using the same PI error metric for both actuators, the system is able to avoid the chatter and dead bands associated with opposing, single-acting actuators. To realize the commanded brake pressure, the hysteretic brake actuator is controlled through saturated proportional feedback on the brake pressure, as measured by the Touareg, and reported through the CAN bus interface.

1.9.2 Steering Control

The steering controller accepts as input the trajectory generated by the path planner, the UKF pose and velocity estimate, and the measured steering wheel angle. It outputs steering commands at a rate of 20 Hz. The function of this controller is to provide closed loop tracking of the desired vehicle path, as determined by the path planner, on quickly varying, potentially rough terrain.

The key error metric is the cross-track error, $x(t)$, as shown in Fig. 1.24, which measures the lateral distance of the center of the vehicle's front wheels from the nearest point on the trajectory. The idea now is to command the steering by a control law that yields an $x(t)$ that converges to zero.

Stanley's steering controller, at the core, is based on a non-linear feedback function of the cross-track error, for which exponential convergence can be shown. Denote the vehicle speed at time t by $u(t)$. In the error-free case, using this term, Stanley's front wheels match the global orientation of the trajectory. This is illustrated in Fig. 1.24. The angle ψ in this diagram describes the orientation of the nearest path segment, measured relative to the vehicle's own orientation. In the absence of any lateral errors, the control law points the front wheels parallel to the planner trajectory.

The basic steering angle control law is given by

$$\delta(t) = \psi(t) + \arctan \frac{k\, x(t)}{u(t)} \tag{1.6}$$

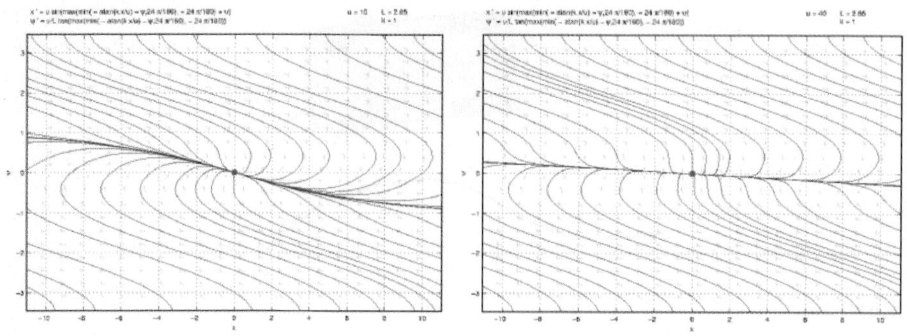

Fig. 1.25. Phase portrait for $k = 1$ at 10 and 40 meter per second, respectively, for the basic controller, including the effect of steering input saturation

where k is a gain parameter. The second term adjusts the steering in (nonlinear) proportion to the cross-track error $x(t)$: the larger this error, the stronger the steering response towards the trajectory.

Using a linear bicycle model with infinite tire stiffness and tight steering limitations (see [Gillespie, 1992]) results in the following effect of the control law:

$$\dot{x}(t) = -u(t) \sin \arctan\left(\frac{kx(t)}{u(t)}\right) = \frac{-kx(t)}{\sqrt{1 + \left(\frac{kx(t)}{u(t)}\right)^2}} \tag{1.7}$$

and hence for small cross track error,

$$x(t) \approx x(0) \ \exp -kt \tag{1.8}$$

Thus, the error converges exponentially to $x(t) = 0$. The parameter k determines the rate of convergence. As cross track error increases, the effect of the arctan function is to turn the front wheels to point straight towards the trajectory, yielding convergence limited only by the speed of the vehicle. For any value of $x(t)$, the differential equation converges monotonically to zero. Fig. 1.25 shows phase portrait diagrams for Stanley's final controller in simulation, as a function of the error $x(t)$ and the orientation $\psi(t)$, including the effect of steering input saturation. These diagrams illustrate that the controller converges nicely for the full range attitudes and a wide range of cross-track errors, in the example of two different velocities.

This basic approach works well for lower speeds, and a variant of it can even be used for reverse driving. However, it neglects several important effects. There is a discrete, variable time delay in the control loop, inertia in the steering column, and more energy to dissipate as speed increases. These effects are handled by simply damping the difference between steering command and the measured steering wheel angle, and including a term for yaw damping. Finally, to compensate for the slip of the actual pneumatic tires, the vehicle is commanded to have a steady state yaw offset that is a non-linear function of the path curvature and the vehicle speed, based on a bicycle vehicle model, with slip, that was calibrated and verified in testing. These terms combine to stabilize the vehicle and drive the cross-track error to zero, when run on the physical vehicle. The resulting controller has proven stable in testing on terrain from pavement to deep, off-road mud puddles, and on trajectories with tight enough radii of curvature to cause substantial slip. It typically demonstrates tracking error that is on the order of the estimation error of this system.

1.10 Development Process and Race Results

1.10.1 Race Preparation

The race preparation took place at three different locations: Stanford University, the 2004 Grand Challenge Course between Barstow and Primm, and the Sonoran Desert near Phoenix, AZ. In the weeks leading up to the race, the team permanently moved to Arizona, where it enjoyed the hospitality of Volkswagen of America's Arizona Proving Grounds. Fig. 1.26 shows examples of hardware testing in extreme offroad terrain; these pictures were taken while the vehicle was operated by a person.

In developing Stanley, the Stanford Racing Team adhered to a tight development and testing schedule, with clear milestones along the way. Emphasis was placed on early integration, so that an end-to-end prototype was available nearly a year before the race. The system was tested periodically in desert environments representative of the team's expectation for the Grand Challenge race. In the months leading up to the race, all software and hardware modules were debugged and subsequently frozen. The development of the system terminated well ahead of the race.

The primary measure of system capability was "MDBCF" – mean distance between catastrophic failures. A catastrophic failure was defined as a condition under which a human driver had to intervene. Common failures involved software problems such as the one in Fig. 1.9; occasional failures were caused by the hardware, e.g., the vehicle power system. In December 2004, the MDBCF was approximately 1 mile. It increased to 20 miles in July 2005. The last 418 miles before the National Qualification event were free of failures; this included a single 200-mile run over a cyclic testing course. At that time the system development was suspended, Stanley's lateral navigation accuracy was approximately 30 cm. The vehicle had logged more than 1,200 autonomous miles.

In preparing for this race, the team also tested sensors that were not deployed in the final race. Among them was an industrial strength stereo vision sensor with a 33 cm baseline. In early experiments, we found that the stereo system provided excellent results in the short range, but lacked behind the laser system in accuracy. The decision not to use stereo was simply based on the observation that it added little to the laser system. A larger baseline might have made the stereo more useful at longer ranges, but was unfortunately not available.

The second sensor that was not used in the race was the 24 GHz RADAR system. The RADAR uses a linear frequency shift keying modulated (LFMSK) transmit waveform; it is normally used for adaptive cruise control (ACC). After

Fig. 1.26. Vehicle testing at the Volkswagen Arizona Proving Grounds, manual driving

carefully tuning gains and acceptance thresholds of the sensor, the RADAR proved highly effective in detecting large frontal obstacles such as abandoned vehicles in desert terrain. Similar to the mono-vision system in Sect. 1.6, the RADAR was tasked to screen the road at a range beyond the laser sensors. If a potential obstacle was detected, the system limits Stanley's speed to 25 mph so that the lasers could detect the obstacle in time for collision avoidance.

While the RADAR system proved highly effective in testing, two reasons led us not to use it in the race. The first reason was technical: During the National Qualification Event (NQE), the USB driver of the receiving computer repeatedly caused trouble, sometimes stalling the receiving computer. The second reason was pragmatical. During the NQE, it became apparent that the probability of encountering large frontal obstacles was small in high-speed zones; and even if those existed, the vision system would very likely detect them. As a consequence, the team felt that the technical risks associated with the RADAR system outweigh its benefits, and made the decision not to use RADAR in the race.

1.10.2 National Qualification Event

The National Qualification Event (NQE) took place September 27 through October 5 on the California Speedway in Fontana, CA. Like most competitive robots, Stanley qualified after four test runs. From the 43 semifinalists, 11 completed the course in the first run, 13 in the second run, 18 in the third run, and 21 in the fourth run. Stanley's times were competitive but not the fastest (Run 1: 10:38; run 2: 9:12; run 3: 11:06; run 4: 11:06). However, Stanley was the only vehicle that cleared all 50 gates in every run, and avoided collisions with all of the obstacles. This flawless performance earned Stanley the number two starting position, behind CMU's H1ghlander robot and ahead of the slightly faster Sandstorm robot, also by CMU.

1.10.3 The Race

At approximately 4:10am on October 8, 2005, the Stanford Racing Team received the race data, which consisted of 2,935 GPS-referenced coordinates along with speed limits of up to 50 mph. Stanley started the race at 6:35am on October 8, 2005. The robot immediately picked up speed and drove at or just

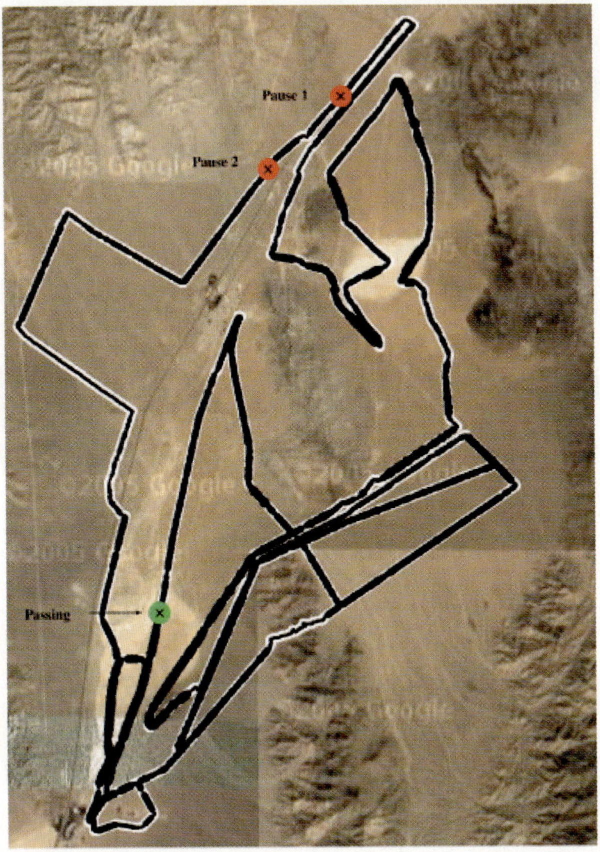

Fig. 1.27. This map shows Stanley's path. The *thickness* of the trajectory indicates Stanley's speed (thicker means faster). At the locations marked by the red 'x's, the race organizers paused Stanley because of the close proximity of CMU's H1ghlander robot. At Mile 101.5, H1ghlander was paused and Stanley passed. This location is marked by a green 'x'.

below the speed limit. 3 hours, 45 minutes and 22 seconds into the race, at Mile 73.5, DARPA paused Stanley for the first time, to give more space to CMU's H1ghlander robot, which started five minutes ahead of Stanley. The first pause lasted 2 minutes and 45 seconds. Stanley was paused again only 5 minutes and 40 seconds later, at Mile 74.9 (3 hours, 53 minutes, and 47 seconds into the race). This time the pause lasted 6 minutes and 35 seconds, for a total pause time of 9 minutes and 20 seconds. The locations of the pauses are shown in Fig. 1.27. From this point on, Stanley repeatedly approached H1ghlander within a few hundred yards. Even though Stanley was still behind H1ghlander, it was leading the race.

5 hours, 24 minutes and 45 seconds into the race, DARPA finally paused H1ghlander and allowed Stanley to pass. The passing happened a Mile 101.5;

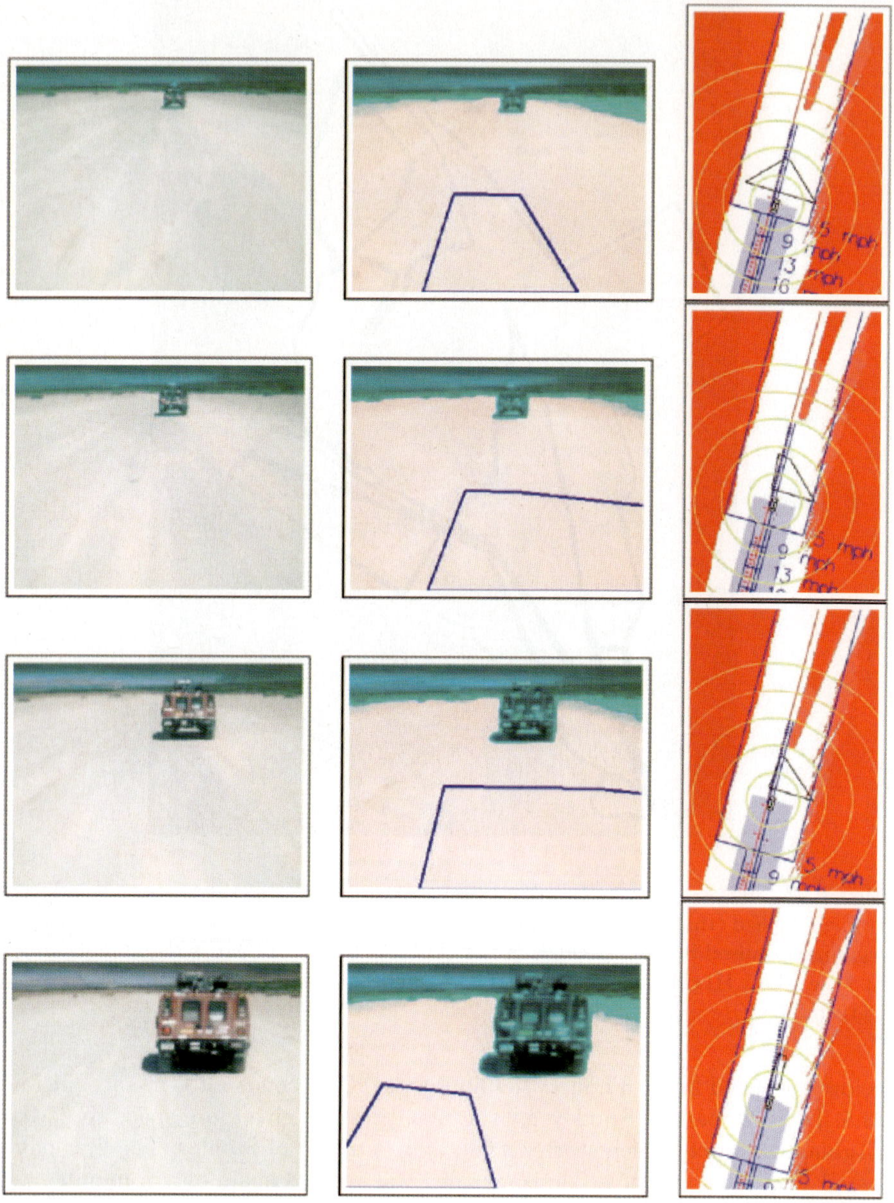

Fig. 1.28. Passing CMU's H1ghlander robot: The left column shows a sequence of camera images, the center column the processed images with obstacle information overlayed, and the right column the 2D map derived from the processed image. The vision routine detects H1ghlander as an obstacle at a 40 meter range, approximately twice the range of the lasers.

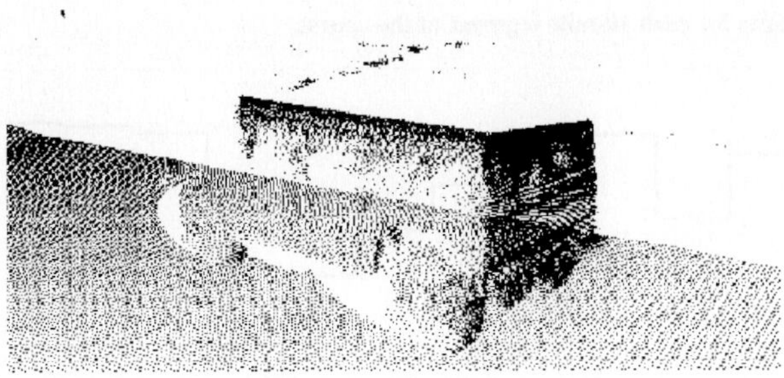

Fig. 1.29. Laser model of CMU's H1ghlander robot, taken at Mile 101.5

the location is marked by a green circle in Fig. 1.27. Fig. 1.28 shows processed camera images of the passing process acquired by Stanley, and Fig. 1.29 depicts a 3-D model of H1ghlander as it is being passed. Since Stanley started in second pole position and finished first, the top-seeded H1ghlander robot was the only robot encountered by Stanley during the race.

As noted in the introduction to this article, Stanley finished first, at an unmatched finishing time of 6 hours 53 minutes and 58 seconds. Its overall average velocity was 19.1 mph. However, Stanley's velocity varied wildly during the race. Initially, the terrain was flat and the speed limits allowed for much higher speeds. Stanley reached its top speed of 38.0 mph at Mile 5.43, 13 minutes and 47 seconds into the race. Its maximum *average* velocity during the race was 24.8 mph, which Stanley attained after 16 minutes and 56 seconds, at Mile 7.00. Speed limits then forced Stanley to slow down. Between Mile 84.9 and 88.1, DARPA restricted the maximum velocity to 10 mph. Shortly thereafter, at Mile 90.6 and 4 hours, 57 minutes, and 7 seconds into the race, Stanley attained its minimum average velocity of 18.3 mph. The total profile of velocities is shown in Fig. 1.30.

As explained in this paper, Stanley uses a number of strategies to determine the actual travel speed. During 68.2% of the course, Stanley's velocity was limited as pre-calculated, by following the DARPA speed limits or the maximum lateral acceleration constraints in turns. For the remaining 31.8%, Stanley chose to slow down dynamically, as the result of its sensor measurements. In 18.1%, the slow-down was the result of rugged or steep terrain. The vision module caused Stanley to slow down to 25 mph for 13.1% of the total distance; however, without the vision module Stanley would have been forced to a 25 mph maximum speed, which would have resulted in a finishing time of approximately 7 hours and 5 minutes, possibly behind CMU's Sandstorm robot. Finally, 0.6% of the course Stanley drove slower because it was denied GPS readings. Fig. 1.31 illustrates the effect of terrain ruggedness on the overall velocity. The curve on the top illustrates the magnitude at which Stanley slowed down to accommodate rugged terrain; the bottom diagram shows the altitude profile, as provided by DARPA. The terrain ruggedness triggers mostly in mountainous terrain. We believe that the ability to adapt the speed to the ruggedness of the terrain was an essential ingredient in Stanley's success.

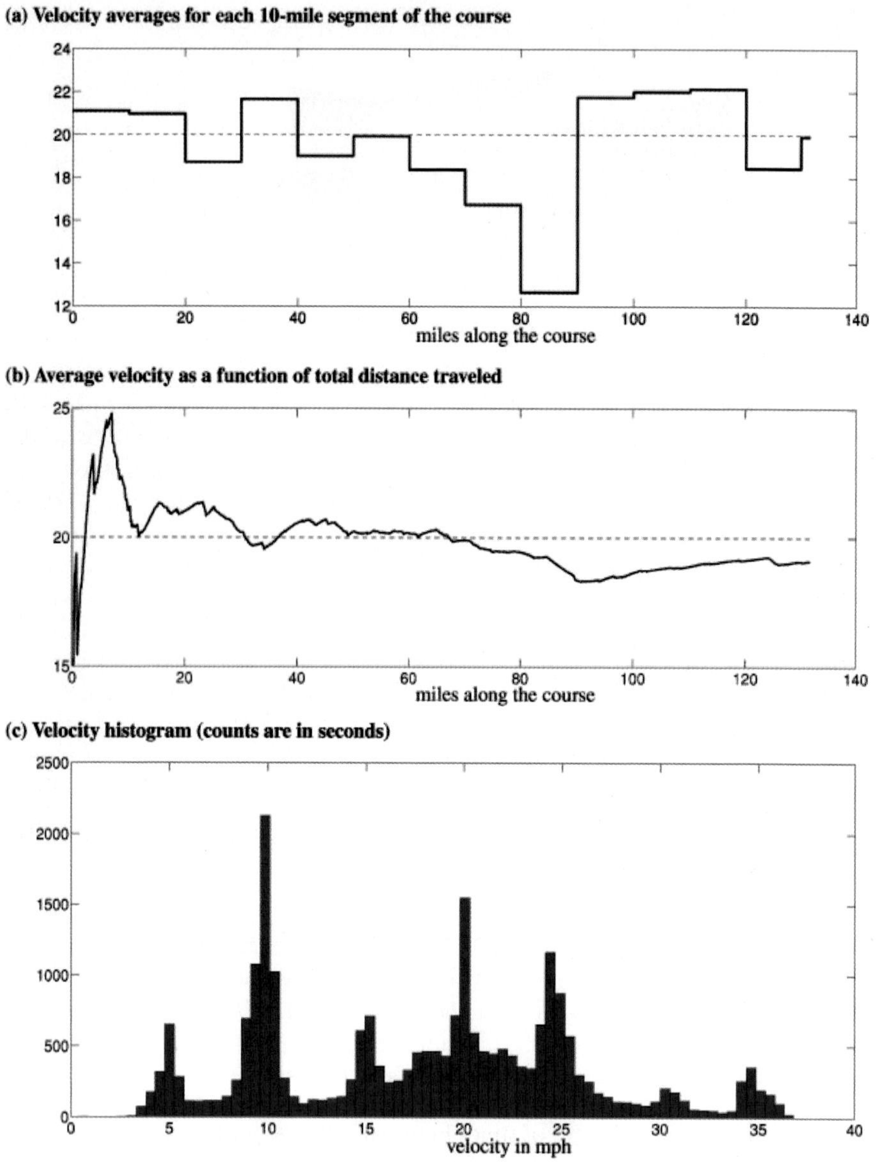

Fig. 1.30. Stanley's cumulative velocity

Stanley also encountered some unexpected difficulties along the course. Early on in the race, Stanley's laser data stream repeatedly stalled for durations of 300 to 1,100 milliseconds. There were a total of 17 incidents, nearly all of which occurred between Mile 22 and Mile 35. The resulting inaccurate time stamping of the laser data led to the insertion of phantom obstacles into the map. In four

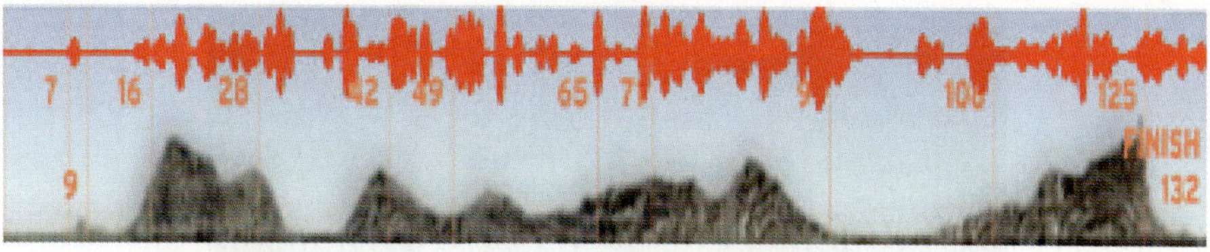

Fig. 1.31. This diagram shows where along the race course the road conditions forced Stanley to slow down. Slow-down predominately occurred in the mountains.

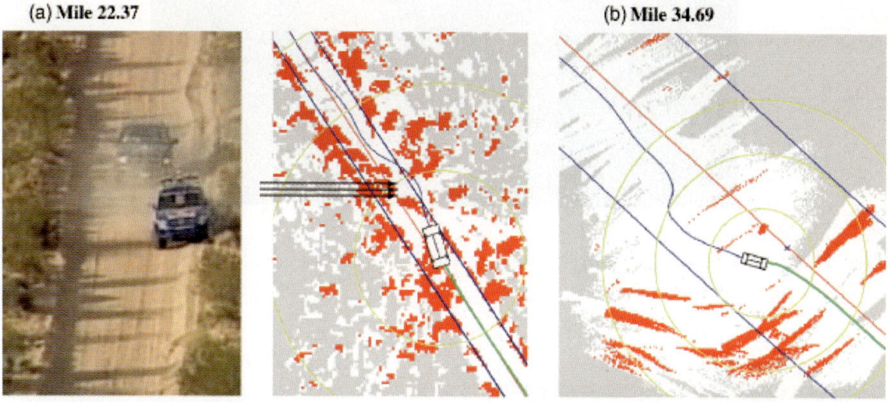

Fig. 1.32. Problems during the race caused by a stalling of the laser data stream. In both cases, Stanley swerved around phantom obstacles; at Mile 22.37 Stanley drove on the berm. None of these incidents led to a collision or an unsafe driving situation during the race.

of those cases, those incidents resulted in a significant swerve. The two most significant of these swerves are shown in Fig. 1.32. Both of those swerves were quite noticeable. In one case, Stanley even drove briefly on the berm as shown in Fig. 1.32a; in the other, Stanley swerved on an open lake bed without any obstacles, as shown in Fig. 1.32b. At no point was the vehicle in jeopardy, as the berm that was traversed was drivable. However, as a result of these errors, Stanley slowed down a number of times between Miles 22 and 35. Thus, the main effect of these incidents was a loss of time early in the race. The data stream stalling problem vanished entirely after Mile 37.85. It only reoccurred once at Mile 120.66, without any visible change of the driving behavior.

During 4.7% of the Grand Challenge, the GPS reported 60 cm error or more. Naturally, this number represents the unit's own estimate, which may not necessarily be accurate. However, this raises the question of how important online mapping and path planning was in this race.

Stanley frequently moved away from the center axis of the RDDF. On average, the lateral offset was ±74 cm. The maximum lateral offset during the race was

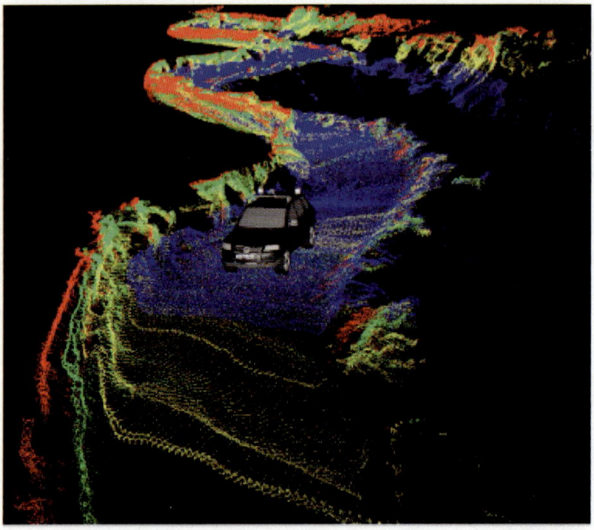

Fig. 1.33. Sensor image from the Beer Bottle Pass, the most difficult passage of the DARPA Grand Challenge

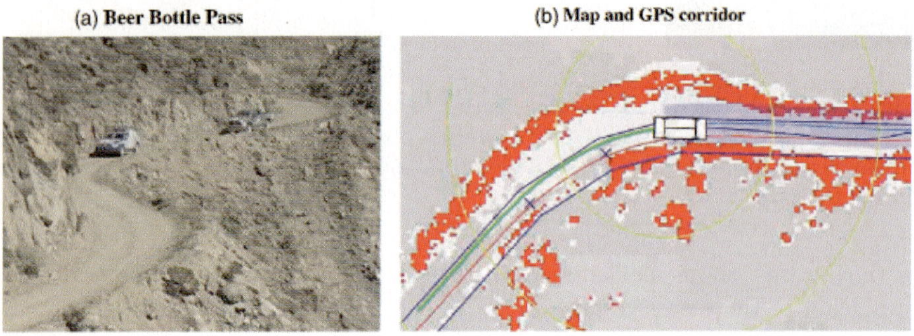

Fig. 1.34. Image of the Beer Bottle pass, and snapshot of the map acquired by the robot. The two blue contours in the map mark the GPS corridor provided by DARPA, which aligns poorly with the map data. This analysis suggests that a robot that followed the GPS via points blindly would likely have failed to traverse this narrow mountain pass.

10.7 meters, which was the result of the swerve shown in Fig. 1.32b. However, such incidents were rare, and in nearly all cases non-zero lateral offsets were the results of obstacles in the robot's path.

An example situation is depicted in Fig. 1.33. This figure shows raw laser data from the Beer Bottle Pass, the most difficult section of the course. An images of this pass is depicted in Fig. 1.34a. Of interest is the map in Fig. 1.34b. Here the DARPA-provided corridor is marked by the two solid blue lines. This image

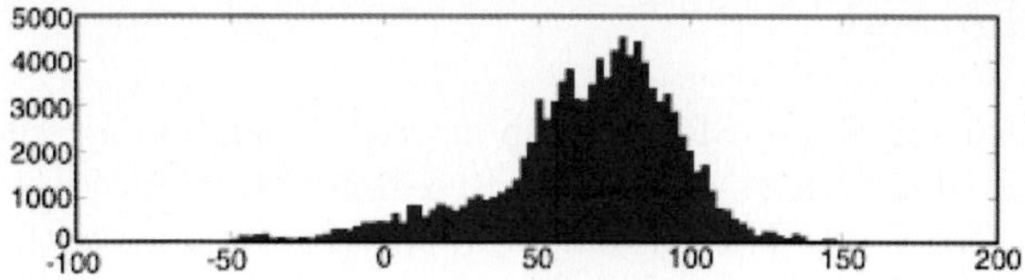

Fig. 1.35. Histogram of lateral offsets on the beer bottle pass. The horizontal units are in centimeters.

illustrates that the berm on Stanley's left reaches well into the corridor. Stanley drives as far left as the corridor constraint allows. Fig. 1.35 shows a histogram of lateral offsets for the Beer Bottle Pass. On average, Stanley drove 66 cm to the right of the center of the RDDF in this part of the race. We suspect that driving 66 cm further to the left would have been fatal in many places. This sheds light on the importance of Stanley's ability to react to the environment in driving. Simply following the GPS points would likely have prevented Stanley from finishing this race.

1.11 Discussion

This article provides a comprehensive survey of the winning robot of the DARPA Grand Challenge. Stanley, developed by the Stanford Racing Team in collaboration with its primary supporters, relied on a software pipeline for processing sensor data and determining suitable steering, throttle, brake, and gear shifting commands.

From a broad perspective, Stanley's software mirrors common methodology in autonomous vehicle control. However, many of the individual modules relied on state-of-the-art Artificial Intelligence techniques. The pervasive use of machine learning, both ahead and during the race, made Stanley robust and precise. We believe that those techniques, along with the extensive testing that took place, contributed significantly to Stanley's success in this race.

While the DARPA Grand Challenge was a milestone in the quest for self-driving cars, it left open a number of important problems. Most important among those was the fact that the race environment was static. Stanley is unable to navigate in traffic. For autonomous cars to succeed, robots like Stanley must be able to perceive and interact with moving traffic. While a number of systems have shown impressive results [Dickmanns et al., 1994, Hebert et al., 1997, Pomerleau and Jochem, 1996], further research is needed to achieve the level of reliability necessary for this demanding task. Even within the domain of driving in static environments, Stanley's software can only handle limited types of obstacles. For example, the present software would be unable to distinguish tall grass from rocks, a research topic that has become highly popular in recent years [Dima and Hebert, 2005, Happold et al., 2006, Wellington et al., 2005].

Acknowledgment

The Stanford Racing Team (SRT) was sponsored through four *Primary Supporters*: Volkswagen of America's Electronics Research Lab, Mohr Davidow Ventures, Android, and Red Bull. The Primary Supporters together with the Stanford team leaders form the *SRT Steering Committee*, which oversees the SRT operations. The SRT has also received support from Intel Research, Honeywell, Tyzx, Inc., and Coverity, Inc. Generous financial contributions were made by David Cheriton, the Johnson Family, and Vint Cerf. A huge number of individuals at Stanford, Volkswagen, and related facilities helped us in developing this robot, which is gratefully acknowledged.

The SRT also thanks DARPA for organizing this great race. Our final gratitude goes to the many friends we made during the event, and the many people who helped us and other teams along the way.

Finally, we are most grateful for all the friends we made during the race and the events leading up to the race, and the many journalists who covered this event.

References

Brooks and Iagnemma, 2005. Brooks, C. and Iagnemma, K. (2005). Vibration-based terrain classification for planetary exploration rovers. *IEEE Transactions on Robotics*, 21(6):1185–1191.

Crisman and Thorpe, 1993. Crisman, J. and Thorpe, C. (1993). SCARF: a color vision system that tracks roads and intersections. *IEEE Transactions on Robotics and Automation*, 9(1):49–58.

DARPA, 2004. DARPA (2004). Darpa grand challenge rulebook. On the Web at http://www.darpa.mil/grandchallenge05/Rules_8oct04.pdf.

Davies and Lienhart, 2006. Davies, B. and Lienhart, R. (2006). Using CART to segment road images. In *Proceedings SPIE Multimedia Content Analysis, Management, and Retrieval*, San Jose, CA.

Dickmanns, 2002. Dickmanns, E. (2002). Vision for ground vehicles: history and prospects. *International Journal of Vehicle Autonomous Systems*, 1(1):1–44.

Dickmanns et al., 1994. Dickmanns, E., Behringer, R., Dickmanns, D., Hildebrandt, T., Maurer, M., Schiehlen, J., and Thomanek, F. (1994). The seeing passenger car VaMoRs-P. In *Proceedings of the International Symposium on Intelligent Vehicles*, Paris, France.

Dima and Hebert, 2005. Dima, C. and Hebert, M. (2005). Active learning for outdoor obstacle detection. In Thrun, S., Sukhatme, G., Schaal, S., and Brock, O., editors, *Proceedings of the Robotics Science and Systems Conference*, Cambridge, MA.

Duda and Hart, 1973. Duda, R. and Hart, P. (1973). *Pattern classification and scene analysis.* Wiley, New York.

Ettinger et al., 2003. Ettinger, S., Nechyba, M., Ifju, P., and Waszak, M. (2003). Vision-guided flight stability and control for micro air vehicles. *Advanced Robotics*, 17:617–640.

Farrell and Barth, 1999. Farrell, J. and Barth, M. (1999). *The Global Positioning System.* McGraw-Hill.

Gat, 1998. Gat, E. (1998). Three-layered architectures. In Kortenkamp, D., Bonasso, R., and Murphy, R., editors, *AI-based Mobile Robots: Case Studies of Successful Robot Systems*, pages 195–210. MIT Press, Cambridge, MA.

Gillespie, 1992. Gillespie, T. (1992). *Fundamentals of Vehicle Dynamics*. SAE Publications, Warrendale, PA.

Happold et al., 2006. Happold, M., Ollis, M., and Johnson, N. (2006). Enhancing supervised terrain classification with predictive unsupervised learning. In Sukhatme, G., Schaal, S., Burgard, W., and Fox, D., editors, *Proceedings of the Robotics Science and Systems Conference*, Philadelphia, PA.

Hebert et al., 1997. Hebert, M., Thorpe, C., and Stentz, A. (1997). *Intelligent Unmanned Ground Vehicles: Autonomous Navigation Research at Carnegie Mellon University*. Kluwer Academic Publishers.

Iagnemma et al., 2004. Iagnemma, K. and Dubowsky, S. (2004). *Mobile Robots in Rough Terrain: Estimation, Motion Planning, and Control with application to Planetary Rovers*. Springer Tracts in Advanced Robotics (STAR) Series, Berlin, Germany.

Julier and Uhlmann, 1997. Julier, S. and Uhlmann, J. (1997). A new extension of the Kalman filter to nonlinear systems. In *International Symposium on Aerospace/Defense Sensing, Simulate and Controls*, Orlando, FL.

Kelly and Stentz, 1998. Kelly, A. and Stentz, A. (1998). Rough terrain autonomous mobility, part 1: A theoretical analysis of requirements. *Autonomous Robots*, 5:129–161.

Ko and Simmons, 1998. Ko, N. and Simmons, R. (1998). The lane-curvature method for local obstacle avoidance. In *Proceedings of the IEEE/RSJ International Conference on Intelligent Robots and Systems (IROS)*, Victoria, Canada.

Pomerleau and Jochem, 1996. Pomerleau, D. and Jochem, T. (1996). Rapidly adapting machine vision for automated vehicle steering. *IEEE Expert*, 11(2):19–27.

Pomerleau, 1991. Pomerleau, D. A. (1991). Rapidly adapting neural networks for autonomous navigation. In Lippmann, R. P., Moody, J. E., and Touretzky, D. S., editors, *Advances in Neural Information Processing Systems 3*, pages 429–435, San Mateo. Morgan Kaufmann.

Pomerleau, 1993. Pomerleau, D. A. (1993). Knowledge-based training of artificial neural networks for autonomous robot driving. In Connell, J. H. and Mahadevan, S., editors, *Robot Learning*, pages 19–43. Kluwer Academic Publishers.

Simmons and Apfelbaum, 1998. Simmons, R. and Apfelbaum, D. (1998). A task description language for robot control. In *Proceedings of the Conference on Intelligent Robots and Systems (IROS)*, Victoria, CA.

van der Merwe, 2004. van der Merwe, R. (2004). *Sigma-Point Kalman Filters for Probabilistic Inference in Dynamic State-Space Models*. PhD thesis, OGI School of Science & Engineering.

van der Merwe and Wan, 2004. van der Merwe, R. and Wan, E. (2004). Sigma-point kalman filters for integrated navigation. In *Proceedings of the 60th Annual Meeting of The Institute of Navigation (ION)*, Dayton, OH.

Wellington et al., 2005. Wellington, C., Courville, A., and Stentz, A. (2005). Interacting markov random fields for simultaneous terrain modeling and obstacle detection. In Thrun, S., Sukhatme, G., Schaal, S., and Brock, O., editors, *Proceedings of the Robotics Science and Systems Conference*, Cambridge, MA.

A Robust Approach to High-Speed Navigation for Unrehearsed Desert Terrain

Chris Urmson, Joshua Anhalt, Daniel Bartz, Michael Clark, Tugrul Galatali, Alexander Gutierrez, Sam Harbaugh, Josh Johnston, Hiroki "Yu" Kato, Phillip Koon, William Messner, Nick Miller, Aaron Mosher, Kevin Peterson, Charlie Ragusa, David Ray, Bryon Smith, Jarrod Snider, Spencer Spiker, Josh Struble, Jason Ziglar, and William "Red" Whittaker

Robotics Institute, Carnegie Mellon University, Pittsburgh, PA 15213

Summary. This article presents a robust approach to navigating at high-speed across desert terrain. A central theme of this approach is the combination of simple ideas and components to build a capable and robust system. A pair of robots were developed which completed a 212 kilometer Grand Challenge desert race in approximately seven hours. A path-centric navigation system uses a combination of LIDAR and RADAR based perception sensors to traverse trails and avoid obstacles at speeds up to 15m/s. The onboard navigation system leverages a human based pre-planning system to improve reliability and robustness. The robots have been extensively tested, traversing over 3500 kilometers of desert trails prior to completing the challenge. This article describes the mechanisms, algorithms and testing methods used to achieve this performance.

2.1 Introduction

Autonomously navigating at high-speeds for long distances necessitates a robust and capable robot. While there have been numerous examples of highly capable autonomously navigating robots (Kelly & Stentz, 1997), (Coombs, Lacaze, Legowik & Murphy, 2000), (Wettergreen et al., 2005) none of them have been able to drive challenging desert roads and trails at high speeds for multiple hours (Urmson et al., 2004). Achieving this combination of robotic skill and stamina was a goal of the DARPA Grand Challenge. In meeting the Grand Challenge two principles emerged as the keys to robustness & success: keep the components simple and test as frequently and as aggressively as possible. Sandstorm and H1ghlander (Figure 2.1) are the embodiment of this strategy.

In simplifying the overall robotic systems, a balanced approach using mechanical and software solutions was adopted to take advantage of existing technologies where possible. For example, the selection of the High Mobility Multi-purpose Wheeled Vehicle (HMMWV) & Hummer H1 chassis instead of a more readily available SUV chassis meant that the onboard navigation software need not be as sensitive to terrain features (i.e. an H1's ground clearance is much larger than that of a conventional SUV and so the perception system can ignore rocks that are irrelevant to the H1 but would cause significant damage to a smaller vehicle).

M. Buehler, K. Iagnemma, and S. Singh (Eds.): DARPA'05, STAR 36, pp. 45–102, 2007.
springerlink.com

Fig. 2.1. Sandstorm and H1ghlander were developed to navigate at high-speed in desert terrain

Similarly, human input during a pre-planning process is used to reduce the complexity of the onboard navigation system. While the Grand Challenge precluded any intervention while robots were operating on a course, the two hours before the challenge could be leveraged with human input to increase the likeliness of success. During this time human editors worked to modify the route to account for dangerous terrain that might be difficult for a robot to detect in real time. The process provided enough information for preemptive action to be taken to avoid some dangerous situations and helped reduce the complexity of the onboard navigation system. By balancing simplicity with the necessary complexity required to complete the challenge, a robust system was developed.

The development process consisted of short development cycles interleaved with periods of intensive field testing. While this took a toll on both personnel and equipment, it enabled the discovery of problems and weaknesses with both implementations and ideas simultaneously.

2.1.1 Format of the Grand Challenge

The 2005 Grand Challenge was a 212 kilometer race through the Mojave Desert. To win the challenge, a team's robot had to complete the course in less time than any other robot, and do so within 10 hours. Information about the route and exact distance of the challenge was withheld until two hours prior to race start, precluding pre-running or recording of the race route.

The route description contained a series of waypoints marking the corridor and speed limits within which the robots were required to travel. Once away from the starting line, the robots were required to be completely autonomous. The only communications allowed to a robot were publicly available GPS signals and a safety kill system used by challenge personnel to ensure safety.

The Red Team developed two robots and used a combination of autonomous and human pre-planning to complete the Grand Challenge. Once underway, the robots used onboard sensors to adjust a pre-planned route, to avoid obstacles and to correct for errors in position estimation.

The approach presented in this article proved successful; both Sandstorm and H1ghlander completed the Grand Challenge and demonstrated a new level of robust high-speed navigation.

2.1.2 Overview

This article describes the Red Team vehicles, software and testing process in depth. Section two begins the discussion with a detailed description of the electro-mechanics that make up both robots, and explains the significant differences between them. Section three describes the software architecture and components that operate onboard the robots including algorithms for perception, navigation and tracking. Both the benefits and limitations of the utilized approaches are described. Section four provides a short description of the preplanning system while section five provides a detailed discussion of the various testing processes employed by the team during the development cycle. In chapter six, an analysis of how the Red Team robots performed during the Grand Challenge event is presented. Chapter seven closes the discussion with a number of lessons learned and some ideas for future work building from the Grand Challenge.

2.2 The Robots

Sandstorm and H1ghlander are substantially different robots; each has its own unique actuation and control challenges. While much hardware and software is shared between the platforms, each vehicle has distinct performance characteristics.

The decision to develop two robots was not made lightly. The extra time, effort and personnel necessary to build and maintain two similar, but not identical robots represented a significant concern. However, operating two robots provides two principal advantages. During testing, the potential availability of two robots helps ensure that there is at least one operational for software testing. Increased testing time is one of the keys increasing robustness and capability. During the challenge, two robots provide an increased level of reliability since even if one of them fails, the other will continue. In the team's estimation, the advantages of having two robots for testing and racing outweighed the concerns.

2.2.1 Chassis

Sandstorm is a modified 1986 Model 998 HMMWV. The vehicle has been highly customized for autonomous control. The roof and passenger compartments have been removed in favor of a large electronics enclosure. Pictures of Sandstorm before and after the vehicle modifications are included in figure 2.2. The electronics box is suspended atop the vehicle platform on twelve coil over damper struts. Suspension lowers the natural frequency of the electronics enclosure so that sensors can be rigidly mounted. It also minimizes the shock dose to the

Fig. 2.2. Sandstorm before and after modifications

computers and electronics so they are protected from severe accelerations. Sandstorm's control frame floats with the electronic enclosure, causing the navigation system to drive the floating electronics box, without having a full understanding of the position and orientation of the chassis. This is one of the main causes of Sandstorm's characteristic smooth, but slightly sloppy driving style.

H1ghlander is built from a 1999 commercial H1 truck chassis with an upgraded 2001 electrical system. This upgrade includes a new vehicle harness, engine controller, and transmission controller. These were installed to allow easy reprogramming of engine and transmission functions. H1ghlander also uses an upgraded hydraulic steering system which provides quick, high accuracy steering response. All other vehicle components remain stock with the exception of a race quality suspension similar to Sandstorms. Unlike Sandstorm, H1ghlander retains three of its four passenger seats. This allows team members to ride in the vehicle for development, and also gives other people the unique experience of riding in an autonomous vehicle. Figure 2.3 shows H1ghlander before and after its vehicle modifications. Because H1ghlander does not have a floating electronics enclosure, the navigation system has a better estimate of the true pose of the vehicle. This helps H1ghlander drive more crisply than Sandstorm.

Fig. 2.3. H1ghlander before and after modifications

2.2.2 Controls Strategy

Throughout the vehicle system, feedback controllers are used in order to regulate systems and position actuators. In most cases the method used is a variant of the Proportional Integral Derivative (PID) controller. While PID control is not always the most accurate or highest performance controller, it is easy to use and robust. Equation 2.1 represents the general PID controller.

$$u = K_p e + K_i \int e dt + K_d \frac{d(-PV)}{dt} \tag{2.1}$$

In general, there are not accurate models available (or the time needed to develop them) for most of the vehicle systems, which makes more complicated control difficult. All PID algorithms are implemented in discrete-time through the use of real-time processes running with fixed time steps. The use of a simple, and easy to tune control strategy across the entire vehicle helped ensure the reliability and robustness of these systems.

The vehicle electronics systems for both Sandstorm and H1ghlander utilize multiple ECMs. ECMs are automotive or commercial grade components that contain one or more embedded processors, including the input/output circuitry, and memory necessary to carry out a given task or function. ECMs are distributed throughout the vehicle system, and are inter-connected by a series of automotive grade data-links.

Each ECM runs a Real Time Operating System (RTOS) in order to ensure that the different control and communications loops occur at deterministic rates, and that safety critical tasks are performed reliably. Software routines such as boot code, communications, and input/output functions were developed using a combination of hand coded C or assembly language. Control routines were developed using Simulink models, which were then auto-coded into functions that were run on the control modules. This allowed rapid development and testing of controllers in simulation, before auto-coding them to run on the vehicles.

There is a clean break in functionality between vehicle system and navigation software, which allowed the use of a very simple interface between the two subsystems. This interface, as shown in Figure 2.4, allows a navigation process to command a vehicle velocity and curvature. The vehicle system reacts to those commands and provides measured velocity and steering position back to the navigation system. The vehicle system also safeguards the vehicle via emergency-stop radios and deactivation buttons.

2.2.3 Power and Cooling

One of the significant differences between Sandstorm and H1ghlander is their power generation and cooling systems. Neither vehicle has the factory alternator installed, but both require about 4 kW of auxiliary power for all of the necessary computing, sensing and actuation components. Sandstorm employs an auxiliary generator mounted on the aft of the vehicle. The 24V generator is turned by

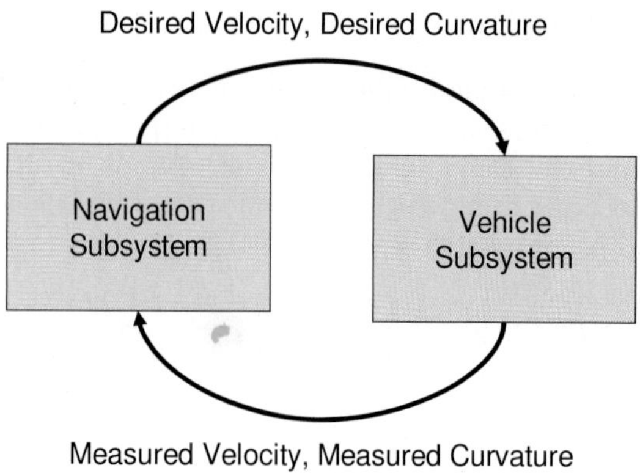

Fig. 2.4. Vehicle to navigation system interface

a small diesel engine which shares the propulsion engine's fuel supply. A compressor mounted to the propulsion engine provides cooling for an evaporator mounted in the electronics enclosure.

H1ghlander's power system is more complex. It incorporates a switch-reluctance generator driven by the main propulsion engine that generates 4 and 7 kW (depending on engine RPM) at 340 VDC. This high-voltage bus is used to efficiently power H1ghlander's electric air conditioning compressor, and is down-converted to both 12V and 24V in order to supply the necessary vehicle and electronics power.

Both vehicles' power systems are controlled by Electronic Control Modules (ECM) which contain embedded processors and input and output circuitry to monitor and control their respective power components. The power control algorithms are very similar. They use a modulated-voltage to control current technique that was developed in order to maintain acceptable power levels even at times when current draw exceeds available power. This control scheme, as outlined in figure 2.5, uses predominantly Integral (I) or "follower" control with offset gains and feedback derived from logic based on measured current and voltage. The controller is designed to operate with batteries in parallel with the output voltage busses. A dual/switching controller is needed because at steady state the generator is able to provide more than enough power to supply the components, but when the batteries are at a low state of charge, the current draw required to charge the batteries can exceed the available power, causing an over-current or "stall" condition at the generator. To compensate for this condition, the power controller is designed to change the output voltage of the generator in order to keep the current draw to acceptable levels. By lowering the voltage difference between the generator and batteries can be decreased. While in this "max-current" case, the output voltage is modulated to keep current draw at about 90% of available current. As the batteries charge, the required current drops and the controller switches to voltage control and maintains an ideal voltage level. This basic power control method has proven to be robust,

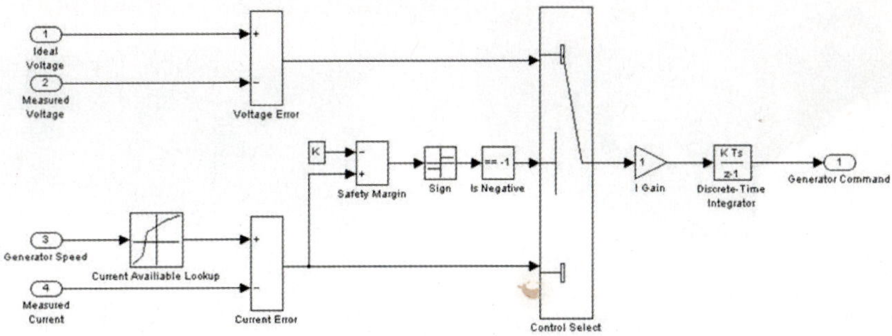

Fig. 2.5. H1ghlander's power controller

and has been extended in H1ghlander to control the power consumption of two down-converters and the cooling unit.

For both Sandstorm and H1ghlander, the power systems are designed to seamlessly function through minor power generation glitches. For example, during testing Sandstorm has traveled over 125 kilometers without any power production from the auxiliary generator, demonstrating the robustness of this design.

2.2.4 Steering

Electronic actuation of steering is a fundamental part of autonomous vehicle control. Sandstorm and H1ghlander have vastly different steering actuation strategies, and in turn have very different steering performance characteristics. In both cases, the systems respond to steering curvature commands from a tracker in the navigation software. The commanded curvature is linearly mapped to a steering angle in the controller, which is then maintained. There is no feedback control around actual curvature, only around steering angle. This proved a challenge to the vehicle's tracking system since it does not account for wheel slip, caused by differing ground conditions, or mechanical sensor slip which inherently changes the mapping between curvature and steering angle.

Mechanically, Sandstorm retains its complete stock steering system. To electronically steer the wheels, a large, driven gear is mounted to the top of the steering column, behind the steering wheel. A drive gear, attached to a DC motor and harmonic drive gear-set (shown in Figure 2.6 (a)) is mated with the steering column gear. The harmonic drive gearing provides a very high gear ratio with zero backlash and large amounts of torque. The downside of this high-reduction gearing is that it limits steering speed.

The motor is controlled through a drive amplifier by an ECM, which runs a closed loop control algorithm around steering angle. Controller feedback is provided by a rotational sensor mounted to the output shaft of the power-steering gearbox, which outputs a PWM signal proportional to steering position. For robustness, there is also a multi-turn sensor that measures position at the motor. A PID controller is used to maintain wheel steering position by outputting motor

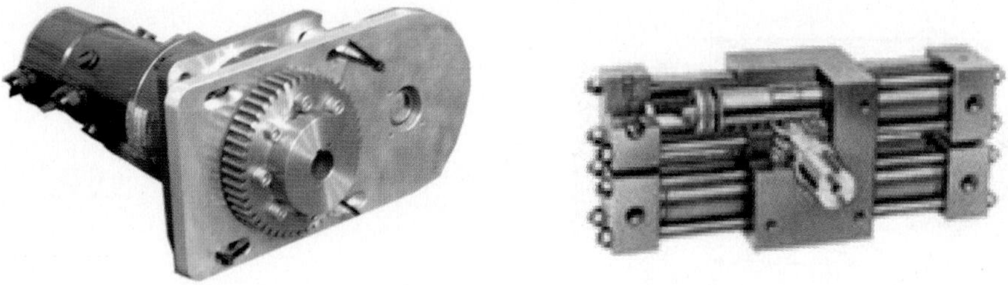

(a) (b)

Fig. 2.6. (a) Sandstorm's steering actuator and (b) H1ghlander's hydraulic steering actuator

torque and reading steering angle. This steering approach retains a majority of the stock steering system, which makes the system simple and robust. The downsides include the limited steering actuation speeds and limited accuracy due to a large mechanical dead-band in the power steering linkage, which causes hysteresis in the controller.

H1ghlander employs a different steering strategy; all of the stock steering components were removed and replaced with a full hydraulic steering system. The hydraulic system is composed of a dual-cylinder rotary hydraulic actuator (shown in Figure 2.6 (b)), a fixed displacement hydraulic pump, and an electro-hydraulic valve to control the hydraulic flow.

Electronics in the valve maintain a closed-loop control of the valve's spool position. Spool position is directly proportional to hydraulic flow (which can be mapped to cylinder velocity) and is commanded by an ECM. Steering angle is measured in the rotary actuator both by measuring the rotary output shaft position, and the linear position of one of the hydraulic cylinders. The ECM reads these positions, selects which one to use for feedback, and outputs a desired spool position based on a PID control algorithm. The advantage of this steering strategy is very responsive steering, and the ability to hold a very precise steering angle. The downside is the complexity of a hydraulic system, which is prone to leaks, heat, and filtration issues, each of which was encountered during the development.

Figure 2.7 illustrates the differences in steering response between Sandstorm and H1ghlander.

2.2.5 Velocity Control

The control of vehicle velocity is an important aspect of high performance driving. In a racing atmosphere, speed control must be accurate and responsive as it is constantly being adjusted to ensure vehicle stability. Velocity also poses a controls challenge, since it involves two different mechanical systems (propulsion engine and brakes) to maintain speed in any number of environmental conditions.

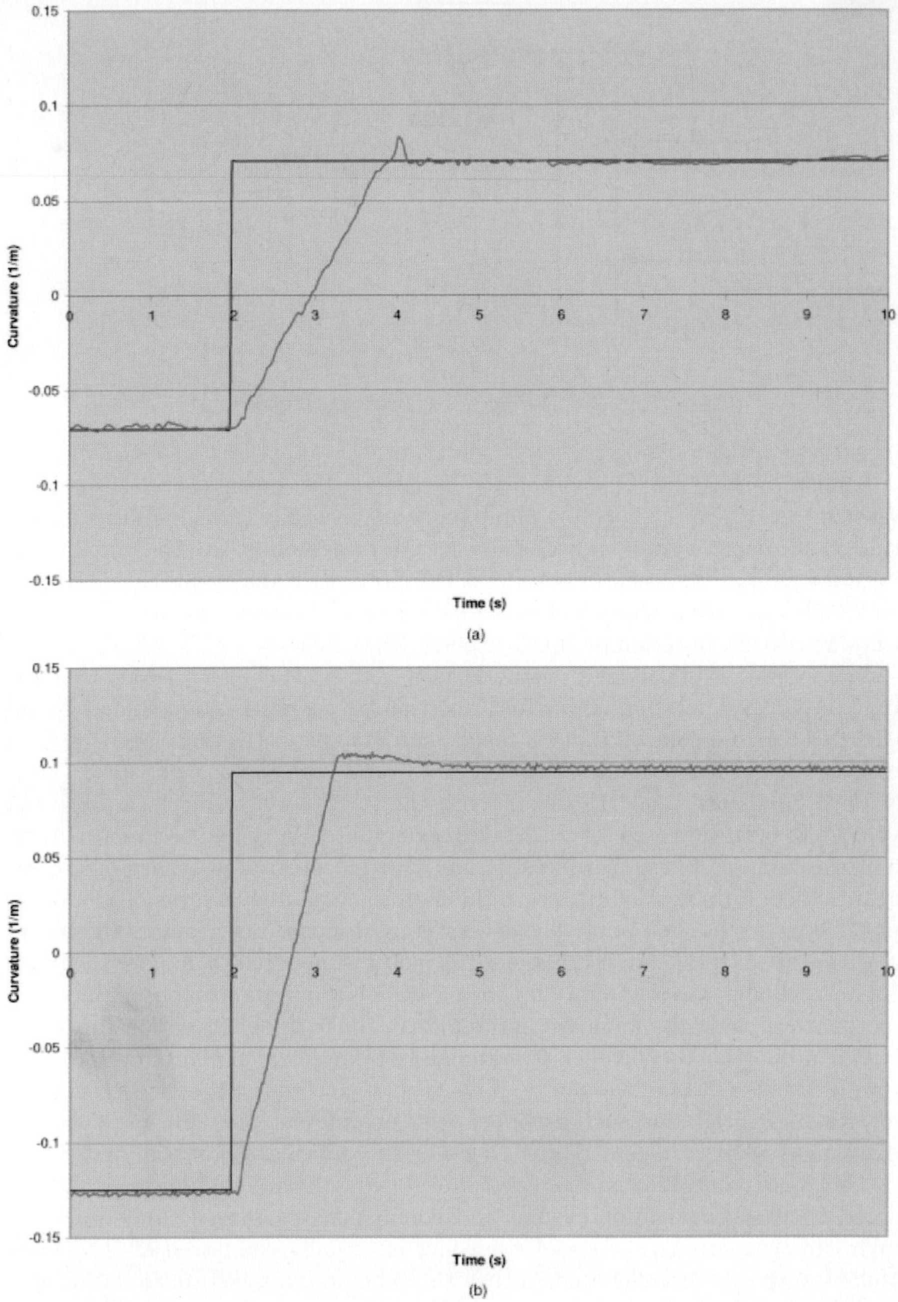

Fig. 2.7. Plots of Sandstorm (a) and H1ghlander (b) steering response

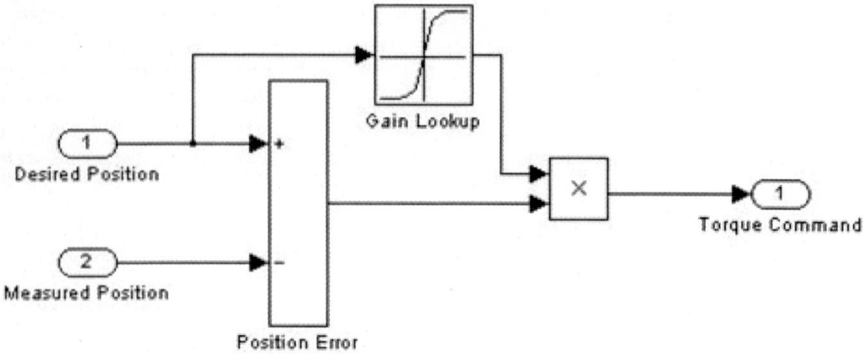

Fig. 2.8. Model of the brake controller

Sandstorm has a mechanically controlled engine. This means that to actuate the throttle, a valve on the injection pump must be physically turned. To accomplish this, an automotive-grade throttle body actuator was modified and mounted to the injection pump. The actuator is a simple DC motor with analog position feedback. An ECM reads this position and runs a PID closed loop control algorithm in order to command the injection pump to a specific throttle level.

In contrast, H1ghlander's engine is fully electronically controlled, meaning that its entire operation, from fuel injection to timing is commanded by an electronic engine controller. This makes autonomous activation very simple; a message is sent across a data-link and acted on by the engine controller.

Both Sandstorm and H1ghlander use the stock service brakes to slow the vehicle. In both cases the service brakes are actuated by an electric motor. Both motors are three phase brushless design with an integral 50:1 harmonic drive gear reduction. In Sandstorm's case the motor is mounted to press on the brake pedal. This results in a relatively slow braking response but provides significant mechanical advantage. In H1ghlander the motor is mounted to actuate the brake master cylinder directly. This mounting achieves quicker response, since less motor travel accounts for more braking force. In both configurations an ECM runs a proportional controller to command braking, which effectively provides torque-based control of the motor. This type of control inherently compensates for system degradation such as brake wear or different pressure line losses. A diagram of Sandstorm and H1ghlander's brake controller is given in figure 2.8.

The speed controller is a piece of embedded software that receives desired speed from the navigation system and commands the throttle and brakes to maintain that speed. The speed controller is actually comprised of three controllers: a speed controller using a throttle command, a speed controller using a brake command, and transition logic to determine which one to use. The control strategy mimics a layman human driver, where only the engine or the brakes will be actuated at any time. This is contrary to the driving strategy of most racers,

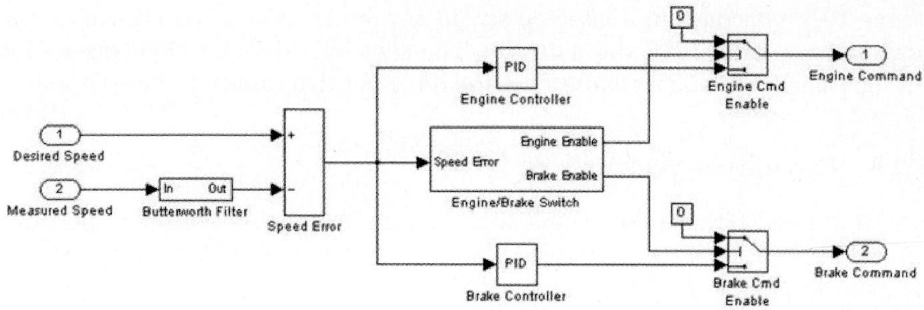

Fig. 2.9. Model of the speed controller

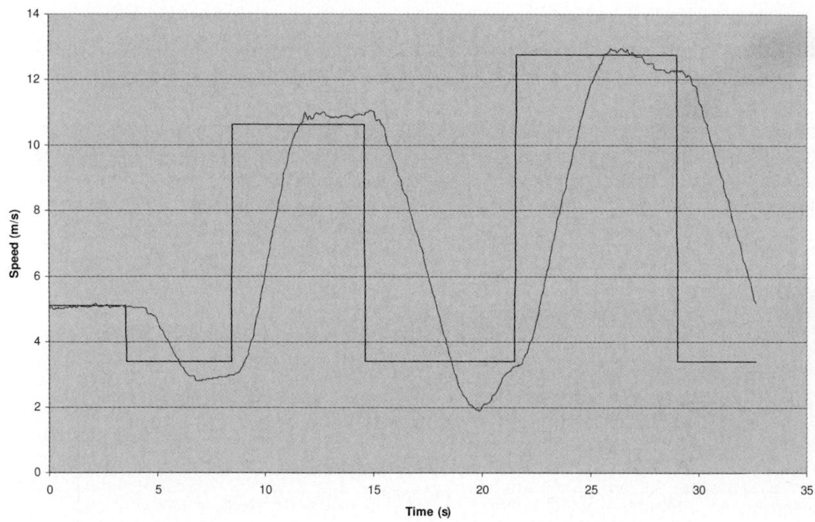

Fig. 2.10. A plot of speed controller performance

who often times actuate the throttle and brakes simultaneously. Figure 2.9 illustrates this speed controller graphically. The throttle and brake controllers each use a PI control scheme. Transition logic integrates the speed error and uses that as a time-delay switch with limits depending on commanded speed and whether the system is switching from the throttle to the brakes or vice versa.

This method allows more error at higher commanded speeds to minimize changeover between throttle and brakes, but also generates a quick response for sharp deceleration before turns. The speed feedback used in the control algorithms is obtained from the transmission and smoothed using a third-order Butterworth filter.

The speed controller maintains speed to within 0.5 m/s on average, with a 1-2 m/s undershoot when braking (response shown in figure 2.10). This undershoot is due to a nonconstant contact point in the brake system which causes the braking force to increase dramatically. The controller does not compensate for this non-linear braking response since the error was determined to be acceptable.

2.2.6 Perception Sensors

Sandstorm and H1hghlander combine data from a variety of sensors to perceive the world. In selecting sensors, the team evaluated monocular cameras, stereo cameras, LIDAR, and RADAR systems to find modalities amenable to generating terrain evaluations under the difficult conditions of the Grand Challenge. Table 2.1 outlines the sensing modalities considered and the advantages and disadvantages for this problem.

Table 2.1. A qualitative comparison of sensors considered for inclusion in the navigation system

Sensor	Advantages	Disadvantages	Selected
Conventional Camera	- wide vertical field of view - large swath of dense data	- data quality decreases due to lighting changes and glare - conventional processing techniques are challenged off-road	No
Stereo Camera	- wide vertical field of view - generates dense three dimensional data sets	- data quality decrease due to lighting changes and glare - measurement accuracy decreases as the square of range	No
LIDAR	- accurate range measurements - straightforward to integrate - wide horizontal field of view	- provides only a single plane of data -correlating between line scans may be challenging when moving over rough terrain	Yes
Automotive RADAR	- commercial grade hardware - integrated signal processing to identify targets - operates through most visual obscurants	- off-road performance is not characterized - output provides only limited information for external processing	No
Navigation RADAR	- dense, raw signal strength output - operates through most visual obscurants	- understanding raw RADAR signal returns is complicated	Yes

These considerations led to a perception strategy based on a set of five LIDAR and a navigation RADAR. Three of the LIDAR operate to characterize terrain, using overlapping field of view to provide redundancy. The two remaining LIDAR and the RADAR are used to detect obvious obstacles at long ranges. Figure 2.11 illustrates the sensor fields of views while Figure 2.12 shows the sensor locations on the robots. This design provides a robust perception suite, with multiple sensors observing the significant portions of terrain in front of the robots. The

Table 2.2. Characteristics of Navigation Sensors

Sensor	Characteristic
Narrow field of view LIDAR	Horizontal field of view: 60° Instantaneous field of view: 0.17° x 0.17° Dots per scan: ~240 Scan Rate: 50 Hz Manufacturer: Riegl Model: Q140i
Wide field of view LIDAR	Horizontal field of view: 180° Instantaneous field of view: 0.86° x 0.86° Dots per scan: 181 Scan Rate: 75Hz Manufacturer: SICK Model: LMS 291
Navigation RADAR	Horizontal field of view: 360° Instantaneous field of view: 1.2° x 4.0° Measurements per scan: 300 Scan Rate: 2.5 Hz Manufacturer: NavTech Model: DS2000

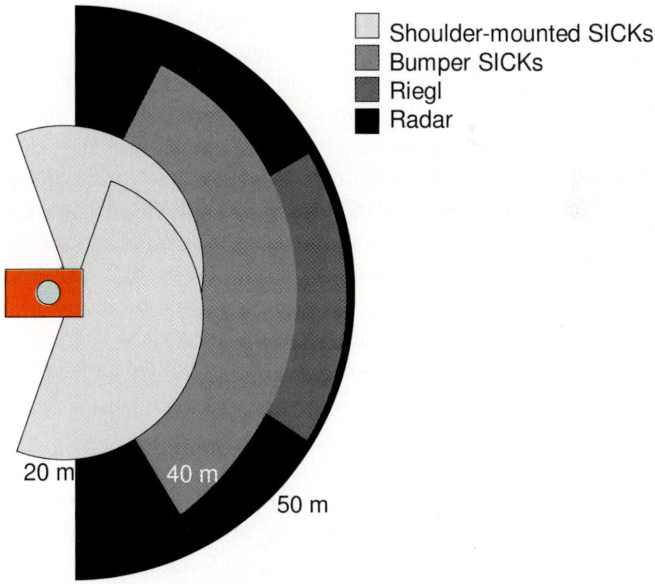

Fig. 2.11. Sandstorm and H1ghlander have multiple sensors with overlapping fields of view

remainder of this section describes the specifics of the sensors selected for the robots. Table 2.2 presents the specifications of the sensors used on the robot.

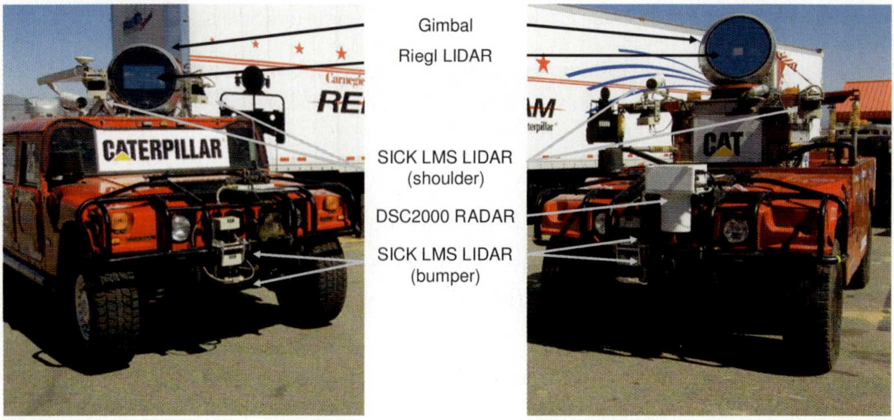

Fig. 2.12. H1ghlander (left) and Sandstorm with their sensors labeled

2.2.6.1 LIDAR

A Riegl Q140i scanning laser range finder is used as the primary terrain perception sensor for both robots due to its long sensing range, ease of integration and few, well understood, failures modes. A limitation of scanned LIDAR is that it is generally only possible to collect dense point data in a single plane. Flash LIDAR does not suffer this limitation but is still too range-limited to be useful at Grand Challenge speeds. Two-axis mechanically-scanned LIDAR have reasonable range, but cannot scan rapidly in both axes and thus do not provide significant benefits over a single scanning plane for this application.

In addition to the long range LIDAR, four SICK LMS 291 laser scanners are used to provide short range supplemental sensing. Two are mounted in the front bumper, providing low, horizontal scans over a 120° wedge centered in front of the robot. These sensors can be used to detect obvious, large, positive obstacles. The other two SICK LMS laser scanners are mounted to the left and right of the vehicle body. These sensors perform terrain classification.

2.2.6.2 RADAR

While LIDAR may have difficulties sensing in dusty environments, RADAR operates at a wavelength that penetrates dust and other visual obscurants but provides data that is more difficult to interpret. Because of its ability to sense through dust the NavTech DS2000 Continuous Wave Frequency Modulated (CWFM) radar was used as a complimentary sensor to the LIDAR devices.

2.2.6.3 Pose Estimation

Reliable and robust position sensing is essential since it central to performing reliable control and building usable world models. The implementation of position sensing is a major undertaking that can drain valuable development resources. To avoid this problem, Sandstorm and H1ghlander use an off-the-shelf pose estimation system. The Applanix M-POS provides position estimates by fusing inertial and differential GPS position estimates through a Kalman filter. The output estimate is specified to have sub-meter accuracies, even during extended periods of GPS dropout. The M-POS system also provides high accuracy angular information, through carrier differencing of the signal received by a pair of GPS antennas, and the inertial sensors. The M-POS system outputs a pose estimate over a high speed serial link at a rate of 100 Hz. This constant stream of low-latency pose information simplifies the task of integrating the various terrain sensor data sources.

2.2.6.4 Stabilization and Pointing

The ability to interpret data from long range sensors, such as a Riegl LIDAR scanner, can be severely hampered by pitching and rolling induced by robot motion over terrain. The performance of the single axis scanning LIDAR is particularly affected by mechanical excitation in the pitch axis. When sensing at reasonably long ranges, even small-scale pointing error can result in a dramatic change in where a sensor's beam intersects terrain. Because of this, range data associated with small obstacles or terrain details at distance become ambiguous, and the overall perception performance is severely degraded. Active attenuation of the terrain excitations can reduce this effect, yielding interpretable terrain range data.

The Gimbal sensor mounting design aligns the Riegl LIDAR's optical aperture with the center of the gimbal and balances the mass distribution around the rotational center. Each axis includes minimal-mass components and the simplest possible gimbal support structure with the design goal of minimizing the moment

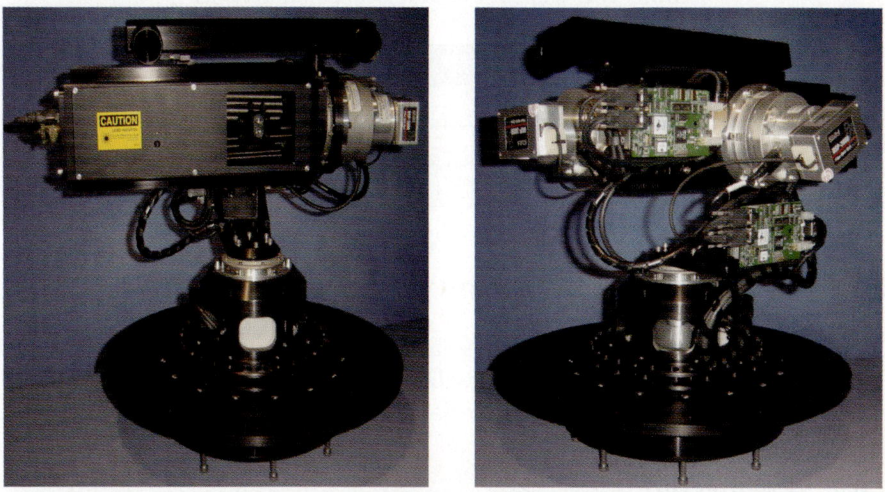

Fig. 2.13. Front and back view of the gimbal

of inertia. By minimizing the moment of inertia, the overall responsiveness of the gimbal is increased.

Harmonic drive actuators are used to point the gimbal based on feedback from a combination of incremental and absolute position encoders, and fiber-optic gyros. Each gimbal axis assembly is designed for electrical and mechanical simplicity. To achieve this goal common design and components are used on each of the axes. The aluminum bracket components are designed to have minimal mass and moment of inertia about their respective rotational axes while still maintaining sufficient strength and stiffness.

The entire gimbal mechanism is enclosed within a protective carbon fiber shell. The shell prevents water and dust from damaging the mechanism and electronics and includes a front window with specially coated optical glass for the sensors to operate through. The fully assembled gimbal is shown in figure 2.13. Table 2.3 describes the gimbal characteristics while Table 2.4 describes its performance.

Table 2.3. Gimbal Characteristics

Parameter	Value
Payload Compliment	High-resolution LIDAR line scanner
Payload Dimensions	240mm x 250mm x 500mm
Payload Weight	12+ Kg
Platform Weight	≈ 25Kg
Peak Power	550 W

Table 2.4. Gimbal Performance Characteristics

Axis	Range of Motion (°)	Angular Velocity (°/s)	Acceleration (°/s^2)
Pitch	± 40	360	49,500
Roll	±40	360	4,200
Yaw	±90	360	1,450

2.3 Onboard Navigation Software

Onboard navigation software combines incoming sensor data with a pre-planned route to generate a new safe and traversable route. In the following sections the architecture and algorithms used to drive Sandstorm and H1ghlander are presented.

2.3.1 Architecture

The navigation software-architecture was designed with the infrastructure to support high-speed navigation while being robust to sensor failures and adaptable

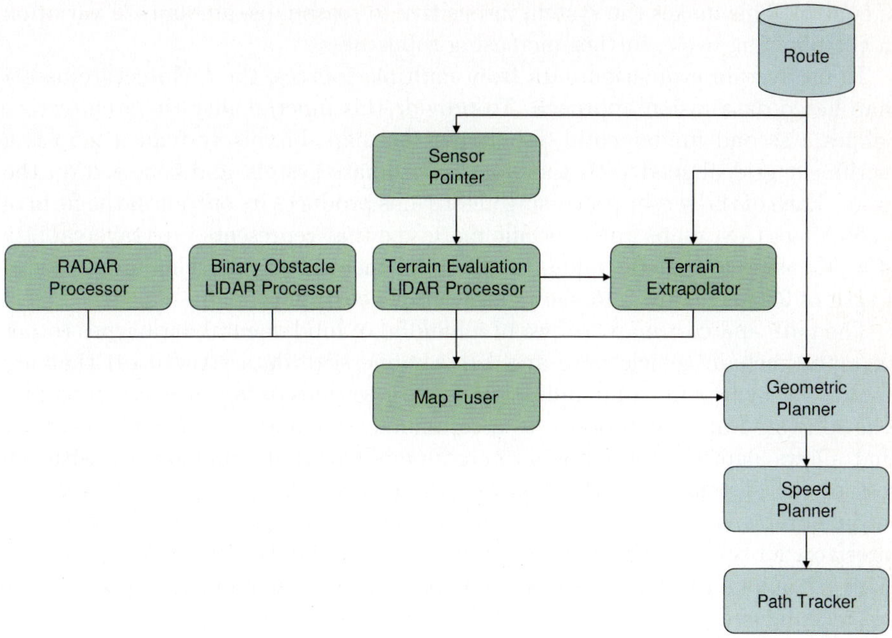

Fig. 2.14. The path-centric architecture of the onboard navigation software

enough to support a rapid, relatively unstructured development process. These design goals led to a path-centric navigation architecture, built around a set of well defined, rigid data interfaces.

In this path-centric architecture (see figure 2.14), the fundamental command action is to execute a path. This differs from a majority of autonomous navigation architectures which use an arc as the fundamental action. The path data structure is pervasive through out this approach; the pre-plan is provided as a path, path-planning acts as a filter on the path, and the perception system uses it to steer sensor focus and account for incompletely sensed terrain.

The path-centric architecture has several advantages that improve performance and robustness over arc-centric architectures (Simmons et al., 1995), (Kelly & Stentz, 1997), (Betulla, Manduchi, Matthies, Owens & Ranki, 2000), (Biesiadecki, Maimone & Morrison, 2001), (Urmson, Dias & Simmons, 2002). It provides a simple method for incorporating human input through a pre-planned route. Given a route, planning can be performed in a fixed width corridor around the pre-planned route, thus reducing the search space for a planning algorithm from the square of the path length to linear in the path length. The path-centric approach avoids problems with arc-based arbitration such as discontinuities in steering commands (due to contradictory information) and jerky control (due to discrete arc-sets). Furthermore, since the navigation system commands the execution of paths rather than discrete arcs. This effectively decouples the steering control frequency from the planning frequency, further increasing the smoothness of control. This makes the system insensitive to reasonable amounts of variation in the planning cycle, further increasing robustness.

To use terrain evaluation data from multiple sources, the architecture uses a map based data fusion approach. To provide this functionality the architecture defines a second fundamental data type; the map. In this system, a map is a rectilinear grid aligned with the world coordinate system and centered on the robot. Each of the sensor processing algorithms produces its output in the form of a cost map. Cost maps are a specific map type that represents the traversability of a cell using a numeric value. In this implementation, the cells have an edge length of 25cm. Figure 2.15 shows an example cost map.

The path and cost map are two of a handful of fundamental data types (other examples include vehicle pose and LIDAR line scan data structures) that are used as the syntax for communication between various data processing modules. The software implementation uses a communication and infrastructural toolset that allows algorithm developers to create modules that communicate with the rest of the system using the specified data types through a set of abstract, reconfigurable interfaces (Gowdy, 1996). During development and debugging, the interfaces for an algorithm can be configured to read data from time-tagged files using a common set of data access tools. As an algorithm matures, the interfaces are reconfigured to communicate with the rest of the navigation system. This approach helped reduce the required up-time and availability of the robot.

By using a common set of carefully defined and strictly controlled data types as the syntax for communication, it is possible to quickly develop new features for either path or map processing. While the syntax is defined and controlled, the

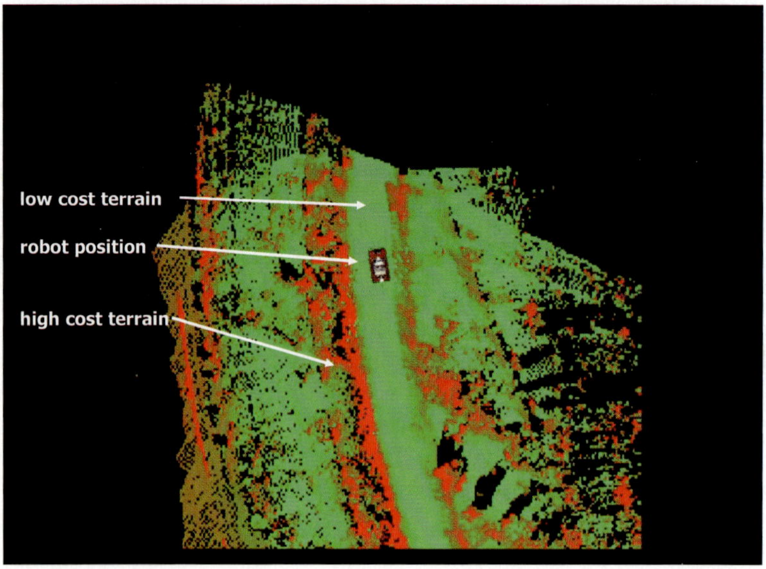

Fig. 2.15. An example cost map showing low (light) and high (dark) cost terrain

semantics, or meaning, of the data being passed between modules is free to be adapted as new ideas evolve and algorithms are developed. This flexibility makes the overall system robust and adaptable to ever-evolving ideas that develop as the navigation problem is explored.

2.3.2 Sensor Fusion

The map fusion process is critical to the robustness of the navigation system, as it enables the system to cope with sensor failures and missing data. To use the data from the various sensor processing algorithms it is necessary to combine it into a composite world model (either implicitly or explicitly). In this system the data is combined in the sensor fusion module by generating a composite map using a weighted average of each of the input maps.

Each of the processing algorithms specifies a confidence for the output map it generates. The fusion algorithm then combines the maps with these weightings to generate the composite expected cost map. This design allows the sensor processing algorithms to adjust their contribution to the composite map if they recognize that they are performing poorly. In practice a set of static weights, based on a heuristic sense of confidence in the algorithms ability to accurately assess the safety of terrain, worked well. With calibrated sensors, this approach produces usable composite terrain models. Figure 2.16 shows various input maps and the resulting, fused composite map.

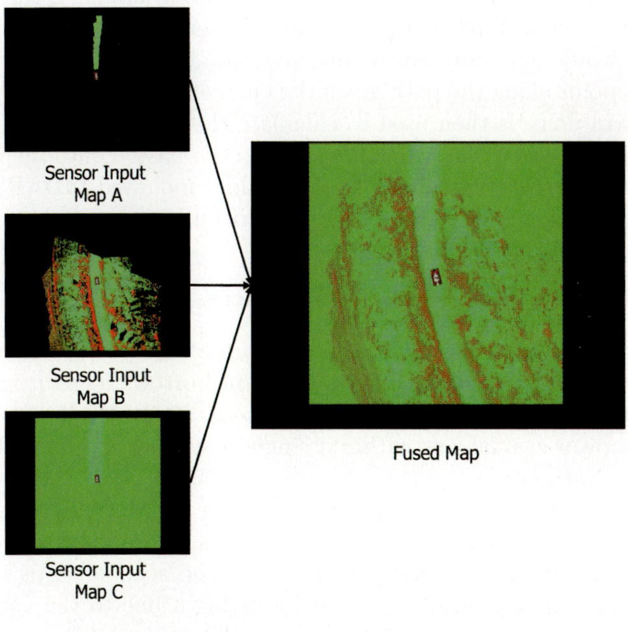

Fig. 2.16. An illustration of fused sensor maps

2.3.3 Perception

In this approach to high-speed navigation, three principal risks are considered-hitting large, obvious obstacles that can destroy a vehicle, driving on avoidable rough terrain that will damage a vehicle over prolonged periods of time, and dynamic effects such as sliding and roll-overs which cause a loss of control and can also potentially destroy a vehicle. The perception algorithms presented here are designed to address these risks. The Binary Obstacle LIDAR and RADAR processors are designed to quickly detect obvious obstacles at range. The Terrain Evaluation LIDAR processor is designed to generate a continuous valued classification of terrain, ranging from safe and smooth to intraversable. The slope calculations in this algorithm are used to steer the robot away from terrain with a likelihood of causing static tip-over, but falls short of estimating dynamic tip-over. Instead, the risk from dynamic effects is mitigated in a speed planning algorithm.

The various perception algorithms provide an overlapping (both geometrically and in terms of capability) set of models that reduce the likelihood of missing the detection of any obstacles while also providing robustness in the face of a sensor or algorithmic failure.

2.3.3.1 Sensor Pointing

The sensor pointing algorithm uses a pre-planned path as a guide as to where to point the Riegl LIDAR. A priori knowledge of the path enables the algorithm to point the sensor around corners, prior to the robot making a turn, and helps the perception system build detailed models of terrain in situations where the fixed sensors would generate limited information. A simple algorithm calculates a look-ahead point along the path given the current pose and speed of the robot. The look ahead point is then used to calculate the pitch, roll and yaw required to point at this location. These commands are then passed onto the gimbal. The data generated by the pointed and fixed shoulder mounted LIDARS is used by the terrain evaluation LIDAR processing algorithm.

2.3.3.2 Terrain Evaluation LIDAR Processing

Terrain classification and obstacle detection are at the core of high-speed outdoor navigation. The terrain evaluation system borrows ideas from Kelly and others (Kelly & Stentz, 1997), (Batavia & Singh, 2002), (Kelly et al., 2004) in performing terrain evaluations within a single line scan to reduce the effects of imperfect pose estimation.

The terrain evaluation approach is derived from the Morphin algorithm (Simmons et al., 1995), (Golberg, Maimone & Matthies, 2002), (Urmson et al., 2002) but has been adapted to operate on a single line scan of data instead of a complete cloud. The algorithm operates by fitting a line to the vertical planar projection of points in vehicle width segments. The slope and chi-squared error over this neighborhood of points provide the basis for evaluation. The operation is performed at each LIDAR point in a scan. If a point does not have a minimum number of points within a support distance or the surrounding points are not sufficiently dispersed, the point is not classified. The traversability cost is calculated as a weighted maximum of the slope and line fit residual.

Once traversability costs have been determined, each point is projected into a cost map, with independent cost maps maintained for each sensor. The terrain evaluation from each sensor is periodically combined into a composite output map. The traversability cost for each cell in the composite map is computed as the weighted average of the costs from each sensor, with weights proportional to the number of points used in generating the traversability cost for each sensor map.

While this basic algorithm works well, it blurs small obstacles over a large area since it does not separate foreground obstacles from background terrain (see Figure 2.17). The foreground obstacle pixels cause the line fit for the background pixels to have significant residual errors. To address this problem, a filter is used to separate foreground features from background terrain. During the evaluation process, any point at a significantly shorter range than the point being evaluated is ignored. This has the effect of removing discrete foreground obstacles (and their contribution to the line fit residual) from the evaluation of background terrain, while still correctly detecting obstacles. Figure 2.17 illustrates the effect of this filtering on a scene consisting of four cones in a diamond configuration. Without filtering, each of the four cones is represented as obstacles the size of a car, with the filtering applied the cones are represented as obstacles of the approximately the correct size.

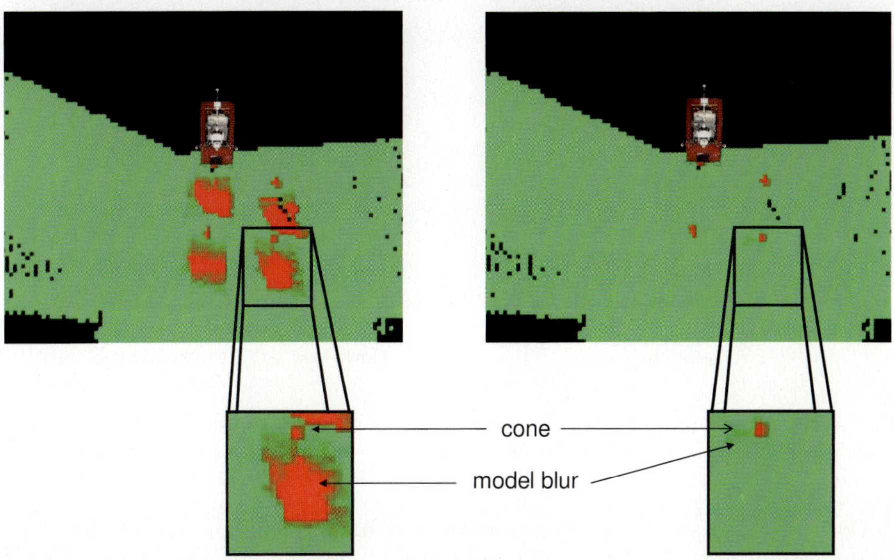

Fig. 2.17. Obstacle blur before (left) and after foreground separation filter is applied

Limitations

There are two situations where the terrain evaluation produces incorrect output: when LIDAR scans graze a surface, and when traveling through narrow tunnels or canyons.

In grazing scenarios, the algorithm will misclassify grazing returns as obstacle locations. The grazing returns are classified as obstacles because the shape of the LIDAR return is indistinguishable from a return with significant obstacles. Fortunately this occurs rarely and is mitigated through the use of multiple sensors. Applying a nearest neighbor clustering approach (Batavia & Singh, 2002) may be one solution to this problem.

The second problem of operation in very constrained environments (such as narrow tunnels) results in areas with sensor data not being classified. This problem stems from the requirement that the terrain evaluation only occur at LIDAR that have neighbors with a minimum dispersion. This requirement causes a half vehicle width of data at each end of the LIDAR scan to be ignored. In normal outdoor navigation scenarios, this defect is insignificant. In situations where the LIDAR returns all occur within a narrow area (e.g. a tunnel) important terrain features are not classified since the points that make up the walls of the tunnel do not have sufficiently dispersed neighbors. This problem is illustrated in Figure 2.18. Note that the approach to the tunnel and tunnel walls near the opening are classified correctly but as soon as the sensor measurements become constrained within the tunnel, only the center of the tunnel is classified while the outer half vehicle width on either edge of the scan are not classified, including the walls.

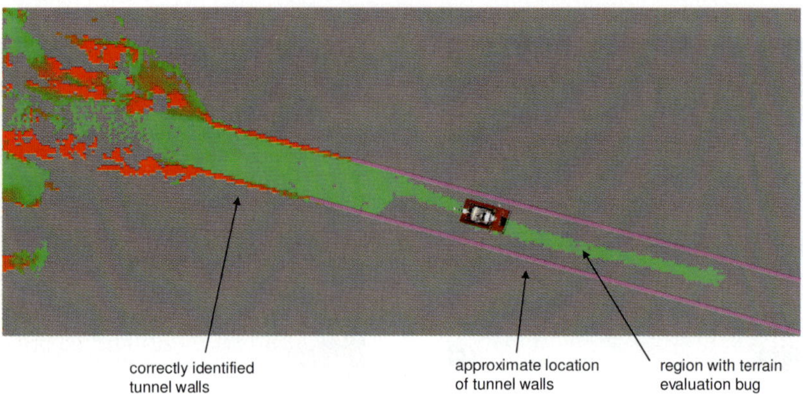

Fig. 2.18. An example of where a constrained environment causes the LIDAR terrain evaluation algorithm to behave poorly

Both of these shortcomings are mitigated by the combination of the other terrain evaluation algorithms and the terrain extrapolation modules described in the following sections.

2.3.3.3 Binary Obstacle LIDAR Processor

The Binary Obstacle LIDAR Processors is designed to quickly and robustly detect obstacles by collecting points over short time periods while a vehicle drives over terrain. The algorithm uses the fact that LIDAR points cluster on vertical faces to detect obstacles.

Using geometric information to determine a binary measure of traversability is common and has been a topic of research for decades. Typically this information is gleaned from images using classification techniques (Ulrich & Nourbakhsh, 2000) or geometric analysis of 3D terrain data (Singh & Keller, 1991), (Talukder, Manduchi, Rankin & Matthies, 2002). In recent work (Roth, Hammer, Singh & Hwangbo, 2005) have extended the RANGER algorithm to use LIDAR point clouds accumulated as a vehicle drives. The algorithm derives roughness, roll, and pitch by fitting planes to patches of point clouds.

As a LIDAR is moved through space, it sweeps the terrain and a point cloud representing this terrain is built by registering each scan using the vehicle and sensor pose. Selected pairs of points from this cloud are compared to compute the slope and relative height of the terrain.

Traversability is determined by performing a point-wise comparison of points within a region surrounding the point in question. If the slope and vertical distance between the two points is determined to be greater than a threshold value, both points are classified as obstacles. Given two points to compare, slope is computed as:

$$\theta = \tan^{-1}(|\Delta z|, \sqrt{\Delta x^2 + \Delta y^2}) \tag{2.2}$$

If Δz and θ are greater threshold values, an obstacle is inserted into the cost map.

Comparison of full rate LIDAR data is computationally expensive (Lalonde, Vandapel & Hebert, 2005). To make comparison rates reasonable, points are binned into 2D (x, y) cells and hashed by 2D cell location. Each hash location contains a list of all points within a cell. When a new point hashes to a hash location containing a list of points far from the new point, the list of points is cleared and the new point is inserted. This data structure allows near constant time comparison of nearby points by doing a hash lookup in the region of a point of interest.

With long range sensors, small errors in attitude of the sensor cause large errors in point registration. The comparison of two measurements of the same terrain patch from two different view points can falsely generate an obstacle if the vehicle pose estimate is erroneously pitched or elevated. Furthermore, the likelihood of inconsistent pose estimates is increase with time and distance traveled. Thus it is important to delete old points. To accommodate fast deletion, points are inserted into a ring buffer in the order that they are received. Once the ring buffer is full each new point overwrites the current oldest point in the buffer and the hash table.

Limitations

This algorithm relies on excellent registration of sensors to the world. The calibration of sensors relative to the vehicle center is a significant cause of error when multiple sensors are compared. Relative errors in pose estimation can cause false positives between scans of a single sensor. These two considerations limit the effectiveness of this algorithm at range.

Because of the reliance on registration, sensors that are not rigidly mounted relative to the vehicle coordinate frame cannot be processed using this algorithm. On H1ghlander, a considerable amount of capability is gained by comparing the bumper mounted SICKs. In flat terrain, these sensors can detect large obstacles (fence posts, large rocks, cliff walls, etc) at up to 50 meters. Because of its floating electronics box, Sandstorm cannot use the bumper mounted sensors with this algorithm.

Since this algorithm is designed to complement the LIDAR terrain evaluation algorithm it is not sensitive to terrain features like roughness and gentle slopes. The algorithm misses small ruts and washouts and produces false positives when tuned to attempt to detect such small features.

2.3.3.4 Radar Obstacle Detection

Radar sensing has several advantages for off-highway autonomous driving. It provides long range measurements and is not normally affected by dust, rain, smoke, or darkness. Unfortunately, it also provides little information about the world. Resolution on most small antennas is limited to 1 or 2 degrees in azimuth and 0.25m in range. Radar scanning is generally performed in 2D sweeps with a vertical beam height of <5 degrees. More narrowly focused beams are difficult to achieve and terrain height maps cannot be extracted from so wide a beam because objects of many heights are illuminated at the same time. This prevents using geometric or shape algorithms like those commonly used with LIDAR.

Attempts at using electromagnetic effects to gain information, such as correlating polarization with object density, have met with little success (Yamaguchi, Kajiwara & Hayashi, 1998). This leaves intensity of backscatter returns, binned by range and azimuth, as the sole identifier. Most previous systems use constant (Kaliyaperumal, Lakshmanan & Kluge, 2001) or adaptive thresholding (Jiang, Wu, Wu, & Sun, 2005), but achieve only marginal performance on good paved roads and are insufficient for offhighway driving. Many obstacles have surfaces that reflect energy away from the radar antenna, returning very low backscatter returns. Other objects that pose little risk to a large vehicle, such as brush, gentle inclines, and small rocks have large radar crosssections. Thus, the intensity of backscatter returns is a poor measure of the risk posed by an object.

Context Filtering

The primary challenge of processing radar data is separating dangerous or interesting objects from pervasive clutter. Early experimentation with thresholding and energy filtering failed in desert environments, generating too many false positives from innocuous objects. Thus another feature was therefore required for classification.

In the desert, clutter tends to occur over wide areas. Vegetation, inclines, and rough road sections all produce backscatter returns distributed over a significant region. Conversely, obstacles, like telephone poles, fence posts, and cars are generally isolated from each other and surrounded by road or clear dirt. Smooth ground such as this is specular because the low angle of radar beam incidence tends to reflect energy away from the antenna. Therefore, a potential classification feature is isolation, or the quality of a moderately low cross-section object of small footprint surrounded by clear, specular areas. While this limits the class of obstacles detected, it defines an important class of obstacles that can be detected with a small rate of false positives.

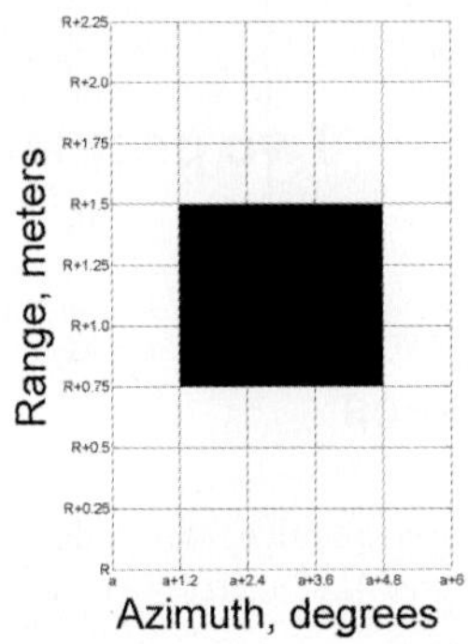

Fig. 2.19. The kernel used in context filtering. The black pixels are multiplied by positive one and the white pixels by negative one. The center pixel is set to the sum of these values.

To implement an algorithm exploiting this classifier, radar data is organized into a 2D image consisting of range and azimuth bins (Figure 2.19). A kernel consisting of two radii is convolved with this image. While the kernel is centered on a pixel, the energy between the inner and outer radii is subtracted from the energy contained within the inner radius. The value for this pixel is compared to a threshold and then reported as obstacle or not. The strength of this filter is dictated by the ratio of negative to positive space, i.e. the ratio of the two radii. The size of the inner radius determines the footprint size for which the filter is tuned. Results from a scene in the Nevada desert are presented in Figure 2.20.

Radar Map Hysteresis

As a vehicle drives forward, obstacles drop below the radar beam or become obscured, causing them to fade in and out of individual radar images. Thus, some form of memory is required to preserve their size and location. A simple approach is to add any newly classified obstacle pixel to a map combined with the previously reported locations of other obstacles. Due to the conversion from

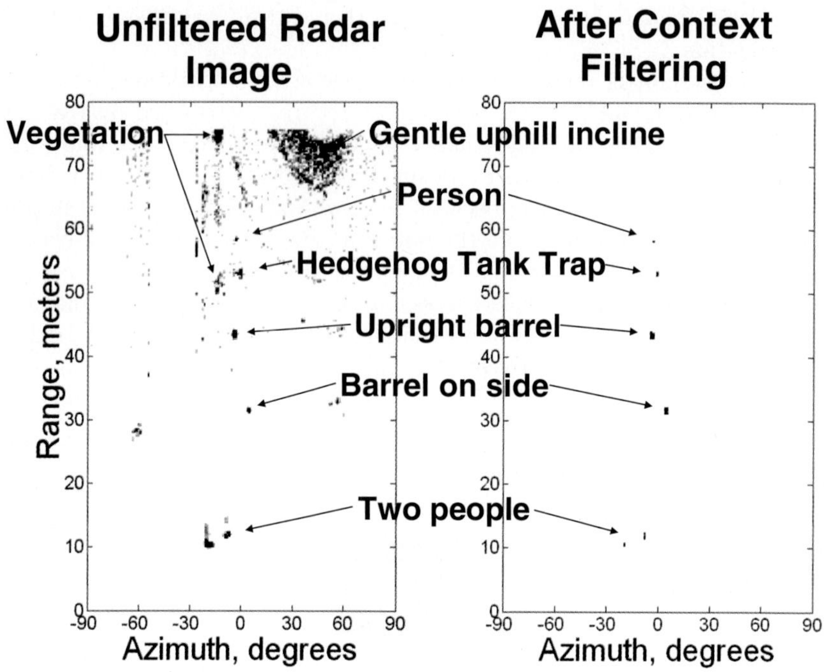

Fig. 2.20. Example context-filtered results from a desert scene. White pixels are empty and darkness represents strength of backscatter return.

polar to Cartesian cost maps, the location and size of an obstacle is refined as a vehicle approaches it.

Hence, a modification is required to the simplistic method that stores the union of all reported obstacle pixels. If a new obstacle blob is reported and overlaps with an old blob, then the old blob is removed and replaced with a new, more refined model of the obstacle location. This is implemented as a recursive algorithm that searches the neighboring pixels in Cartesian space to identify the extent of the old blob.

The results of this hysteresis are shown in Figure 2.21. As the vehicle approaches a fence post, the reported width is reduced from over 1.5m to a more reasonable 0.5m as better azimuth data becomes available. Obstacles that the vehicle has already passed are stored in the map even though they are not in the radar antenna's 180 degree field of view.

Limitations

The radar system was tested on Sandstorm and H1ghlander for over 3000 kilometers of off-highway driving. It effectively detected the narrow range of obstacles it was designed to detect, but does not generalize beyond those. As a complement to the LIDAR processing algorithms, it makes no attempt to detect road edges or other areas of rough terrain, making radar-only off-highway naviga-

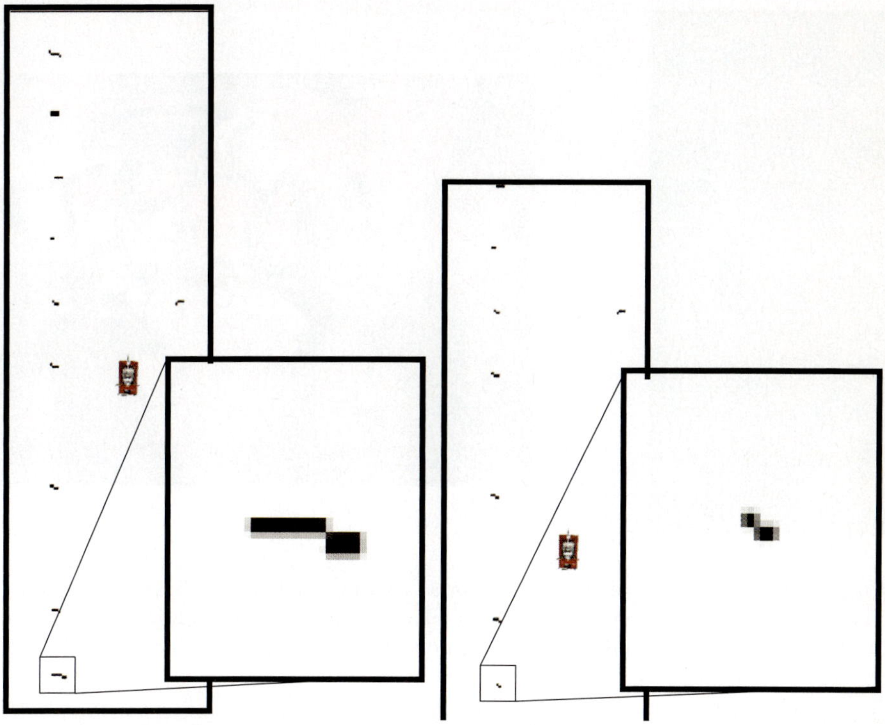

Fig. 2.21. Two maps recorded driving along a barb-wire fence during the 2005 Grand Challenge. The left map shows the magnified fence post at 35m range. The right map shows the same post, this time at 15m range. As the obstacle approaches, its size and position are refined, while obstacles behind the vehicle and out of the antenna's field of view are retained.

tion risky. While limited, this implementation demonstrates that, with proper scoping, RADAR can provide valuable long-range obstacle data to complement LIDAR and potentially vision based sensing.

2.3.3.5 Terrain Extrapolation

Perception in high-speed outdoor navigation often suffers from incomplete data due to a combination of occlusion and vehicle motion which can cause sensors to skip over terrain. Without a method for inferring reasonable traversability values for unseen terrain, this problem can result in catastrophic failures, such as the example illustrated in Figure 2.22. To address this problem, the onboard navigation system incorporates a Terrain Extrapolation Module (TEM). While appropriately interpreting the traversability of unsensed terrain is not new, there is very little published work in the literature (Nabbe, Kumar & Hebert, 2004).

The terrain extrapolation algorithm infers terrain costs based on three assumptions: (1) the characteristics (width, traversability) of a trail or road do

Fig. 2.22. An example of incomplete data (a) which led to significant damage (b). The red continuous line represents the pre-planned path. The data shows that the pre-planned path is on the road edge and veers off the road, which combined with the with the hill crest to cause the robot to leave the road.

not change rapidly over short distance along a path (2) the cost of traversability can change rapidly in the direction perpendicular to the path (i.e. trails may be sharply defined), and (3) the preplanned path is parallel to the actual traversable path. Figure 2.23(a) gives an example of input cost data for the TEM. As a path is driven, the algorithm samples a cross-section of costs perpendicular to the pre-planned path, represented by the black line in Figure 2.23(a). Each sample is combined with a running, alpha-blended, average of costs from the same lateral offset. The result is a constantly evolving, "learned" profile of the costs across the path. The alpha blending average filters out high frequency terrain features such as perpendicular fences while allowing the TEM to adapt to new trail conditions over a few tens of meters.

The calculated trail cost-profile is used to generate a "hallucinated" cost map by painting the profile along the pre-planned route. The map is fused with the sensor evaluation cost maps with a low confidence. The result is a cost map with TEM data only in locations in where there is no other available cost data, as shown in Figure 2.23(b).

Limitations

The primary limitation of the TEM is that it requires the true traversable path to be approximately parallel to the pre-planned path. Since the profile data is calculated as cross-sections of the pre-planned path, the profile of the pre-planned path matches the profile of the traversable path only when the two paths are approximately parallel, as shown in Figure 2.24(a). If the pre-planned

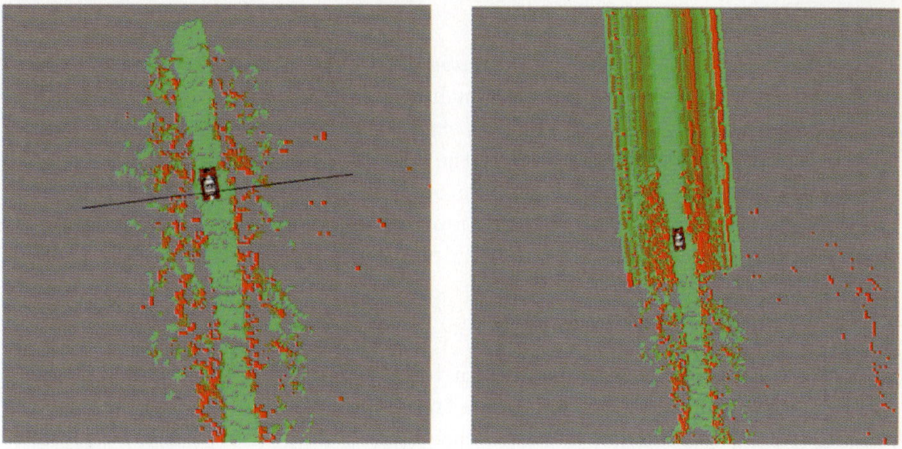

Fig. 2.23. (a)Example of typical cost map from H1ghlander. The black line represents a crosssection sample of costs the TEM uses to produce a cost profile. (b) Same scene as (a), but with the TEM's output fused into the upper half of the cost map.

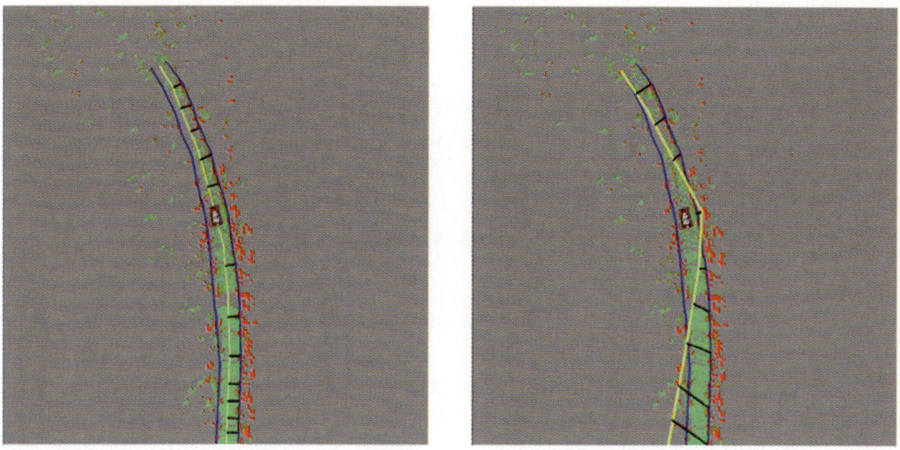

Fig. 2.24. (a) An example of the pre-planned path (light line) running approximately parallel to the traversable path (dark lines). The black lines are part of the cross-section sampled by the TEM. In this case the profile generated by the TEM produces an accurate profile of the path. (b) An example of the pre-planned path not running parallel to the traversable path. In this case, the cross-sections sampled do not correlate to the cross-section of the traversable path.

and traversable paths are at significantly different angles, as shown in Figure 2.24(b), the profile generated by the TEM will become blurred as the cross-sections of the traversable path do not project consistently into the cross-section of the pre-planned path.

2.3.4 Planning

The planning portion of the online navigation system is broken into a pair of modules that adjust the pre-planned path based on terrain evaluation generated by the perception algorithms. The first stage, the geometric planner, adjusts the path to avoid obstacles and minimizes the cost of traversability of the terrain the robot will drive over. The speed planner operates on the output of the geometric planner and preemptively slows the robot for any sharp turns that may result when the geometric planner generates a plan to avoid obstacles.

2.3.4.1 Geometric Planning

Trajectory planning algorithms attempt to find an optimal path from a starting point to a goal point. There has been a tremendous amount of work in this area ranging from deterministic, heuristic-based algorithms (Hart, Nilsson & Rafael, 1968), (Nilsson, 1980) to randomized algorithms (Kavraki, Svestka, Latombe & Overmars, 1996),(Lavalle, 1998), (Lavalle & Kuffner, 1999), (Lavalle & Kuffner 2001). There has also been significant research in algorithms to set speeds and curvatures reactively (Coombs, Lacaze, Legowik & Murphy, 2000), (Roth et al., 2005), (Shimoda, Kuroda & Iagnemma, 2005). In the Grand Challenge, a prescribed route consisting of a centerline with a set of bounds was provided. The bounds and centerline did not necessarily define a road, but did constrain where a robot may travel. While the details of the terrain were unclear, the route guaranteed that there was a traversable path within the corridor. This information can be exploited to significantly reduce the computational requirements of path planning. While the approach presented here was designed for the Grand Challenge, there are many scenarios in which an autonomous robot can be given a reasonable preplan of the route it must traverse.

Conformal Search

Because the path is assumed to be traversable, search can be limited to expansion near and in the direction of the path. A search graph is constructed relative to the pre-planned path that conforms to the shape of the path and constrains the motion of the vehicle. The spacing of the graph along the path is varied to increase stability as speed increases. The graph is searched using A* and the nodes comprising the solution are connected by straight-line segments.

Possible expansion nodes are grouped in linear segments, oriented normal to the direction of travel of the path, similar to railroad ties (Figure 2.25). Nodes are spread with a fixed spacing across each of the segments. Each node is allowed to expand to neighboring nodes in the next segment. A node is considered to be a neighbor of another node if its lateral offset is within one step of the current node. Expansion opposing the direction of travel or within a segment is disallowed.

The cost at each node is calculated from the composite fused cost map. A rectangular window is centered on the node and aligned with the direction of travel of the path. Costs in the cost map within the window are aggregated using a weighted averaged to produce a C-space expanded estimate of cost of traversability at that node. To encourage the planner to produce paths which are

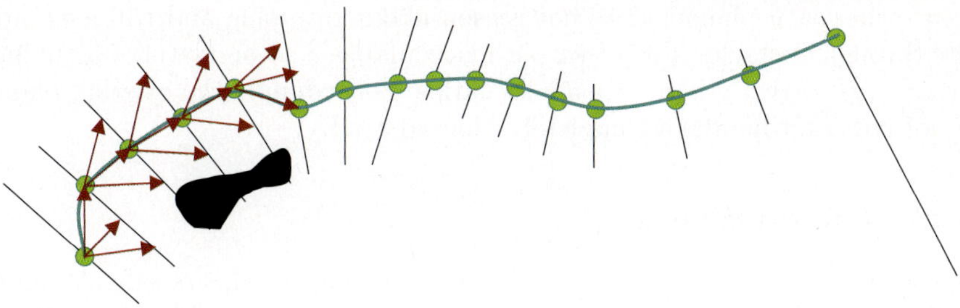

Fig. 2.25. The conformal planner operates between cells on adjacent segments normal to the preplanned path

shaped like the pre-planned path, traversal cost to the left and right neighbors is increased by a factor $\sqrt{2}$.

Each cycle, the graph is regenerated and searched using A* to produce an optimal path given the most recent sensor data. The search starts at the point closest to the vehicle on the last path output by the planner. The first few meters (proportional to speed) of the previous path are used as the starting point to generate the new path. This starting path is used to account for vehicle motion during the search.

The raw output path tends to have sharp turns; A* chooses to either maintain a fixed offset from the pre-planned path or to avoid an obstacle as hard as possible. These sharp turns slow the vehicle considerably, as the speed planner will reduce speed to ensure safety. In order to remove these sharp turns, a greedy smoothing operator is applied to the path. The smoothed path is only accepted if it has a cost approximately equal to the non-smooth path.

In most cases, the search operates quickly; faster than 20 Hz on the navigation computers. Occasionally, the search space is too complicated for the search to complete within a reasonable amount of time. Because the vehicle is a real-time system traveling at high speed on rough terrain, planner lockup is unquestionably bad. To prevent lockup, the search times out after a 20th of a second, returning the best path found at that point. In practice this path has been acceptably drivable.

Limitations

In situations where sharp turns are required, two of the perpendicular segments can overlap resulting in path that has a small knot in it. The knots rarely cause a problem but can cause the tracker to become confused and either reverse or drive poorly.

The planner does not explicitly consider speed and the speed of the output path is assumed to be close to the speed of the pre-planned path. While the speed planner will slow the output path to account for obstacle avoidance, explicitly reasoning about speed and geometry at the same time could improve performance and generate safer paths.

Since this path planner does not reason about reversing and will generate a route through obstacles if no other path is available, a second level of planning is necessary to correctly handle some scenarios. For example the existing planner can not correctly handle a completely blocked road.

2.3.4.2 Speed Planning

The speed planner is responsible for ensuring driving speeds are safe. As vehicle speed increases, dynamics become important. Speed induces side-slip (Gillespie, 1992) and can cause rollover (Diaz-Calderon & Kelly, 2005) in vehicles with a high center of gravity. Considerable research has been done to characterize vehicle performance at low speed (Golda, Iagnemma & Dubowsky, 2004) considering both roughness (Catelnovi, Arkin & Collins, 2005) and compressibility (Talukder, 2002) for speed setting and is primarily reactive.

Maximum tractive force is limited by friction, thus speeds must be planned such that they gradually decrease as turns are approached. While obstacle avoidance and controller error can cause executed curvatures to be larger than the initial plan calls for, an estimate of maximum safe vehicle speed can be determined in advance as a function of path curvature and maximum deceleration.

Modeling

The vehicle is modeled as a point mass with rigid wheels on a flat surface. Mass is equally distributed over the wheels and does not shift. Under these assumptions, sliding occurs when static friction cannot counter longitudinal or lateral acceleration.

$$\frac{v^2}{R} < \mu g \tag{2.3}$$

$$|\dot{v}| > d \tag{2.4}$$

Where μ is the coefficient of friction, g is gravity, v is the speed of the vehicle, d is the maximum deceleration of the vehicle, and R is the radius of curvature of the path. While d is limited by the tire-soil interaction, in practice maximum deceleration is slightly smaller because the velocity controller does not operate at the traction limit.

Numerous unmodeled dynamics including suspension effects, changes in friction and tire stiffness cause inaccuracies in this model. In practice it is important to incorporate a safety margin, k_{safety}, to decrease lateral acceleration and maximum deceleration.

$$\mu_{eff} = \mu k_{safety} \tag{2.5}$$

$$d_{eff} = d k_{safety} \tag{2.6}$$

$$v_{lat} < \sqrt{\mu_{eff} g R} \tag{2.7}$$

Equation (2.4) can be reformulated to represent a maximum speed, v_{lat} of a point P_i relative to the following point, P_{i+1}.

$$v_{lon} = \sqrt{2d_{eff} \mid P_{i+1} - P_i \mid + P_{i+1}.v} \quad (2.8)$$

Radius of curvature is defined as $ds/d\theta$ where $d\theta$ is differential heading change of the path and ds is arc length. Because the path is represented as discrete points, $d\theta$ and ds cannot be measured directly and must be estimated by $\Delta\theta$ and Δs. For a point P_k, define two headings θ^+ and θ^- as the heading of two points, $P_{k+spacing}$ and $P_{k-spacing}$.

$$\Delta\theta = \theta^+ - \theta^- \quad (2.9)$$

Arc length is approximated by the sum of lengths between the points between $P_{i-spacing}$ and $P_{i+spacing}$.

$$\Delta s = \sum_{i=-spacing}^{sapcing-1} \mid P_{k+i+1} - P_{k+i} \mid \quad (2.10)$$

$$R = \Delta s / \Delta\theta \quad (2.11)$$

A two pass process is applied to determine a maximum speed at each point in the path. The first pass walks the path in the forward direction and sets a maximum speed at each point to the lower of a maximum overall speed for the vehicle and the result of equation (2.7). The second pass walks the path from the last point to the first point and limits the change in velocity so that it is constrained by equation (2.4).

The algorithm runs two ways. Before executing a path, the algorithm processes the entire path once setting speeds well beyond the sensing horizon of the vehicle. While the vehicle is driving, the algorithm runs on the output of the planner (a small subset of the overall path) allowing the vehicle to slow for unexpected obstacles.

In practice, this algorithm performs well enough to drive safely on desert trails. As the vehicle approaches turns, it smoothly decelerates to a safe speed. As the vehicle approaches obstacles, it slows as necessary to swerve around them.

Limitations

Traction is limited by the sum of two vector components: lateral acceleration required to turn plus longitudinal acceleration of the tires (Gillespie, 1992). Because this model uses the maximum of each of these components, the maximum deceleration and coefficient of friction must be tuned for the worst case turning while decelerating. This results in a slightly lower maximum speed through turns when the vehicle is not decelerating.

Tires are not rigid and consequently do not suddenly break away as this model suggests. In reality, as lateral acceleration increases, tires walk sideways inducing a side slip angle (Gillespie, 1992). The side slip angle is a function of many parameters including tire pressure, tire stiffness, and the coefficient of friction. When vehicles operate with high slip angles, they are hard to control. Additionally, if slip angle is not countered by the control layer, it will induce error.

Many effects which are functions of the terrain and environment decrease tractive force. A wheel bouncing on washboard terrain has less contact with the ground and as a result cannot apply as much force. On side slopes and in banked turns, gravity and the "up force" generated by the curvature of the terrain changes the maximum possible speed before rollover and breakaway.

Additional system effects should also be considered. Controllers have greater error at high speed. The planner operates poorly when it has very little information. An analysis of these errors could provide a function that predicts errors at particular speeds and curvatures. A cap on these errors would determine maximum speed through turns and in straights.

2.3.4.3 Path Tracking

At the lowest level of the onboard navigation system a modified conventional pure pursuit path tracking algorithm (Amidi, 1990), (Coulter, 1992) steers the vehicle. As is common, the look-ahead distance of the tracker is adjusted dynamically based on speed. The control gains are configured to provide a balance between good performance at both low speed in tight maneuvering situations, and at high speed on straight-aways and soft corners.

The basic pure pursuit algorithm works well if the output arcs are executed faithfully by the underlying vehicle controllers. Errors in the mapping between steering angle and curvature in the low level control scheme induce systematic tracking errors. For example, the steering angle sensor on Sandstorm would intermittently shift relative to its mechanical ground, this resulted in a moving zero point for the mapping between steering angle and curvature that could not be corrected by the vehicle control system. The effect of this shift was to cause the robot to track with a significant steady-state lateral offset from a desired path.

To correct for this problem, the basic pure-pursuit tracker is augmented with an integral correction function. The error term is calculated as proportional to the lateral offset between the vehicle and the path when the commanded steering angle is near zero curvature. This causes the integral term to accumulate on straight-aways, but not in corners where pure-pursuit tracking would normally have significant errors. The scaled, integrated curvature correction term is then added to the desired curvature generated by the basic pure-pursuit algorithm before it is passed on to the vehicle control system.

This solution worked well and effectively removed the tracking bias problems induced by the shifting steering sensor. The addition of the integral term also made the vehicle robust to some mechanical damage to the steering system that would have otherwise degraded driving performance.

2.4 Pre-planning

The pre-planning system creates a path, including its associated speeds and estimated elapsed time, prior to robot operation. It incorporates aspects common to mission scaled schedulers and planners for space missions (Tompkins, Stentz & Whittaker, 2004) and provides the critical input that allows the navigation system to make assumptions about the world it is operating in. Pre-planning helps a robot by allowing it to anticipate and slow down for obstacles and conditions without sensing them directly. Good human drivers use similar foreknowledge to adjust radii, favor lanes and set speeds to slow down for harsh terrain features.

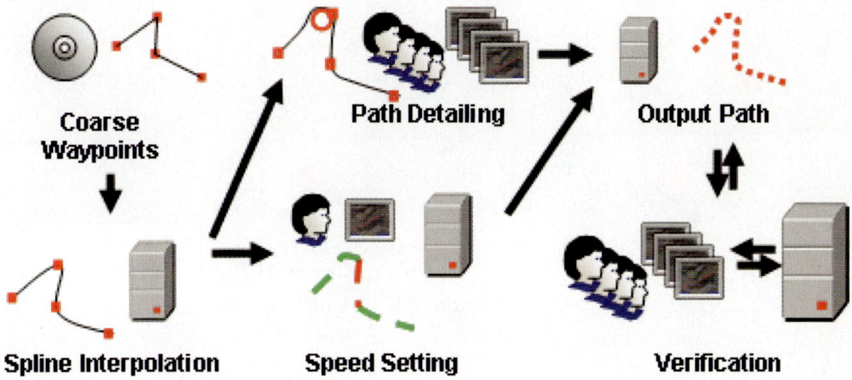

Fig. 2.26. An illustration of the pre-planning process

As illustrated in Figure 2.26, the pre-planning system involves the initial generation of a path using splines to smooth an initially coarse list of waypoints. The smoothed path is then modified by human editors to widen turns, and identify high risk areas in the path. In parallel, risk is assessed throughout the course and an automated speed setter uses this information to achieve a desired elapsed time. This resulting path is then output in the form of a series of fine waypoints which are used by the robot to traverse the course. As part of this process, several rounds of verification are performed, to find and remove problems with the path. Periodically the current best path is transferred to a robot to ensure that there is always a route available, in case of some unexpected failure.

Human editors perform feature extraction that is beyond the state-of-the-art in automated image understanding. Image extraction algorithms are limited to items such as road extraction from imagery (Harvey, McGlone, Mckeown & Irvine, 2004) and object detection and delimiting (Flores-Parra, Bienvenido & Menenti, 2000) for large scale features such as buildings. Even though delimiting and detecting objects would be useful for the identification of underpasses, overpasses and gates, the algorithms are only semiautomated, reducing their value for this application. Furthermore, even if these technologies were implemented, they are still unable to detect subtle obstacles such as road washouts which can be potentially fatal for a robot.

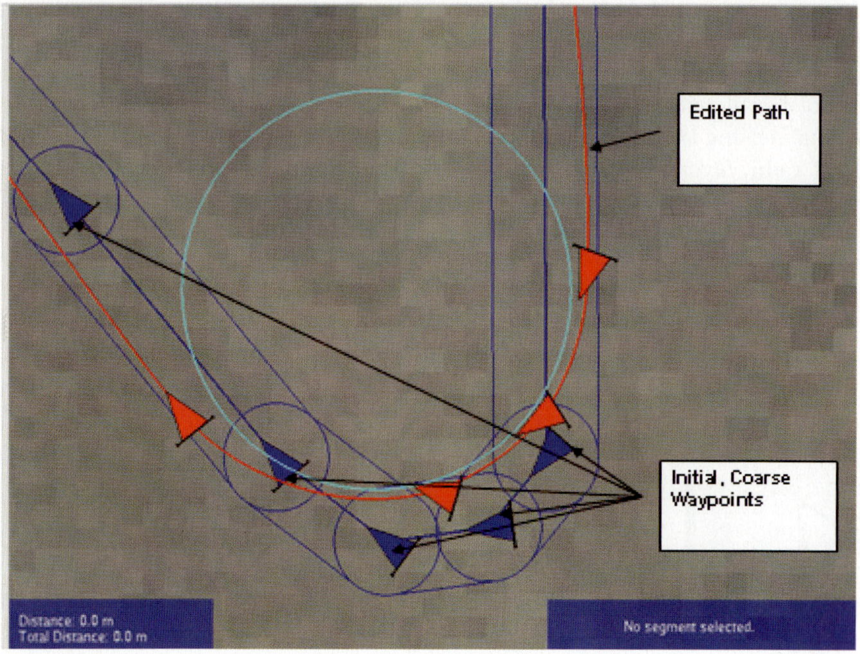

Fig. 2.27. An example of path detailing: adjusting a turn to achieve a minimum turning radius

2.4.1 Path Editing

Path editing is a process that transforms a set of coarse waypoints and speed limits into a pre-planned path with one meter spaced waypoints. The smoothing process produces a route of curved splines from waypoint to waypoint defined by a series of control points and spline angle vectors. Human editors can alter splines by shifting a control points, and spline angle vectors that specify the location and orientation of a path. The generated splines are constrained to ensure C2 continuity which helps ensure a drivable path.

The human editing process removes unnecessary curvature from the smoothed path. Smooth paths are also generally faster since decreasing the amount of curvature in a path reduces concerns for dynamic roll-over and side slip. Figure 2.27 shows an instance where an editor has widened a turning radius in a path.

2.4.2 Speed Setting

During pre-planning, a speed setting process specifies the target speeds for an autonomous vehicle given a target elapsed time to complete a pre-planned path. Speed setting is performed by assessing the risk for a given robot to traverse a section of terrain based on available information. An automated process then uses a speed policy generated by combining the risk assessment with any speed limits imposed on the course to assign speeds to each waypoint in the path.

The risk estimation process discretizes risk into four levels (dangerous, moderate, safe and very safe) in classifying terrain. Each risk level maps to a range of safe robot driving speeds. Risk is first assigned regionally, over multi-kilometer scale terrains, using general characteristic of the terrain under consideration. Once the entire route has risk assigned at a coarse level, a first order approximation of the ease/difficulty of that route, as well as an estimate of the overall elapsed time can be generated.

In addition to classifying risk at a macro level, risk is also assigned to local features of importance. This processing step characterizes and slows the path down for washouts, overpasses, underpasses, and gates. In this way human editors provide a robot with a set of "pace notes", similar to the information used by rally race drivers. These details allow a robot to take advantage of prior knowledge of the world to slow preemptively. This is a critical part of the process for increasing the robustness of the onboard navigation system.

2.4.3 Verification

The verification step helps ensure that each pre-planned route is free from errors prior to a robot executing it. The verification process is performed in three ways: (1) in-line as an automated method which operates periodically while the route is edited, (2) through multiple reviews by human editors, and (3) through an automated external independent check on the final route.

2.4.3.1 Automated Inline Verification

The inline verification process provides human editors periodic updates of locations where the path being edited violates any constraints. The specific constraints considered are: (1) exceeding of corridors boundaries, (2) path segments with radii tighter than a robot's turning radius, and (3) areas where the route is very narrow and warrants extra attention. Each of these potential problems is flagged for review by a human editor. These flags are then used as focal points for interpreting the path.

2.4.3.2 Human Review

Editors review each segment multiple times to ensure the route final route is safe. While detailing a route, each segment undergoes an initial review by editors which fixes major problems. The first review looks for any errors in the output of the automated planner, and attempts to identify areas of high risk for a robot, such as washouts. These high risk areas are then flagged, to be confirmed in a second review. A second review takes the output of the first review, and refines the route, confirming marked "flags" and adding additional "flags" for any high risk areas missed in the first review. The expectation is that after completion of the second review there will be no need for additional editing of the geometry of the route. In the 3rd and 4th reviews the main focus is to verify that all problems identified by the automated inline verification process have been cleared, as well as to confirm that any problems identified by the automated external verification algorithm are addressed.

2.4.3.3 Automated External Verification

An automated external verification process operates on the final output to the robot and checks heading changes, turning radius, speeds, and boundary violations. In addition, the verification process outputs warnings where there are areas of high slopes and sections of narrow corridors along the path. These warnings are used to identify areas for the human editors where extra care should be used. The verification process also produces a number of strategic route statistics such as a speed histogram for time and distance, a slope histogram, and a path width histogram. These statistics are used in determining the target elapsed time for the route and in estimating the risk for the route. This process is repeated several times as the path detailing progresses until the route is deemed safe for the robots to use.

2.5 Testing and System Performance

To succeed in the Grand Challenge, robots must demonstrate both performance and reliability. To achieve these two requirements, the team identified that a system of tests and performance milestones were required. Performance milestones were set to ensure driving skills and reliability would meet near term program goals such as passing the site visit demonstration, winning pole position at the National Qualifying Event (NQE) and completing the Grand Challenge. Examples of performance milestones included, shakedown cruises, NQE skills tests and race length autonomous runs at race pace and in race environment. In addition to milestones the team developed test methods that enabled quantitative evaluation of the effects of changes to the robots' hardware and software on performance.

2.5.1 Controlled Tests

The impetus for a controlled test methodology was to monitor development progress and measure driving skill. The team also used the tests to evaluate the effects of changes in hardware and software on the robots' overall ability to drive. The ability to drive was defined as three major skills. The first skill was the ability of the robots to follow a preplanned path based on position sensing only. The second skill was the ability of the robots to track a pre-planned path while assisted by perception sensors. The third skill was the ability of the robots to dynamically make significant modifications to a preplanned path to avoid sensed obstacles.

Fig. 2.28. H1ghlander (and Sandstorm) were tested extensively on desert terrain

2.5.1.1 Test Formulation

There are few, if any, documented standardized tests to measure autonomous driving skills at speed. A review of technical reports of DARPA Grand Challenge 2005 finishers Stanford Racing Team (Stanford Racing Team, 2005), Team Gray (Trepagnierl, Kinney, Nagel, Donner & Pearce, 2005) and Team Terramax (Oshkosh Truck Corp, 2005) found that they all placed great value on testing but did not mention specific tests to measure driving skill. While a method for characterizing tracking performance has been described (Roth & Batavia, 2002), the authors are unaware of literature that describes a standardized process for evaluating perception based navigation.

The literature on the subject of automotive dynamic testing includes International Organization for Standardization standard ISO-3888-1, Passenger cars — Test track for a severe lane-change maneuver Part 1 – Double lane-change (International Standards Organization, 1999) (Figure 2.29). This test was designed as a means to subjectively evaluate vehicle dynamic performance. The test is subjective because it only quantifies a small part of a vehicle's handling characteristics and is highly dependent on the input from the driver. This dependence on driver skill is what makes the test attractive for adaptation to autonomous ground vehicle driving skill testing.

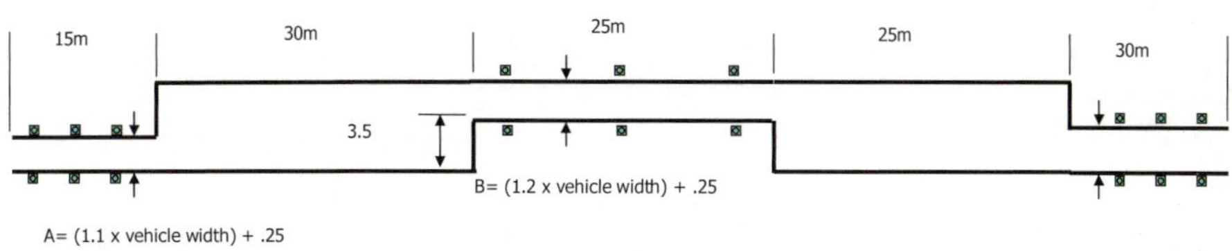

Fig. 2.29. ISO-3888-1 Test track for a severe lane change maneuver

Table 2.5. Test steps for autonomous ground vehicle path tracking skill assessment

1.	Create a route file through the test course.
2.	Create a path definition file for the route created in step 1 setting the corridor width slightly wider than the test track's lane width and the speed to a constant (e.g., 5 meters/sec). The path definition file must include an area before the test course begins for the robot to achieve the required constant velocity.
3.	Load the path definition file into the robot.
4.	Command the robot to drive the route described in the path definition file.
5.	Record the time the robot is on the test track entry to exit.
6.	Record the number of times the robot touches or exits the test track's boundaries.
7.	Repeat steps 2 through 6 increasing the speed by an incremental value (e.g., 2 meters/second) until the robot can no longer successfully traverse the course or the operation is deemed to be unsafe. Multiple runs at each speed increment are required to demonstrate consistency.

The original ISO-3888-1 course was modified, adding 1.5 meter to lane width in all sections to account for nominal pose estimator and tracking errors. This additional lane width enabled a quantitative measurement of performance considering total system error. Table 2.5 describes the basic steps used in the autonomous ground vehicle path tracking skills assessment.

2.5.1.2 Blind Path Tracking Test

The blind path tracking test uses the modified ISO-3888-1 to perform a quantitative evaluation of path tracking performance. A route file for this test is created by recording a smooth, "best line" route through the test course. The recorded path is then transferred to the robot and used as the baseline path to be tracked. Figure 2.30 is a graphical representation of the path definition file used in the blind and perception assisted path tracking tests. When conducting the blind path tracking test the robot is configured such that only the pose sensor is considered in path planning. The absence of all perception sensor input limits path tracking error to that induced from the pose sensor, path tracking algorithm and drive by wire actuation control errors.

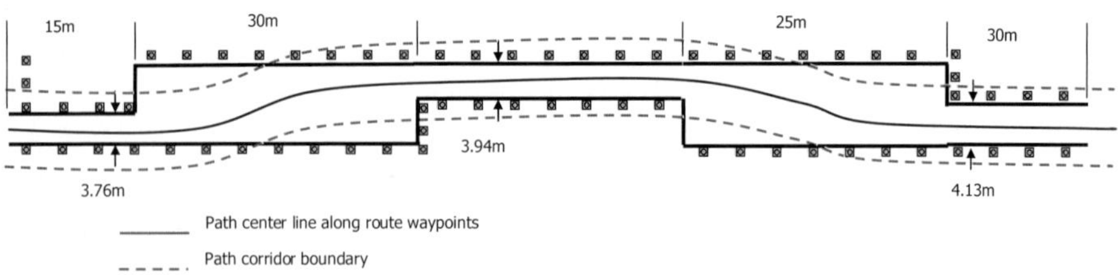

Fig. 2.30. Bind and perception assisted path tracking route through modified ISO-3888-1 test track

2.5.1.3 Perception Assisted Path Tracking Test

The perception assisted tracking test is conducted using the same path definition file used in the blind path tracking test and shown in Figure 2.30. At the entrance to each lane segment a perpendicular boundary wall has been added to the original ISO-3888-1 test track. These boundary walls are intended to ensure a robot will plan a through the desired lanes. While performing this test, robots are configured to use all of their perception and pose sensor data. The test measures driving skill when given a nominal path through an area constrained by boundary obstacles.

2.5.1.4 Perception Planning Test

The perception planning test is conducted on the same modified ISO-3888-1 test track but uses a route file that follows the center line of the test track (see Figure 2.31). This test measures an autonomous ground vehicle's ability to significantly modify the pre-planned path based on perception.

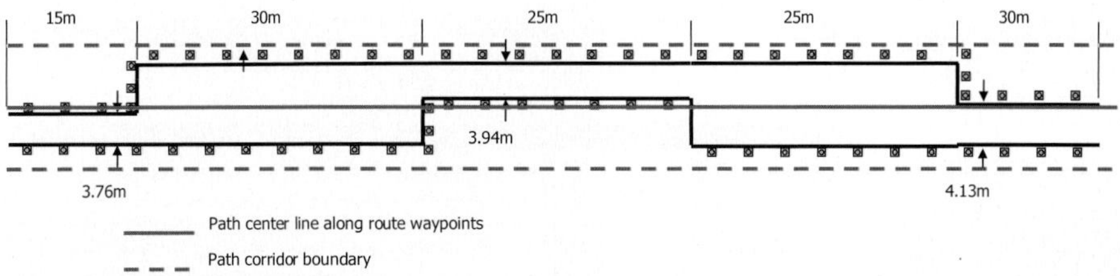

Fig. 2.31. Perception planning route through the modified ISO-3888-1 test track

2.5.1.5 Test Execution

Field tests were conducted at a brown site in Pittsburgh, Pennsylvania and at the Nevada Automotive Test Center (NATC) in Silver Springs, Nevada. While testing at the Pittsburgh site small cardboard boxes wrapped in plastic bags were used as lane boundaries. As perception capability improved cones were substituted for lane boundaries and obstacles. Figure 2.32 shows Sandstorm and H1ghlander operating on the test tracks at LTV Steel and NATC.

During development and testing a full battery of tests was performed on the modified ISO-3888-1 test track on eight separate occasions. Test personnel included test conductor, robot operator, chase car driver and timers. In general a complete set of blind path tracking, perception assisted tracking and perception planning tests were performed during each session. Table 2.6 is an example of data collected during a typical test session on the modified ISO-3888-1 test tracks.

Sessions on the modified ISO-3888-1 track generally started with the blind path tracking test at an initial speed of 5 meters per second. The team would execute a minimum of three runs then increase the speed by 2 meters per second.

Fig. 2.32. Sandstorm (a) and H1ghlander (b) on the LTV and NATC test courses

Table 2.6. Example data collected during testing

Test #	TOD	Type	200m (s)	Speed (m/s)	Notes
1	2:35	BT-5	40.98	4.88	Cone 9L brushed
2	2:40	BT-5	41.20	4.85	Cone 4L brushed
3	2:44	BT-5	41.13	4.86	Good
4	2:49	BT-7	29.20	6.85	Cone 4L hit hard
5	2:52	BT-7	29.45	6.79	Cone 4L hit hard
6	3:09	BT-7	31.70	6.31	Cone 4L hit more hard
7	3:12	BT-7	31.67	6.32	Cone 4L hit more hard
8	3:15	BT-9	**ABORT**		**E-stop because robot leaving course at 4L**
9	3:40	PT-5	45.58	4.39	Good
10	3:58	PT-5	45.26	4.42	Good
11	4:04	PT-5	43.64	4.58	Good
12	4:09	PT-7	36.67	5.45	Good
13	4:11	PT-7	37.29	5.36	Good
14	4:15	PT-7	38.70	5.17	Cone 7R hit
15	4:17	PT-9	29.70	6.73	Cone 7R hit
16	4:21	PT-9	31.27	6.40	Cone 7R hit
17	4:25	PP-5	45.89	4.36	Center 3rd wall hit
18	4:28	PP-7	39.72	5.04	Corner 1st wall brushed, 3rd wall corner crushed
19	4:30	PP-9	37.60	5.32	Corner 1st wall , brushed, 3rd wall corner crushed
20	4:32	PP-9	N/A		2nd outer and center wall hit, 3rd corner crushed
21	4:37	PP-9	33.61	5.95	3rd wall corner, 2 wall center punch & corner, 1st wall box crunched

At each speed increment three runs were executed. Speed was increased until the vehicle left the course and had to be stopped via the emergency stop link or the test team deemed operations at higher speeds were unnecessary. The blind path tracking test was never conducted at speeds above 13 meters per second. As confidence in the robot's blind tracking ability increased the number of blind tracking tests were decreased and eventually were not included in the test routine. The test was held in reserve for regression testing after hardware or software changes were made to a robot that would affect the basic path tracking.

Like the blind tracking test, the initial speeds for perception tracking test was 5 meters per second. Multiple runs at the slower speeds were eventually found to be unnecessary. As in blind path tracking, incremental speed changes were of 2 meters per second and the maximum speed was 13 meters per second. As confidence in the perception sensing systems' ability to correctly sense and localize obstacles increased, emphasis on conducting perception tracking was diminished.

The perception planning test was performed in the same way as the perception tracking test with the exception of using a different path.

2.5.1.6 Evaluation

The initial blind tracking test of Sandstorm (see figure 2.33) conducted on June 17, 2005, revealed that the robot did not have a robust path tracking ability. The observed quality of driving was poor with significant overshoot when cornering. The path tracking control algorithm was modified, adding an integral term, and when the test was repeated on August 17, 2005, performance was notably improved. The team was satisfied with Sandstorm's blind tracking performance at this point and did not conduct the test on Sandstorm again. Similar performance improvement trends were observed withH1ghlander as well.

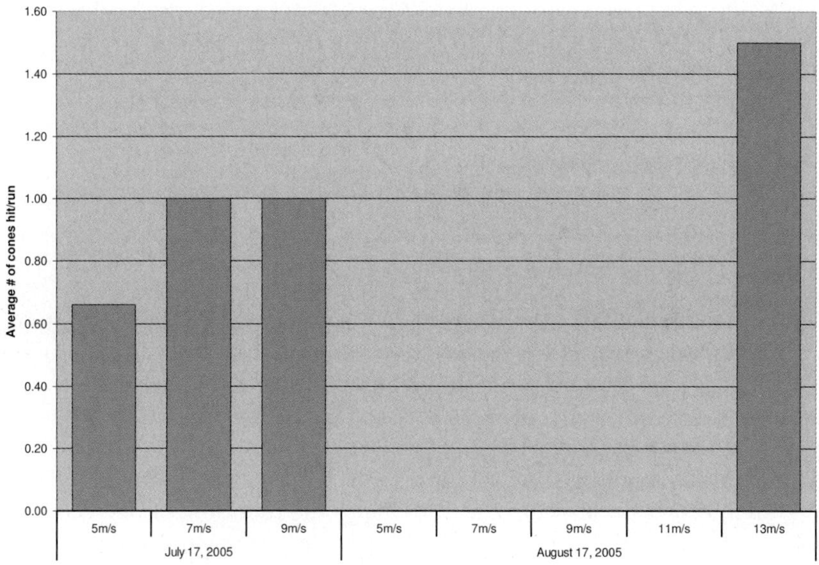

Fig. 2.33. Blind tracking performance for Sandstorm

Analysis of the data collected during the perception tracking and planning tests is inconclusive of overall performance gains due to changing hardware and software configurations but it did help detect system anomalies. As an example, performance of the vehicles was observed to decline in September after the short range sensors were removed and replaced during addition of a sensor washing system. This decline was directly attributable to the lack of adequate calibration of the sensors for object localization.

2.5.1.7 Perspectives on Controlled Tests

The blind tracking, perception tracking and perception planning tests are an effective tool for measuring autonomous ground vehicle driving skill. The tests are relatively easy to set up and inexpensive to conduct. While straight forward, the tests are an effective means to evaluate hardware and software configuration changes.

The blind tracking test is an excellent tool for measuring an autonomous ground vehicle's ability to blindly follow waypoints and is broadly applicable to all automotive and truck class autonomous ground vehicles.

The perception tracking test is a good tool to measure the effects of perception on path tracking and also as a subjective qualitative assessment of driving quality. The perception planning test is a good tool to measure the effectiveness of an autonomous ground vehicle's ability to dynamically adapt to obstacles impeding the pre-planned path. The perception planning test is limited; while it measures a robot's ability to rapidly adapt to obstacles, it does not measure if a vehicle reacts in a smooth manner. For example, if a human is driving and sees an obstacle in its path they will generally react as early and smoothly as possible to change their trajectory to avoid the obstacle. The perception planning test could be adapted for this purpose by elongating the length of the track segments between lane change barriers.

2.5.2 End-to-End Tests

Red Team deployed to Nevada in late August of 2005. Once there, the team began conducting a series end-to-end system tests designed to simulate race day conditions. These tests followed the following process:

- System test team identifies a route and creates a route data definition file (RDDF) (This could take several days to complete).
- The RDDF is delivered to the pre-planning team who have 1 hour and 45 minutes to create a path file (Ideally this was accomplished on the morning of the systems test).
- The path file is reviewed by the field team to ensure it is safe and viable.
- A system test team provides the path file to the robot operators who have 15 minutes to load and launch the robot (This step was timed in order to validate the operators' ability to reliably launch the robot in a timely manner).
- Robots were then chased and performance monitored until either run completion or intervention.

Identifying routes that were race length including a wide collection of terrain and free from undue risk proved to be a formidable task. No non-looping, race length route was identified near the test area. Figure 2.34 shows two of the routes used for systems tests. The "Pork Chop" route is a 48 kilometer loop featuring pavement, gravel road, dry lake bed, sand, wash out, dirt road/trail, cattle guard, gates, parallel fences, high voltage power lines, rail road crossings and elevations ranging from 4300 to 4750 feet. The "Hooten Wells Extended" route is 85 kilometers one way. This route was looped by adding tight circle

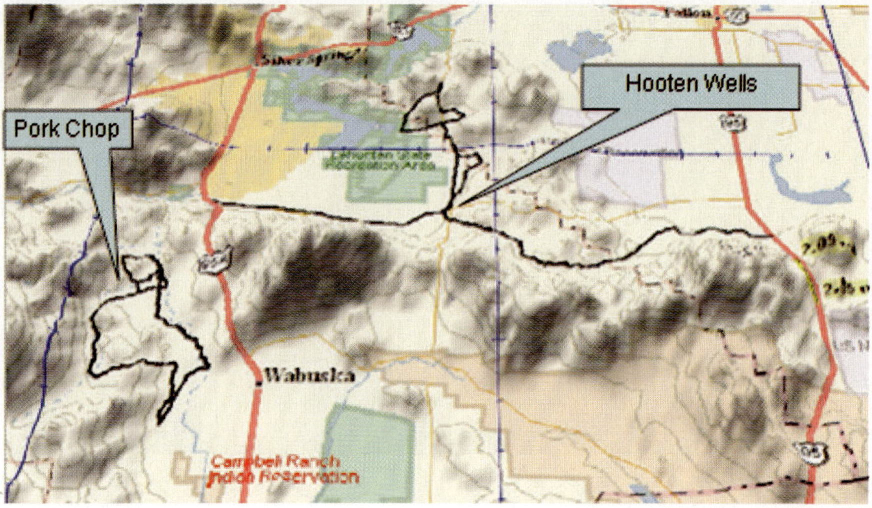

Fig. 2.34. The "Pork Chop" route is a 48km loop while the "Hooten Wells" route is 85km each way

turns at both ends. The Hooten Wells route features gravel road, dry lake bed, large wash outs, dirt road/trail, cattle guard, gates, parallel fences, high voltage power lines, a canyon and elevations ranging from 4000 to 4700 feet.

During tests along these routes Sandstorm and H1ghlander each drove over 1600 kilometers autonomously. Each robot completed challenge length runs at race pace on routes more difficult than the 2005 DARPA Grand Challenge race route. Figure 2.35 shows the daily total number of meters driven during testing at the NATC. The figure shows dates of the three significant failures incurred during testing: H1ghlander sheering off its front right wheel, Sandstorm being "clothes lined" by a tree and H1ghlander rolling.

2.5.3 Test Conclusions

Over the course of this development, the technology readiness level (TRL) (Mankins, 1995) of the robots presented in this paper improved from TRL 3 (analytical or experimental characteristic proof of concept) in the summer of 2003 to TRL 5 (Technology component demonstration in a relevant environment). This change in readiness was driven by the rigorous test program. In order to achieve TRL 9 (actual technology system qualified) a much wider set of systems tests must be developed. Although the testing described above effectively prepared Sandstorm and H1ghlander to compete in the DARPA Grand Challenge, it is not adequate for fielding an autonomous ground vehicle for everyday use.

Standardized tests must be developed that measure a robot's ability to sense and accurately localize obstacles of varying size. These tests should account for differing perception sensing modes. Standard tests that measure an autonomous

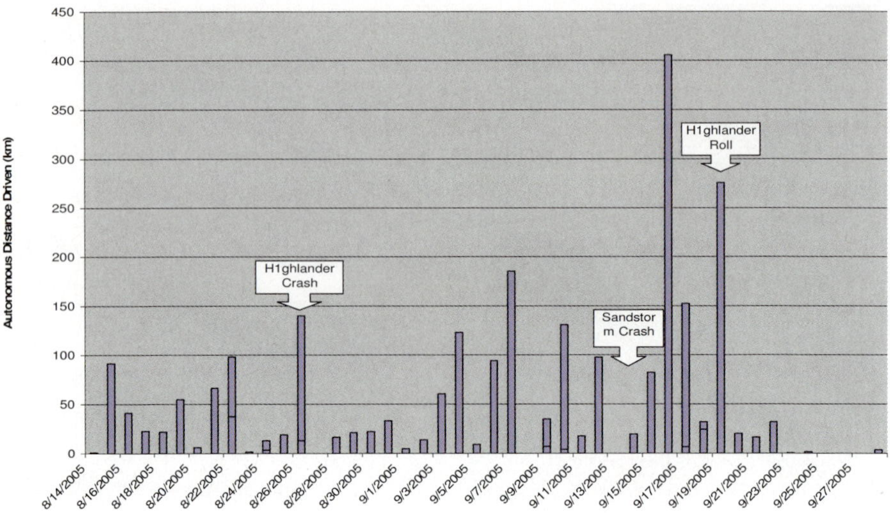

Fig. 2.35. Daily autonomous driving distance during testing at NATC

vehicle's ability to safely and reliably interact with other vehicles and humans are needed. These tests and others are required in order to move autonomous ground vehicles from technological curiosities to common tools used by people everywhere.

2.6 The Grand Challenge

The 2005 DARPA Grand Challenge began at 6:40AM on October 8, 2005 with H1ghander and Sandstorm departing the starting chutes 1st and 3rd respectively. Both robots completed the challenge, finishing 3rd and 2nd (Figure 2.36). Figure 2.37 illustrates the race day route. The route consists primarily of wide,

Fig. 2.36. Sandstorm (a) and H1ghlander (b) cross the Grand Challenge finish line 2nd and 3rd

Fig. 2.37. The 2005 Grand Challenge Route with the approximate times at which Sandstorm encountered challenging terrain

straight dirt roads. There are a few technical areas including 3 underpasses and a narrow, winding descent through Beer Bottle canyon. Overall, the route is less difficult than the system test courses that Sandstorm and H1ghlander had operated on prior to the grand challenge.

2.6.1 Strategy

To increase the chance of success, the two robots were run with different speed ranges for each of the risk levels used by the pre-planning system. Table 2.7 describes the speeds used for each risk level. H1ghlander ran at near full speed while Sandstorm operated with a more conservative speed plan. Both speed policies were within the operational ranges for which the robots had been tested. The speed policies resulted in pre-planned expected completion times of six hours and nineteen minutes for H1ghlander and seven hours and one minute for Sandstorm.

The dual speed strategy was selected to increase the likelihood of at least one Red Team robot correctly dealing with some challenging, unforeseen obstacle on the Grand Challenge route. While the strategy was successful in that both

Table 2.7. Risk level to speed range mapping during the Grand Challenge

Risk Level	Speed Ranges for Sandstorm (m/s)	Speed Ranges for H1ghlander (m/s)
Dangerous	5	5
Moderate	7.5	7-9
Safe	10-10.5	10-12
Very Safe	12	13-13.5

robots completed the challenge, it limited Sandstorm below its ability and in retrospect prevented it from winning the Grand Challenge.

2.6.2 Sandstorm

Sandstorm completed the 212 kilometer course in approximately seven hours and four minutes to finish 2nd, eleven minutes behind the winning team from Stanford. At approximately 6:50 AM, Sandstorm launched on-time and without incident. Upon departure all mechanical, sensing, and navigation systems were operating nominally. In general, Sandstorm drove the route well. It completed the course within 1% of the prepanned time and came within 11 minutes of winning the grand challenge. Sandstorm drove cleanly through most of the course, only touching an obstacle during the tricky Beer Bottle pass descent.

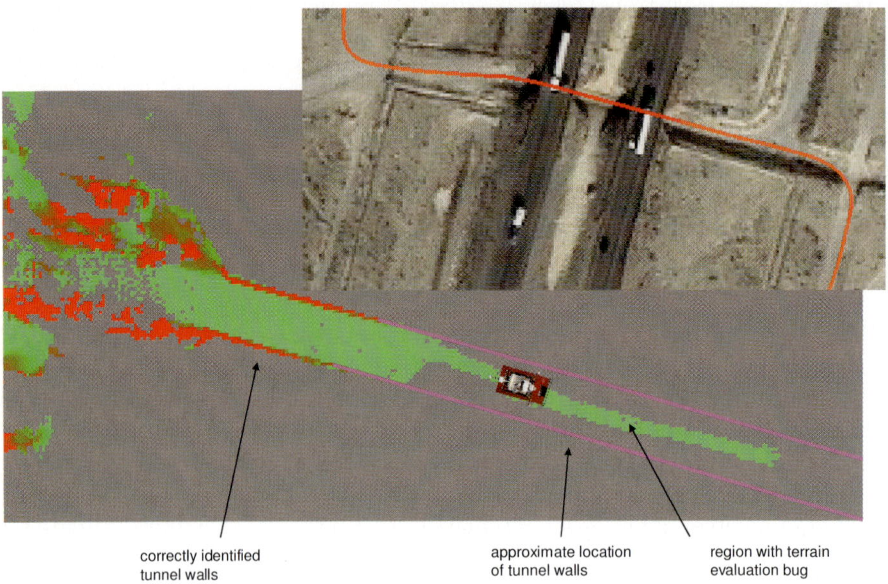

Fig. 2.38. The third underpass and sensor data illustrating the terrain evaluation bug

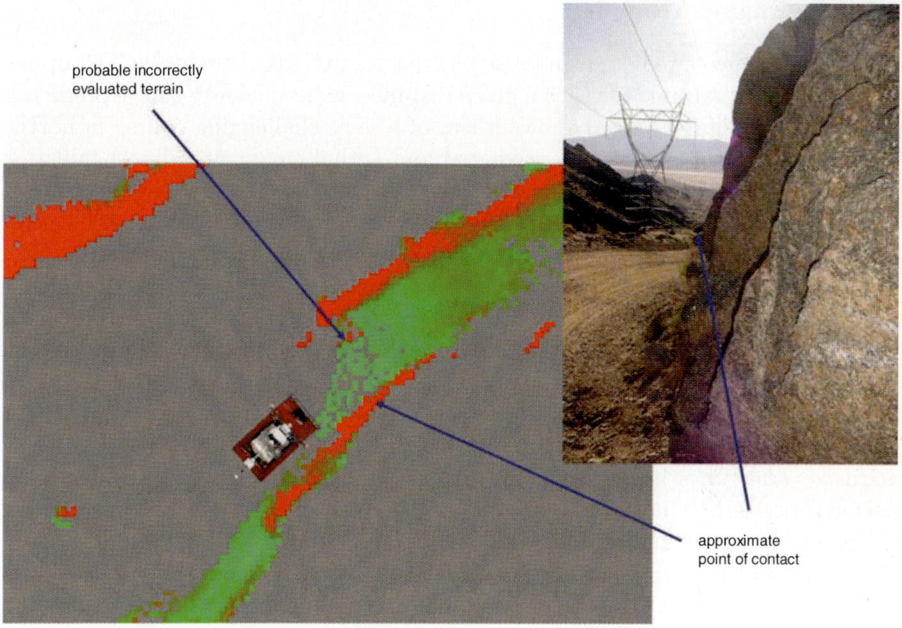

Fig. 2.39. Sensor map illustrating the point of contact

Sandstorm cleanly navigated through 3 underpasses. Both underpass 2 and 3 were sufficiently long and constrained that the LIDAR terrain evaluation algorithm failed to detect the walls (see figure 2.38). Fortunately the redundancy in perception algorithms made the navigation system robust to this failure and Sandstorm cleanly navigated all three tunnels without contact.

The only contact Sandstorm made with an obstacle during the Grand Challenge occurred during the descent through Beer Bottle pass. It is impossible to perform a complete reconstruction of the incident since the log file containing the pose of the gimbal is corrupt and the documentation camera stopped operating prior to entering the canyon. Figure 2.39 shows a cost map constructed from only the shoulder, short range LIDARs, and thus provides an incomplete picture of what the robot saw. It is likely that the robot attempted to cut close to the wall to avoid the incorrectly evaluated terrain to the left of the trail. Despite this minor contact, Sandstorm emerged from the canyon relatively unscathed.

Sandstorm drove well and completed the grand challenge course but was on average 0.25m/s too slow to claim victory.

2.6.3 H1ghlander

H1ghlander completed the challenge in seven hours and fourteen minutes, 55 minutes longer than its intended completion time. This failure to achieve intended speeds and elapsed time was unprecedented in H1ghlander's testing history.

2.6.3.1 Engine Trouble

Figure 2.40 shows predicted progress plotted versus actual progress. The upper curve shows actual time to reach a given distance while the lower shows predicted performance. Figure 2.40(a) shows a run of a very challenging course in northern Nevada. In this figure, the predicted and actual curves are almost indistinguishable. Figure 2.40(b) shows performance during the Grand Challenge course. Early in the race (27 kilometers) the predicted progress separates from the actual progress. The actual progress continues to separate throughout the race.

Figure 2.41 compares velocity controller performance on various grades during the race and during the last long distance traverse before the race. The pre-race plot shows a slight decline in performance at higher grades while the in-race plot shows a significant decline at all speeds.

Analysis of velocity performance shows seven locations where H1ghlander came to a stop. Two of these stops were on hills where the vehicle rolled backwards several times before being able to crest the hills. Significantly degraded performance is first noticeable after 27 kilometers and continues for the entire challenge. Onboard audio indicates that the engine was stalling regularly while H1ghlander drove. Immediately following the race, the engine idle speed was oscillating violently. Data was logged while this phenomenon was occurring, but since the race the problem has not reoccurred. Samples of engine and transmission fluids show no anomalies and the onboard engine diagnostics showed no faults during and after the race. At this time, the cause for this engine problem remains unexplained.

2.6.3.2 Gimbal Failure

H1ghlander's gimbal stopped responding to commands 87 kilometers into the race, failing after a hard left turn. Examination of vehicle speed indicates that this is an area where H1ghlander was having engine trouble.

The gimbal is sensitive to power fluctuations, which may have been the cause of its failure. While H1ghlander's power is typically stable, the engine conditions discussed above may have caused significant power generation problems. When engine RPMs are low, the generator stops producing power and the power system switches to using batteries. This condition is rare, as engine idle speed is set to be significantly higher than the generator cutoff speed. Power system switching was tested repeatedly early in the development of H1ghlander and had not shown any problems leading up to the race.

It is conjectured that while the engine was not running normally, the generator shut off temporarily, causing a brown out in the gimbal power system. The gimbal stopped responding to commands and never recovered. Normally, when processes fail to respond, they are restarted. In this case, the gimbal control computer would have also browned out and as there is no capability to restart failed computers, the restart system did not attempt to restart the gimbal processes.

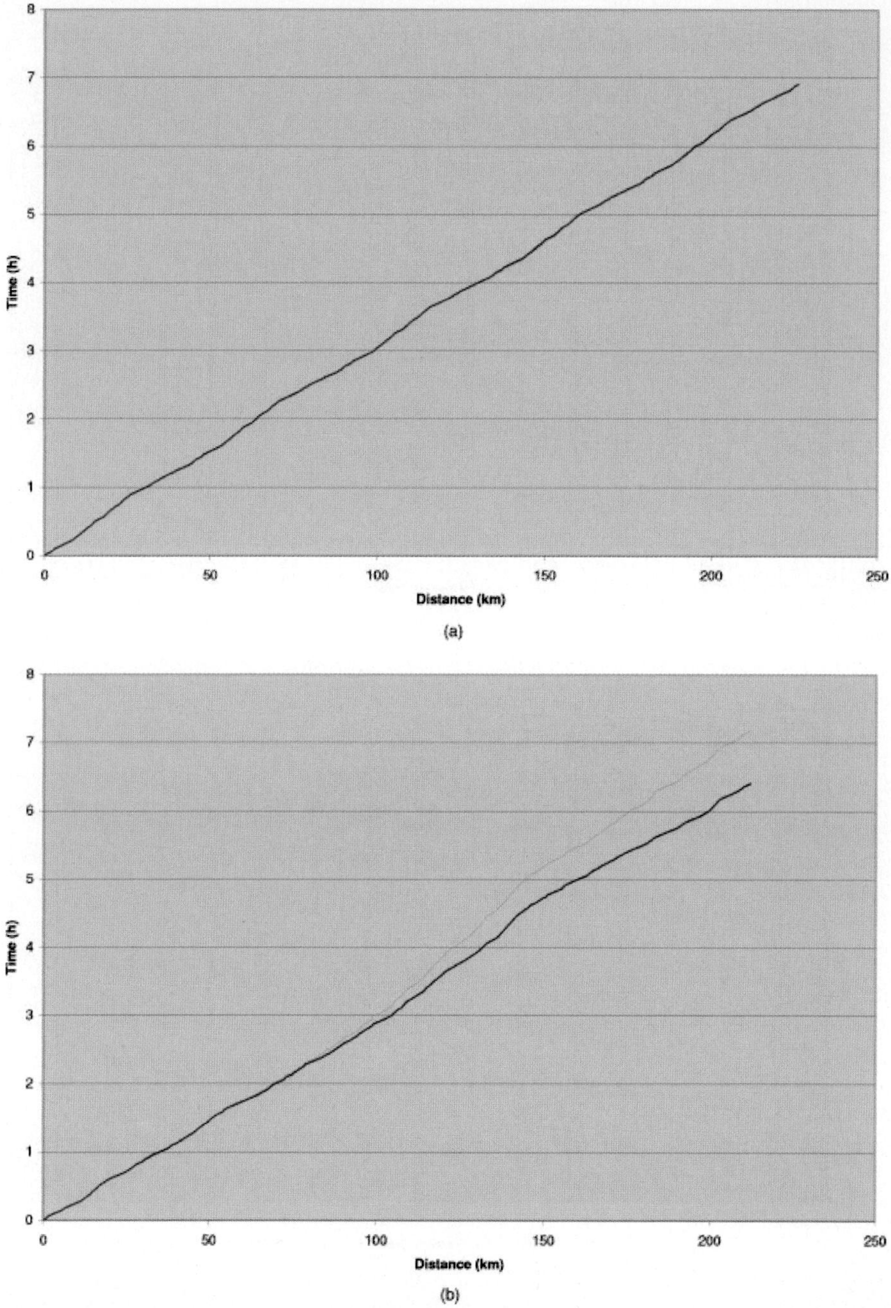

Fig. 2.40. A comparison of H1ghlander's actual and pre-planned distance traveled for (a) a representative test and (b) during the Grand Challenge

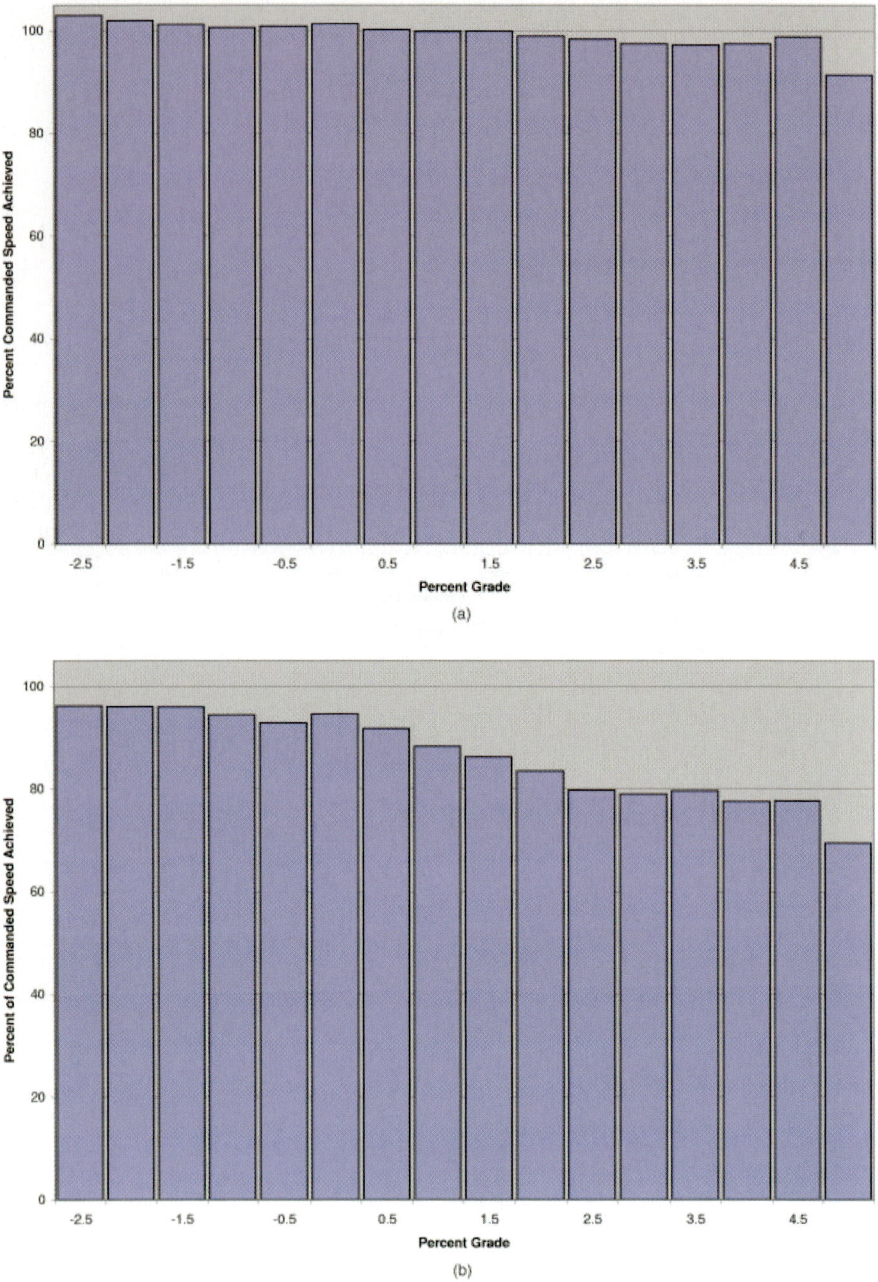

Fig. 2.41. A comparison of H1ghlander's ability to maintain commanded speed between (a) testing and (b) performance during the Grand Challenge

While this hypothesis seems the most logical explanation for the gimbal failure, there is insufficient data to determine the root cause of failure as power diagnostics are only logged into a short rolling buffer. Under normal testing conditions, system failures are examined immediately and information about power and state of components involved in the failure is recorded. Due to the nature of the race, the vehicle could not be stopped for examination, thus this data was lost.

Despite these two failures, H1ghlander completed the Grand Challenge, illustrating the ability of the robot to survive a number of significant faults.

2.7 Conclusions

Through the application of simple, well implemented ideas, it is possible to achieve robust, high-speed desert navigation. At each point in the development process we attempted to use the simplest possible approach, be it the first order approximations of physical constraints in the speed planning algorithm or the expectation based map merging algorithm. The key to success was performing testing and analysis to understand where these approaches would fail, and determining whether a more complex solution was warranted.

The robustness of this approach was clearly demonstrated by Sandstorm and H1ghlander's performances at the 2005 Grand Challenge. Despite a failing engine and the loss of a perception sensor, both robots completed the course within 20 minutes of the winning time. Without the careful design and implementation of the navigation system, and the selection of a robust vehicle chassis, this feat would not have been possible.

2.7.1 Next Steps

While the DARPA Grand Challenge represented a significant step forward for high-speed navigation in desert terrain, the robotic navigation problem is far from solved. Before it will be possible to deploy, full-autonomous robots, techniques for dealing with dynamic environments must be developed. The Grand Challenge was carefully designed to keep the competitors separated. Passing in the challenge was constrained such that slower vehicle were stopped well before, and throughout the time, faster vehicles encountered them. While this was reasonable for the Grand Challenge, it is not a viable way for robots to deal with traffic in general.

While the desert provides numerous challenges, the terrain is relatively straight-forward to model. Most desert terrain is well characterized by its geometry since there is relatively little vegetation. This simplifies the perception and terrain evaluation tasks. The same cannot be said for jungles, forests or even farm land. In these environments, a purely geometric understanding of the world would be misleading since there is vegetation that can and must be driven over and through. Completing the same challenge in heavily vegetated terrain would be a tremendous next step.

The Grand Challenge forced teams to develop reliable, near turn-key solutions by requiring a long duration performance on a specific day. The race format meant it was not possible to develop a technology that worked some of the time, it was necessary for the robots and supporting systems to work when called on. While this level of reliability and readiness was met, it was achieved with large teams with very specialized knowledge and training. To move this technology from a compelling demonstration to a production ready capability will require significant effort to both harden and quantify performance.

2.7.2 Lessons for Development of Autonomous Vehicles

While Sandstorm and H1ghlander performed well, and the experience of developing and testing the robots was rewarding, there is much room for improvement in both the performance and the process used to develop them. Over the course of this research we learned several lessons:

System testing is essential but results can be misleading- Testing was critical to the success of Sandstorm and H1ghlander but at times also provided a false sense of confidence. Because of the tight development and testing schedule, component testing was often short and incomplete, due to an emphasis on integrated testing. Since the overall system was relatively robust to sensor failures and bad data, it was possible for intermittent and subtle problems to go undetected for long periods of time. A combination of better unit/component testing or a more in-depth analytical evaluation of developing algorithms would have reduced this problem.

Reliability is critical, and can be achieved- One of the keys to success in the Grand Challenge was balancing innovation with engineering. We used a programmatic approach that effectively annealed ideas to strike this balance. In the first phase we cultivated a broad set of potentially good ideas, hoping to uncover those that would be the keys to success. In the second phase we focused these ideas while achieving internally and externally imposed milestones (e.g. complete a 320 kilometer traverse, or perform obstacle avoidance with a given sensor). These milestones allowed us to judge the viability of competing technical approaches while working to also increase reliability. In the third and final phase, we accepted only ideas that were required to fix problems with the existing system. We used a strict process to report any problems identified during testing and then worked to quickly rectify them in a way that prevented recurrence of the same fault. During this final phase, the robots began to realize their potential for reliable operation.

Use commercial, off-the-shelf (COTS) components, but only with support- A common mantra in developing new systems is to buy not build, and in general this is true, but, buyer beware. Without support, COTS components can be worse than in-house, custom built components. With in-house built components there is at least someone who designed and built the system who can debug it. An unsupported COTS component that behaves unexpectedly may force a significant redesign of a subsystem, potentially incurring more cost and time than if the component had been in-house custom engineered.

Know the problem- much of the technical approach described in this paper was excessive given the final form of the Grand Challenge. The groomed roads and carefully detailed route provided by the organizers greatly reduced two of the competitive advantages (namely the H1 & HMMWV chassis and the pre-planning system) applied by the team. Furthermore, the team put an excess of wear-and-tear on the vehicles during testing operating on more rugged terrain than that encountered during the challenge. Had the final race conditions been known ahead of time, it would have been possible to shed a significant amount of technical complexity.

Correctly scoping a problem is a key to success- One of the programmatic successes was a process to cut technical ideas if they were not progressing or showing results. This helped keep the team focused on the overall goal of completing the Grand Challenge rather than chasing after interesting ideas that were irrelevant to the task at hand. Without this focus we would not have been able to achieve the reliability and robustness necessary to complete the Grand Challenge.

Acknowledgements

This work would not have been possible without the tremendous support of our team sponsors including Caterpillar, SAIC, Boeing, Intel, HD Systems, Harris, Applanix, Chevron, Phillips, AMGeneral, TTTech and the numerous other contributors.

References

Amidi O. (1990). Integrated mobile robot control (Tech. Rep. CMU-RI-TR-90-17). Pittsburgh, PA: Carnegie Mellon University, Robotics Institute.

Batavia P., & Singh S. (2002, May). Obstacle detection in smooth high curvature terrain. Paper presented at the IEEE International Conference on Robotics and Automation (ICRA 2002), Washington, D.C..

Bellutta, P., Manduchi, R., Matthies, L., Owens, K., & Rankin, A. (2000, October). Terrain perception for DEMO III. Paper presented at the IEEE Intelligent Vehicles Symposium (IVS 2000), Dearborn, MI.

Biesiadecki, J., Maimone, M., & Morrison, J. (2001, June). The Athena SDM rover: a testbed for Mars rover mobility. Paper presented at the International Symposium on Artificial Intelligence (iSAIRAS 2001), St-Hubert, Canada.

Castelnovi, M., Arkin, R.C., & Collins, T.R. (2005, April). Reactive speed control system based on terrain roughness detection. Paper presented at the IEEE International Conference on Robotics and Automation (ICRA 2005), Barcelona, Spain.

Coombs, D., Lacaze, A. D., Legowik, S. A., & Murphy, K. N. (2000, October). Driving autonomously offroad up to 35 km/h. Paper presented at the IEEE Intelligent Vehicles Symposium (IVS 2000), Dearborn, MI.

Coulter, R. (1992). Implementation of the pure pursuit path tracking algorithm (Tech. Rep. CMU-RI-TR-92-01). Pittsburgh, PA: Carnegie Mellon University, Robotics Institute.

Diaz-Calderon, A., & Kelly, A. (2005). On-line stability margin and attitude estimation for dynamic articulating mobile robots. International Journal of Robotics Research, 24(10), 845-866.

Flores-Parra, I., Bienvenido, J., & Menenti, M. (2000, July). Characterization of segmentation methods by multidimensional metrics: application to the delimitation of structures. Paper presented at the IEEE National Geoscience and Remote Sensing Symposium (IGARSS 2000), Honolulu, HI.

Gillespie T. D. (1992). Fundamentals of vehicle dynamics. Warrendale, PA, Society of Automotive Engineers.

Golda, D., Iagnemma, K., Dubowsky, S. (2004, April). Probabilistic modeling and analysis of high-speed rough-terrain mobile robots. Paper presented at the IEEE International Conference on Robotics and Automation, New Orleans, LA.

Golberg, S., Maimone, M., & Matthies, L. (2002, March). Stereo vision and rover navigation software for planetary exploration. Paper presented at the IEEE Aerospace Conference, Big Sky, MT.

Gowdy, J. (1996). IPT: An object oriented toolkit for interprocess communication (Tech. Rep. CMU-RI-TR-96-07), Pittsburgh, PA: Carnegie Mellon University, Robotics Institute.

Hart, P., Nilsson, N., & Rafael, B. (1968). A formal basis for the heuristic determination of minimum cost paths. IEEE Transactions on Systems Science and Cybernetics, 4(2), 100–107.

Harvey, W., McGlone, J., McKeown, D., & Irvine, J. (2004). User centric evaluation of automated road network extraction. Photogrammetric Engineering and Remote Sensing, 70(12), 1353–1364.

International Standards Organization (1999). Passenger cars test track for a severe lane-change maneuver part 1: double lane-change first edition 1999-10-01, Geneva, CH, International Standards Organization.

Jiang, T.Z., Wu, H., Wu K., & Sun X.W. (2005). Threshold design method of CFAR for millimeter-wave collision warning radar. Journal of Infrared and Millimeter Waves, 24(3), 217–220.

Kaliyaperumal, K., Lakshmanan, S., & Kluge, K. (2001). An algorithm for detecting roads and obstacles in radar images. IEEE Transactions on Vehicle Technology, 50(1), 170–182.

Kavraki, L., Svestka, P., Latombe, J., & Overmars, M. (1996). Probabilistic roadmaps for path planning in high-dimensional configuration spaces. IEEE Transactions on Robotics and Automation, 12(4),566–580.

Kelly, A. & Stentz A. (1997, January). An Approach to Rough Terrain Autonomous Mobility. Paper presented at the International Conference on Mobile Planetary Robots, Santa Monica, CA.

Kelly, A. & Stentz, A. (1997). An Analysis of Requirements for Rough Terrain Autonomous Mobility. Autonomous Robots, 4(4).

Kelly, A., Amidi, O., Bode, M., Happold, M., Herman, H., Pilarski, T., Rander, P., Stentz, A., Vallidis, N., & Warner, R. (2004, June), Toward reliable off-road autonomous vehicle operating in challenging environments. Paper presented at the International Symposium on Experimental Robotics (ISER 2004), Singapore.

Lalonde, J., Vandapel, N., & Hebert, M. (2005, June). Data structures for efficient processing in 3-D. Paper presented at Robotics: Science and Systems 1, Cambridge, MI.

LaValle, S. (1998). Rapidly-exploring random trees: a new tool for path planning (Technical Report). Ames, IA: Iowa State University, Computer Science Department.

LaValle, S., & Kuffner, J. (1999, May). Randomized kinodynamic planning". Paper presented at the IEEE International Conference on Robotics and Automation (ICRA 1999), Detroit, MI.

LaValle, S., & Kuffner, J. (2001). Rapidly-exploring Random Trees: Progress and prospects". In B. R. Donald, K. M. Lynch, and D. Rus, (editors), Algorithmic and Computational Robotics: New Directions, (pp. 293-308). Wellesley, MA: A K Peters.

Mankins, J. (1995). Technology readiness levels: A white paper. http://www.hq.nasa.gov/office/codeq/trl/trl.pdf.

Nabbe, B., Kumar, S., & Hebert, M. (2004, October). Path planning with hallucinated worlds. Paper presented at IEEE/RSJ International Conference on Intelligent Robots and Systems (IROS 2004), Sendai, Japan.

Nilsson, N. (1980). Principles of Artificial Intelligence. Palo Alto, CA: Tioga Publishing Company.

Oshkosh Truck Corp, Oshkosh, WI Rockwell Collins, Cedar Rapids, IA Rockwell Scientific, Thousand Oaks, CA University of Parma, Parma, Italy (2005). Team Terramax DARPA Grand Challenge 2005 (Technical Paper). OshKosh, WI.

Roth, S. A. & Batavia, P. (2002, July). Evaluating path tracker performance for outdoor mobile robots. Paper presented at the Conference on Automation Technology for Off- Road-Equipment, Chicago, IL.

Roth, S.A., Hamner, B., Singh, S., & Hwangbo, M. (2005, July). Results in combined route traversal and collision avoidance. Paper presented at the International Conference on Field & Service Robotics (FSR'05), Port Douglas, Australia.

Shimoda, S., Kuroda, Y., & Iagnemma, K. (2005, April). Potential field navigation of high speed unmanned ground vehicles on uneven terrain. Paper presented at the IEEE International Conference on Robotics and Automation (ICRA 2005), Barcelona, Spain.

Simmons, R., Krotkov, E., Chrisman, L., Cozman, F., Goodwin, R., Hebert, M., Katragadda, L., Koenig, S., Krishnaswamy, G., Shinoda, Y., Whittaker, W., & Klarer, P. (1995, August). Experience with rover navigation for lunar-like terrains. Paper presented at IEEE/RSJ International Conference on Robots and Systems (IROS'95), Pittsburgh, PA.

Singh, S., & Keller, P. (1991, April). Obstacle detection for high speed autonomous navigation. Paper presented at the IEEE Proceedings International Conference on Robotics and Automation (ICRA'91), Sacramento, CA.

Stanford Racing Team (2005). Stanford racing team's entry in the 2005 DARPA grand challenge (Technical Paper). Stanford, CA.

Talukder, A., Manduchi, R., Rankin, A., Matthies, L. (2002, June). Fast and reliable obstacle detection and segmentation for cross-country navigation. Paper presented at the IEEE Intelligent Vehicle Symposium (IVS 2002). Versailles, France.

Talukder, A., Manduchi, R., Castano, R., Owens, K., Matthies, L., Castano, & A., Hogg, R. (2002, October). Autonomous terrain characterization and modeling for dynamic control of unmanned vehicles. Paper presented at the IEEE/RSJ International Conference on Intelligent Robots and Systems (IROS 2002). Lausanne, Switzerland.

Tompkins, P., Stentz, A., & Whittaker, W.L. (2004, March). Field experiments in mission-level path execution and re-planning. Paper presented at the Conference on Intelligent Autonomous Systems (IAS-8). Amsterdam, Netherlands.

Trepagnier, P., Kinney, P., Nagel, J., Dooner, M. & Pearce, J. (2005). Team gray technical paper DARPA grand challenge 2005 (Technical Paper).

Ulrich, I., & Nourbakhsh, I. (2000, July). Appearance-based obstacle detection with monocular color vision. Paper presented at the AAAI National Conference on Artificial Intelligence. Austin, TX.

Urmson, C., Dias, M., & Simmons, R. (2002, October). Stereo vision based navigation for sun-synchronous exploration. Paper presented at the IEEE/RSJ International Conference on Intelligent Robots and Systems (IROS 2002), Lausanne, Switzerland.

Urmson, C., Anhalt, J., Clark, M., Galatali, T., Gonzalez, J.P., Gowdy, J., Gutierrez, A., Harbaugh, S., Johnson-Roberson, M., Kato, H., Koon, P.L., Peterson, K., Smith, B.K., Spiker, S., Tryzelaar, E., & Whittaker, W.L. (2004). High speed navigation of unrehearsed terrain: Red Team technology for Grand Challenge 2004 (Tech. Rep. CMURI- TR-04-37). Pittsburgh, PA: Carnegie Mellon University, Robotics Institute.

Wettergreen, D., Cabrol, N., Baskaran, V., Calderon, F., Heys, S., Jonak, D., Luders, R.A., Pane, D., Smith, T., Teza, J., Tompkins, P., Villa, D., Williams, C., & Wagner, M.D. (2005, September). Second experiments in the robotic investigation of life in the Atacama Desert of Chile. Paper presented at the 8th International Symposium on Artificial Intelligence, Robotics and Automation in Space (iSAIRAS 2005), Mnchen, Germany.

Yamaguchi, H., Kajiwara, A., & Hayashi, S. (1998, May). Empirical study of polarization techniques for clutter reduction. Paper presented at the IEEE International Radar Conference. Dallas, TX.

3

KAT-5: Robust Systems for Autonomous Vehicle Navigation in Challenging and Unknown Terrain

Paul G. Trepagnier[1], Jorge Nagel[2], Powell M. Kinney[2], Cris Koutsougeras[2], and Matthew Dooner[2]

[1] The Gray Insurance Company, Metairie, LA 70002
[2] Tulane University, New Orleans, LA 70118

Summary. Kat-5 was the fourth vehicle to make history in DARPA's 2005 Grand Challenge, where for the first time ever, autonomous vehicles were able to travel through 100 miles of rough terrain at average speeds greater than 15 mph. In this paper, we describe the mechanisms and methods that were used to develop the vehicle. We describe the main hardware systems with which the vehicle was outfitted for navigation, computing, and control. We describe the sensors, the computing grid, and the methods that controlled the navigation based on the sensor readings. We also discuss the experiences gained in the course of the development and provide highlights of actual field performance.

3.1 Introduction

The DARPA Grand Challenge aimed at fostering the development of a completely full-scale autonomous vehicle that would live up to the challenge of high-speed navigation through challenging and unknown terrain. Organized by DARPA (Defense Advanced Research Projects Agency), the Grand Challenge was first held in 2004, but by the conclusion of the event, no vehicle had completed more than 7 miles of the prescribed course. Even with such poor results, DARPA sponsored the Grand Challenge again in 2005. This time their hopes were justified as not one, but five vehicles finished the 132-mile race. Out of the original 195 vehicles entered, Team Gray's Kat-5 was one of five vehicles to complete the course and one of only four to accomplish this feat within the 10 hour time limit. The vehicle as entered in the race is shown in Figure 3.1.

This paper describes the main challenges posed by this endeavor, and specifically for Kat-5, the choices that were made and the designs that were developed to address them.

3.2 Challenges

According to the rules published by DARPA, the autonomous vehicle had to be able to traverse a route of up to 175 miles across desert terrain in under 10 hours while avoiding ditches, berms, rocks, boulders, fences, and other natural

M. Buehler, K. Iagnemma, and S. Singh (Eds.): DARPA'05, STAR 36, pp. 103–128, 2007.
springerlink.com

Fig. 3.1. Team Gray's Kat-5 as entered in the 2005 Grand Challenge

or man-made obstacles [DARPA, 2005b]. The route to be followed was unknown by the team until two hours before the start of the race. It consisted of a series of GPS waypoints which were an average of 275 feet apart. For each waypoint, an acceptable lateral boundary around the waypoint was also specified. The sequence of waypoints and the lateral boundaries specified a corridor that the vehicle had to stay within or the vehicle could be disqualified by officials (see Figure 3.2). The route was only given as a sequence of GPS waypoints but otherwise it was completely unknown; not just unrehearsed. The rules did not prevent normalization of DARPA's data before they were fed to the vehicles, neither did they prevent elevation map databases, however, Kat-5 did not make use of any information other than its sensor readings and DARPA's waypoint data given to it in raw form.

These conditions presented many challenges that needed to be met by innovative hardware and software design. The following subsections present an overview of each of the main challenges provided by the DARPA Grand Challenge.

3.2.1 Endurance and Reliability

The actual race route was a substantial distance of 132 miles which had to be completed without refueling or pit stops. Having to endure this length of distance

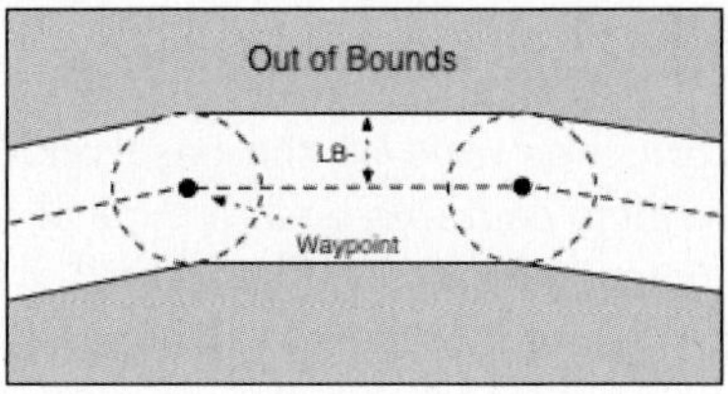

Fig. 3.2. The corridor as defined by DARPA. See [DARPA, 2005a].

in rough terrain at relatively high speed is a massive endurance test for any vehicle, let alone a vehicle operating autonomously. Every piece of hardware on the vehicle could become a failure point due to the excessive shock and vibration caused by the rough terrain.

3.2.2 Environment Sensing

All decision making for steering and other controls had to be based solely on the data collected by the vehicle's onboard sensors. This introduces a major problem as collecting and fusing sensor data can be a computationally expensive process and the sensor data itself can be difficult to interpret [Coombs et al., 2001]. In addition, the vehicle must be able to accurately identify its own position through an onboard sensor. The data readings then have to be coordinated with position readings. Timestamping is complicated by delays of the onboard computer networks.

3.2.3 Artificial Intelligence

Based solely on the sensory data collected, the vehicle needed to independently decide how to generate control signal sequences for steering, braking, and acceleration so that it could avoid obstacles and stay within the given course. In addition, it needed to adjust speed to make sure that the vehicle would execute turns safely.

3.2.4 Control Systems

The combination of all of the aforementioned navigational constraints along with the requirement for high vehicle speed created the need for a very agile set of physical actuators and control algorithms. These algorithms and actuators had to respond to the need for abrupt changes in speed and steering quickly in order to handle the rapidly changing environment.

In addition to the above, this team had to work on a short development timeline. Unlike many of the other teams that competed in the 2005 Grand Challenge, this team entered the competition at a late date. As a result, we effectively had only six months to create and test the complete autonomous vehicle.

3.3 System Design

A 2005 Ford Escape Hybrid was used as the base vehicle platform for the autonomous vehicle. This vehicle contains both a typical four-cylinder gas engine and a 330-volt electric motor/generator. The hybrid system allows for the vehicle to adapt intelligently to many different driving situations with maximum efficiency and performance by automatically using a combination of the two. [Ford Motor Company, 2005]. This ensures maximum fuel economy, but still acceptable performance. This vehicle was chosen because of the system that it was already outfitted with in order to power its electric motor. This system was capable of supplying enough clean power for the additional onboard electronics without an additional alternator or generator. It also had sufficient ground clearance for rough terrain and a wheel base that offered a good balance between agile maneuverability and stability.

During this project, the physical vehicle was viewed as a "black-box" and was only interfaced with at the highest level. The following sections describe the hardware components that were used to make the vehicle capable of operating autonomously and handling the challenges presented in the previous section. The hardware architecture for Kat-5 is outlined in Figure 3.3.

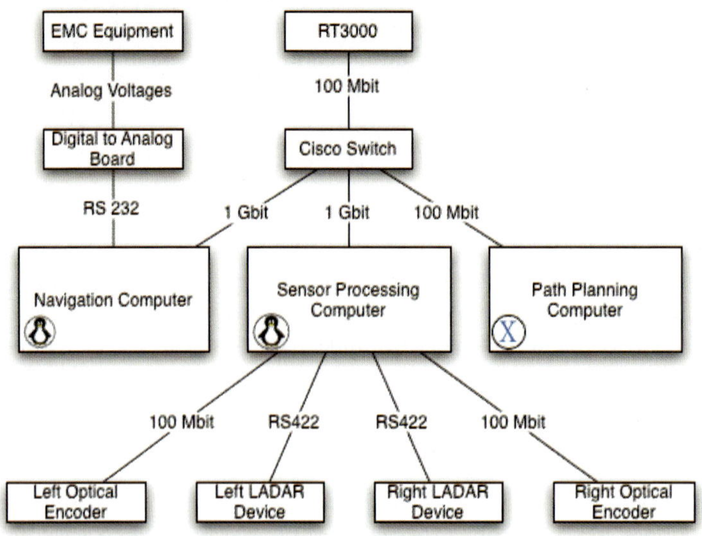

Fig. 3.3. The communication architecture in Kat-5

3.3.1 Computing Devices

The computing capacity for the autonomous vehicle was provided by a computing grid consisting of four networked computers. The computers included two identical marine computers running Linux and two identical Apple Mac Mini computers running OS X. Marine computers (made for use on boats) were used as the two main computers because they would run on 12 volts of direct current without modifications to their power supplies and also because they already had integrated shock mounts. One of the marine computers was dedicated to processing data from the sensors, while the other was dedicated to producing navigation control signals. The Mac Mini performed all of the path planning.

A key challenge associated with the computing devices was making them hardened against shock and vibration. The two marine computers were designed for a hostile environment, so they came with integrated shock mounts and were capable of withstanding shocks of up to 30 Gs. Also, each of their connections (both internal and external) came reinforced with silicon glue and other restraints to ensure that they could not come loose due to shocks and vibrations encountered in rough seas. A similar shock mount system was also created in house for the Mac Minis. An image of the shock mount system is shown in Figure 3.4. The computing devices were linked together by a rugged Cisco 2955 Catalyst switch capable of handling shocks of up to 50 Gs. A mix of UDP and TCP protocols were used for the ethernet communications, with UDP being used for communications where reliability was not as important as speed and TCP for communications where reliability was paramount.

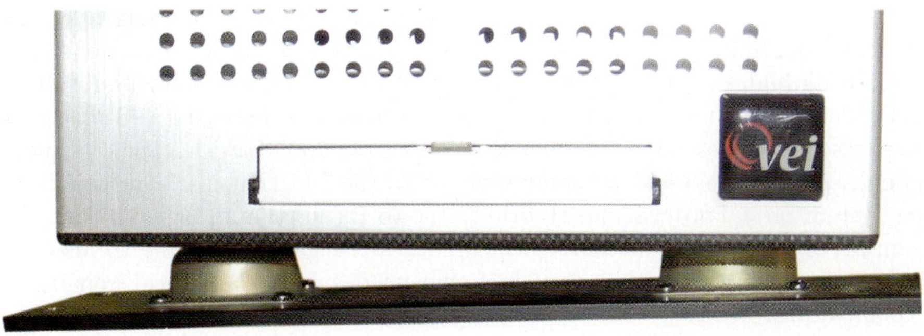

Fig. 3.4. The shock mount system for the navigation computer could withstand shocks of up to 30 Gs

3.3.2 Vehicle Actuators

Interfacing with the primary vehicle controls (steering wheel, accelerator, and brake) was accomplished by a custom drive-by-wire system designed by Electronic Mobility Controls (EMC). This is a custom solution designed to outfit vehicles for handicapped drivers. It consists of actuators and servos mounted on the steering column, brake pedal, throttle wire, and automatic transmission. It is primarily controlled by a console with a touch-sensitive screen and an evaluator's console which contains a throttle/brake lever and a miniature steering wheel. We bypassed these controls and connected the equipment's wire harness to a computer via a digital to analog board. Thus, the electrical signals that

Table 3.1. Input voltages and corresponding digital ranges for interfacing with EMC equipment

Input	Voltage Range	Digital Steps	Software Range
Steering	0.4V to 4.6V	512	-250 (left lock) to 250 (right lock)
Accelerator/Brake	0.4V to 4.6V	256	-124 (full brake) to 124 (full accelerator)

the manual controls would normally produce were actually produced by the computer (which later will be referred to as the navigation computer). These electrical signals took the form of the analog voltages shown in Table 3.1.

The EMC equipment was chosen because it inherently satisfied many of the previously mentioned challenges. Since the equipment is designed for handicapped drivers and must meet Department of Transportation standards, it is fully redundant and very rugged. It also has a very fast reaction time when given a command, so it can quickly turn the steering wheel or apply the brake if it needs to. It was also able to be installed and tested very quickly by trained experts, so our short development timeline was not adversely affected.

After initial testing during a hot summer day, we noticed that the computing equipment was overheating and then malfunctioning due to the high temperatures in the cabin of the car. This revealed an issue between having proper fuel efficiency and having an acceptable cabin temperature. If the air conditioner was kept on its highest setting, the equipment did not overheat, but the resulting fuel economy was projected to be too low to finish the expected 175 mile race (projections were based on the fuel economy of the 2005 Ford Escape 4 cylinder model). This lowered fuel economy was due to the fact that if the air conditoning system on a Ford Escape Hybrid is set to its maximum setting, then the compressor must run constantly, which causes the gasoline engine to also run constantly. This defeats the whole fuel efficient design of the hybrid's engine as explained previously.

As a result of this problem, we created a simple on/off mechanism for the air conditioning system that was suited to the cooling needs of the equipment rather than the passenger's comfort. The device consisted of a temperature sensor, a BASIC stamp, and a servo motor. We mounted the servo to the air conditioning system's control knob so that the servo could turn the air conditioner on and off. The BASIC stamp is a simple programmable micro-controller with 8 bidirectional input and output lines and a limited amount of memory which can hold a small program. We programmed the BASIC stamp to monitor the temperature of the cabin near the equipment. If the temperature dropped below a certain threshold, the air conditioner was turned off. If the temperature rose above a certain temperature, the air conditioning system was turned to its maximum setting. This simple system solved our temperature problems while not adversely affecting our fuel efficiency, yet still only interfacing with the vehicle at its highest level.

3.3.3 Positioning System

The position and pose of the car is reported by an Oxford Technical Solutions RT3000, an integrated Global Positioning System (GPS) with two antennas and an Inertial Navigation System (INS). This system ensures that the vehicle is aware of its position on the Earth with a best-case accuracy of less than 10 centimeters. This accuracy is possible due to its use of the Omnistar High Performance (HP) GPS correction system. Nominal accuracies of the different parameters available from the RT3000 are shown in Table (3.2).

Table 3.2. Accuracies of primary vehicle parameters given by the RT3000 [Oxford Technical Solutions, 2004]

Parameter	Accuracy
Position	10 cm
Forward Velocity	0.07 km/hr
Acceleration	0.01 %
Roll/Pitch	0.03 degrees
Heading	0.1 degrees
Lateral Velocity	0.2 %

The RT3000 uses a Kalman filter to blend all of its inputs so as to derive clean unbiased estimates of its state. A Kalman filter is a method of estimating the state of a system based upon recursive measurement of noisy data. In this instance, the Kalman filter is able to much more accurately estimate vehicle position by taking into account the type of noise inherent in each type of sensor and then constructing an optimal estimate of the actual position [Kelly, 1994]. In the standard RT3000, there are two sensors (GPS and INS). These two sensors complement each other nicely as they both have reciprocal errors (GPS position measurements tend to be noisy with finite drift while INS position measurements tend to not be noisy but have infinite drift) [Bruch et al., 2002].

The RT3000 also accepts additional custom inputs to reduce drift in its estimate of vehicle position when GPS is not available. This is important since when GPS is not present, the estimate of position will begin to drift due to the Kalman filter's heavy reliance on INS measurements. One of these custom inputs is a wheel speed sensor which provides TTL pulses based upon an encoder placed on a single wheel on the vehicle. When a wheel speed sensor is added to the RT3000, it initially uses GPS data to learn how these TTL pulses correspond to actual vehicle movement. Then when GPS data is not available due to tunnels, canyons, or other obstructions, the RT3000 is able to minimize the positional drift by making use of the wheel speed sensor and its latest reliably known correspondence to the vehicle's movement. [Oxford Technical Solutions, 2004].

The wheel speed sensor consisted of a digital sensor capable of detecting either ferrous metal or magnets that are in motion. We mounted it in the wheel well adjacent to the stock Antilock Brake System (ABS) sensor, which allowed the wheel speed sensor to read the same magnets mounted on the wheel that the ABS sensor did.

This level of accuracy allowed the car to precisely know its location on the earth, and therefore made the artificial intelligence algorithms much more accurate. Its 100 Hz refresh rate also notified the control systems of positional error very quickly, which allowed for immediate corrections in course due to bumps from rough terrain and other sources of error.

3.3.4 Vision Sensors

Two Sick LMS 291 Laser Detecting and Ranging (LADAR) devices provided the autonomous vehicle with environmental sensing. Each LADAR device scans a two-dimensional plane using a single pulsed laser beam that is deflected by an internal rotating mirror so that a fan shaped scan is made of the surrounding area at half-degree intervals [Sick AG, 2003]. Rather than pointing the LADAR devices at the ground horizontally, we mounted the LADAR devices vertically. We chose to align them vertically because it made obstacle detection much easier. In the simplest case, by analyzing the measurement data beam by beam in angular order, obstacles were easy to locate as either clusters of similar distance or gaps in distance [Coombs et al., 2001].

A set of two 12 volt DC batteries connected in series supplied the required 24 volts of DC current required for the two Sick LADAR devices. These two batteries were then charged by an array of six solar panels that were placed on the top of the vehicle's aluminum rack. These solar panels were then divided into two sets of three panels each, and each set was connected to a solar regulator which monitored the status of the corresponding battery that it was connected to, and provided charge when necessary. The solar panels were the only source of power supplied to the two batteries that comprised the 24 volt power system.

Next, we built a platform that oscillated back and forth, so that the LADAR units would scan all of the terrain in front of the vehicle repeatedly. To ensure that we knew the precise angle at which the LADAR devices were pointed at any time, an ethernet optical encoder from Fraba Posital was placed on the shaft which was the center of rotation. The optical encoder provided both the current angle and the angular velocity of the shaft. To decrease the delay between reading a value from the sensor and reading a value from the encoder, a separate 100 MBit ethernet connection with its own dedicated ethernet card was used to connect the I/O computer with the encoder. This delay was assumed to be the result of the TCP protocol's flow control algorithms [Jacobson and Karels, 1988] and their handling of congestion in the ethernet switch. After placing each encoder on its own dedicated ethernet connection, communications delays between the encoder and the I/O computer were relatively consistent at approximately .5 ms.

Testing revealed that the actual LADAR scan was taken approximately 12.5 ms before the data was available at the I/O computer. When this time was added to the .5 ms of delay from the encoder communications, we had a 13 ms delay from the actual scan to the actual reading of the encoder position and velocity. To counteract the angular offset this delay created, we multiplied

Fig. 3.5. A sample 3D elevation map (a) created by the vision system and the actual terrain (b) it was created from

the velocity of the encoder times the communications delay of .013 seconds to calculate the angular offset due to the delay. This angular offset (which was either negative or positive depending on the direction of oscillation) was then added to the encoder's position, giving us the actual angle at the time when the scan occurred. This extra processing allowed us to accurately monitor the orientation of the LADAR platform to within .05 degrees.

Although the vehicle's vision system was only comprised of these two LADAR devices, their oscillating platforms and vertical orientation allowed them to sense the environment with very fine detail and as precise an accuracy as possible. Figure 3.5 shows an actual picture of a parking lot and the resulting maps from the sensor readings. Several holes are visible in the 3D elevation map, especially along the right side of the image. These are artifacts created by the fact that the lasers cannot see behind obstacles. Therefore, the section behind an obstacle will have no elevation data associated with it. We designed both oscillating mounts to cover a thirty degree range and mounted each of them on different sides of the vehicle. This allowed us to have much finer detail in the center of the path, as both LADAR devices were able to scan this area. It also offered redundant coverage in the center of the path so that if one sensor failed, the vehicle could still sense obstacles most likely to be directly in its path.

All sensor readings had to be converted to a common coordinate system, which was chosen to be the geospatial coordinate system because the GPS reports vehicle position in the geospatial coordinate system. The two Sick LMS 291 LADAR units captured a two-dimensional scan of the terrain in front of the vehicle along with the exact time at which the scan took place. Using the highly accurate data from the RT3000, these two-dimensional scans were then transformed into the vehicles coordinate frame and then into a geospatial coordinate frame via a set of coordinate transformation matrices. This is accomplished in two similar steps. In each step a transformation matrix is defined so that

$$\mathbf{P_2} = \mathbf{T_{1\rightarrow 2}}\mathbf{P_1} + \mathbf{\Delta_1} \tag{3.1}$$

where $\mathbf{T_{1\rightarrow 2}}$ is the transformation matrix for going from coordinate frame 1 to coordinate frame 2, $\mathbf{\Delta_1}$ is the vector representing the position of the origin of coordinate frame 1 with respect to the origin of coordinate frame 2, and $\mathbf{P_1}$ and $\mathbf{P_2}$ are the same point in coordinate frames 1 and 2, respectively.

The first step converts from the sensor coordinate frame to the vehicle's coordinate frame. The vehicle's coordinate frame is located on the center of the rear axle of the vehicle and has the standard X (longitudinal), Y (lateral) , and Z (vertical) axes orientation. The sensor's coordinate system is centered on the sensor with the sensor's native X, Y, and Z axes. Therefore, the vehicle's coordinate system is related to the sensor's coordinate system via rotations and translation. Thus, a simple linear transformation of the form (3.1) can convert one to the other. This transformation is defined through a matrix, $\mathbf{T_{s \to v}}$ (to be used in place of $\mathbf{T_{1 \to 2}}$ in Equation (3.1)), which is defined as

$$\mathbf{T_{s \to v}} = \begin{bmatrix} \cos\psi_s & -\sin\psi_s & 0 \\ \sin\psi_s & \cos\psi_s & 0 \\ 0 & 0 & 1 \end{bmatrix} \begin{bmatrix} \cos\theta_s & 0 & \sin\theta_s \\ 0 & 1 & 0 \\ -\sin\theta_s & 0 & \cos\theta_s \end{bmatrix} \begin{bmatrix} 1 & 0 & 0 \\ 0 & \cos\phi_s & -\sin\phi_s \\ 0 & \sin\phi_s & \cos\phi_s \end{bmatrix} \tag{3.2}$$

where ψ_s, θ_s, and ϕ_s are the yaw (around the z-axis), pitch (around the y-axis), and roll (around the x-axis) of the sensor coordinate frame relative to the vehicle's coordinate frame. This transformation takes into account deviations in yaw, pitch, or roll caused by the mounting of the sensor. For example, if the sensor were mounted pointed slightly downward, it would have a negative pitch that would need to be countered by setting θ_s to its inverse (or positive) value. In addition, the angle of deflection caused by the oscillation is processed here by adding it to ϕ_s.

The same basic transformation and translation was done again in order to translate from the vehicle's coordinate system to the common geospatial coordinate system. Yet another transformation matrix, $\mathbf{T_{v \to g}}$, was constructed for this purpose.

$$\mathbf{T_{v \to g}} = \begin{bmatrix} \cos\psi_v & -\sin\psi_v & 0 \\ \sin\psi_v & \cos\psi_v & 0 \\ 0 & 0 & 1 \end{bmatrix} \begin{bmatrix} \cos\theta_v & 0 & \sin\theta_v \\ 0 & 1 & 0 \\ -\sin\theta_v & 0 & \cos\theta_v \end{bmatrix} \begin{bmatrix} 1 & 0 & 0 \\ 0 & \cos\phi_v & -\sin\phi_v \\ 0 & \sin\phi_v & \cos\phi_v \end{bmatrix} \tag{3.3}$$

where ψ_v, θ_v, and ϕ_v are the heading (around the z-axis), pitch (around the y-axis), and roll (around the x-axis) of the vehicle relative to the geospatial coordinate system. These heading, pitch, and roll values are generated by the GPS/INS navigation sensor that is mounted on the vehicle.

After taking into account both of these transformation matrices, the full equation for transformation from sensor coordinate system to geospatial coordinate system is

$$\mathbf{P_g} = \mathbf{T_{v \to g}}(\mathbf{T_{s \to v}} \mathbf{P_s} + \mathbf{\Delta_s}) + \mathbf{\Delta_v}. \tag{3.4}$$

where $\mathbf{\Delta_s}$ is the vector representing the position of the sensor with respect to the center of the vehicle and $\mathbf{\Delta_v}$ is the vector representing the position of the center of the vehicle with respect to the center of the GPS/INS navigation sensor.

At this point, each of the measurement values from the LADAR units now has a latitude, longitude, elevation, and timestamp. These elevation values are then placed into the same elevation grid, where the number of scans and time since last scan are used to derive the probability that the vehicle can drive over that geospatial location. We now describe how this probability is derived.

The data from the two sensors are correlated by placing data from both sensors into the same elevation grid (see Figure 3.5 for an example of this). The routines that this team developed to build internal representations for maps do not need to account for which sensor saw the obstacle, but only the number of times any sensor saw the obstacle and how recently did a sensor see the obstacle. As a result of this, any number of LADAR sensors can be used without having to change the algorithms at all.

The timestamp is very important to the algorithms due to the fact that LADAR scan data can have anomalies in it. Highly reflective surfaces can cause the LADAR devices to register incorrect distances for obstacles. To counteract these anomalies, scans are only kept in memory for a certain amount of time. After that time has passed, if no more obstacles have been registered in that same area, the obstacle is removed. This also ensures that moving obstacles are handled correctly. If a vehicle or other object crossed the path perpendicular to the autonomous vehicle, the sensors would effectively register a sequence of obstacles in front of the autonomous vehicle. These obstacles would appear as a complete obstruction of the vehicle's path, forcing the vehicle to immediately stop. After enough time had passed, the obstacles would expire, and the autonomous vehicle would be able to start moving again.

A persistent obstacle, on the other hand, will not expire. Consider, for example, a large boulder that is located in the center of the path. At a distance of approximately forty to fifty meters, the LADAR devices will start to register parts of the boulder. As the autonomous vehicle gets to within ten to twenty meters of the vehicle, the previous scans would begin to approach the obstacle expiration time. Since the sensors are still registering the boulder, the previous scans will not expire. Instead, the previous scans along with the current scans would all be retained, giving the boulder a higher count for total number of scans. This high count for the number of scans causes the boulder to have an extremely high probability of being an obstacle.

3.4 Software Design

Software played a critical role in the design of the autonomous vehicle. Rather than using the C programming language to build the software for Kat-5 like virtually all of the other teams competing in the DARPA Grand Challenge, we decided to use the Java programming language instead. While many detractors might say that Java offers slower performance and lacks conventional parallel programming models [Bull et al., 2001], we decided that its benefits outweighed its deficiencies.

It was imperative in the design of Kat-5 that all systems, physical or software, be completely modular and independent of platform or peers. This requirement was met fully from the beginning of development within the software realm by the use of the Java programming language. Java's simple language semantics, platform independence, strong type checking, and support for concurrency have made it a logical choice for a high integrity system [Kwon et al., 2003] such as an autonomous vehicle.

Because of the platform-independance of Java, we were able to use multiple hardware solutions that fit different niches in our overall design. All of the interface systems were deployed on Linux systems while the math-intensive processing was performed primarily on Apple systems. Despite this disparity in operating systems, all computers were given the same exact executables. A block diagram of our process architecture is shown in Figure 3.6.

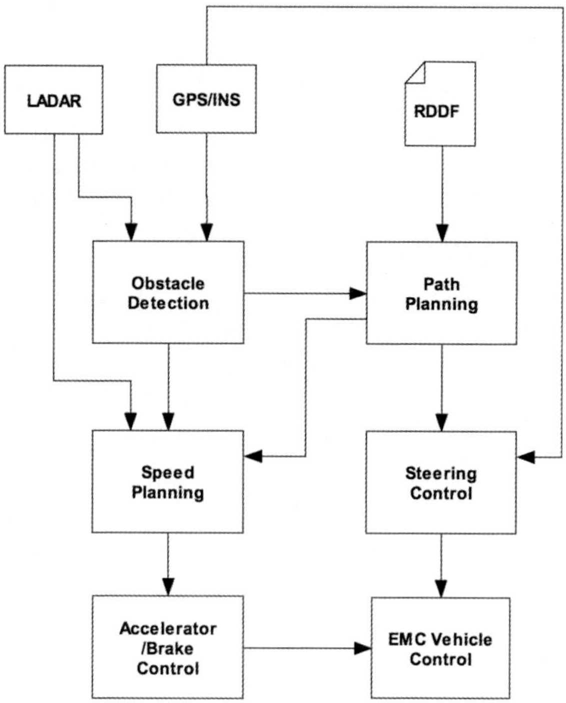

Fig. 3.6. The software used in driving the vehicle is a highly modular system [Trepagnier et al., 2005]

To ensure the quality of our code base, we followed the practices of Test Driven Development which is a software development practice in which unit test cases are incrementally written prior to code implementation [George and Williams, 2003]. As a result of this, we created unit tests for all critical software modules on the vehicle. This allowed us to run the unit tests each time we deployed our code to detect if we had introduced any bugs. This alone saved us enormous amounts of time and helped to provide us with the stable code base that allowed us to finish the Grand Challenge.

For example, geospatial operations such as projecting a point a given distance along a heading (or bearing) or converting between different coordinate systems were some of the most commonly used pieces of code on the vehicle. These geospatial operations were also consistently modified and improved throughout development. To ensure that this ongoing development did not introduce errors into this critical part of the code base, unit tests were created to test each geospatial operation. These unit tests were created by performing the operations manually, then using these results to craft a unit test that ensured that the given operation produced the same results. Each geospatial operation had several of these tests to ensure complete testing coverage of each operation.

The following subsections describe our use of Java to develop the software necessary to meet the challenges described in the previous section.

3.4.1 Path Planning and Obstacle Avoidance

Path Planning is accomplished through the use of cubic b-splines [de Boor, 1978] designed to follow the center of the route while still ensuring that the path they create is not impossible for the vehicle to navigate. This assurance means that the curvature at any point along the path is below the maximum curvature that the vehicle can succesfully follow. In addition, the curvature is kept continuous so that it is not necessary to stop the vehicle in order to turn the steering wheel to a new position before continuing.

B-splines were chosen for use in the path planning algorithms primarily because of the ease in which the shape of their resulting curves can be controlled [Berglund et al., 2003]. After an initial path is created that follows the center of the corridor, the path is checked against the obstacle repository to determine if it is a safe path. If the path is not safe, a simple algorithm generates and adjusts control points on the problem spots of the curve until the spline avoids all known obstacles while still containing valid maximum curvature. At this point, the path is both safe and drivable.

Once a safe path is designed that avoids obstacles and yet remains drivable, the next step in the planning process is to decide on the speed at which the vehicle should take each section of the path. The speed chosen is based on the curvature at that point in the path and upon the curvature further down the path. The speed is taken as the minimum of speed for the current curvature and the speed for future path curvature. The future path curvature is defined by a simple function that multiplies the curvature at a given future point by a fractional value that decreases towards zero linearly based upon the distance from the current path point to the given future path point.

3.4.2 Speed Controller

In order to design intelligent controls suitable for a given system, that system, in this case the car's engine and the external forces, must be understood. System identification is a method by which the parameters that define a system can be determined by relating input signal into a system with the system's response

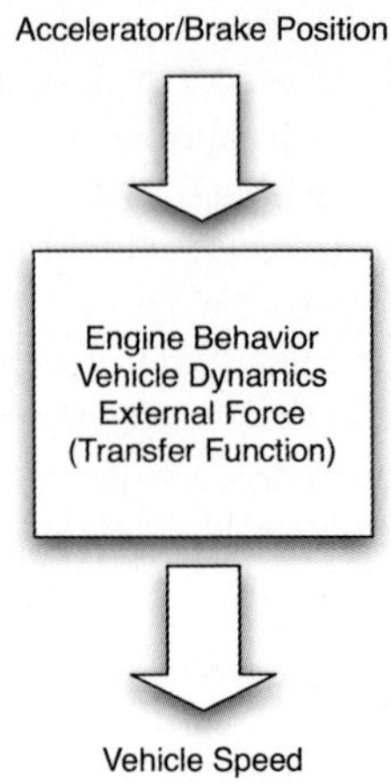

Fig. 3.7. The transfer function developed through system identification is intended to characterize the reaction of the system to the inputs

[Aström and Wittenmark, 1995]. It is the goal of this method to develop a transfer function that behaves in the same way as the actual system. For instance, when attempting to control the speed of a vehicle, the inputs are the brake and accelerator position and the output is the vehicle's speed (see Figure 3.7). If it is assumed that the transfer function, $H(s)$, is first-order, it can be written as

$$y(s) = H(s)u(s) \tag{3.5}$$

where H(s) is the transfer function of a system, $u(s)$ is the input to the system, and $y(s)$ is the output from the system. System identification, as described in [Aström and Wittenmark, 1995], was applied to real world data from the propulsion system to come up with the transfer function of the system.

As far as the speed control of the vehicle was concerned, it seemed like a simple control system could be designed to handle the accelerator and brake. However, as it turned out, there were many factors in the physical engine system that made for a fairly complex transfer function. Being a gas-electric hybrid engine, the coupling of the two propulsion systems was controlled by an intelligent computer tuned for fuel efficiency, a computer that we had no information about. In addition, the mapping of the requested pedal position and the actual position achieved was not linear and had to be remapped in the software layer.

It was eventually decided that the speed of the vehicle would be controlled by an integrated proportional-derivative (PD) controller [Aström and Wittenmark, 1995]. This controller bases its output on the previous output and on the current error and derivative of the error. In the time domain, the controller can be written as

$$u(t_2) = (t_2 - t_1)(K_p e(t_2) + K_d e'(t_2)) + u(t_1) \tag{3.6}$$

where K_p and K_d are tunable coefficients, $u(t)$ is the output of the controller at time t, and $e(t)$ is the error at time t. The error is defined in a conventional manner: actual output subtracted from target output. Actual output is reported by the RT3000 and target speed is derived from the path planning algorithms. Since the actual output is limited to $u(t) \in [-1, 1]$, no windup of the controller is experienced. There was some overshoot and oscillation inherent in the system but with tuning it was reduced to a manageable level (see Figure 3.8).

The integrated PD controller was designed and tuned against the transfer function that we had arrived at through the system identification process mentioned above. It was a simple matter to arrive at the weights needed for optimal performance against the computational model; however, the computational model was a far cry from the true real-world system. The modeled values were used as a baseline from which to work from when tuning the real controller.

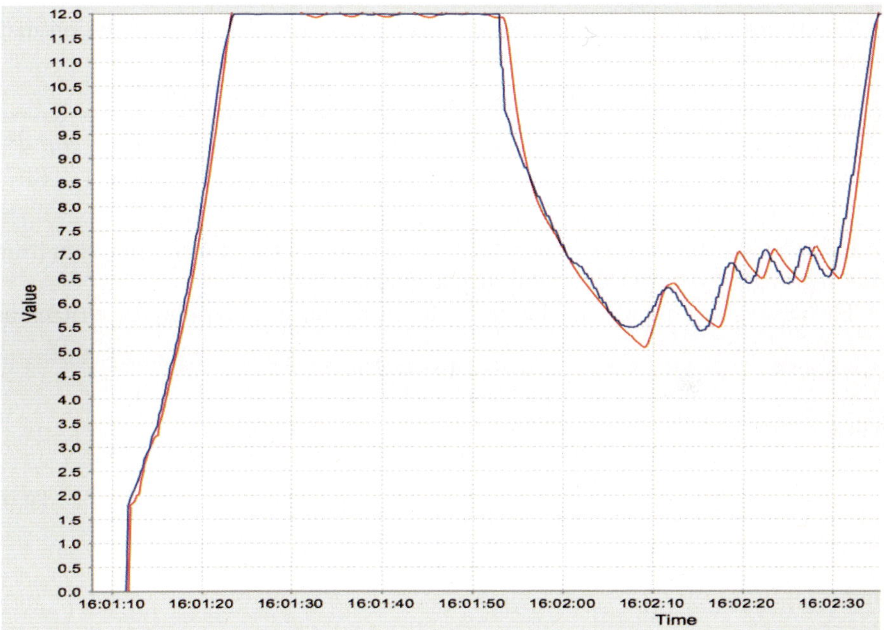

Fig. 3.8. The PID comes very close to matching current speed to target speed during a run at the NQE

3.4.3 Steering Controller

The steering controller for the vehicle was a lead-lag controller based on the classical single-track model or bicycle model developed by Riekert and Schunck [Riekert and Schunck, 1940].

Table 3.3. Variables used in the steering controller's vehicle model

Symbol	Description
M	Vehicle mass
v	Vehicle velocity
ψ	Vehicle yaw with respect to a fixed inertial coordinate system
I_ψ	Yaw moment of inertia about vertical axis at CG
$l_{f/r}$	Distance of front/rear axle from CG
l	$l_f + l_r$
y_S	Minimum distance from the virtual sensor to the reference path
d_f	Front wheel steering angle
d_S	Distance from virtual sensor location to CG
$c_{f/r}$	Front/rear tire cornering stiffness
μ	Road adhesion factor

The transfer function from steering angle to vehicle lateral acceleration may be written as

$$\frac{\mu c_f v^2 \left(M l_f d_s + I_\Psi\right) s^2 \; + \; \mu^2 c_f c_r l v \left(d_s + l_r\right) s \; + \; \mu^2 c_f c_r l v^2}{I_\Psi M v^2 s^2 + \mu v \left(I_\Psi \left(c_f \; + \; c_r\right) + M \left(c_f l_f^2 \; + \; c_r l_r^2\right)\right) s + \mu M v^2 \left(c_r l_r \; - \; c_f l_f\right) + \mu c_f c_r l^2} \tag{3.7}$$

The introduced variables are defined in Table 3.3.

By applying the Laplace integrator twice we obtain the transfer function from steering angle $d_f(s)$ to the lateral displacement $y_S(s)$.

The state space representation as found in [Hingwe, 1997] may be written as

$$\dot{x} \; = \; Ax \; + \; Bu \tag{3.8}$$

where

$$x \; = \; \begin{bmatrix} y_s \\ \dot{y}_s \\ \Psi \\ \dot{\Psi} \end{bmatrix} \qquad u = \delta_f \tag{3.9}$$

and

$$A \; = \; \begin{bmatrix} 0 & 1 & 0 & 0 \\ 0 & -\left(\frac{(\phi_1 + \phi_2)}{v}\right) & \phi_1 \; + \; \phi_2 & \left(\frac{\phi_1 (d_S - l_f) + \phi_2 (d_S + l_r)}{v}\right) \\ 0 & 0 & 0 & 1 \\ 0 & -\left(2 \frac{(l_f c_f - l_r c_r)}{I_\Psi v}\right) & 2\left(\frac{(l_f c_f - l_r c_r)}{I_\Psi}\right) & -2\left(\frac{\left(c_f \left(l_f^2 - l_f d_S\right) + c_r \left(l_r^2 - l_r d_S\right)\right)}{I_\Psi v}\right) \end{bmatrix} \tag{3.10}$$

$$B \; = \; \begin{bmatrix} 0 & 0 \\ \phi_1 & \left(\frac{\left(\phi_2 l_r - \phi_1 l_f - v^2\right)}{v}\right) \\ 0 & 0 \end{bmatrix} \tag{3.11}$$

where

$$\phi_1 = 2c_f \left(\frac{1}{M} + \frac{l_f d_S}{I_\Psi} \right) \tag{3.12}$$

$$\phi_2 = 2c_r \left(\frac{1}{M} + \frac{l_r d_S}{I_\Psi} \right) \tag{3.13}$$

The outputs of the vehicle model as shown in Equation 3.14 are the lateral error at the virtual sensor and the yaw.

$$y = \left[1 \; 0 \; d_S \; 0 \right] x \tag{3.14}$$

Lead and lag compensators are commonly used in control systems. A lead compensator can increase the responsiveness of a system; a lag compensator can reduce (but not eliminate) the steady state error [Bernstein, 1997]. The lead-lag compensator was designed using the frequency response of the system. The lead compensator used is stated in (3.16) and the lag compensator is stated in (3.15). The resulting controller is the convolution of the two functions multiplied by the low frequency gain, which was 0.045. The parameters used in (3.16) and (3.15) were produced using rough estimates which were then tuned by trial and error.

$$F_{lag}(s) = \frac{850s + 1}{900s + 1} \tag{3.15}$$

$$F_{lead}(s) = \frac{2s + 4}{0.2s + 1} \tag{3.16}$$

The step response of the closed-loop system is shown in Figure 3.9.

The input to the controller is defined as $[y_S]$ where y_S refers to the minimum distance from the virtual sensor to the reference path. The virtual sensor is a point projected a given distance ahead of the vehicle along the vehicle's centerline. This point is commonly referred to as the look-ahead point, and the distance from the look-ahead point to the RT3000 is referred to as the look-ahead distance. The output of the controller is the steering angle measured at the tire with respect to the centerline.

Control gains were not scheduled to the speed of the vehicle. Gain scheduling was evaluated, but not implemented, as the steering controller by itself offered more than enough stability and accuracy. No additional knowledge, such as whether the vehicle was going straight or in a turn, was given to the steering controller.

Several assumptions were made in implementing this model. The relationship between the steering wheel angle and the resulting tire angle was assumed to be linear. The measurements made of this relationship showed that the actual variation was negligible. Also, the location of the vehicle's center of gravity was assumed to be at the midway point between the front and rear axles.

The steering controller managed to provide incredible accuracy, even at speeds of over thirty miles per hour. An analysis was made of the steering controller's accuracy from the first 28 miles of the 2005 DARPA Grand Challenge. An unfortunate error that is described later in Section V prevented the team from being

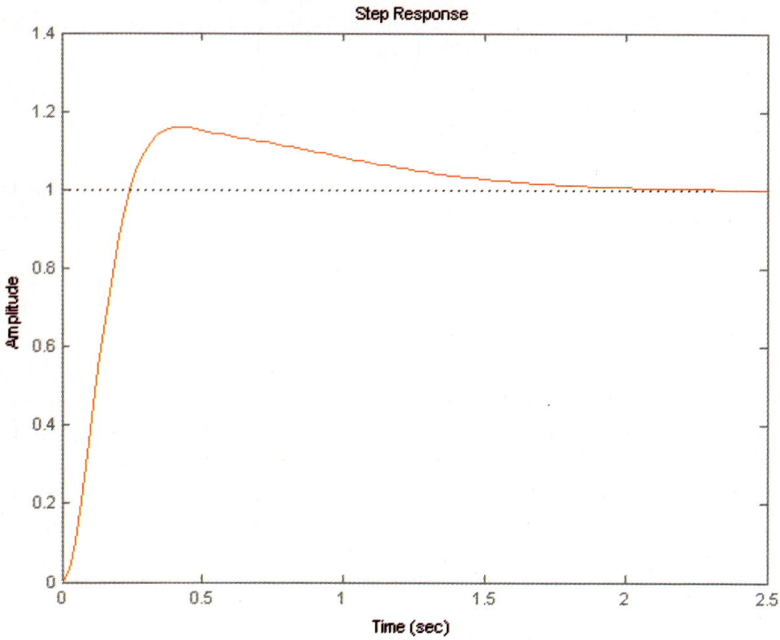

Fig. 3.9. Step response of closed-loop system

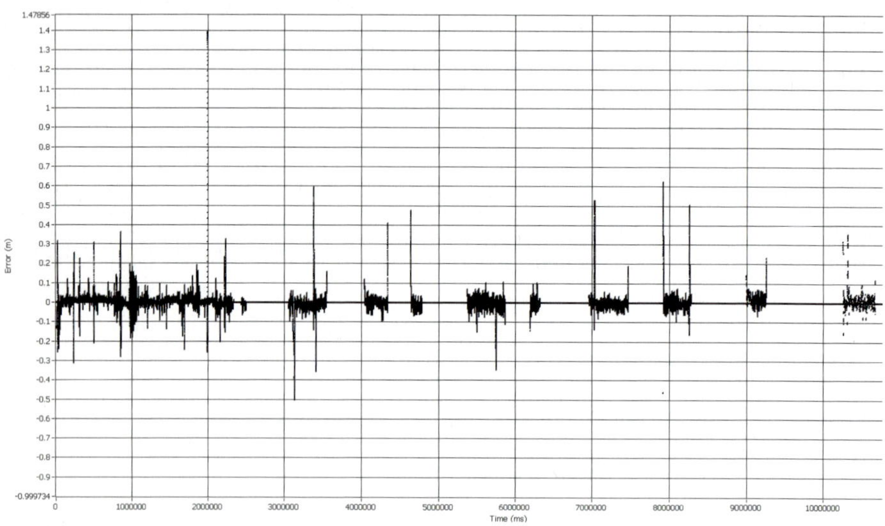

Fig. 3.10. The steering controller proved to have an extremely high accuracy during the 2005 DARPA Grand Challenge. The standard deviation for the 28 miles shown here was just 5 cm from the desired path.

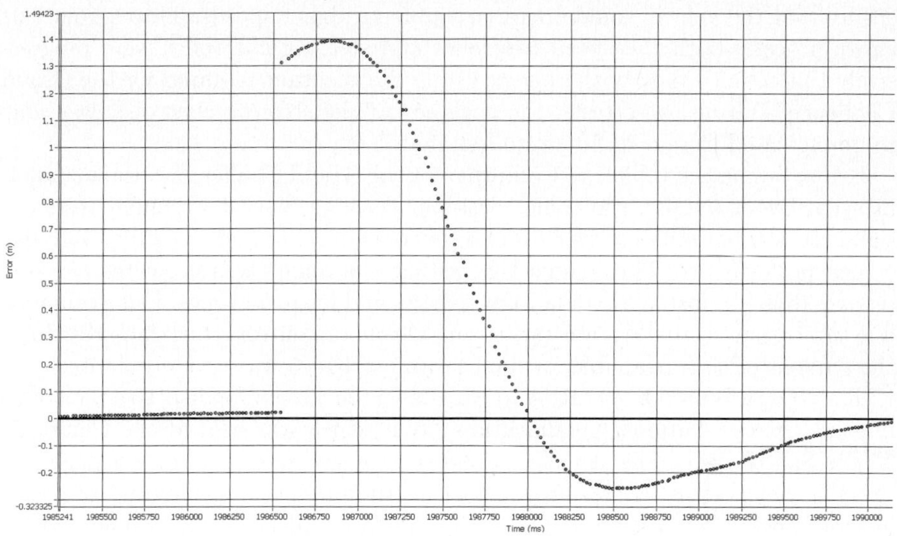

Fig. 3.11. The steering controller handled this GPS jump by reducing speed to allow the vehicle to safely steer back onto the planned path

able to perform an analysis on more data from the Grand Challenge. As shown in Figure 3.10, the steering controller was able to maintain a standard deviation of 5 centimeters in regards to the desired path. This is an excellent performance, especially when considering the roughness of the terrain and a top speed of over 35 mph for Kat-5.

As a measure of safety the magnitude of the y_s signal was monitored to prevent the vehicle from becoming unstable. If y_s were ever to cross a given threshold, meaning the vehicle is severely off path, the speed was instantly reduced to 2 mph. This allowed the vehicle to return onto the desired path and prevented a possible rollover. The algorithm was repeatedly tested by manually overriding the steering controller and taking the vehicle off path, then allowing it to regain control.

This algorithm proved to be very effective. Figure 3.10 shows the path error from the first 28 miles of the 2005 Grand Challenge, and a 1.5 meter spike at around the 2000 second mark. At this point, a GPS jump caused a spike in path error from 5 cm to 1.5 meters in a span of just 10 ms. As shown in Figure 3.11, the steering controller was able to handle this GPS jump safely and quickly, and return Kat-5 back to the proper planned path.

3.5 Field Testing

Initial field testing concentrated on improving the accuracy of both the speed controller and steering controller at low to medium speeds. After extensive testing at these speeds, the team rented out a local racetrack, and proceeded to ensure that the vehicle could operate reliably at higher speeds. Throughout this entire process, both the speed controller and steering controller were progressively tuned, until they both operated with the accuracy required for the Grand Challenge. During these tests, the accuracy of the steering controller was able to be increased from over 50 cm to less than 5 cm.

Before taking the vehicle to California for the Grand Challenge National Qualification Event (NQE), the vehicle was put through several strenuous tests designed to test not only hardware and software endurance and reliability, but also vehicle performance. Throughout this testing, the vehicle was subjected to harsh terrain, difficult obstacle avoidance scenarios, and long run times. Unfortunately, due to weather-related circumstances outside of our control (Hurricane Katrina), the amount of time available for final testing of Kat-5 was severely limited.

The NQE was used by DARPA to pare down the 43 semifinalists to a pool of 23 finalists. At the National Qualification Event which was held at the California Motor Speedway in Los Angeles, California, Kat-5 ran into several problems initially which were the result of a lack of thorough testing before the event. Several quick changes were made to the software and hardware, which allowed Kat-5 to produce several successful runs at the end of the NQE and advance to the Grand Challenge Event. The results of Kat-5's seven qualifying runs are listed below:

Run 1 – Did Not Finish

At the end of the fast portion of the course, Kat-5's geospatial system encountered a failed assertion in its geospatial processing algorithm and shut down. Without a course to follow, the vehicle continued straight ahead, missing a right turn and hit the side-wall at approximately 20 miles per hour (8.94 m/s) causing some damage to the front body, but no major structural damage. The assertion was caused by a bug in the code that created a composite polygon from the individual polygonal segments of the corridor. Apparently, the composite polygon created in one section of the NQE course was not a valid polygon, and this caused a failed assertion error. To get around this bug, the algorithm was changed to use the individual segments directly rather than the composite polygon composed of the individual segment polygons.

Run 2 – Did Not Finish

The GPS position of the vehicle was reported as two to three meters north of its true location. While attempting to enter the tunnel, the vehicle saw the entrance, but thought that the entrance was off of the course, and thus it could not navigate so as to enter the tunnel without going out of bounds. The front left of the vehicle hit the corner of the entrance. Major damage was done to the front end, but it was repaired quickly. The RT3000 has a basic GPS component but it is also using a ground OmniStar Correction Service to further refine the readings of the GPS component and thus provide finer accuracy than is possible on GPS alone. Upon investigation, it was discovered that the ground station frequency used by the RT3000 for OmniStar Correction Service was still set to the one corresponding to the eastern United States rather than the western United States and thus during the previous runs only basic GPS was actually used.

Run 3 – Did Not Finish
: After approximately 100 meters, the circuit breaker in the engine compartment of the vehicle overheated and popped open, cutting power to all of the vehicle's robotic systems. The drive-by-wire system used its reserve power to apply the brake and then shut off the vehicle. The circuit breaker was replaced with a heavy-duty wire and locking disconnect plug.

Run 4 – Finished in 16:42
: Kat-5 completed the entire 2.2-mile course. Upon exiting the tunnel, the vehicle gunned its accelerator causing it to leave the corridor before slamming on its brakes. After reacquiring a GPS lock, it navigated back onto course and continued to the end where it finished the course.

Run 5 – Finished in 15:17
: Kat-5 had a perfect qualifying run until the very end of the course where it completely failed to avoid the tank trap. It hit the obstacle, but continued its forward motion, pushing the tank trap across the finish line. Data analysis indicated that the position of the sun on the extremely reflective tank trap caused both LADAR devices to be blinded.

Run 6 – Did Not Finish
: Upon exiting the tunnel, the vehicle behaved as it did in run 4, but when it recovered GPS lock, it began driving in circles. It had to be shut down remotely. There was a logical error in one of our recovery systems that did not allow the vehicle to draw a path back onto the course after recovering from failure in the RT3000. There were several bugs in the RT3000 itself that caused its integrated INS to lag almost five full seconds behind its integrated GPS. Although the RT3000 is designed to handle GPS outages with little performance degradation in normal working conditions, these bugs created significant problems that caused a drift of over 5 meters when GPS signal was lost. After an hour of work on-site with engineers from Oxford Technical Solutions, the problems were quickly diagnosed and fixed, then incorporated into an upgraded firmware release for the RT3000 by Oxford Technical Solutions.

Run 7 – Finished in 15:21
: Kat-5 completed its first and only perfect run, securing its spot as a finalist.

Several images of the output of the path planning systems during actual runs from the NQE are shown in Figures 3.12, 3.13, and 3.14. In this visualization, obstacles are represented by clusters of points and the path is represented by the sparse sequence of dots.

The Grand Challenge was held on October 8 in Primm, Nevada. Kat-5 was one of the twenty-three finalists that was able to participate in the race. She left the starting line with a burst of speed and never looked back. Kat-5 finished the race in an elapsed time of seven hours and thirty minutes with an average speed of 17.5 mph to claim fourth place. Several issues were discovered after analyzing the vehicle during a post-race inspection.

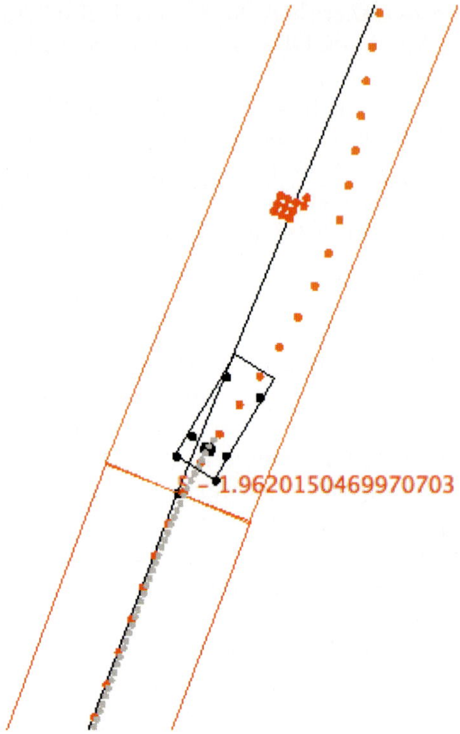

Fig. 3.12. Kat-5 is planning a path to avoid an obstacle in the center of the path at NQE. The vehicle is moving towards the top right.

- The vehicle's steering was severely out of alignment. The team assumed this was due to Kat-5's attack of the rough terrain at relatively high speed. Amazingly, the steering algorithm was able to easily handle this issue, as evidenced by the fact that Kat-5 was able to complete the race.
- The Antilock Braking System was displaying intermittent failures that caused the brakes to behave in an erratic fashion. This issue was also assumed to be the result of the rough terrain. Like the steering controller's ability to handle the problem steering alignment, the speed controller also was able to handle the erratic brakes.
- The logging system crashed after 28 miles. This unfortunate issue means that a full analysis of the Grand Challenge for Kat-5 is impossible. The cause for this crash was apparently a crash of the logging server Java program running on the logging computer. The root cause of this is currently undetermined, but the primary suspect is that diagnostic information produced by the error discussed in the next item caused some piece of executing code to throw an unchecked exception. This unchecked exception caused the Java Virtual Machine to exit.

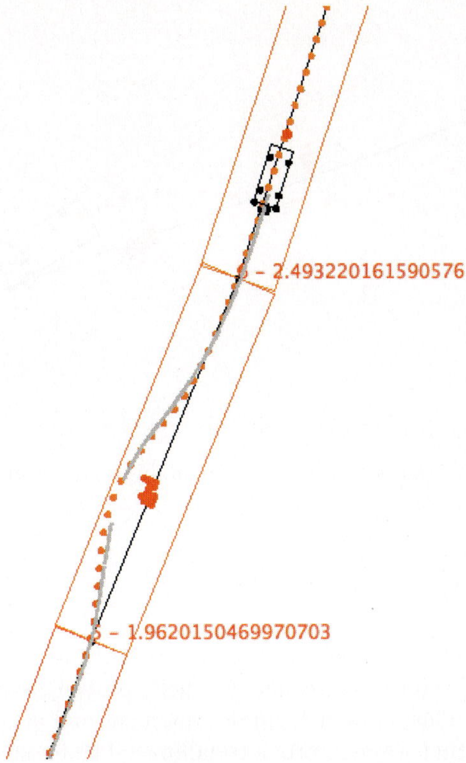

Fig. 3.13. The path planning algorithms identified an obstacle at NQE and avoided it. The vehicle is moving towards the top right.

- The only major flaw in Kat-5's performance on the day of the Grand Challenge was an error in the path planning algorithms that caused them to time out when faced with sections of the route with extremely wide lateral boundaries.

  We had anticipated that the path planning algorithms might occasionally time out, and therefore we had programmed Kat-5 to slow down to 3 mph for safety reasons until the algorithms had a chance to recover. However, whenever Kat-5 encountered sections with an extremely wide lateral boundary, the algorithms timed out continuously due to the error until a section with a narrower lateral boundary was encountered. This caused Kat-5 to drive the dry lake bed sections of the race, which were considered the easiest, at 3 mph instead of 40 mph. Calculations by both DARPA and Team Gray about the time lost due to this bug have shown that if this error had not occurred, Kat-5 would have posted a much better finishing time. This bug has since been fixed.

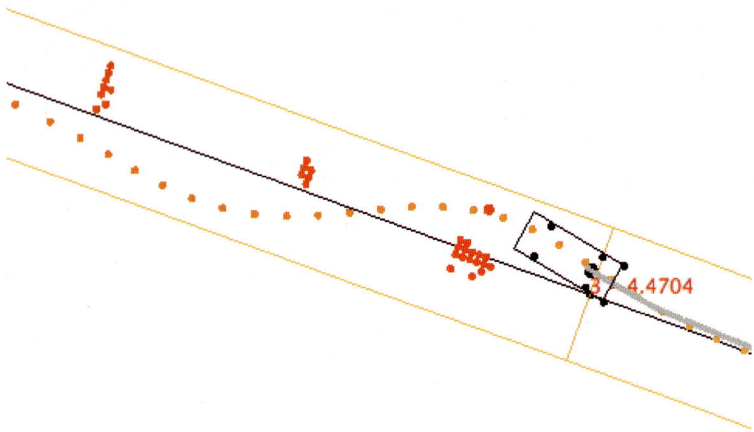

Fig. 3.14. Kat-5 shows that she is capable of planning a path that can handle slaloming around obstacles. The vehicle is moving towards the top left.

3.6 Conclusion

The recipe of success in this effort was rugged hardware, simple software, and good teamwork. We chose to use off-the-shelf, proven hardware that we customized as needed. This allowed rapid development and also provided a reliable hardware platform that was not prone to failures. The harsh terrain could easily dislodge a wire or render a sensor or communications link useless and would therefore mean the end of the race. This was the motivation for choosing the EMC drive-by-wire system, the shock resistant equipment mounts, the rugged 50G tolerant Cisco equipment, and all of the other equipment. This was also the reasoning for choosing the RT3000 which not only was built to provide supreme accuracy but also was designed to take additional inputs from custom-made sensors and seamlessly incorporate them into its adaptive mechanism. The RT3000 produced better than 10 cm accuracy at an incredibly fast 100 Hz. As an off-the-shelf, industrial system, it is much more reliable than one that we would conceivably build ourselves.

The simplicity of the design also resulted in an agile system. The sensor vision system of the vehicle was just two LADAR devices. All of the I/O bandwidth of the system was dedicated to reading just these two sensors. So the system was able to read massive data streams very quickly without much overhead for synchronization and fusion of the sensor streams. Thus, processing of the data was efficient and the computation of the control sequences was fast, resulting in a very agile overall system. While other sensors were considered for inclusion into the vision system, their contributions were not considered useful enough to warrant the complications they would have added to the sensor fusion algorithms.

The choice of Java was also a good one since the development, monitoring, and testing were all profoundly quick and produced code that was thoroughly validated before it even went into testing. On another note, because of the portability of Java byte code, if we had to change our computing grid architecture or even switch a PC with a Mac or add a new one of either platform to the overall system it would not be any problem. The choices we made were well validated since the vehicle endured the entire course well within the time limit and with only about 7 gallons of gas. We did not win the race but participating in it was very rewarding as it validated our choices and methods. Indeed, we like to think that reaching the finish line after 132 miles of autonomous driving in the desert was not just beginner's luck but rather the result of our simple design methods, good decisions, and good system integration.

Acknowledgment

This project was entirely financed and supported by The Gray Insurance Company, Inc.

References

[Aström and Wittenmark, 1995] Aström, K. J. and Wittenmark, B. (1995). *Adaptive Control.* Addison Wesley Lognman, Reading, Massachusetts, second edition.

[Berglund et al., 2003] Berglund, T., Jonsson, H., and Soderkvis, I. (2003). An obstacle-avoiding minimum variation b-spline problem. In *2003 International Conference on Geometric Modeling and Graphics (GMAG'03)*, page 156.

[Bernstein, 1997] Bernstein, D. (1997). A student's guide to classical control. *IEEE Control Systems Magazine*, 17:96–100.

[Bruch et al., 2002] Bruch, M. H., Gilbreath, G., Muelhauser, J., and Lum, J. (July 9-11, 2002). Accurate waypoint navigation using non-differential gps. In *AUVSI Unmanned Systems 2002.*

[Bull et al., 2001] Bull, J. M., Smith, L. A., Pottage, L., and Freeman, R. (2001). Benchmarking java against c and fortran for scientific applications. In *Java Grande*, pages 97–105.

[Coombs et al., 2001] Coombs, D., Yoshimi, B., Tsai, T.-M., and Kent, E. (2001). Visualizing terrain and navigation data. Technical report, National Institute of Standards and Technology, Intelligent Systems Division.

[DARPA, 2005a] DARPA (2005a). 2005 darpa grand challenge rddf document.

[DARPA, 2005b] DARPA (2005b). 2005 darpa grand challenge rules.

[de Boor, 1978] de Boor, C. (1978). *A practical guide to splines.* Springer-Verlag.

[Ford Motor Company, 2005] Ford Motor Company (2005). Escape hybrid: How it works.

[George and Williams, 2003] George, B. and Williams, L. (2003). An initial investigation of test driven development in industry. In *SAC '03: Proceedings of the 2003 ACM symposium on Applied computing*, pages 1135–1139, New York, NY, USA. ACM Press.

[Hingwe, 1997] Hingwe, P. (1997). *Robustness and Performance Issues in the Lateral Control of Vehicles in Automated Highway Systems.* Ph.D Dissertation, Dept. of Mechanical Engineering, Univ. of California, Berkeley.

[Jacobson and Karels, 1988] Jacobson, V. and Karels, M. J. (1988). Congestion avoidance and control. *ACM Computer Communication Review; Proceedings of the Sigcomm '88 Symposium in Stanford, CA, August, 1988*, 18, 4:314–329.

[Kelly, 1994] Kelly, A. (1994). A 3d state space formulation of a navigation kalman filter for autonomous vehicles. Technical report, CMU Robotics Institute.

[Kwon et al., 2003] Kwon, J., Wellings, A., and King, S. (2003). Assessment of the java programming language for use in high integrity systems. *SIGPLAN Not.*, 38(4):34–46.

[Oxford Technical Solutions, 2004] Oxford Technical Solutions (2004). *RT3000 Inertial and GPS Measurement System.*

[Riekert and Schunck, 1940] Riekert, P. and Schunck, T. (1940). Zur fahrmechanik des gummibereiften kraftfahrzeugs. In *Ingenieur Archiv*, volume 11, pages 210–224.

[Sick AG, 2003] Sick AG (2003). *LMS 200 / LMS 211 / LMS 220 / LMS 221 / LMS 291 Laser Measurement Systems Technical Description.*

[Trepagnier et al., 2005] Trepagnier, P. G., Kinney, P. M., Nagel, J. E., Dooner, M. T., and Pearce, J. S. (2005). Team Gray technical paper. Technical report, The Gray Insurance Company.

4

The TerraMax Autonomous Vehicle

Deborah Braid[1], Alberto Broggi[2], and Gary Schmiedel[3]

[1] Rockwell Collins, Cedar Rapids, Iowa 52498, USA
[2] VisLab, University of Parma, 43100 Parma, Italy
[3] Oshkosh Truck Corp., Oshkosh, Wisconsin 54902, USA

Summary. The TerraMax vehicle is based on Oshkosh Truck's Medium Tactical Vehicle Replacement (MTVR) truck platform and was one of the 5 vehicles able to successfully reach the finish line of the 132 miles DARPA Grand Challenge desert race. Due to its size (30,000 pounds, 27'-0" long, 8'-4" wide, and 8'-7" high) and the narrow passages, TerraMax had to travel slowly, but its capabilities demonstrated the maturity of the overall system. Rockwell Collins developed, integrated, and installed the intelligent Vehicle Management System (iVMS), which includes vehicle sensor management, navigation, and vehicle control systems. The University of Parma provided the vehicle's vision system, while Oshkosh Truck Corp. provided project management, system integration, low level controls hardware, modeling and simulation support and the vehicle.

4.1 Introduction

TerraMax™, a completely autonomous vehicle, was developed by Oshkosh Truck Corporation in cooperation with its partners Rockwell Collins and University of Parma in response to Congress' goal that one third of military vehicles be unmanned by 2015. The Oshkosh TerraMax™ was one of only five vehicles that successfully completed the 132-mile DARPA Grand Challenge course in October 2005 (5th place), and it was the only vehicle whose mission is to provide medium- to heavy-payload logistic support to the battlefield. During the race, the fully-autonomous vehicle was successful in demonstrating obstacle avoidance, negotiating tunnels, narrow roads and cliffs, GPS waypoint following and 28 hours of non-stop continuous operation - all applicable to military missions.

4.2 The Vehicle

The TerraMax vehicle shown in Figure 4.1 is based on Oshkosh's Medium Tactical Vehicle Replacement (MTVR) MK23 truck platform. The MTVR was designed with a 70% off-road mission profile. It can carry a 7-ton payload off-road or a 15-ton payload on-road. All-wheel drive, TAK-4™ independent suspension, and central tire inflation make rocks, dips, holes and crevasses easier to handle.

M. Buehler, K. Iagnemma, and S. Singh (Eds.): DARPA'05, STAR 36, pp. 129–153, 2007.
springerlink.com

Fig. 4.1. The TerraMax vehicle

And the truck can handle 60% grades and 30% side slopes. A 425-hp Cat C-12 engine powers the truck. This kind of vehicle was chosen for the DARPA Grand Challenge (DGC) because of its proven off-road mobility, as well as for its direct applicability to potential future autonomous missions. The TerraMax team participated also to the 2004 DARPA Grand Challenge (Ozguner et al., 2004) with the same vehicle. Two significant vehicle upgrades for the 2005 DGC were the addition of rear-wheel steering and integrated sensor structure/roll cage. Rear steer has been added to TerraMax to give it a tighter 29-foot turning radius. Although this allows the vehicle to negotiate tighter turns without needing frequent back ups, the back up maneuver is required to align the vehicle with narrow passages. The sensor mounting structure/roll cage provided added protection to the sensors as well as key vehicle components.

4.2.1 Autonomous System Integration

The Autonomous System consists of Computers, Communication Network, Sensors, Vehicle Control Interface and the supporting mounting and protection structures. The Autonomous System utilized in the 2004 DGC was completely removed and upgraded for the 2005 DGC.

4.2.2 Computers and Communication Network Integration

The Computers and Communication Network hardware was packaged in a modular shock absorbing rack located inside the base of the passenger seat as shown in Figure 4.2. The video monitor, keyboard, and mouse were securely mounted on the dashboard. This arrangement allowed the sensitive computer equipment to survive the high-G shock and vibration experienced on the trail.

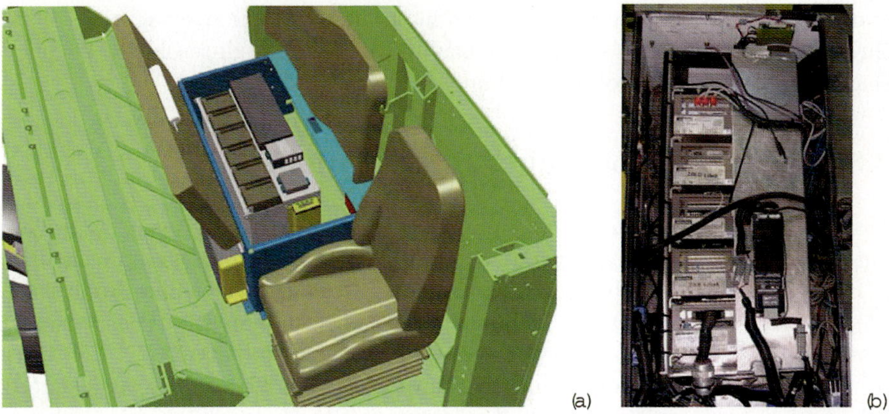

Fig. 4.2. Computers mounted under passenger seat: (a) CAD simulation; (b) real installation

4.2.3 Sensor Installation

The Sensors were mounted to modular adjustable mounts that were integrated into the roll cage. The roll cage also serves as a protective conduit, through which, the vital sensor communication and power cables are routed. The adjustable mounts used were selected for their ability to retain the set position regardless of the pounding taken from the trail. The location of each sensor was optimized for functionality while maintaining a high level of protection from the environment i.e. rain, sun, engine heat, brush, and, yes, even bridge supports. A sensor cleaning system was also developed to keep the lenses of the TerraMax sensors free of debris such as dust, water, and mud. The main components of this system are: Cleaning Controller, Valve Array, and Washer Tank. The Cleaning Controller controls the sequence and duration the sensors are dusted, washed, and dried. The Valve Array has electrically controlled valves that pass pressurized water and air through pattern nozzles to the sensor lenses.

4.2.4 Vehicle Control Actuator Integration

The Vehicle Control Integration was comprised of four key areas: Brake, Throttle, Gear Selection, and Steering. The Brake was controlled via Ethernet by a proportional Voltage to Pressure valve. This method was utilized, in-part, because of safety concerns of mechanical systems interfering with a driver while on the road. The Throttle was controlled via an I/O card in the Vehicle Manager computer. A Pulse Width Modulated (PWM) signal allowed precise control of engine throttle level. The Gear Selection was controlled via a relay card in the Vehicle Manager computer. A binary pattern applied to the transmission control harness allowed the ability to select the desired gear required by the trail conditions. The Steering was controlled via serial communications to a servo drive and motor. The servomotor is connected in parallel with the steering wheel shaft through a gearbox. The servomotor has an integrated high-resolution encoder that allows precise control of wheel angle.

4.3 Terramax Modeling and Simulation

Modeling and Simulation efforts supported the controls development by providing information such as underbody clearance, steer angles and lateral stability. A full vehicle model of the truck was created in Advanced Dynamic Analysis of Mechanical Systems (ADAMS) by assembling subsystem models of suspensions, steering, chassis and tires. A typical NATO Reference Mobility Model (NRMM) obstacle course with over 70 different obstacles of different sizes and shapes was used to evaluate the underbody clearance (See Figure 4.3). The results of this simulation gave an idea about the truck's capability to maneuver through different obstacles at low speeds.

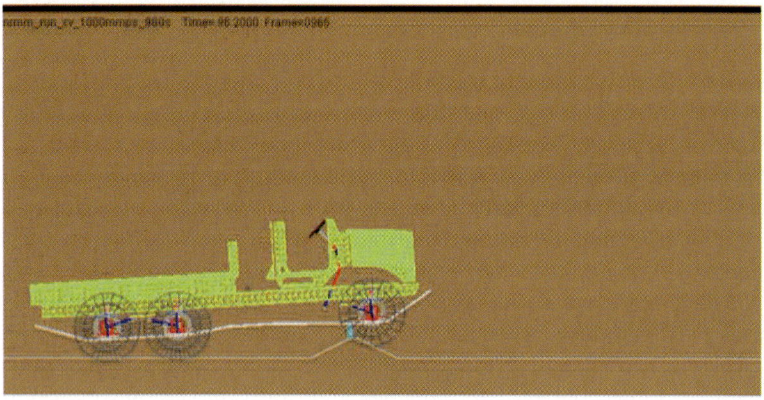

Fig. 4.3. ADAMS model of the MTVR negotiating a simulated obstacle

The steering model was used to predict the front and rear steer angles (See Figure 4.4) for a given steering wheel input. The rear steer model included a dwell and had different gear ratios than the front.

The lateral stability of the truck was evaluated through constant-radius tests. Tire forces were monitored to detect tire lift-offs. The results of these simulations

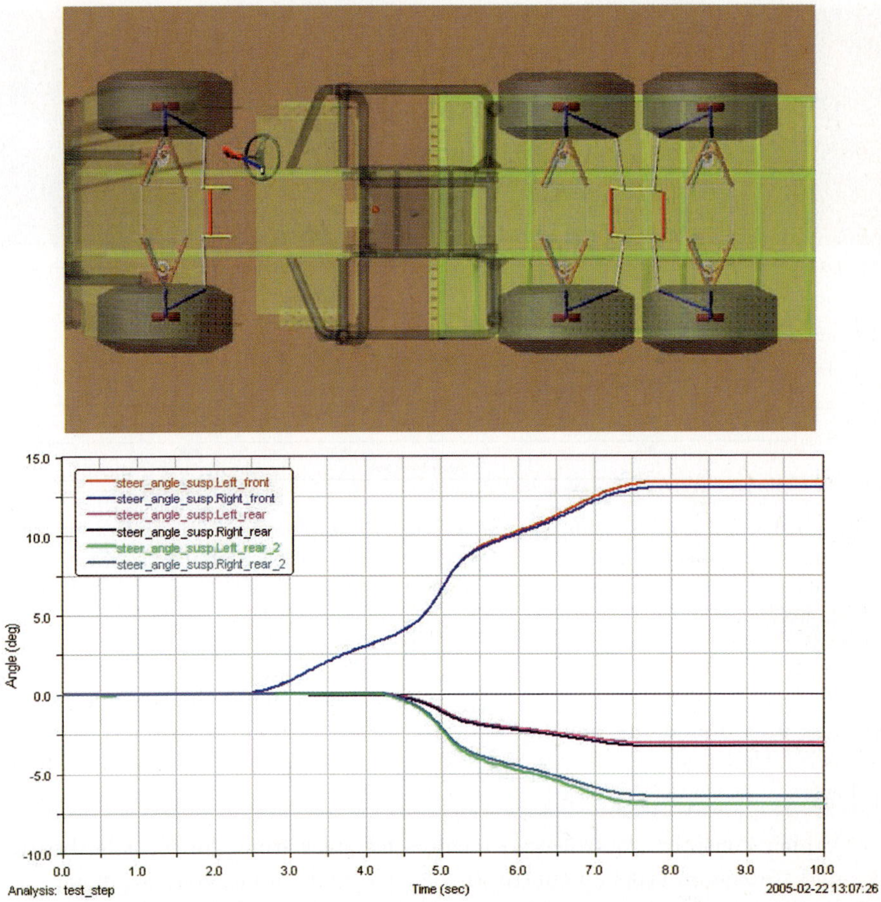

Fig. 4.4. Steering angle behavior of each wheel end throughout a 360° rotation of the steering wheel originating from a straight ahead position

as shown in Figure 4.5 were used to evaluate the capability of the truck to take a particular turn at different speeds without rolling-over.

4.4 Vehicle Management System

The Rockwell Collins intelligent Vehicle Management System (iVMS) (Braid et al. 2006) consists of hardware and software components that together provide an extensive set of autonomous capabilities. In order to accomplish this, the iVMS interfaces with the vehicle systems and all onboard sensors. The primary commands to the vehicle interface are throttle, brake, steering, and transmission.

The general architecture for the iVMS software is a set of applications that communicate to each other over a 100BaseT Ethernet network utilizing Transmission Control Protocol (TCP) and User Datagram Protocol (UDP) protocols

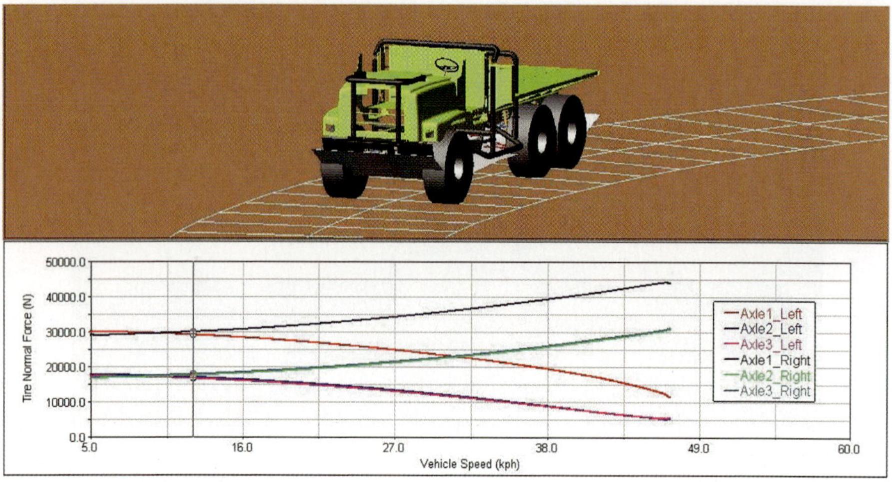

Fig. 4.5. Lateral stability simulation of an MTVR traveling on a constant radius path with increasing speed. Graph depicts weight transfer on each of the six wheel ends. Simulation was conducted at the maximum gross vehicle weight rating (GVWR) of 58,000 pounds.

and a commercial Ethernet switch. The iVMS software has the key role of performing all autonomous behavior and interfacing to numerous LRUs (Line Replaceable Units) and the key vehicle systems. The software applications are as follows:

- Vehicle control – controls and receives feedback from the throttle, brakes, and steering in order to control the vehicle while in autonomous mode.
- Real time path planner – computes the real time path utilizing the desired path while avoiding the obstacles along the desired path
- Obstacle detection – uses LIDAR and Vision to detect positive and negative obstacles. Obstacle data coming from the various sensors are merged into a single obstacle database used by the real-time path planner.
- Behavior management – decides what mode the vehicle should be in based on the current conditions of the other functions
- Navigation – computes present position and provides a dead reckoning function.

A Graphical User Interface (GUI) provides multiple functions to the user including data visualization, recording, and playback. The GUI is primarily a development tool and is not considered to be an integral part of the real-time iVMS system. Figure 4.6 shows the GUI displayed on a monitor in the cab.

A system management function is also implemented that provides a user interface for execution control and status display for the iVMS applications. Once the system has been initialized, the system manager performs a health management function that continuously monitors the status of the application and

Fig. 4.6. The driving cabin, with the monitor showing the graphical user interface

automatically stops and restarts applications as necessary to maintain normal functionality. The iVMS can continue to operate normally without the system manager once initialized so it is not included as one of the iVMS applications. The system architecture can be viewed in Figure 4.7.

The following sections of this paper will go into further detail on each of the iVMS functions.

4.4.1 Vehicle Control

The vehicle control function of the iVMS provides the TerraMax control actions that emulate the actions a human would perform when driving the truck. The controls provided by the iVMS are steering, throttle, brake, and transmission control. Steering control is provided through an electronic servo connected directly to the MTVR steering gearbox. The standard MTVR steering gearbox has dual inputs so the steering servo for autonomous operation and hand wheel are both connected to the steering gear allowing the steering control to be switched between manual and autonomous operations without changing mechanical linkages. The steering control function is responsible for providing wheel angle commands that guide the vehicle to the path defined by the real-time path planner. This is accomplished by computed deviations from the desired path and converting the deviations to steer angle commands that are sent to the steering servo. The steering control uses capture and track steering control modes. Capture steering control is used for initial course capture and track steering control is used during normal operation. Capture and track control modes are automatically selected based on current conditions.

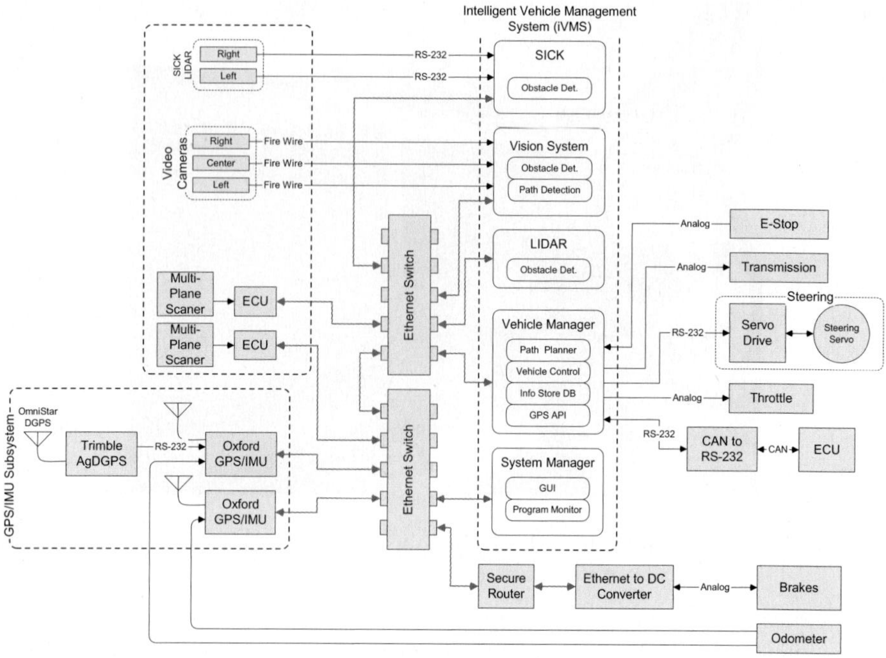

Fig. 4.7. TerraMax iVMS System Architecture

The capture controller uses course error as the control parameter. The controller creates a steer angle command that aligns the ground track of the vehicle with the direct bearing to the active (TO - "next") waypoint. This type of control is sometimes referred to as homing control since the path followed is uncontrolled and the path to the TO waypoint may not be a straight line. Capture conditions occur during initial course capture so the capture controller is only active if certain conditions exist at the time the autonomous mode is first activated.

The track controller uses linear cross track deviation and cross track deviation rate to align the vehicle's path along the ground with the active TO waypoint course. Track angle error and steer angle command limiters are used to limit the commanded steer angle to values that are achievable by the vehicle. The command limiters incorporate vehicle dynamic limits with margins built in to ensure the vehicle does not get into an unsafe condition. This also means that the vehicle operates at levels below its maximum dynamic capability when in autonomous mode. Turn anticipation for waypoint sequences is also used so the transition onto the new course is accomplished without overshoots.

The throttle controller interfaces directly to the electronic engine control unit (ECU) through a digital PWM interface. The throttle controller is responsible for controlling the vehicle's speed to the desired speed specified by the path planner. This is accomplished primarily through throttle position control but engine and service brakes are also used in certain situations to manage the speed.

The throttle position control uses proportional and integral control. Reset conditions to the throttle position are provided for transmission up shift and down shift and to activate the engine brake. Engine brakes are activated during engine idle so throttle position overrides are used when engine brakes are required. Throttle position faders are used to reactivate the throttle position control when the engine brake is disabled. Engine and service brakes are used primarily to control speed on steep grades and for speed management during deceleration.

The brake controller provides an analog signal to a pressure actuator connected to the air brake system (service brakes). The throttle and behavior control functions provide brake actuation parameters to the brake controller and the brake controller determines the pressure actuator signal. The brake control parameter provided by the throttle control function is speed deviation which is used by the brake controller to provide a brake application that is proportional to the speed deviation. Behavior control provides brake override signals for emergency stop (e-stop), e-stop pause, and other special situations requiring speed control or position holding. The emergency stop condition results in a full brake command. Brake modulation to limit slipping in full brake conditions are provided by the Anti-lock Brake System (ABS) system that is part of the basic MTVR.

The MTVR has a seven speed automatic transmission. The transmission control function provides forward, neutral, and reverse gear control for the automatic transmission. The selection of the transmission gear is through a digital signal to the transmission control unit. The transmission controller receives a desired gear signal from the behavior control function and converts the desired gear into the digital interface to the transmission. Behavior control uses the actual gear position to determine allowable state transitions and to prevent transmission faults due to incorrect gear selection sequences.

4.4.2 Real-Time Path Planner

The real-time path planner (See Figure 4.8) is responsible for deriving the desired trajectory of the vehicle and providing that trajectory to the vehicle control function. The trajectory includes a desired path along the ground as well as the desired speeds and boundary area. The desired trajectory is derived using the path and speed constraints contained in the DARPA Route Data Definition File (RDDF), which contains a list of waypoints that define a path along the ground, a path boundary, and maximum speed for each leg of the path. The real-time path planner provides reactive path corrections to this nominal path to account for current conditions, such as vehicle dynamic limits, obstacles, road edges, terrain grade, etc.

The path planner implements a tree algorithm that branches from the base at the current TO waypoint. Constraints for path boundary and speed are applied to the tree build function so the tree size is bounded by the constraints. Branches

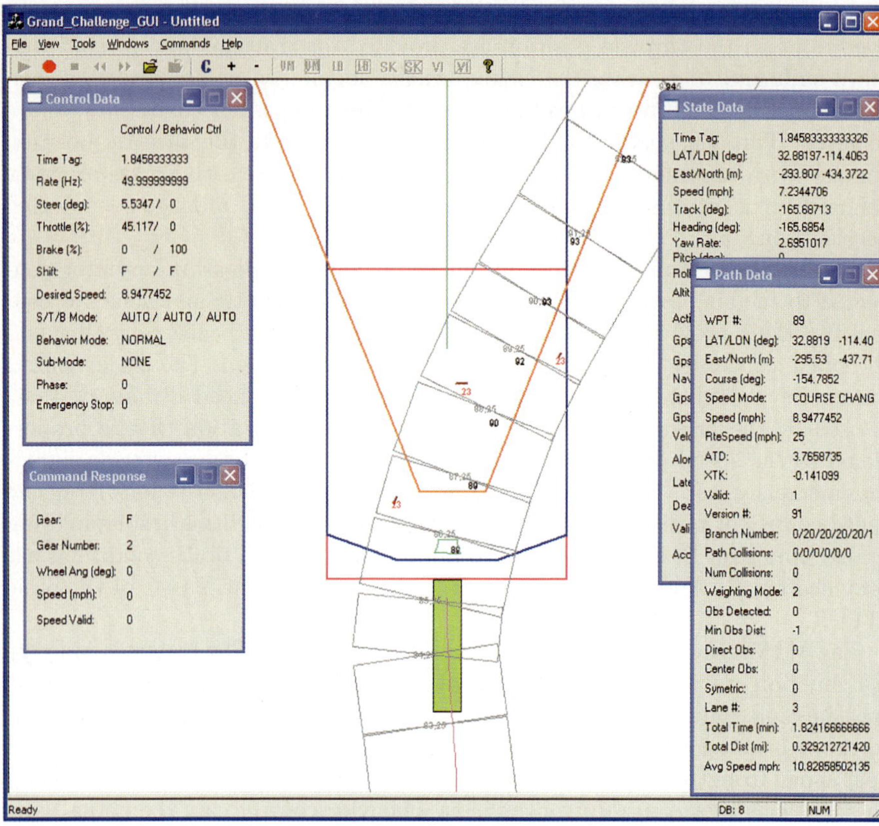

Fig. 4.8. iVMS Graphical User Interfaces Depicting Real Time Path Planning Along a Defined Route

of the tree are computed using a model of the steering system and vehicle dynamics to insure that the candidate paths are drivable. The tree algorithm was derived from the Rapidly-exploring Random Tree (RRT) path planner (Kuffner and LaValle, 2000) where the growth of the tree was limited to a fixed number of branches (levels).

Once built, the tree represents a series of candidate paths, one of which is selected as the path to be used by the vehicle control. Selection of the best path from the candidate paths is based on a scoring algorithm that considers distance from the route centerline, path curvature, obstacle avoidance, boundary area constraints, and other factors. Over 2000 candidate paths are evaluated each planning cycle to determine the best path.

The real-time path planner also contains a speed management function that adjusts the speeds as necessary to account for path geometry and current conditions. The initial desired speed is set to the RDDF speed constraint for the leg and the speed management function reduces the speed as necessary.

Figure 4.8 shows a graphical representation of the RDDF data and the resulting real-time path generated by the path planner. In the main window of the illustration, the central box represents TerraMax vehicle, light grey boxes represent the desired path defined in the RDDF file, and the black numbers near the center of the boxes represent the real-time path generated by the path planner. The width of the grey boxes defines the lateral boundary of the path and the length of the boxes is defined by the waypoints in the file. The short red lines in the diagram are representations of obstacles that have been detected by the perception sensors. Dialog boxes along the sides of the main window show data associated with the path, vehicle state, and control state. As shown in the example diagram, the real-time path is adjusted to the right of the RDDF path center in order to avoid the obstacles in the turn.

4.4.3 Obstacle Detection

LIDAR and vision sensors are used to detect obstacles in front of the vehicle. Obstacles detected by the sensors are registered to the vehicle navigation position and stored in an obstacle database. The real-time path planner queries the database to determine if obstacle collisions occur on the proposed paths.

Several different types of obstacle clearance information are provided to the path planner to aid in path selection. Obstacle collision information is reported by the database in terms of the closeness of the object collision to the proposed path. Buffer regions of various sizes are used to determine the collision proximity relative to the path.

Bearing and distance to the nearest collision is provided by the obstacle database that is an indication of the proximity of the obstacles to the proposed path. Obstacle distance is used primarily in the speed manager function to lower the speed if an obstacle is in close proximity to the vehicle's planned path.

Road and cliff edges are handled as special cases by the obstacle database. Since the consequences to the vehicle of breaching a cliff edge are very severe, additional weight to negative road/cliff edges are used. The database also reports if any negative road/cliff edges are in the immediate area that is used by the speed manager to reduce speeds accordingly.

4.4.4 Behavior Management

The behavior management module is the central "brain" of the system. Its purpose is to monitor and react to dynamically changing conditions. This module receives input from the real-time path planner, obstacle database, navigation sensors and the vehicle interface module.

Several behaviors have been designed into the behavior module, using a state transition architecture. When a specific event or a change from normal operating conditions is detected, one of the behaviors is activated to handle the situation at hand. Each behavior executes an ordered list of instructions, providing a set of commands to the vehicle controller.

Some of the conditions the behavior module will react to are as follows:

- Transition in E-Stop state: When the e-stop is in Pause mode, a behavior will command the vehicle to come to a stop. When e-stop transitions to Run, another behavior is initiated to begin normal operation.
- No valid path ahead: The behavior initiated in this condition commands the vehicle to come to a stop and wait for a valid path. If no valid path is found, it will command the vehicle to back up and try again.
- Obstacle detected behind the vehicle while backing up: Another behavior will stop the vehicle and command it back into normal operation to try to find a valid path ahead.
- A large course change requiring a backup maneuver: The switchback behavior guides the vehicle around a 3-point turn.
- Narrow tunnel condition: The tunnel behavior will guide the vehicle through a narrow tunnel, using the LIDAR scan data.
- Stuck between obstacles:

If the vehicle cannot make progress along the route because it continues to go back and forth, getting stuck between obstacles, the stuck behavior will take over. It will first try to position the vehicle at different angles to search for a valid path. If no valid path is found, it then commands the system to ignore low confidence obstacles, in an attempt to eliminate false obstacles. The last resort is to go forward toward the DARPA route, ignoring all obstacles.

The real-time path planner, behavior management, and vehicle control functions work together to determine the actual path the vehicle follows. Nominally, the vehicle follows the path generated by the real-time path planner but that real-time path can be overwritten by the behavior manager based on the current conditions. This design approach is similar to the Distributed Architecture for Mobile Navigation (DAMN) (Rosenblatt, 1997) developed by Carnegie Mellon University where the behavior management functions as the DAMN Arbiter and Behaviors. Unlike the DAMN architecture, the behavior manager uses rulesbased decision logic to determine control behavior modes rather than a voting scheme to select the control mode. This approach was chosen over the more complex voting scheme since it is deterministic and more robust, which were considered to be important attributes given the nature of the competition. This approach is lends itself to fleet applications, which was also an important consideration in the iVMS design.

4.4.5 Navigation

Two Oxford Technical Solutions (OXTS) RT3100s (www.oxts.co.uk) supply GPS position information to the iVMS system. The RT3100 is a combined GPS/IMU sensor that provides real-time data even in the absence of GPS signal. The high 100 Hz update rate has a very low latency to insure that the system is using the most accurate position possible. One RT3100 is configured to use DGPS corrections transmitted via RS-232 from an external GPS receiver subscribed to the Omnistar correction service. The other RT3100 is configured to use WAAS corrections.

In the case of loss of GPS signal, such as driving through a tunnel, the IMU portion of the RT3100 takes over and begins dead reckoning. In order to aid the INS solution in dead reckoning mode, a wheel speed sensor on the vehicle provides input to the RT3100. Tests have shown that the wheel speed input helps to keep the IMU solution stable and extends the time the RT3100 is able to dead reckon.

In the case of a failure or short-term loss of the RT3100's, a second dead reckoner is implemented using sensed wheel speed and wheel angle. This represents an independent backup navigation function. Because of the potentially large errors that can build up when it is in a dead-reckoning mode, the RDDF boundary area checks in the path planner are disabled so the vehicle can continue navigation relative to the terrain and terrain obstacles for short periods of time.

The RT3100's were capable of dead reckoning after loss of GPS using the IMU but, during field testing, it was found that the error characteristics of the wheel speed/wheel angle dead reckoner were more desirable than the IMU error characteristics. The conditions where the dead reckoner was most likely to be used was while traveling through railroad tunnels and highway overpasses where the GPS satellite signal would be masked. This meant that the vehicle would be moving in a straight line and only in the tunnel for a short time (typically less than 30 seconds). Under these conditions, the wheel speed/wheel angle dead reckoner was able to maintain a predictable accuracy of less than 1 meter. The IMU was also capable of similar accuracies but the error characteristics were less predictable. The predictability of the error was important since the obstacles while in the tunnel (i.e. the tunnel walls) were very close to the vehicle and small, abrupt changes in position error caused the collision detection function to stop the vehicle.

The wheel speed/wheel angle dead reckoner relied on an analytical model to relate wheel angle and speed to heading (yaw) rate so the accuracy deteriorated significantly during turns. The large heading error build up during turns makes this implementation practical only under a very narrowly defined set of conditions.

4.5 Sensors

The sensors were carefully selected to provide the required navigation and perception capability. Figure 4.9 shows the sensor locations on TerraMax. The sensors selected for the DARPA Challenge 2005 are as follows:

- Oxford GPS/INS
- Trimble GPS
- Single-plane LIDAR
- Multi-plane LIDAR
- Forward-looking Vision System

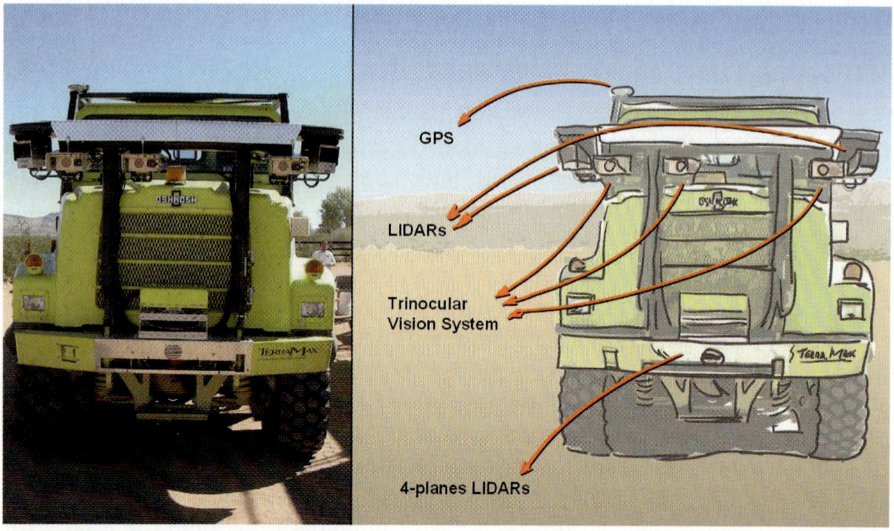

Fig. 4.9. The TerraMax sensor suite: the picture shows the GPS position, the three front looking cameras, the two SICKs, and the two multiplane laserscanners

4.5.1 Oxford GPS/INS

The OXTS RT3100s are mounted on the floor of the cab on the approximate centerline of the vehicle. In order to obtain a more accurate position solution and eliminate any errors over time, the position solutions from the two RT3100s were averaged together. In the case of a failure of one of the RT3100s, the system will switch to using the remaining RT3100 as the sole GPS source.

4.5.2 Trimble GPS

The Trimble GPS (www.trimble.com/aggps132.shtml) is an agriculture GPS unit used to receive differential corrections used by the GPS receivers embedded in the Oxford RT3100s. The Trimble receiver outputs differential corrections at 1 Hz through RS232. In order to output the differential corrections the Trimble receiver is placed in base station mode and must also have a subscription.

4.5.3 Single-Plane LIDAR

There are two SICK LMS-291 LIDARs used for positive and negative obstacle detection (See Figures 4.10 and 4.11). They are mounted on the outermost edges of the front rollbar. They are pointed 10 degrees down and 25 degrees outward from the truck so that there is good coverage on extreme turns. The two LIDARs are configured to scan a 100-degree scan area with a 1-degree resolution.

The orientation of the SICK LIDARs was chosen to gain visibility to positive obstacles near the vehicle and detect negative road edges. Positive obstacle

Fig. 4.10. The SICK LIDARs and the cameras are placed on a rigid bar onto the vehicle hood

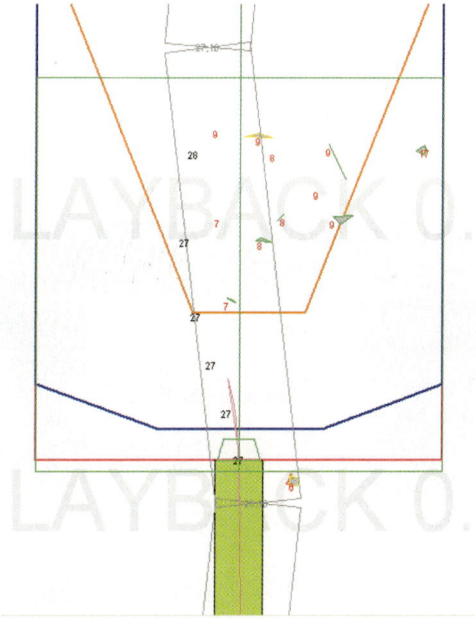

Fig. 4.11. iVMS Graphical User Interface depicting SICK obstacles as green and yellow polygons. In the figure the vehicle is located at waypoint 27 with iVMS calculated micro-waypoints extending ahead of the vehicle.

detection was accomplished be translating the range returns into local level coordinates and comparing the relative heights of neighboring scan points. A history of scan returns were maintained that effectively mapped the surface directly in front of the vehicle. Detection thresholds were set so obstacles below a specific height would not be detected as an obstacle. The minimum obstacle height was set based on the capability of the vehicle. A convex hull algorithm was used to defined the outermost edge of the obstacle.

Negative road edge detection followed a similar approach as for the positive obstacle detection. A specific search algorithm was used to find any negative height discontinuities. Each discontinuity that was detected was further evaluated to determine if the true edge and if that discontinuity was a continuation of the previously detected edge.

4.5.4 Multi-plane LIDAR

The IBEO ALASCA LIDAR (Lages, 2004) is a 4-plane scanner that is used for positive obstacle detection. The LIDAR is mounted level in the front bumper (See Figure 4.12) and has two planes that scan toward the ground and two planes that scan toward the sky. With a range of 80 meters and a resolution of 0.25 degrees it can detect obstacles accurately at long and close range. The 170-degree scan area allows seeing obstacles around upcoming turns.

The LIDAR sends scan data via Ethernet to the LIDAR PC via a TCP connection. An algorithm then transforms the raw scan data into obstacles by looking for large positive slopes in the scan data (See Figure 4.13).

The IBEO obstacle detection algorithm followed a similar approach as the SICK obstacle detection algorithms. The scan returns were translated from the

Fig. 4.12. The front of the TerraMax vehicle. Two 4-plane laserscanners are visible; The one inside the bumper is the one used during the race, the other (over the bumper) is a backup.

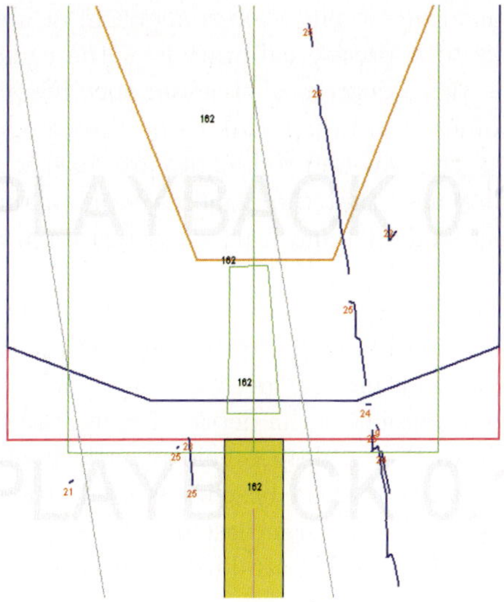

Fig. 4.13. iVMS Graphical User interface depicting IBEO obstacles as blue lines. In the figure the vehicle is located at waypoint 162 with iVMS calculated micro-waypoints extending ahead of the vehicle.

sensor coordinate frame to a local level coordinate frame. The scan returns were then compared to neighboring returns and discontinuities were detected. Since the IBEO LIDAR was a multi-planes, the obstacle detection algorithm was able to compare the scans from each plane to aid in the obstacle detection.

The SICK and IBEO LIDARs were configured so the data provided by the sensors were complementary. The SICK LIDARs were configured to detect obstacles from 0 to 20 meters in front of the vehicle. The IBEO LIDAR was configured to detect obstacles from 20 to 60 meters in front of the truck. Obstacles detected from both sensors were put into the obstacle database and used for path planning.

Field testing indicated that, although the two LIDAR sensors provided similar data, the characteristics of the data were somewhat different. For example, the SICK LIDAR data was very consistent but more susceptible to reflections than the IBEO LIDAR. The IBEO LIDAR provided more accurate data but would occasionally return spurious data spikes. Putting both sets of data into the database without prior filter resulted in multiple copies of some obstacles database. A fusion of the LIDAR sensors to eliminate the multiple copies reduced database utilization and sped up the execution of the collision detection algorithms.

4.5.5 Trinocular Vision System

The vision system is based on multi stereoscopic vision (forward looking trinocular system). It consists of three identical cameras mounted on a rigid bar on top of the hood. The two lateral cameras lay at a distance, which is about 1.5

meters, while the central one is placed asymmetrical at about 0.5 meters from the right one. Thanks to a precise calibration of the cameras - performed on a graduated grid- the three degrees of freedom specifying cameras orientation are fixed to known values, and in particular -in order to ease and speed-up the subsequent processing- the yaw and roll angles are fixed to zero for all cameras. The pitch angle is chosen so that the cameras frame a small portion over the horizon (to limit direct sunlight) and frames the terrain at about 4 meters from the vehicle.

The trinocular system sends three video streams at 10 Hz (640x480, color with Bayer pattern) to the vision PC via a firewire connection. The PC selects which stereo pair to use depending on the speed of the vehicle. Since the baseline of the stereo vision system influences the depth of view, the large baseline is used at high vehicle speeds so that a deeper field of view is obtained, the medium one at medium speeds, and the short baseline is used at low speeds. This is one of the very few examples of very large baseline stereo systems (1.5 m) used on rough off-road terrain and delivering a robust environmental perception at more than 50 m, regardless of terrain slope.

The rationale behind the design is mainly the need for mechanical robustness: three non-moving cameras have been preferred with respect to a pan-tilt solution such as the one used by other teams, for example the Red Team (Whittaker, 2005). A few other considerations were the basis for this choice: vision must be able to sense obstacles at large distances (more than 50 meters away on rough terrain), therefore a stereo vision system was the only choice. Furthermore, the baseline (distance between the stereo cameras) had to be large enough to guarantee depth perception at large distances. Systems based on moving cameras when both stereo and a large baseline are used, are subject to a number of mechanical problems such as -for example- the non negligible momentum caused by vehicle vibrations which has to be compensated for. As a result, the experience with multiple cameras providing different video streams to choose from turned out to be a winning solution, capable of removing most of the mechanical problems of a gazing system.

Vision provides sensing for both obstacle detection and path detection (see Figure 4.14 and 4.15).

1. Image disparity is first used to estimate the average terrain slope in front of the vehicle (Labayrade et al., 2002). Slope information is then used for both obstacle detection and path detection. Any significant deviation from the average smooth slope detected previously is then identified as an obstacle. The exact location of obstacles is then obtained via stereo triangulation between the two views of the object. A fairly precise localization is obtained, but nonetheless it can be further refined via sensor fusion with raw data coming from the multi-plane lidar. In this way it is possible to detect thin vertical posts and fence poles. The system is able to detect even small obstacles (Broggi et al., 2005), but -due to both the size and capabilities of the vehicle and to the team strategy- it was tuned with very high thresholds, so that the number of false positives was reduced to a minimum. In other

Fig. 4.14. Images showing left and right images of different situations; on the right images, colors show the presence of detected obstacles: different colors mean different distances. The additional horizontal lines represent the 5m, 50m, and horizon position. Posts and thin poles as well as fences posts are correctly detected.

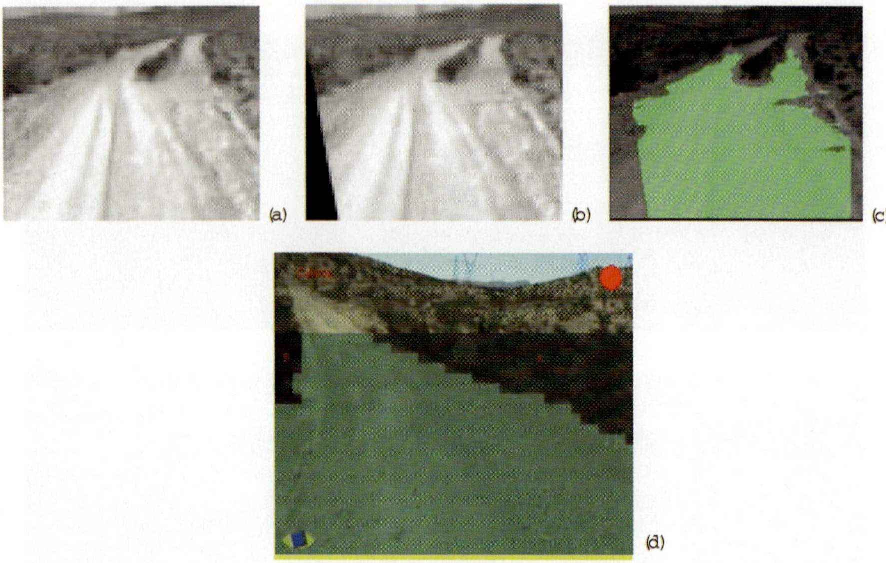

Fig. 4.15. Image showing different steps of path detection: first the free-space is determined then the path is localized. Right image (a), warped left image (b), free space (c), the final result of path detection (d).

words, the capability of detecting small obstacle was traded for a higher robustness of the detection. Nevertheless, the system was demonstrated to be able to detect small construction cones used during both the tests and the qualification runs. Anyway, since the vehicle is able to negotiate 60cm steps, obstacles smaller than 60 cm needs to be detected primarily for speed management issues.

2. Image disparity is also used to compute the area in front of the vehicle which features a smooth slope, the so-called freespace. The free-space is one of the features that concur to construct a representation of the path to be followed by the vehicle: also similarity in texture, similarity in color, and shape information are taken into account, fused together, and delivered to the following path planning module. Free space is obtained using a standard image warping (Bertozzi et al., 1998) in order to localize deviations from a smooth road surface: figure 4.15 shows the right image (a), the warped left image (b), and in green- the matching cluster representing free space (c). Figure 4.15(d) shows the final result of path detection. This algorithm also signals the presence of a straight segment of road, in order to increase vehicle speed. When a curved path is present (the red dot at the top right of figure 4.15 shows the presence of a non-straight road), vehicle speed is reduced.

Vibrations are automatically filtered out since the slope detection algorithm (Broggi et al., 2005), which is the first to be performed, also extracts information that is used to electronically stabilize the oncoming images. Different light levels

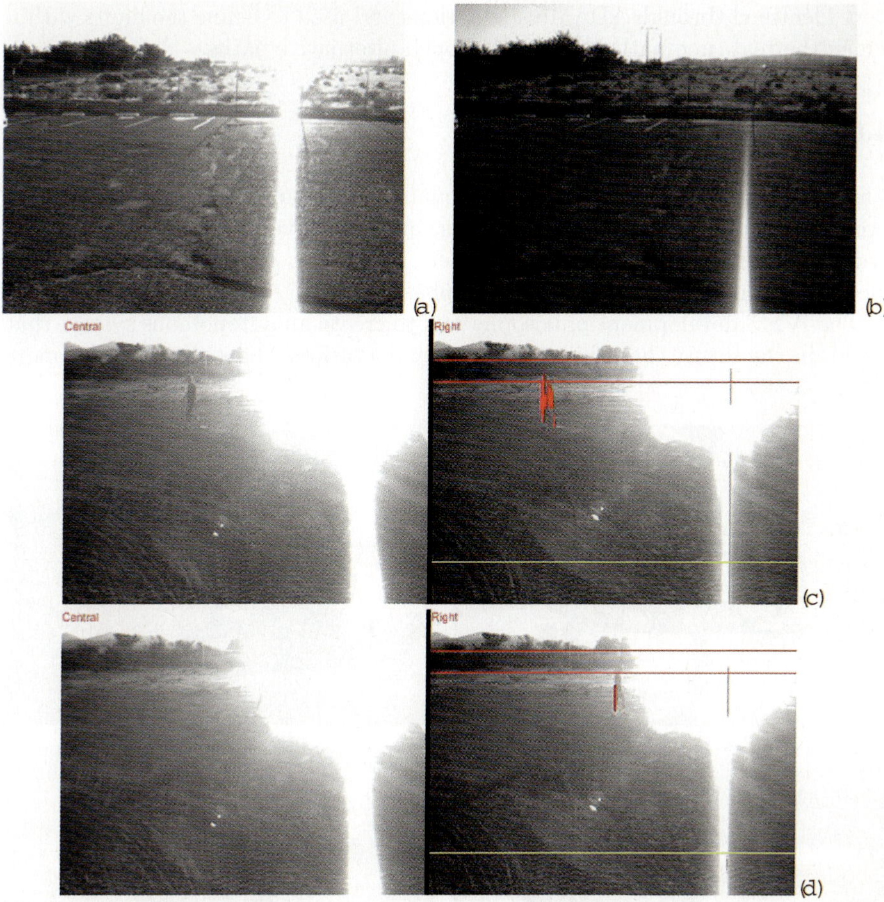

Fig. 4.16. Images captured at sunrise, representing the view with (a) and without (b) the developed camera gain control scheme; (c) and (d) show the result of obstacle detection in bad illumination conditions: although a part of the images are oversaturated, the terrain is visible and the algorithm can be run; obstacles can be detected until they enter the oversaturated area.

are compensated for by an automatic gain control scheme, which allows to sense the environment even with direct sunlight into the camera. Figure 4.16 shows some examples of the custom gain control mechanism.

The camera boxes have a sun shade aimed at reducing to a minimum the quantity of direct sunlight hitting the cover glass, in order to avoid over saturation and reflections due to dirty glass.

The MTVR is a production vehicle with thousands of units produced and in service with the US Marine Corps, US Navy and other services throughout the world. Performance of the MTVR was not specifically tracked as part of the TerraMax development efforts, rather the dynamic capabilities of the vehicle were identified through ADAMS simulations and used to define the limits within which the real time path planner developed alternative paths.

4.6 Vehicle Performance

The vehicle was able to conclude the qualification runs with excellent results, avoiding obstacles, passing into tunnels, and maneuvering (with back-ups) in order to align itself with narrow barriers and gates. Figure 4.17 shows some pictures taken during the qualification phase.

The iVMS development philosophy was to create an autonomous system that could, in the future, be utilized in military operations. This allowed for a more

Fig. 4.17. Some phases of the qualification runs

Fig. 4.18. Two pictures taken during the race (courtesy of DARPA)

rugged implementation of the iVMS for real time navigation across unknown terrain. As a result, Team TerraMax was one of only five teams to traverse the 132-mile course and the only vehicle to overcome an overnight "pause" of the autonomous system. During the race, TerraMax reached a maximum speed of 42 mph. This is impressive not only due to the size and weight of the TerraMax, but due to the fact that true obstacle avoidance was achieved at these speeds.

Figure 4.18 shows a few pictures taken during the last part of the race.

During the race the TerraMax was paused 13 times by DARPA officials to maintain a minimum distance between the competing vehicles or for passing stopped vehicles. The TerraMax automatically stopped and realigned its path approximately 52 times during the race. The majority of the path resets occurred while traversing beer bottle pass where the road was very narrow and the turns were very tight compared to the size of the vehicle (see figure 4.18).

An automatic reversion mechanism was implemented to manage redundant sensors. The reversionary logic was activated due to sensor abnormality twice during the race. The reversionary logic correctly selected the operational sensor and continued to operate normally after the reversion.

A software application health monitor was implemented to monitor the health of the system and start and stop applications as necessary to keep the system executing normally. During the race the health monitor function was activated and correctly reset applications to keep the system operating normally with only minor interruptions in service. A periodic database integrity check was also performed to prevent fatal errors from corrupting the database and to recover data if a data error was found.

During the race the TerraMax struck the edge of one of the concrete underpass barriers. The impact of the tunnel caused the IBEO LIDAR and passenger vision camera to be severely misaligned. The misalignment caused a geo-registration error of the detected obstacles. This caused the path planner to offset the path to compensate for obstacles, slowing the progress of the vehicle for the last half of the race. TerraMax completed the 132-mile race with an official time of 12 hours, 51 minutes and over 28 hours of continuous operation.

4.7 Conclusions

This technology demonstrated by TerraMax has the potential of improving soldier survivability in the battlefield by removing soldiers from harms way, especially during convoy operations – the ultimate outcome of the Congressional goal. The development of fully autonomous systems has allowed Oshkosh and its partners to fully understand the requirements related to both leader-follower and autonomous operation. The opportunity exists to develop and deploy this technology to allow for robotic replacement of convoy personnel, allowing the personnel to be refocused on more pressing duties, and ultimately reduce the convoy personnel exposure to enemy threats.

On the technical side, the choice of the (i) sensors suite delivered the sufficient amount of information for the successful conclusion of the race and demonstrated to be robust enough to deal with the extreme conditions of a desert environment in summer. The experience with multiple cameras providing different video streams to choose from turned out to be a winning solution, capable of removing most of the mechanical problems of a gazing system. The 28 hours of uninterrupted service provided by the (ii) processing systems and software architecture demonstrated the stability and robustness. Finally the (iii) algorithmic solutions proved to be fast and reliable, reducing the number of wrong detections to a minimum.

The TerraMax partners of Oshkosh Truck, Rockwell Collins, and University of Parma have demonstrated the scalability and portability of the autonomous technology by installing and operating the system on an Oshkosh Palletized Loading System (PLS). The PLS is a 10 x 10 vehicle with a gross vehicle weight

Fig. 4.19. Palletized Loading System (PLS) operating autonomously near Barstow, CA

of 84,000 pounds and is capable of delivering a 33,000 pound payload. The vehicle was successfully demonstrated at the Yuma Test Center in January 2006, exhibiting the same autonomous capabilities as the TerraMax. The project was completed in approximately 75 days. Figure 4.19 shows the PLS vehicle during an autonomous run.

References

Ozguner, U., Redmill, K.A., Broggi, A., (2004, June), Team TerraMax and the DARPA grand challenge: a general overview, Intelligent Vehicles Symposium, 2004 IEEE, Page(s):232–237, Parma, Italy.

D. Braid, C.W.Johnson, (June 2006), DARPA Grand Challenge Intelligent Vehicle management System (iVMS) and the Transition to Leader/Follower, 6th Annual Intelligent Vehicle Systems Symposium & Exhibition, Traverse City, MI, USA, in press.

J. Kuffner and S. LaValle, (April 2000) RRT-Connect: An Efficient Approach to Single-Query Path Planning, In Proceedings 2000 IEEE International Conference on Robotica and Automation, San Francisco, pages 995–1001, vol.2.

J. Rosenblatt, (January 1997), DAMN: A Distributed Architecture for Mobile Navigation, doctoral dissertation, tech. report CMU-RI-TR-97-01, Robotics Institute, Carnegie Mellon University.

U. Lages, (June 2004), Laser Sensor Technologies for Prevent Safety Functions, ATA EL 2004, Parma, Italy. Labayrade, R.; Aubert, D.; Tarel, J.-P., (2002, June), Real time obstacle detection in stereovision on non flat road geometry through "v-disparity" representation, Intelligent Vehicle Symposium, 2002. IEEE, Page(s):646–651 vol.2, Paris, France.

Whittaker, W. (2005, August), The Red Team technical paper, submitted to DARPA for the Grand Challenge 2005.

Broggi, A.; Caraffi, C.; Fedriga, R.I.; Grisleri, P. (2005, June), Obstacle Detection with Stereo Vision for Off-Road Vehicle Navigation, 2005 IEEE Computer Society Conference on Computer Vision and Pattern Recognition Volume 3, Page(s):65–65, San Diego, USA.

Bertozzi, M., Broggi, A., Fascioli, A., (1998, June) Stereo Inverse Perspective Mapping: Theory and Applications, Image and Vision Computing Journal, 16(8):585–590.

5

Virginia Tech's Twin Contenders: A Comparative Study of Reactive and Deliberative Navigation

Brett M. Leedy, Joseph S. Putney, Cheryl Bauman, Stephen Cacciola, J. Michael Webster, and Charles F. Reinholtz

Unmanned Systems Group, Virginia Tech, Blacksburg, Virginia 24060

Summary. There are two commonly accepted paradigms for organizing intelligence in robotic vehicles, namely, reactive and deliberative. Although these paradigms are well known to researchers, there are few published examples directly comparing their development and application on similar vehicles operating in similar environments. Virginia Tech's participation, with two nearly identical vehicles in the DARPA Grand Challenge, afforded a practical opportunity for such a case study. The two Virginia Tech vehicles, Cliff and Rocky, proved capable of off-road navigation, including road following and obstacle avoidance in complex desert terrain. Under the conditions of our testing, the reactive paradigm developed for Cliff produced smoother paths and proved to be more reliable than the deliberative paradigm developed for Rocky. The deliberative method shows great promise for planning feasible paths through complex environments, but it proved unnecessarily complex for the desert road navigation problem posed by the Grand Challenge. This case study, while limited to two specific software implementations, may help to shed additional light on the tradeoffs and performance of competing approaches to machine intelligence.

5.1 Introduction

The 2005 DARPA Grand Challenge was a 132 mile race of autonomous ground vehicles through the Mojave Desert. Virginia Tech produced two off-road autonomous vehicles (Figure 5.1) to compete for the $2 million prize. From an initial field of 195 teams, both Virginia Tech vehicles passed a series of qualifying events and ultimately qualified for the main Grand Challenge Event, along with 21 other teams. Although they were built on two similar base vehicle platforms, one vehicle was developed using a reactive paradigm, while the other vehicle was developed using a deliberative navigation paradigm (Murphy, 2000). These competing strategies were developed and evaluated independently for the Challenge. This paper discusses the strategy, capability, and performance of both of the Virginia Tech Grand Challenge entries.

M. Buehler, K. Iagnemma, and S. Singh (Eds.): DARPA'05, STAR 36, pp. 155–182, 2007.
springerlink.com

Fig. 5.1. Virginia Tech's entries to the 2005 DARPA Grand Challenge, Cliff (left) and Rocky (right)

5.2 Base Platform

Both Virginia Tech Grand Challenge vehicles were initially designed as interchangeable platforms on which to develop two very different navigation strategies. However, the terrain mapping for Rocky's deliberative path planning required additional terrain mapping LADAR units, resulting in some hardware differences between the two vehicles. This section includes the details of the base vehicles, power system, drive-by-wire conversion, and network architecture (Leedy, 2006).

5.2.1 Base Vehicle

The Virginia Tech Grand Challenge base vehicles are Club Car XRT 1500s, utility vehicles produced by Ingersoll-Rand. This base platform may seem like an unlikely choice for a desert race due to its diminutive size, but it has proven to be a capable off-road vehicle. The XRT 1500 is extremely agile with a turning radius of 3.5 m. The vehicle also provides a top speed of 40 km per hour and a minimum ground clearance of 16.5 cm under the rear differential skid plate. Stock vehicle weight is 567 kg with the capability of carrying a 454 kg payload.

Cliff, a redesign of Virginia Tech's entry to the 2004 DARPA Grand Challenge, is built on a prototype XRT 1500 which had not yet gone into production at the time of the vehicle's donation. An aircooled 20 horse power (hp) Honda GX620 gasoline engine supplies power to the drive train. The vehicle's roll cage was customized to provide protection for electronic equipment located in the payload area as well as mounting locations for vision and laser sensors.

Rocky's platform is also a Club Car XRT 1500. However, Rocky is a production vehicle powered by a Kubota D722 20 hp liquid-cooled diesel engine. The roll cage on Rocky was replaced with a custom built cage constructed of thicker wall tubing. Rocky also makes use of Club Car's optional heavy-duty suspension upgrade.

5.2.2 Drive-by-Wire Conversion

To enable full computer control of the vehicle actuation systems, the throttle, brake, and steering were converted to drive by wire. The drive-by-wire systems on Cliff and Rocky are nearly identical. Both vehicles actuate the throttle using

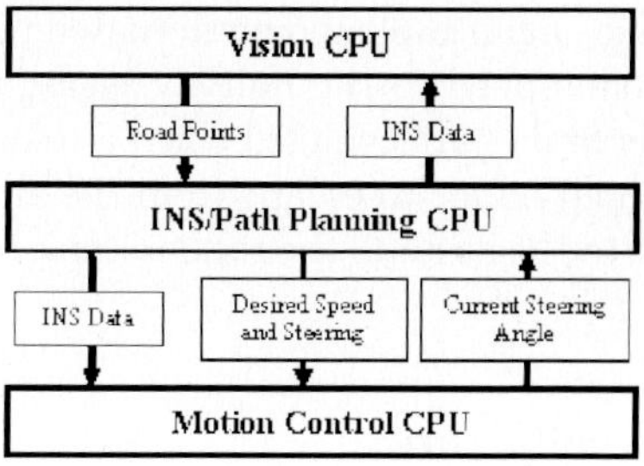

Fig. 5.2. System-level data flow diagram for Cliff

Fig. 5.3. Novatel Propak LBplus positioning system. This system consists of a Novatel Propak LBplus GPS receiver (left) and a Honeywell HG1700 IMU in a Novatel IMUG2 enclosure (right).

a dc gear motor with integrated encoder feedback. The throttle cable is wrapped around a pulley mounted to the output shaft of the motor.

The steering wheel and column were removed from the vehicles to make space for the drive-by-wire system. Both vehicles use right-angle gear motors fitted with quadrature encoders to actuate the steering. The output shaft of the right-angle gear head is directly coupled to the input shaft of the stock steering rack with a chain coupling.

The drive-by-wire braking is accomplished by replacing the master cylinder and brake pedal assembly with an electronically controlled hydraulic pump. The braking system uses a Hydrastar HBA-16 actuator from Carlisle Industrial Brake. Operator control of this electrohydraulic brake is accomplished using a Teleflex-Morse ECFP electric foot pedal connected to the vehicle's Motion Control computer.

In addition to the drive-by-wire brake, a manual/emergency brake can be applied by the onboard operator or emergency stop system. The stock parking brake consists of a pedal assembly that pulls a steel cable directly connected to the rear brake calipers. After removing the ratcheting mechanism, the modified parking brake assembly is used as a manual auxiliary brake. This brake is also automatically activated in the event of a power loss to the vehicle, such as a DARPA "disable" emergency stop condition. To use this braking system as an autoengaging safety brake, an air tank and air actuated piston were installed. Air hold-off pressure must be applied to the unit to release the brake. In the event of vehicle power loss, the brakes are engaged when a solenoid valve opens the air passage and allows the hold-off air to escape from the air-actuated piston. Once power has been restored to the system, the reserve tank recharges the piston and disengages the brake.

5.2.3 Computing Architecture

Both vehicles are equipped with National Instruments PXI-8176 controllers. These controllers are high-performance compact personal computers containing Pentium processors with up to 1 GB of random access memory. The controllers can run at speeds ranging from 1.2 to 2.6 GHz. Due to their high shock resistance, the PXI computers are rigidly mounted to the electronics enclosure without additional shock isolation. The Windows XP operating system provides a familiar visual user interface.

The three computers on Cliff each perform a specific task: Vision, INS/Path Planning, and Motion Control. The INS/Path Planning computer determines Cliff's current position and target location while monitoring for obstacles in front of the vehicle. The Vision computer uses monocular vision to look for roads in the vehicle's field of view and stereovision to localize points along the road. The information from the Vision computer is passed to the INS/Path Planning computer, which determines the appropriate behavior for perceived orientation and surroundings. The Motion Control computer executes speed and steering commands from Path Planning by handling the closed-loop control of all vehicle actuators. Figure 5.2 illustrates the computing architecture of Cliff's three computers.

Rocky uses the same basic architecture as Cliff except with four computers: Vision, Path Planning, INS/Local Mapping, and Motion Control. The INS/Local Mapping computer creates a map of the perceived obstacles and terrain. The Vision computer passes a map of perceived roads to the Path Planning computer. The Path Planning computer then determines the optimal path to take through the surrounding area using data from local map and road map data. These decisions are passed to the Motion Control computer and handled exactly as on Cliff.

5.3 Sensors

Cliff and Rocky both share the same basic sensor suite: GPS/INS for positioning, horizontal LADAR for obstacle detection, and stereovision for road following. Rocky is also equipped with a set of two downward-looking scanning LADAR units for local terrain mapping. These LADAR units are used only with the map-based deliberative scheme to provide terrain information and to detect negative obstacles, such as holes and ditches. This section describes in detail the physical arrangement, purpose, and data format of the main exteroceptive sensor components.

5.3.1 Positioning

Both vehicles used Novatel Propak LBplus positioning systems (Figure 5.3) in the Grand Challenge and in the comparative studies described later in this paper. The system consists of a Novatel Propak LBplus GPS receiver and a Novatel IMUG2 enclosure housing a Honeywell HG1700 inertial measurement unit (IMU). The Propak LBplus unit provides singlepoint position accuracy of 1.5 m CEP. As previously mentioned, this accuracy is increased to 10 cm CEP by L-band differential corrections through the subscription service, OmniSTAR. The position, velocity, and heading from the Propak LBplus are collected at the maximum output rate of 20 Hz. In the event that the global positioning system (GPS) signal becomes occluded, the inertial measurements from the IMU take over seamlessly to provide position and heading.

5.3.2 Obstacle Detection

A single SICK LADAR is used to detect obstacles by scanning a horizontal plane in front of Cliff and Rocky. Anything detected by this scanner is marked as an obstacle. Unfortunately, this includes false obstacles, such as hills and other nonobstacle objects that may pass in front of the scanner. The unit is mounted to the front of each vehicle directly below the brush guard approximately 0.38 m above the ground. By angling the sensor up approximately 1.5° from horizontal, problems related to false hill detection are minimized. The serial data output of the LADAR returns a near-instantaneous twodimensional (2D) polar coordinate array of the range and angle to any solid objects in the sensor's viewing plane. Only the most recent scan from the LADAR is used for both navigation strategies.

5.3.3 Vision

The monocular/stereovision system allows Cliff and Rocky to perceive roads ahead of the vehicle and mark them as preferred areas of travel. The vision system examines the monocular image of the scene, extracts areas that look like roads, then finds the relative position of the road areas using the camera's stereo capabilities. Finally, it passes the road information to the path planning computer.

For the Grand Challenge Event, it was assumed that most of the competition course would follow desert service paths and that these roads would be less likely to contain obstacles than the surrounding terrain. For this reason, a vision system was designed to identify roads and adjust the path of the vehicle to be down the center of the road. A Point Grey Bumblebee stereovision camera, mounted to the top center of the vehicle's roll cage, is used to observe the area in front of the vehicle. Each of the Bumblebee's cameras is capable of outputting progressively scanned 640×480 stereoimages at 30 Hz. The stereoimage processing algorithm operates at approximately 5 Hz.

Before each image is processed, it undergoes a number of modifications to reduce processing time. The image resolution is reduced to 160×120 pixels to lessen the number of recognition operations required on each frame. The image is also converted from the red-green-blue (RGB) image representation to hue-saturation-luminosity representation. The HSL representation allows simpler color definitions in a variety of lighting conditions.

To further reduce the required processing per image, each frame is run through a k-clustering (Green, Yao & Zhang, 1998) algorithm that separates every pixel in each color plane into eight categories defined by the center of the cluster. This clustering method reduces the colors in an image from 16 million to only 512. Instead of using rigidly defined color windows, this method reduces the colors using dynamic logical segments. This operation ensures that each pixel will be converted into colors that are as close to the original color as possible. The reduced-color image is still stored as a 24-bit image to preserve the original color differences.

After simplifying the image, the software searches for the basic geometric characteristics that define a road. This search is done using only one of the 2D images provided by the stereocameras. The software determines if a uniform texture and color pattern form a shape close to that of a desert road (Rasmussen, 2004). Once a road is found, its color patterns are logged to a color look-up table. This table correlates a specific color to a road certainty value.

For each subsequent 2D image frame, every pixel is compared to the color look-up table to create a confidence value for that pixel. This confidence value represents the level of certainty that the specific pixel is part of a road. To simplify further processing, the pixels corresponding to the lowest 30% of the confidence values are removed. Pixels that are not adjacent to or near high confidence value pixels are also removed. Using morphological techniques to connect and separate the areas of an image, the software attempts to create a single area of the image that may be a road (Rasmussen, 2004). The suspected road area must be confirmed as a road by checking its vertical narrowing, edge continuity, and size relative to the image.

If an area of the 2D image is determined to be a road, the image is sent to the stereo-processing software where it is combined with its synchronous image to convert the pixel locations to coordinates in the vehicle reference frame. These coordinates are rechecked to ensure that there are no discontinuities, and that they lie on the same plane as the vehicle. If confirmed, the road points are transformed into UTM global coordinates and sent to the Path Planning computer to be used in the navigation algorithm. The color look-up table for the confirmed road is averaged with the previous ten tables to form the new table for future iterations. To ensure that the look-up table contains enough values to operate accurately, five successful iterations must occur before road points are confirmed and sent to the Path Planning computer.

5.4 Motion Control and Vehicle Safety

Although the two vehicles were designed to use different navigation algorithms and sensor configurations, the drive-by-wire systems were designed to be the same on both vehicles. This standardization allows any navigational algorithm to be implemented on either vehicle, as long as it is compatible with the standard interface. The Motion Control system provides the necessary software to turn desired steering and speed commands into vehicle movement. The motion of both Cliff and Rocky is controlled by three actuators: Steering motor, throttle motor, and brake actuator. The desired commands are generated by the Path Planning software in autonomous operation or a human driver in manual mode.

5.4.1 Speed Control

The vehicle acceleration is controlled by a closed-loop manipulation of an electric gear motor attached to the vehicle's throttle. Since the throttle motor cannot slow the vehicle, a parallel brake control is needed for controlling the speed of the vehicle. The braking system uses an open-loop control to translate a desired reduction in speed to the appropriate brake percent command for the hydraulic brake driver. If the commanded speed is greater than the current speed, a proportional integral differential (PID) control loop handles throttle inputs. If the commanded speed is less than the current speed, brakes are applied based on the commanded urgency of deceleration. Even though the brake and throttle control the speed in parallel, the vehicle will never attempt to increase throttle when braking.

5.4.2 Rollover Prevention and Vehicle Safety

After experiencing two vehicle rollovers, one during Rocky's DARPA site visit, attention was focused on preventing another rollover. A simple dynamic model of the vehicle, that considers gravity and centripetal force, was developed. The basis for this model is shown in Figure 5.4. To account for the rollover effects of unpredictable terrain, a factor of safety is implemented in each calculation.

A rollover condition exists when the resultant of the centripetal force and the weight vector point outside the footprint of the vehicle. Stability can be achieved by slowing the vehicle's forward velocity and reducing the magnitude of the steering angle.

Since the stability calculation depends on velocity and steering angle feedback, the stability calculation described above is not foolproof. Rocky's rollover in qualifying was due to LABVIEW DataSocket communication failures between the INS computer and Motion Control computer. When the failure occurred, the Motion Control computer falsely perceived a zero-speed value from the GPS/INS DataSocket. When the vehicle attempted to accelerate to the commanded speed, the GPS/INS feedback speed remained zero. As a result, the PID controller continued to apply full throttle, and the vehicle rolled in its first turn. The

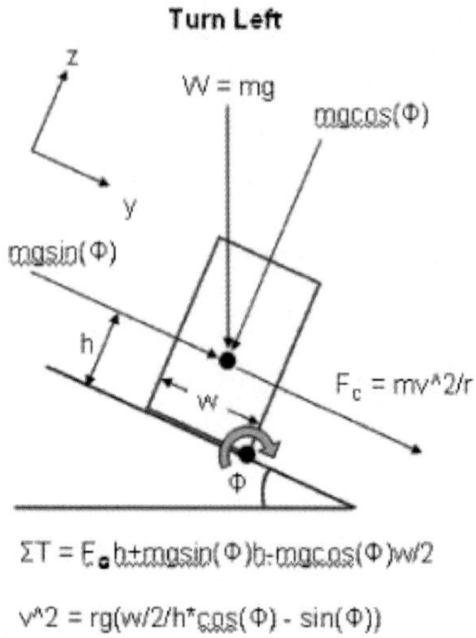

Fig. 5.4. Model of the vehicle on a side slope and in a turn

team replaced the less reliable communication with simple UDP messages. In addition, Motion Control monitors GPS/INS data for communication failures, ensuring that the data are being updated on every iteration. The software will pause the vehicle if a failure occurs, which prevents it from driving without speed feedback. Similar safety systems monitor other potential failures, such as problems with steering.

5.5 Navigation Strategies

An important objective in developing the two Virginia Tech Grand Challenge vehicles was to compare the reactive navigation strategy used on Cliff with the deliberative path planning strategy used on Rocky. These two approaches are usually considered to be opposite ends of the spectrum of navigation strategies (Murphy, 2000). Both approaches were given equal attention during the design and development phase in preparation for the competition. This section provides an explanation of each of the two navigation strategies.

5.5.1 Reactive Navigation with Dynamic Expanding Zones (DEZ)

Waypoint navigation, road following, and obstacle avoidance on Cliff all use a reactive scheme (Murphy, 2000). Reactive algorithms only use the most recent sensor information to make navigational decisions. A technique called the Dynamic Expanding Zones algorithm was developed by Virginia Tech as the main obstacle avoidance strategy for the reactive approach. A set of zones around the vehicle dictate the behavior that the vehicle will exhibit. If obstacles are not detected within these zones, the vehicle will proceed with waypoint/road following.

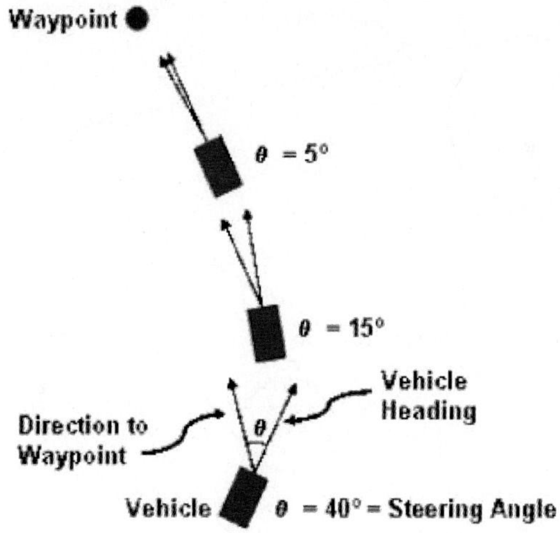

Fig. 5.5. An illustration of the vehicle's commanded steering angle converging toward zero during waypoint navigation

Otherwise, the vehicle will take appropriate action to avoid the obstacles. These zones vary in size and shape, depending on vehicle speed, steering, and sensor status.

5.5.1.1 Waypoint Navigation

A critical component of the 2005 Grand Challenge Event was successful navigation through globally defined waypoints. As with all decision-making software on Cliff, waypoint navigation does not generate a planned path to reach a desired waypoint. Instead, an instantaneous steering angle, equal to the difference between the current heading and direction to the waypoint, is commanded (Figure 5.5). This waypoint navigation strategy acts as a closed-loop feedback control that requires very little computation to calculate the commanded steering angle.

The vehicle reaches a desired waypoint when the vehicle enters a radius defined by the distance between the corridor intersection point and waypoint (Figure 5.6). Whether the vehicle is saving time or avoiding an obstacle, the waypoint radius eliminates the need to travel directly over the waypoint. Traveling over a waypoint was not required for the Grand Challenge Event, as long as the vehicle stayed within the lateral boundary offset (LBO).

5.5.1.2 Road Following

Since roads are generally easier to traverse and have fewer obstacles than unstructured desert terrain, road following is a desirable behavior. If a road exists that leads the vehicle in the general direction of the waypoint, the vehicle ignores waypoint navigation to follow the road. Road data are received from the vision

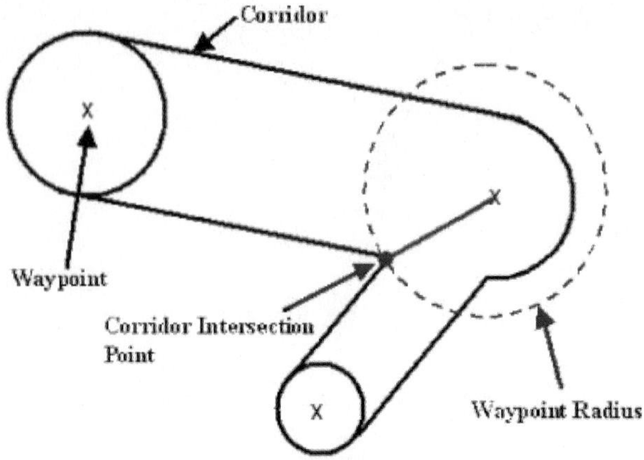

Fig. 5.6. A waypoint radius is created using the LBO of the intersecting corridors

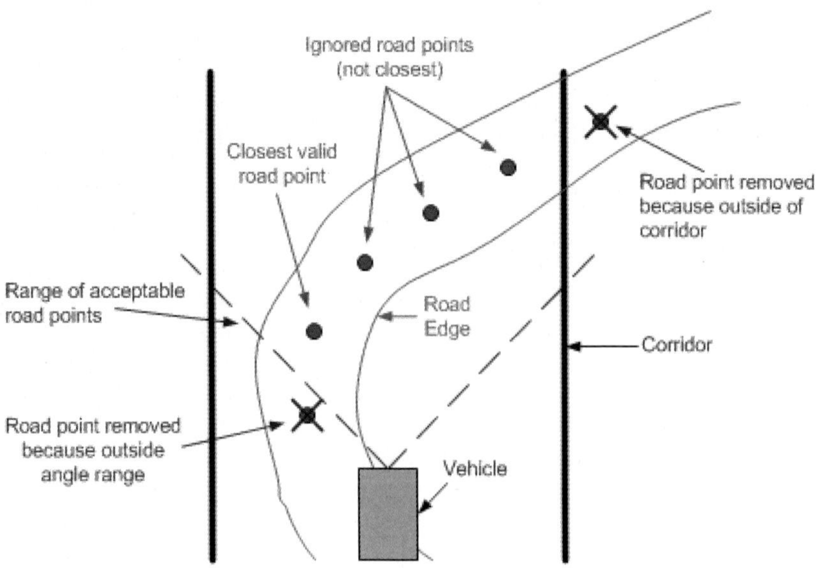

Fig. 5.7. An illustration of how a road point is selected from the road point array

computer as an array of perceived road center points. Points outside of the LBO and points not within 30° of the current heading are ignored, and the closest valid road point is chosen to be the desired road point (Figure 5.7).

Using this desired road point, fuzzy logic control is used to determine if the road point is in the general direction of the desired waypoints. Fuzzy logic is able to substitute numerical variables with linguistic variables to solve ill-defined problems (Zadeh, 1965, 1973). For example, if the vehicle is heading *somewhat*

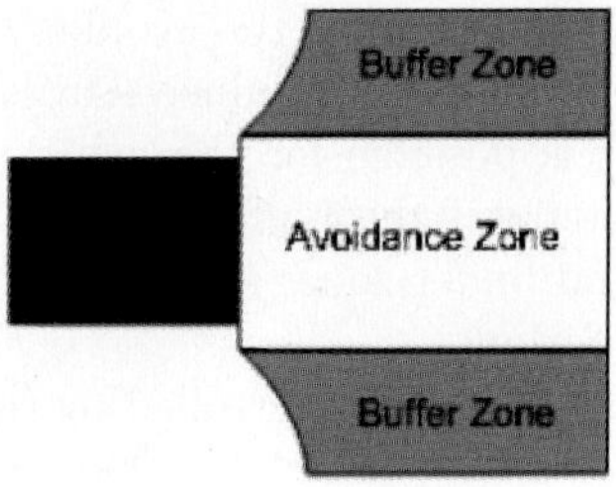

Fig. 5.8. Dynamic Expanding Zone layout

toward the waypoint and away from the corridor boundary, fuzzy control will determine that road following is appropriate. If road following is desired, Cliff steers toward the road point in the same manner as waypoint navigation.

5.5.1.3 Obstacle Avoidance

If a perceived obstacle prevents the vehicle from driving directly to a waypoint or following a road, Cliff ignores the waypoint navigation and road following behaviors to avoid the obstacle. A reactive obstacle avoidance approach, called Dynamic Expanding Zones, has been developed for robust obstacle avoidance. This algorithm is not limited to a specific sensor configuration, and can use any type of instantaneous obstacle map with Boolean elements (obstacle or no obstacle).

5.5.1.3.1 Obstacle Zones. The Dynamic Expanding Zones algorithm uses two zones to determine the avoidance behavior when an obstacle is present (Figure 5.8). The avoidance zone is located directly in front of the vehicle (Reynolds, 1999). If an obstacle is in the avoidance zone, the vehicle must avoid it to continue safely toward the desired waypoint. This zone has a constant width, slightly larger than the width of the vehicle, which prevents the vehicle from clipping the sides of obstacles. The length of the avoidance zone expands dynamically, hence the name Dynamic Expanding Zones. Dynamic Expanding Zones commands a steering angle and speed to avoid any obstacles in this zone.

The second zone, the buffer zone, is adjacent to and of the same length as the avoidance zone. The purpose of the buffer zone is to prevent the vehicle from turning into an obstacle. It also eliminates oscillatory behavior between waypoint/road following and obstacle avoidance. For example, if the vehicle attempts to make a turn to the left when there is an obstacle in the left buffer, Dynamic Expanding Zones will override the turn command. The vehicle will drive straight forward, until the obstacle exits the buffer zone. Once both the avoidance and buffer zones are clear, waypoint/road following will resume.

5.5.1.3.2 Dynamic Expanding Capability. The length of the avoidance and width of the buffer zone are the key factors in the success of this obstacle avoidance algorithm. The length and width of the zones are adjusted based on the current driving conditions. For example, the avoidance zone is shortening when the vehicle is turning. This keeps it from unnecessarily trying to avoid obstacles that are straight ahead. It is possible for the vehicle to make an unnecessary maneuver to avoid an obstacle if the buffers are too wide. For these reasons, the size of the avoidance and buffer zones are dynamically modified to optimize navigation for different situations.

The length of the avoidance zone is controlled by the projected clothoid path and the speed of the vehicle (Shin & Singh, 1990). Using the current steering angle and steering velocity (assumed to be constant), the corresponding clothoid path is calculated. The avoidance zone shrinks to the most distant intersection of this path with the avoidance zone. This means that the length of the avoidance zone shrinks as the steering angle increases. This length control prevents the vehicle from reacting to obstacles too far ahead, but ensures that the vehicle will have ample time to respond as it approaches an obstacle. In addition, as the vehicle increases its speed, the avoidance zone length must also increase to react to obstacles in the distance (Putney, 2006).

Similar to the avoidance zone, the buffer is also dynamically controlled based on the steering angle. Both buffer zones widen symmetrically as the steering angle increases. A larger steering angle requires the vehicle to look for obstacles farther away in the lateral direction. Again, this zone expansion ensures that the vehicle will only avoid the necessary obstacles.

5.5.1.3.3 Commanded Steering Control. Similar to road following, the steering direction is determined by fuzzy logic control. The controller intelligently decides a steering direction, which is optimal for both avoiding an obstacle and staying on course. The fuzzy input variables include distance to obstacles and obstacle summing, discussed below. When a collision with an obstacle is imminent, the Dynamic Expanding Zones method uses only obstacle summing to choose a safe steering direction. For example, if an obstacle is located on the left side of the avoidance zone near the vehicle, the safest steering direction is to the right. On the other hand, when the obstacle is farther ahead, the vehicle has more decision flexibility. As a result, the vehicle can choose a direction that will avoid the obstacle while keeping the vehicle within the boundaries.

Obstacle summing allows Dynamic Expanding Zones to decide which direction is optimal given current obstacle data. An obstacle window is a defined area of interest that encompasses known obstacles in front of the vehicle. The height and width of the obstacle window is defined by the fixed lengths for lateral and length expansion. Only obstacles detected in this window are considered in the obstacle summing calculation. This window allows the vehicle to respond to multiple objects in close proximity instead of just the closest detected obstacle. Using the obstacle window, a value is determined by summing the distances from each of the obstacles to the centerline of the vehicle. Figure 5.9 illustrates how obstacle summing would work if the obstacle window contained two obstacles. The negative values, left of the centerline, represent the wall obstacle (sensed as three points by the laser scanner); while the positive value represents the round

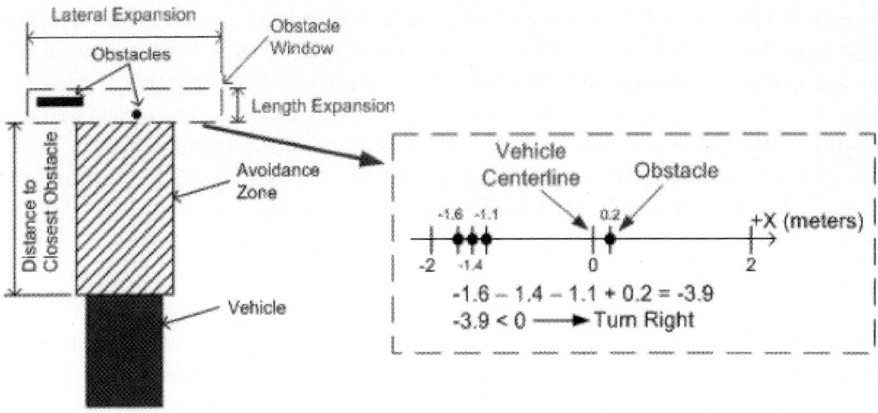

Fig. 5.9. Example calculation for determining a steering direction based on obstacle location

obstacle to the right. This example results in a negative obstacle sum; therefore, a right turn requires a smaller steering maneuver (Putney, 2006).

The magnitude of the commanded steering angle is calculated using the distance to the closest obstacle within the avoidance zone. The steering angle calculation is not an attempt to model the vehicle's actual projected path. Dynamic Expanding Zones varies this approximate path and steering angle with each iteration until the obstacle is avoided. This steering angle calculation eliminates the need for accurate path calculations on each update, which can be computationally expensive. As a result, Dynamic Expanding Zones requires minimal processing power when compared to many deliberative approaches (Putney, 2006).

Once an obstacle has left the peripheral view of the LADAR, it is no longer considered by the reactive strategy. This has proven to be a safe assumption, since only the most extreme maneuvers of the vehicle would cause it to turn back into an obstacle it has already seen without seeing it again.

5.5.1.3.4 Commanded Speed Control. Speed control is critical for properly avoiding obstacles, staying within boundaries, and preventing rollover. The vehicle will always attempt to run at its top speed. However, to prevent rollovers, the speed is limited when the vehicle executes a turn. Though speed control follows the reactive approach, the vehicle can anticipate future maneuvers and take precautionary action. For example, the vehicle slows down when it detects an obstacle in its avoidance zone. The vehicle also anticipates the turn at a waypoint by slowing to a safe speed before it reaches the waypoint radius.

5.5.2 Deliberative Strategy

The Deliberative NonUniform Terrain Search (NUTS) algorithm, developed for Rocky, uses simultaneous sensor fusion and storage to create a homogeneous local terrain traversibility map. This map contains information on discontinuities in terrain height, course boundaries, and roads recognized by the vision system. The map is scanned by an A* graph search (Hart, Nilsson & Raphael, 1968) to determine the desired future path of the vehicle. This operation iterates in real time at a rate of 16 Hz. The goal of the deliberative Terrain Search path planning strategy is to build a continuously updated best path on which to drive. The benefits include planned reaction to perceived future obstacles and holistic driving decisions based on all sensor data (Leedy, 2006).

5.5.2.1 Terrain Mapping

The Terrain search algorithm uses two types of LADAR data to describe the local terrain and obstacles: A two-and-one-half dimensional geometric terrain map, and a binary obstacle map. The geometric terrain map is built using two ground-scanning LADAR units. These can be seen on Rocky in Figure 5.1. The obstacle map is built using a horizontal scanning, front mounted LADAR.

To detect variations in terrain that might affect the planned path of the vehicle, two groundscanning LADAR systems are employed. Figure 5.10 shows the fields of view of Rocky's sensors.

The ground-scanning LADAR units measure ranges to solid objects in a 100° 2D swath about the z axes of the sensors. On a perfectly flat surface, this would allow the scanner beams to reach the ground at a maximum distance of 15 m in front of the vehicle. As the vehicle approaches an obstacle, the scanners record a higher altitude at the location the scan plane intersects the object. Tall obstacles occlude the LADAR; leaving a "shadow" behind the obstacle, which cannot be

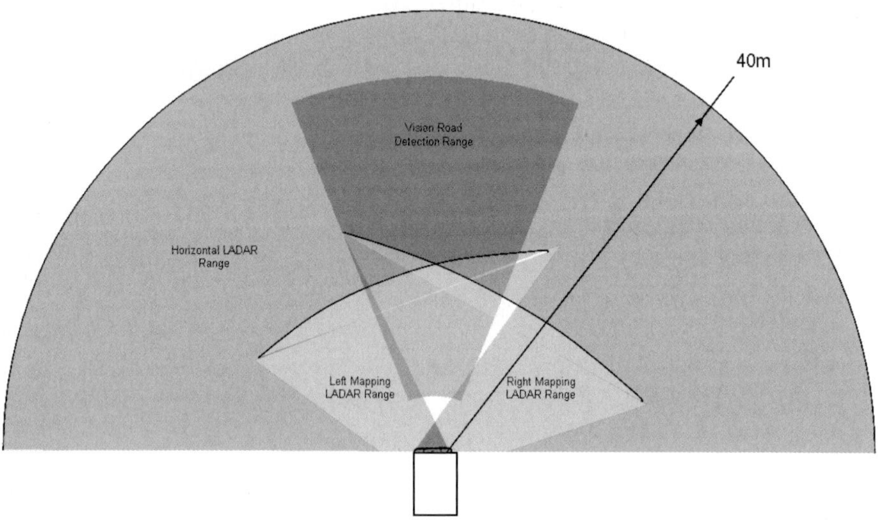

Fig. 5.10. Fields of view of Rocky's ground mapping LADAR, horizontal LADAR, and camera

Fig. 5.11. The image on the left shows a parked car from Rocky's point of view. The image on the right shows scanned terrain colored black. The position and orientation of Rocky is denoted by the grey arrow. A parked car blocks LADAR scans, leaving an unscanned shadow, circled on the right image.

scanned. Figure 5.11 shows the location of scanned points collected as the vehicle approaches car on the left.

The scan planes of the two scanners overlap in front of the vehicle, giving more data directly ahead. This extra information can be used to identify potential obstacles more readily.

Each point collected by the LADAR scanners is transformed from the sensor coordinate frame to the vehicle reference and then rotated into global UTM coordinates. This transformation uses the most recent position, attitude, and heading. The new data are stored as an array of height and position values, then added to the corresponding location in the local map. The new data overwrite any older values stored in the same location. The local map is stored as a 2D array of height values from an arbitrary baseline set at the start of the vehicle's run. The map array is aligned with true north-south and east-west, and represents a 40 m by 40 m field of 20 cm square grid elements. This local terrain map is continuously updated with new LADAR scans at a rate of 16 Hz. As the vehicle moves, the LADAR scanners measure the height of solid objects in their scan plane, and adds these data to the local terrain map, creating a three-dimensional geometric description of the terrain that has been scanned. The data are always represented with the vehicle at a fixed location and varying attitudes and the map grid aligned with the global UTM coordinate frame. This map could readily accept a priori terrain data, if any were available. Figure 5.12 shows a diagram of the vehicle on a local terrain map.

In each program iteration, a 12.5 m by 12.5 m rectangular section of the terrain map is extracted for path planning analysis. This section is transformed back into the vehicle coordinate frame to another grid of 20 cm$\times$20 cm squares. This extracted section is processed using the sigma filter method (Murphy, 2000; Lee, 1984) to find the slope (first derivative) of the perceived terrain. Areas of

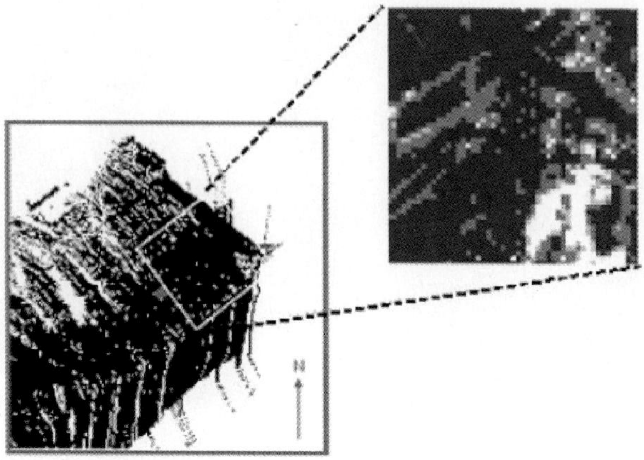

Fig. 5.12. Black areas in the local terrain map (left) indicate scanned points. The local cost map is extracted from this area (right), and sent to the A* decision algorithm. On the terrain map, the vehicle is held at a fixed position on the scrolling globally referenced terrain map, but it may change attitude. The cost map area is held fixed relative to the vehicle.

Fig. 5.13. The local cost map (top-down view, left) generated by the horizontal LADAR for a typical scene with obstacles (vehicle view, right)

high slope are considered to be less passable than areas of little or no slope, so the cost map is scaled by a tuned gain value to return high cost in areas of high slope and low cost in areas of low slope. This map format allows the grid to be searched using standard graph-search algorithms.

Horizontal LADAR scan points, which return a range of less than 40 m, are imported into the local cost map as high-cost obstacles. Any point returned to the scanner is considered impassible, and is marked with an extremely high cost. The data from this sensor are refreshed on every program cycle using only the most recent scan for the cost map. This treatment of the sensor data provides easily interpreted high-cost obstacles wherever the horizontal scanning LADAR detects a solid object. The main drawback of this treatment is the occurrence of "false positives" when the vehicle pitches momentarily or when it approaches a hill. This has the effect of slowing the vehicle speed and causing the vehicle to approach steep hills at an angle. In testing, we found that this rarely created a situation where the vehicle would veer off course. Figure 5.13 shows a local cost map generated by the horizontal scanning LADAR beside a photo of the scene.

Vision Road Mapping – A computer vision road finding algorithm also contributes to the local cost map. The vision processing approach is identical to the one used by the Dynamic Expanding Zones algorithm with data passed as a Boolean map of suspected road points. The scene in front of the vehicle is processed to find points along the road. Figure 5.14 shows one frame of a stereotest scene with the map array of the corresponding road map.

The vision road map is passed to the mapbuilding software as a binary array of the same dimensions as the LADAR local cost maps. If confidence in the vision-recognized road falls below a specific threshold, a blank map is sent to the path planning. When received by the map-building software, all areas marked as road centerline are marked with a lower cost than the surrounding areas. The difference between road and nonroad cost values was tuned through extensive field testing.

5.5.2.2 Deliberative Driving Decision

The NUTS deliberative driving paradigm attempts to drive the optimum path over continuously changing nonuniform terrain perceived by the vehicle. To optimize the path based for both the current and intended future position of the vehicle, the NUTS algorithm computes a new optimum path at each program iteration using an A* graph search. Figure 5.15 is a flow schematic of a single iteration of the NUTS program.

Fig. 5.14. The recognized road from the RGB image (left) is translated into a local road map (right)

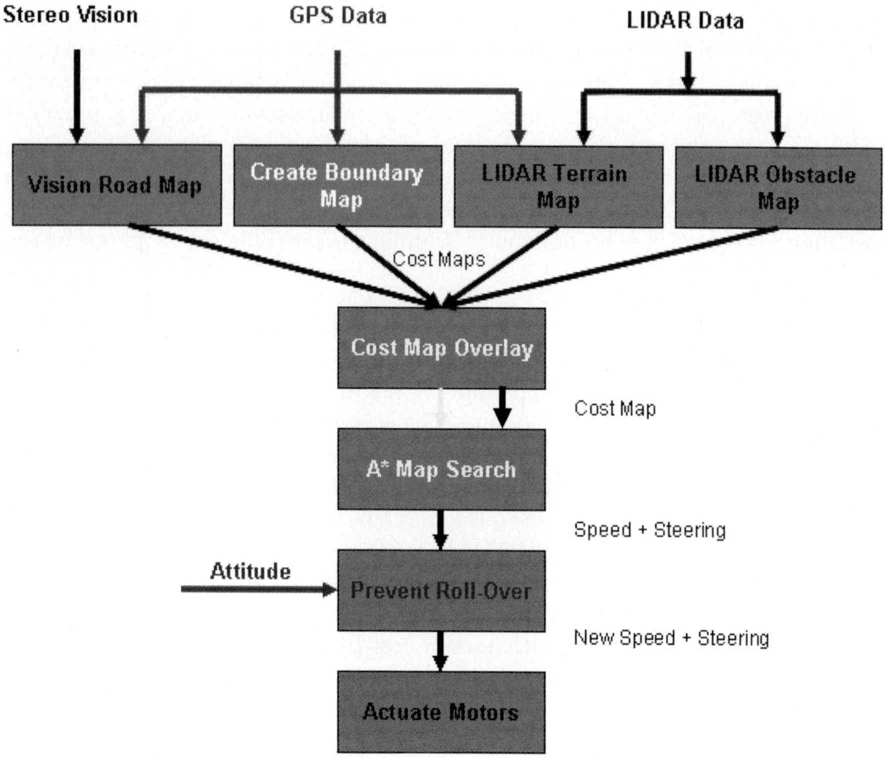

Fig. 5.15. Sensor data for the NUTS algorithm is processed in parallel then combined in the form of a local cost map for the A* search

To generate the final search map, NUTS overlays the four local cost maps generated by the sensor cognition components. Figure 5.16 shows an example of a typical cost map used by NUTS.

Obstacles detected by the horizontal LADAR and areas outside the course boundary are marked with the highest possible value using the obstacle and boundary cost maps. Areas with a geometric change in altitude are assigned a cost based on the "steepness" of the terrain. The road layer adds cost to areas not believed to be a road. The cost overlay values are weighted such that course boundaries and obstacles have the greatest influence over driving decisions. Terrain LADAR and road data are used to guide the vehicle through the optimum path for navigation.

Using the overlaid map, NUTS next attempts to find the best path using an A* least-cost path search (Hart et al., 1968). If the destination waypoint is on the cost map, it is taken as the search goal point. If the destination point is out of the map, NUTS generates a goal point on the border of the map. The path generated by the A* search is passed to the driving component. The driving component generates steering and speed commands based on the vehicle's current pose. This

Fig. 5.16. Typical LADAR overlay map of sensor data for the Virginia Tech Grand Challenge A* graph search (right). Dark shades indicate areas of low cost, while light areas indicate areas of high cost. The area enclosed in the white oval indicated a significant drop off; the area circled in dashes indicates the trees pictured at left.

is accomplished by selecting a point on the path a certain range from the vehicle. The vehicle steers using a pure pursuit algorithm (Coulter, 1992) to head toward the path. Before the steering and speed commands are passed on to the vehicle motion control system, a final check is performed to ensure that the commands are safe.

5.6 Comparative Study

In preparation for the DARPA Grand Challenge, Virginia Tech compared the reactive and deliberative navigation strategies side-by-side. The comparison described in this section attempts to capture the data and lessons learned from applying each navigation strategy. In the future, the team intends to use the results to improve future designs of hybrid paradigms using the best elements of both reactive and deliberative path planning. By implementing both strategies simultaneously with similar developmental teams, this study also sheds light on the nuances of developing and implementing both types of algorithms.

5.6.1 Performance

At the DARPA site visit to Virginia Tech on May 5, 2005, both the reactive and the deliberative algorithms demonstrated their ability to navigate global waypoints, avoid obstacles, and stay within the course boundaries on an off-road obstacle course. This section discusses the differences in performance of the reactive Dynamic Expanding Zones and deliberative NUTS driving algorithms in the areas of waypoint following, driving smoothness/efficiency, obstacle avoidance, and repeatability/reliably.

Fig. 5.17. The lower Plantation Road test field with the RDDF course centerline superimposed

To quantify the ability of each strategy to navigate waypoints, GPS/inertial data were collected during test runs of both vehicles on an open rolling-hill terrain test course set up at the Plantation Road test facility on the Virginia Tech campus (Figure 5.17). These initial tests were run with identical route definition data files (RDDF) paths and with no obstacles to avoid.

Special care was taken to examine the driving algorithms under weather and terrain conditions that were as similar as possible. All test runs were collected on the same day in clear weather with alternating reactive and deliberative runs for measurements at the same speed/weather/lighting combination. All exteroceptive sensors except the high-precision Novatel GPS/INS were shut down on the vehicles. In essence, this test focused on the ability of the algorithm to follow a given path in the absence of obstacles. Both data sets were collected on the same vehicle, using the same sensors and peripheral software. Our goal was to isolate the behavioral differences in the decision-making software.

The RDDF length of this course was 3.173 miles (approximately 5 km) over open field terrain. Each driving algorithm was tested over five laps at maximum commanded speeds of 5, 10, and 15 mph. Position, velocity, actuator state, commanded vehicle state, and the values of many other parameters were collected at a rate of 5 Hz during the tests.

The data, summarized in Table 5.1, clearly show an overall performance edge for the reactive Dynamic Expanding Zones algorithm. Rocky running Dynamic Expanding Zones averaged significantly higher speeds than the same vehicle running the deliberative NUTS path planning. Although the vehicle was fully capable of driving at higher speeds, the rollover safety processes in both algorithms prevented the vehicle from selecting turn/speed combinations that might put it in jeopardy. At 5 and 10 mph, the Dynamic Expanding Zones algorithm averaged speeds near the top speed limit imposed on the software. At 15 mph, however, the serpentine nature of the course triggered safety slowdown procedures for sharp turns, and limited the overall speed.

By attempting to steer a course defined by a square grid, the deliberative method must make more frequent steering adjustments to follow the desired

Table 5.1. Overall performance statistics for non-obstacle avoidance test runs

		Test Run Overall Performance		
Top Set Speed		**5 mph**	**10 mph**	**15 mph**
Reactive (Dynamic Expanding Zones)	**Top Speed (mph)**	6.6	11.4	15.6
	Average Speed (mph)	4.8	7.2	7.9
	Total Time (s)	2365.2	1568.8	1443.6
Deliberative (Non-Uniform Terrain Search)	**Top Speed (mph)**	6.6	10.8	14.4
	Average Speed (mph)	3.5	4.0	4.30
	Total Time (s)	3203.6	2829.4	2594.4

path. The overall course performance highlights one main difference between the reactive and deliberative paradigms: decisiveness. The reactive Dynamic Expanding Zones algorithm is less sensitive to subtle changes in the perceived sensor state. It is important for the vehicle to be flexible and react quickly to a dynamic environment when selecting a desired path, but subtle errors in the vehicle position or orientation can cause the deliberative algorithm to reroute the path, which diminishes the overall performance. While this rerouting may be desirable for long-term planning, it does not seem to be desirable for simple waypoint following. As a result, the purely reactive paradigm had better overall performance on the test course.

Unnecessarily using the steering, brake, or throttle actuators consumes energy and may degrade the dynamic performance of the vehicle, or even result in a rollover. This unnecessary actuation may also cause wear on the involved components, such as the brake pads and steering rack. Unnecessary actuation also burdens the vehicle's power system and reduces the vehicles overall efficiency. The steering actuator consumes 373 W at peak power, and the electrohydraulic brakes consume 240 W at peak power. Brake actuation also takes significant kinetic energy from the vehicle, and dissipates it as waste heat at the brake pads. Hence, unnecessary steering and braking can be significant factors in reducing the efficiency of the vehicle.

While collecting the test run data, it appeared that the reactive algorithm was able to steer more smoothly and efficiently on the course. To measure the efficiency and smoothness of steering on the test runs, the percent of the time the vehicle commands a change in steering angle was examined. The percent of time spent changing steering angle is an indicator of the driving algorithm's smoothness and energy efficiency. If the vehicle constantly seeks a new steering position, it uses a large amount of energy to drive the steering motors. Continuous steering actuation also indicates more weaving of the vehicle.

Table 5.2 shows the percentage of time the steering actuator was running for each algorithm on the same course. The reactive algorithm is nearly an order of magnitude more efficient than the deliberative algorithm under the

Table 5.2. Steering and braking actuation percentage (no obstacle avoidance)

		Steering and Braking Actuation		
		5 mph	10 mph	15 mph
Dynamic Expanding Zones	Travel Distance (mi)	3.22	3.17	3.17
	% Time Turning	6.3%	9.4%	10.0%
	% Time Braking	3.6%	10.6%	13.6%
Non-Uniform Terrain Search	Travel Distance (mi)	3.24	3.21	3.21
	% Time Turning	54.4%	56.3%	57.2%
	% Time Braking	5.1%	10.6%	12.9%

same conditions. The likely cause of this wide disparity in steering efficiency can be traced to the means by which each algorithm generates its path. The reactive driving scheme determines the difference between its current heading and the heading to the next waypoint. As long as the vehicle is heading toward the waypoint, no steering adjustments need to be made. The deliberative approach uses a pure pursuit algorithm to drive the planned path as closely as possible.

In tests with lower maximum speeds, the reactive NUTS algorithm exhibited slightly more frequent braking than Dynamic Expanding Zones (Table 5.2). As the maximum speeds increase, the relationship reverses, but these differences are probably not significant. The Dynamic Expanding Zones algorithm is able to achieve higher average speeds on the course. If the vehicle is allowed to accelerate up to full speed in some sections, it will have to brake for turns. On the other hand, the NUTS algorithm has a high frequency of turning, which means the vehicle must first to slow to a safe speed to execute the maneuver. These effects seem to require roughly the same amount of braking effort.

The mission of the Grand Challenge requires the vehicles to navigate reliably throughout the 134 mile off-road course. To accomplish this reliably, extensive testing and tuning was performed prior to the event. It was found during testing that the reactive algorithm was more robust and less sensitive to small variations in sensor data than the deliberative algorithm. The graphs in Figure 5.18 are overlays of repeated runs on a practice course for the two algorithms. The reactive algorithm clearly produces more repeatable paths than the deliberative algorithm at all speeds.

5.6.2 Application

During development of both navigation strategies, the Virginia Tech Grand Challenge team made several notable observations that provide insight into the practical application of the reactive and deliberative driving schemes. The vehicle's

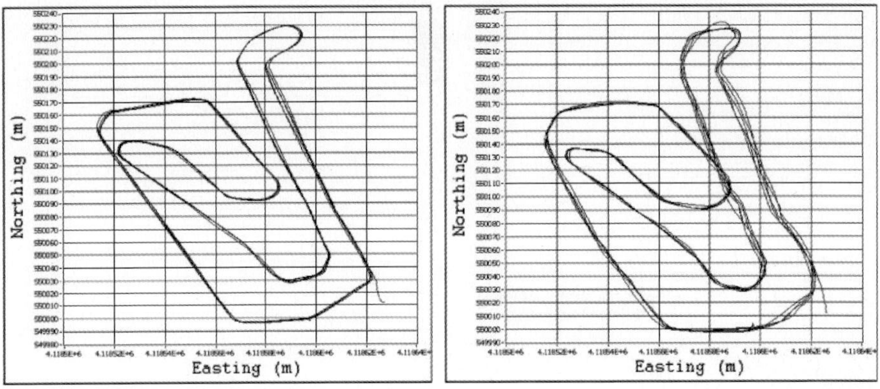

Fig. 5.18. Overlays of repeated runs for the reactive (left) and deliberative (right) algorithms

reactions to sensor stimuli and errors, such as high grass and error in GPS position, show key differences between the strategies. The measures taken to address these and other issues shed light on the characteristics of development and practical application of these strategies. Overall, the Virginia Tech Grand Challenge team found the reactive approach to be the most conducive to upgrades and gradual improvements based on field testing. The "general solution" approach of the deliberative scheme promises higher intelligence in navigation, but requires fundamental changes in navigation strategy to influence small changes in behavior.

5.6.2.1 GPS Error

A common experience for the Virginia Tech Grand Challenge GPS and inertial-based positioning systems is the "GPS pop" (Figure 5.19). This occurs when, after running on inertial-only positioning, the GPS/ INS regains the GPS signal. The perceived position of the vehicle instantaneously jumps from the INS-computed location to the GPS-based position. This "jump" has been measured as up to 2.5 m, in an arbitrary direction, based on the error in INS and GPS.

The reactive approach handles this type of sensor aberration without issue. There is no change in obstacle avoidance or waypoint following performance, except for a small heading adjustment based on the new perceived position with respect to the waypoint. This adjustment is inversely proportional to the vehicle's distance from the waypoint. Because obstacle avoidance is based entirely on the instantaneous measurement of obstacle position relative to the vehicle, no change is affected by noise or error in the positioning system.

The deliberative algorithm is significantly more susceptible to problems due to varying position error. Since the deliberative path planner attempts to drive a path based on data collected in a previous time period, it will use the most recent (corrected) position data to drive a path generated using an older, offset position frame. The now-offset path can potentially carry the vehicle through

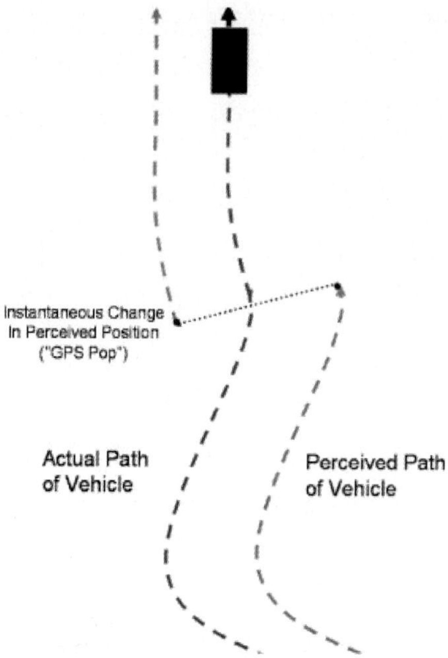

Fig. 5.19. When the GPS/INS reacquires satellites, the system corrects the INS-computed position

obstacles that were detected, but no longer perceived, in the correct location relative to the vehicle.

The general direction and shape of the "correct" path is still valid from the old computed position, but the data are no longer usable for obstacle avoidance. The long-term general planning capability of the NUTS approach could benefit from a reactive driver for more reliable close-in obstacle avoidance.

5.6.2.2 Navigation Range

The Dynamic Expanding Zones navigation approach has proven superior to the NUTS in robustness and reliability for obstacle avoidance in simple situations. The DARPA Grand Challenge is one such case where intelligent planning of complex maneuvers is not required. The reactive navigation strategy is superb at navigation through simple obstacles, such as passing cars and tunnel walls, but lacks the ability to plan through complex situations (Figure 5.20).

Because of the avoidance zone in front of the vehicle, a vehicle running Dynamic Expanding Zones might not be capable of maneuvering through close-quarters situations. In practice, Dynamic Expanding Zones showed a particular weakness in the offset-gate configuration of obstacles (Figure 5.21).

The offset-gate obstacle is traversable by the NUTS deliberative strategy, which is capable of planning a path through any area, regardless of the complexity of the

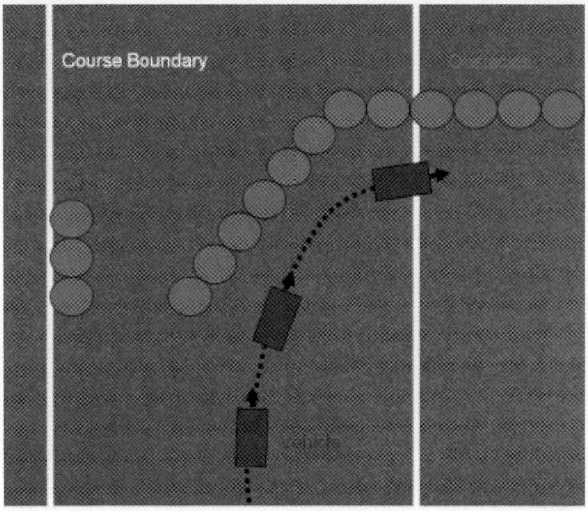

Fig. 5.20. Dynamic Expanding Zones does not take the optimum path in some situations

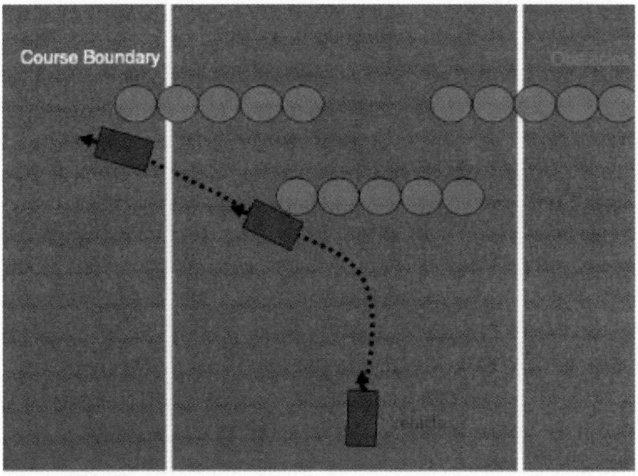

Fig. 5.21. Dynamic Expanding Zones reaction to the offset-gate obstacle

obstacle field (Figure 5.22). This long-range intelligence demonstrates the main attraction of the deliberative approach: The larger the area and complexity of data available for path planning, the greater the advantage to the deliberative approach. The drawback to using this strategy is that, as discussed in this paper, it lacks the adaptability and smooth obstacle avoiding performance of the reactive approach.

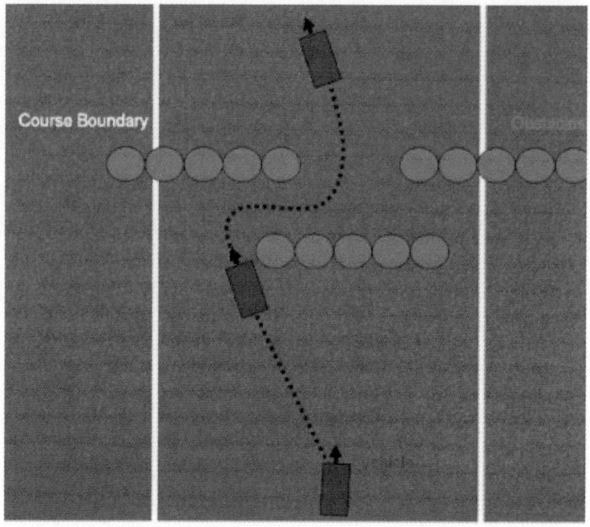

Fig. 5.22. NUTS reaction to the offset-gate obstacle

5.6.2.3 Road Following

Another area in which NUTS excels is road following. The data output from the Virginia Tech Grand Challenge road recognition algorithm to the NUTS is optimal for map-based path following. Rather than simply steer toward a point ahead of the vehicle suspected to be a road, NUTS attempts to find the shortest path onto the low-cost road terrain. Unlike Dynamic Expanding Zones, which ignores road following data in the presence of an obstacle, NUTS is capable of

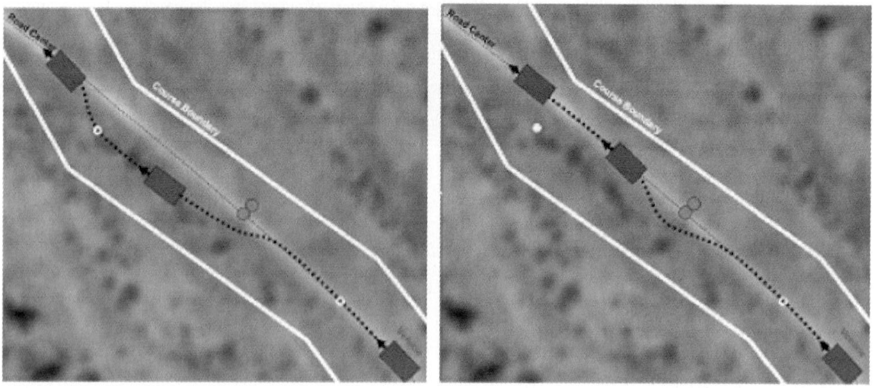

Fig. 5.23. In some situations, Dynamic Expanding Zones (left) can lose a road due to obstacle avoidance while NUTS (right) will maintain the optimal path (aerial photo from Google Local ®)

intelligently planning a path down a road, around an obstacle, and back onto the road again given sufficient sensor data (Figure 5.23).

5.7 Grand Challenge Performance

Neither Cliff nor Rocky were able to finish the 132 mile course. Rocky traveled just over 39 miles along the course, and Cliff traveled just over 44 miles. Both vehicles failed due to mechanical problems, rather than poor navigation decisions. Cliff's drive engine stalled when it briefly slowed to an idle, and Rocky's on-board generator shut down due to a suspected false low-oil reading. Had the base platforms not failed, the Virginia Tech Grand Challenge team is confident that the sensors and navigation systems would have allowed both vehicles to finish the race in just under the 10 h time limit. From the start, the Virginia Tech Grand Challenge strategy was to finish the race in the allotted amount of time and focus on solid navigation rather than higher-speed performance. In overall distance traveled, Cliff and Rocky finished eighth and ninth, respectively.

5.8 Conclusions

A team of dedicated undergraduate and graduate engineering students built the Virginia Tech GrandChallenge vehicles as an exercise in engineering design. Although the conversion to drive-by-wire, power system, motion control, and computing architecture are nearly identical on Cliff and Rocky, these platforms are designed to test and run two very different navigation strategies.

Cliff was designed to use the Dynamic Expanding Zones algorithm, a reactive navigation approach specifically created for the Grand Challenge. This algorithm consists of a number of variable size zones around the vehicle. In these zones, the presence or lack of obstacles dictates the behavior of the vehicle. The zones vary in size depending on the speed and surroundings of the robot in order to only take in the essential information to avoid obstacles. Rocky uses a deliberative navigation approach, making use of a terrain map and an A* algorithm to search through the map for the easiest route to travel. This allows Rocky to navigate more efficiently than Cliff through complicated terrain.

For the actual Grand Challenge Event and the National Qualifying Event, the reactive navigation software was used on both Cliff and Rocky. The deliberative software was not used, simply because the implementation was not mature enough to perform reliably in competition. One of the biggest lessons learned for implementing a deliberative navigation strategy was that it is essential to have a way to translate the optimal grid-based path to a path that is smoothly drivable. Another consideration is the size of terrain data. The deliberative algorithm planned a path from a terrain map of 12.5×12.5 m. This was too small an area to generate a useful plan.

In summary, the Virginia Tech case study describes and emphasizes some of the key design considerations for development of deliberative and reactive navigation.

The reactive strategy is simple and efficient, while the deliberative approach shows the potential to deliver higher navigational intelligence and planning.

References

Coulter, R.C. (1992). Implementation of the pure pursuit path tracking algorithm (Tech. Rep. CMU-RI-TR-92-01). Pittsburgh, PA: Carnegie Mellon University, Robotics Institute.

Gillespie, T.D. (1992). Fundamentals of vehicle dynamics. Warrendale, PA: SAE International.

Green, D., Yao, F., & Zhang, T. (1998). A linear algorithm for optimal context clustering with application to bilevel image coding. Paper presented the International Conference on Image Processing. Chicago, IL.

Hart, P.E., Nilsson, N.J., & Raphael, B. (1968). A formal basis for heuristic determination of minimum path cost. IEEE Transactions on Systems, Science, and Cybernetics 4, 100–107.

Lee, J.S. (1984). Edge detection by partitioning. In E.G. Wegman (Ed.), Statistical image processing (pp. 59–69). New York: Dekker.

Leedy, B.M. (2006). Two minds for one vehicle: A case study in deliberative and reactive navigation. Unpublished Masters thesis, Virginia Tech, Department of Mechanical Engineering, Blacksburg, VA.

Murphy, R. (2000). Introduction to AI robotics. Cambridge, MA: MIT Press.

Putney, J. (2006). Reactive navigation of an autonomous ground vehicle using dynamic expanding zones. Unpublished Masters thesis, Virginia Tech, Department of Mechanical Engineering, Blacksburg, VA.

Rasmussen, C. (2004). Grouping dominant orientations for Ill-structured road following. Paper presented at the IEEE Computer Society Conference on Computer Vision and Pattern Recognition (CVPR'04), San Diego, CA.

Reynolds, C.W. (1999). Steering behaviors for autonomous characters. Foster City, CA: Sony Computer Entertainment America.

Shin, D.H., & Singh, S. (1990). Path generation for robot vehicle using composite clothoid segments (Tech Rep. CMU-RI-TR-90-31). Pittsburgh, PA: Carnegie Mellon University, The Robotics Institute.

Intelligent Off-Road Navigation Algorithms and Strategies of Team Desert Buckeyes in the DARPA Grand Challenge '05

Qi Chen and Ümit Özgüner

Department of Electrical and Computer Engineering, The Ohio State University, Columbus, Ohio 43210, USA

Summary. This paper describes one aspect of our approach in developing an intelligent off-road autonomous vehicle, the Intelligent Off-road Navigator (ION), as team Desert Buckeyes from the Ohio State University for the DARPA Grand Challenge 2005. The real-time navigation is one of the critical components in an autonomous ground vehicle system. In this paper, we focus on the navigation module, whose main responsibility is to generate smooth and obstacle-free local paths with appropriate speed setpoints. For the smooth path generation, we introduce a polynomial interpolation method. To generate obstacle-free paths, a steering controller utilizing a fuzzy obstacle avoidance algorithm is presented. A speed fuzzy controller is introduced to generate the speed setpoints. These two fuzzy controllers collaborate with each other to guide our vehicle ION to the goal safely. The obstacle avoidance algorithm proposed in this paper was also tested in simulations and on small robots successfully. Other issues related to the navigation module are discussed in the paper as well, such as the vehicle's system structure and its finite state machine. As a result, ION achieved great performance in the National Qualification Event (NQE), covered about 30miles in the Nevada Desert with complete autonomous operations, and finished 10th in the Grand Challenge 2005.

Keywords: autonomous ground vehicle, Grand Challenge, real-time, navigation, finite state machine, obstacle avoidance, fuzzy controller.

6.1 Introduction

This paper describes our approach in developing the navigation module of the vehicle ION, the Intelligent Off-road Navigator, as team Desert Buckeyes from the Ohio State University for the DARPA Grand Challenge 2005 (GC05).

GC05 was a competition for off-road autonomous ground vehicles (AGVs). About three thousands waypoints (GPS coordinates) were provided shortly before the race, the AGVs were required to follow the waypoints one by one safely, continuously, smoothly and fast across natural terrain en route to the finish line without any human interference. The challenge was indeed "grand" because the AGVs had to respond to the dynamically changing environment in a timely way

M. Buehler, K. Iagnemma, and S. Singh (Eds.): DARPA'05, STAR 36, pp. 183–203, 2007.
springerlink.com

Fig. 6.1. Vehicle ION: the Intelligent Off-road Navigator. The vehicle is a 2005 Polaris Ranger 6x6. It is 120 inches long and 60 inches wide and its height is 78 inches. Drive by wire capability has been added to the vehicle so that computer control is possible for throttle, brake, steering control, and transmission gear switching.

and the data acquired was both complex and full of erroneous information. The total distance in GC05 was about 132miles.

Developing an autonomous vehicle to traverse the desert means solving a series of problems and developing a series of technologies, such as sensor fusion, navigation, artificial intelligence, vehicle control, signal processing, drive-by-wire technology, reliable software and mapping (Toth et al., 2006), etc. ION was developed in partnership with University of Karlsruhe, which developed the image processing system (Hummel et al., 2006). ION is a 6 wheeled vehicle with full drive-by-wire capability. A set of sensors (LIDARs, radars, cameras and ultrasonic transducers) and a GPS and IMU, the internal measurement unit, provide extensive sensing capability, shown in Figure 6.1. A sophisticated sensor fusion system (Redmill, Martin, & Özgüner, 2006) was developed and used with a complex intelligent analysis, decision and control configuration (ION, 2005). The overall design was quite similar to the one we developed for TerraMax in GC04 (Chen, Özgüner, & Redmill, 2004).

In this paper, we focus on the navigation module whose task is to guide ION toward the goal without colliding with obstacles. Our strategy is to develop a finite state machine that decides the state of ION so that the proper drive mode is selected in different situations. By perceiving both nearby environment and ION's status, the navigation module generates commands for the low-level controller. The command can be, for example, a set of GPS route points called *pathpoints* and the *speed setpoint*, or a sequence of motion specifications called *robotic unit operations*, or some *direct commands* such as "stop-and-wait". The finite state machine design and the obstacle avoidance algorithm utilizing fuzzy logic are introduced.

As the Desert Buckeyes was the team that developed the sensing and intelligence for the 2004 TerraMax, a number of aspects of ION were descendants of technology and approaches used in the 2004 Grand Challenge (Chen et al., 2004), (Yu, Chen, & Özgüner, 2004), (Özgüner, Redmill & Broggi, 2004), (Chen & Özgüner, 2005). We also have developed autonomous vehicles for highway driving, which have shown great performance in the structured environments. (Özgüner, Hatipoglu, & Redmill, 1997), (Redmill & Özgüner, 1998), (Hatipoglu, Özgüner, & Redmill, 2003). As far as off-road is concerned, the environment being unstructured, the navigation module and the obstacle avoidance algorithm proposed in this paper is desired to deal with more complicated situations.

The remainder of this paper is organized as follows. Section 6.2 briefly describes our system structure and general bases. The design of the finite state machine is described in Section 6.3. Section 6.4 introduces navigation module with the path planning algorithm that generates collision-free paths and appropriate speed set-points for the vehicle. ION's performances are presented in Section 6.5. Section 6.6 concludes the paper.

6.2 System Structure

The system structure of ION is illustrated in Figure 6.2. Environment sensor fusion, vehicle ego-state sensor fusion, high-level controller and low-level controller are ION's four major modules. ION is equipped with several cameras, LIDARs, ultrasonic Sonars and a radar. A sophisticated sensor fusion system generates a local grid map that moves with the vehicle. The local map contains the sizes and positions of nearby obstacles. ION's ego states are fused from a GPS and an IMU. The navigation module, also called the high-level controller, obtains information from the sensor fusion modules and generates a series of pathpoints and a speed set point for the low level controller. In some situations, the navigation module sends direct control command to the low level controller, such as "Stop" and specific "robotic unit operations". Obtaining the commands from the high-level controller, the low-level controller enables ION to follow the given route defined by a series of pathpoints at the given speed, as has been done in (Redmill, Kitajima, & Özgüner, 2001).

6.3 Finite State Machine

Since ION was expected to encounter very complicated situations, we developed a finite state machine (FSM) in ION's navigation module to deal with difference situations. Based on the information collected from the sensor fusion modules and the inner state monitoring, the navigator module determines the right machine state for ION to activate the corresponding subsystem/algorithm so that the proper command is generated.

The FSM we implemented in ION is represented conceptually in Figure 6.3 . There are two major states, the "*Path-Point Keeping*" state and the "*Obstacle*

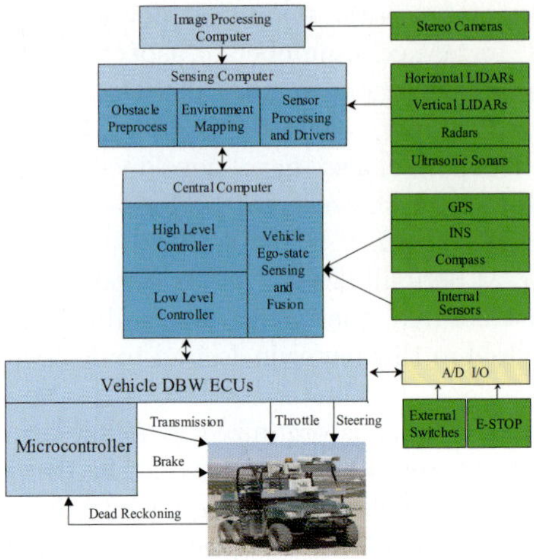

Fig. 6.2. The System Structure of ION

Avoidance" state. ION stays in these two states at the most of racing time, when there are no special events, such as tunnel, sensor failure, and narrow path, etc, detected. When ION is in one of these two states, the pathpoints are generated by the path planning algorithms as described in the following section. The smooth path generation algorithm is for the "*Path-Point Keeping*" state, while the obstacle avoidance algorithm is used in the "*Obstacle Avoidance*" state.

In the FSM, the other states are also very essential for ION to deal with special events that the vehicle may encounter. The "*Tunnel Mode*" state is designed specifically for the situation when ION is in a tunnel where GPS signals are lost temporarily. We assume ION encounters no obstacles in the tunnels and the FSM switches only to the "*Path-Point Keeping*" state from this state. In the "*Tunnel Mode*" state, the distances to both sides of the vehicle are measured by the ultrasonic transducers mounted on ION so as to keep the vehicle in the center of the tunnel. The FSM gets into the "*Robotic Operations*" state when ION needs to adjust its heading or position or both, for example, when narrow paths and very sharp turns, i.e. direct forward path would contact obstacles, are encountered. In this state, a sequence of back-and-forth operations are executed to adjust the ION's status. When there are sensor failures, the FSM transits into the "*Alarm*" state to deal with the malfunction. In this state, the navigation module executes the sensor resetting command to recover the failed sensor or sensors. In the case that the sensor failure is unrecoverable, a flag is received from the sensor fusion module and the FSM switches back to the "*Path-Point Keeping*" state. Carrying these flags, the navigation module reduces the maximum allowed speed to reduce the risk of collision, and the sensor fusion module adjust the coefficients correspondingly as well. The "*Road Following*" state is designed for the situation when the image processing module detects a road. For the situation when ION

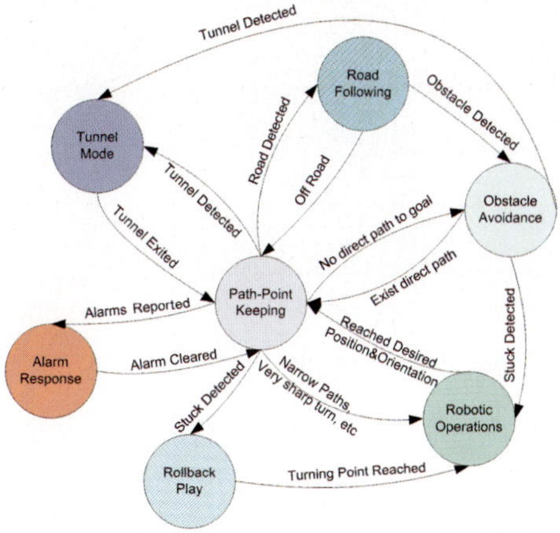

Fig. 6.3. ION's finite state machine of the navigation module

is stuck on the road, the FSM switches to the "*Rollback*" state. In this state, ION drives back along the trajectory it records till the point where another path can be initiated.

We also introduced a watch-dog to prevent the vehicle from being stuck forever. The vehicle might be stuck because the robotic operations couldn't adjust vehicle status amidst obstacles in six tries, or false obstacles block the path totally, or because of some other unexpected events. Anyway, the watch-dog was not designed to deal with normal situations. When ION's position does not change or the FSM rests in a state other than the "*Path-Point Keeping*" or "*Obstacle Avoidance*" state for a certain period of time, the FSM automatically resets to the "*Path-Point Keeping*" state and stays in this state regardless of the obstacles and other events until ION reaches a certain distance. During the FSM resetting process, a high throttle value is sent to the lower level controller so that ION can possibly get out of holes or run over some obstacles like small bushes. With the watch-dog design, ION might be able to get out of stuck in many cases and gain chances to continue.

6.4 Navigation Module

ION's navigation module, also called the high level controller, plans or replans the local path when the machine state is at the "*Path-Point Keeping*" or "*Obstacle Avoidance*" state. The navigation module checks the status of the local path, which consists of 10 pathpoints, at 10Hz. If the path comes across obstacles in the local map, or ION has reached the 5th point of the local path, or ION is indicated passing a waypoint, the navigation module then generates a new local path. Figure 6.4 shows the procedure to generate the local path. There are two path generation algorithms implemented in the navigation module: the smooth path generation algorithm and the obstacle avoidance algorithm.

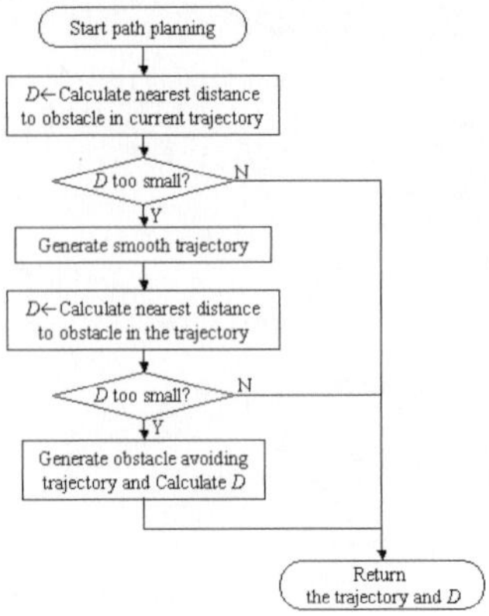

Fig. 6.4. The flow chart for path planning

6.4.1 Smooth Path Generation

A smooth path, whose curvature is continuous, is generated to connect the waypoints. Let $\{P_i,\ i = 0, 1, 2, \cdots\}$ denote the coordinates of the waypoints, where $P_i := (x_i,\ y_i)^T \in \mathbf{R}^2$. To generate a path, $\{P_i(s),\ 0 \le s \le 1\} \subset \mathbf{R}^2$, connecting P_i and P_{i+1}, a polynomial interpolation is applied:

$$P_i(s) = A_0(s)P_{i-1} + A_1(s)P_i + A_2(s)P_{i+1} + A_3(s)P_{i+2}, \quad 0 \le s \le 1 \tag{6.1}$$

where $A_i(\cdot)$'s are scalar polynomial coefficient functions. According to the theory of planar curves (Hsiung, 1998), for a certain $s \in [0,\ 1]$, the vector tangent to the curve and associated at the point $P_i(s)$ is defined by

$$\begin{aligned} T_i(s) =& \lim_{h \to 0} \frac{1}{h}(P_i(s+h) - P_i(s)) \\ =& A_0'(s)P_{i-1} + A_1'(s)P_i + A_2'(s)P_{i+1} + A_3'(s)P_{i+2} \end{aligned} \tag{6.2}$$

where $A_i'(\cdot)$ is the derivative of function $A_i(\cdot)$. Define the unit tangent vector as $\mathbf{u}_i(s) := T_i(s)/\|T_i(s)\|$. $A_i(\cdot)$'s are selected by satisfying

$$\begin{cases} P_i(0) = P_i \\ P_i(1) = P_{i+1} \\ \mathbf{u}_{i-1}(1) = \mathbf{u}_i(0) \end{cases} \tag{6.3}$$

According to (6.1), $\{P_i(s),\ 0 \le s \le 1\}$ is a planar curve determined by four waypoints $(P_{i-1},\ P_i,\ P_{i+1},\ P_{i+2})$. For curve $\{P_{i-1}(s),\ 0 \le s \le 1\}$, it is generated by $(P_{i-2},\ P_{i-1},\ P_i,\ P_{i+1})$ and connects P_{i-1} and P_i. These two curves have one common point, P_i. By satisfying $\mathbf{u}_{i-1}(1) = \mathbf{u}_i(0)$ in (6.3), the unit tangent vector

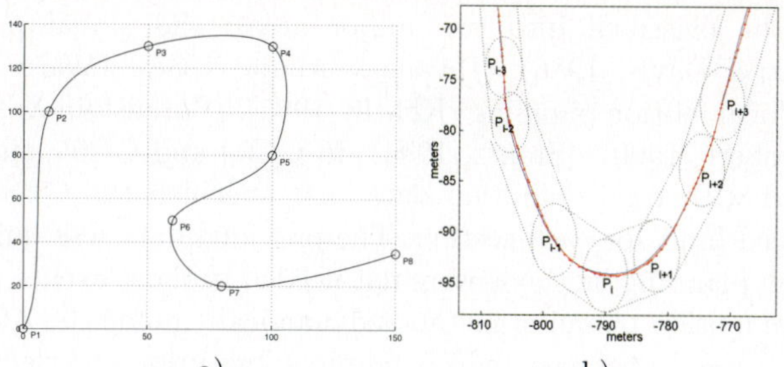

Fig. 6.5. Smooth path generation examples: a) A smooth path generation example: the curve is generated by equation (6.1) with the coefficient polynomials defined in equation (6.4); b) An example of smooth path generation: The solid line with dots (red) is the the planned path that ION generated in the NQE, GC05. The path smoothly connects the waypoints, which are located at the center of each circle. The solid curve (blue) is the trajectory that ION traveled by following the planned path. These two curves almost overlap, which shows ION followed the planned path very well.

of the curve $\{P_{i-1}(s)\}$ is the same as that of the curve $\{P_i(s)\}$ at the point P_i. We say these two curves are smoothly connected.

In ION, we select $A_i(\cdot)$'s in (6.1) as follows:

$$\begin{cases} A_0(s) = (-s + 2s^2 - s^3)/2 \\ A_1(s) = (2 - 5s^2 + 3s^3)/2 \\ A_2(s) = (s + 4s^2 - 3s^3)/2 \\ A_3(s) = (-s^2 + s^3)/2 \end{cases} \tag{6.4}$$

Applying the coefficient polynomials (6.4) to (6.1), the result satisfies the constraints in (6.3) so that the path generated for ION is smooth. For example, given eight points, $\{P_1, \cdots, P_8\}$, arbitrarily in a map, the above method generates a smooth path connecting them. The result is shown in Figure 6.5(a). In ION, we use cubic polynomials for $A_i(\cdot)$'s. The selection of the order of the polynomials is somewhat arbitrary and a third order polynomial satisfies the given constraints. With higher order polynomials that satisfy (6.3), we can obtain similar results. Figure 6.5(b) shows part of the trajectory that ION generated in the NQE and it followed the smooth path very well.

6.4.2 Obstacle Avoidance

In the case that the smooth path comes across some obstacles, the FSM stays in the "*Obstacle Avoidance*" state and a collision-free path is generated by the obstacle avoidance algorithm.

Since the late 1970s, extensive effort has been exerted to develop obstacle avoidance algorithms, see (Latombe, 1991) and references therein. The research can be classified into two major areas: the global path planning (Kambhampati & Davis, 1986), (Hwang & Ahuja, 1988), (Warren, 1989) and the real-time local motion planning (Khatib, 1985), (Borenstein & Koren, 1989), (Adams & Probert, 1990), (Stentz, 1994). In GC04 and GC05, the race routes were described with road definition data that specified the GPS coordinates, width and speed limit for each section. The provided data was quite dense and the global path planning methods were not needed in those events. On the other hand, the local motion planning methods dynamically guide the AGV according to the locally sensed obstacles, which requires less prior knowledge about the environment. The fuzzy controller described in this paper is a real-time local motion planning method, which is more suitable and practical for ION in the Grand Challenge events, since ION senses only the nearby obstacles.

Fuzzy logic navigation methods have been studied and implemented in small robots, rovers and AGVs with small turning radius, see (Saffiotti, 1997), (Hodge & Trabia, 1999), (Seraji, 2000), (Lee, Lai, & Wu, 2005) and references therein. As noted by Saffiotti (Saffiotti, 1997), fuzzy logic has the feature to make it a useful tool to cope with the large amount of uncertainty that is inherent in natural environments. Most of these approaches select the direction by weighting the effort of *target-approach* and the need of *obstacle-avoidance*. However, few fuzzy logic approaches concern the dynamic and kinematic constraints of a big AGV like ION. For ION, if the two consecutive steering decisions are totally opposite, such as left turn and right turn, the steering decisions would neutralize with each other. Therefore, frequent jumps between two consecutive steering decisions should be avoided. Also, since the turning radius of ION is considerably large, it is preferable to respond to obstacles before being too close to them.

In the remainder of this subsection, we introduce two heterogeneous fuzzy controllers for ION's obstacle avoidance path planning system. The steering controller emphasizes goal reaching and plays an important role in obstacle avoidance simultaneously. Fuzzy rules are proposed for ION to approach the goal and avoid obstacles at the same time. The speed controller prevents collisions with obstacles. The rules in each fuzzy controller are defined for reacting in different situations and each rule represents a certain behavior character based on human drivers' knowledge. These two fuzzy controllers collaborate to direct ION to the goal without collision with obstacles.

6.4.2.1 Steering Fuzzy Controller Design

Figure 6.6 shows one example scenario that explains the basic steering rules. Based on the local sensor ability, only the boundary of the obstacle in the local sensor region is detectable. The shape and size of the whole obstacle outside of the sensing region are not known. Practically, we expand the boundary of the obstacle by half the size of the vehicle so that we can consider the vehicle as a mass point. The expanded obstacle boundary is called the "real boundary", henceforth. Furthermore, to ensure the safety of ION, the obstacles is further

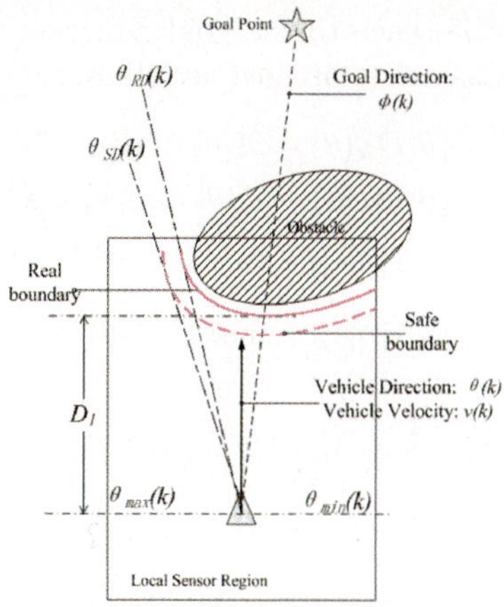

a. Local navigation scenario

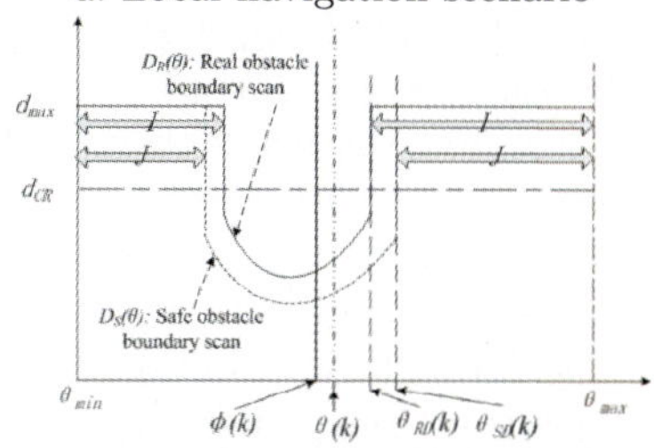

b. The scanning distance to the obstacle's boundaries at time k

Fig. 6.6. An example scenario. In the figures, $[\theta_{min},\ \theta_{max}]$ is the scanning angle range of interest; θ_{RD} and θ_{SD} are the decision angle based on the real boundary and safe boundary, respectively. D_1 is the distance to the obstacle in the current heading direction. d_{CR} is the criterion distance. d_{max} is the maximum scanning distance.

enlarged to create a "safe boundary", so that the real boundaries are enveloped with the safe boundaries. By the two-step obstacle extension, each obstacle has two different boundaries, the "real boundary" and the "safe boundary". Figure 6.6 illustrates the boundaries and the scanning distances.

The obstacle avoidance algorithm to be presented in this subsection originates from a heuristic non-fuzzy algorithm which selects the direction based on the distance to obstacles and the difference to the goal direction. The original algorithm works well when there are few obstacles in the environment and the obstacles are all convex-shaped. However, it has deficiencies in some special situations, such as candidate set being empty, or selecting wide left or right instead of finding a path in the front, or being stuck at some obstacle compositions, or the vehicle swinging toward an obstacle.

In order to overcome these shortcomings, four fuzzy rules are developed one by one and together they work very well. These fuzzy rules are defined to reshape the scanning distances curve and then the steering strategy is to select the direction with the smallest angle difference to the goal direction and enough distance to the obstacles. The formulas are expressed as follows:

$$\begin{aligned} \mathcal{I}(k) &= \{\theta | D_R(\theta) - \Delta(\theta) > d_{CR,R}(k)\} \\ \mathcal{J}(k) &= \{\theta | D_S(\theta) - \Delta(\theta) > d_{CR,S}(k)\} \\ d_{CR,R}(k) &= \min\{d^0_{CR,R}, \max_\theta\{D_R(\theta) - \Delta(\theta)\} - \epsilon\} \\ d_{CR,S}(k) &= \min\{d^0_{CR,S}, \max_\theta\{D_S(\theta) - \Delta(\theta)\} - \epsilon\} \end{aligned} \tag{6.5}$$

and

$$\begin{aligned} \theta_{RD}(k) &= \arg\min_{\theta\in\mathcal{I}}\{|\theta - \phi(k)|\} \\ \theta_{SD}(k) &= \arg\min_{\theta\in\mathcal{J}}\{|\theta - \phi(k)|\} \\ \theta_D(k) &= Select(\theta_{RD}(k), \theta_{SD}(k), \phi(k)) \end{aligned} \tag{6.6}$$

where $\mathcal{I}$ and $\mathcal{J}$ are the candidate sets of angles at which the distance to the obstacle is longer than the criterion distance, $d_{CR,R}(k)$ and $d_{CR,S}(k)$, respectively. $\Delta(\cdot)$ reshapes the obstacle scanning distances curves, $D_R(\cdot)$ and $D_S(\cdot)$, so that different directions have different priorities to be selected from. The rule that decides $\Delta(\cdot)$, called the the "Focus" rule, is described later. $d^0_{CR,R}$ and $d^0_{CR,S}$ are design parameters to prevent the algorithm from being sensitive to obstacles very far away, known as the far-sighted situation, where there may be many false obstacles. ϵ is a small positive constant so that the candidate sets are both nonempty.

Equations (6.5) and (6.6) give the method to find two directions, $\theta_{RD}(k)$ and $\theta_{SD}(k)$, as decisions based on the real boundary and the safe boundary scanning distances, respectively. Both decisions attempt to reach the goal point and keep the obstacle away for certain distances, $d_{CR,R}(\theta, k)$ and $d_{CR,S}(\theta, k)$, respectively. When ION is far away from the obstacles, $d_{CR,R}(k) = d^0_{CR,R}$. Otherwise, when ION is close to the obstacles, the steering controller chooses the direction with the longest distance to the obstacles. In other words, the direction toward open space is selected.

A. The Selection Rule

Let $\theta_D(k)$ be the final decision direction at time k. It is selected from either $\theta_{SD}(k)$ or $\theta_{RD}(k)$ by calculating $|\theta_{SD}(k) - \theta_{RD}(k)|$ and $|\theta_{SD}(k) - \phi(k)|$, where $\phi(k)$ is the goal direction. Thus, the steering strategy reaches the goal and avoids obstacles at the same time. Table 6.1 shows the selection rule noted by "*Select*" in (6.6).

The "*Small*", "*Medium*" and "*Large*" in Table 6.1 are fuzzy descriptions. The defuzzification function picks either $\theta_{SD}(k)$ or $\theta_{RD}(k)$, as the result of $\theta_D(k)$. This ensures that the result would not be some value between $\theta_{SD}(k)$ and $\theta_{RD}(k)$.

Remark 1. *If $\theta_{SD}(k)$ is chosen to be the steering decision, then there is an open space in the direction so that vehicle is allowed to pick up a relatively high*

Table 6.1. The selection rule

		$\lvert\theta_{SD}(k) - \theta_{RD}(k)\rvert$		
	$\theta_D(k) =$	Small	Medium	Large
$\lvert\theta_{SD}(k) - \phi(k)\rvert$	Small	$\theta_{SD}(k)$	$\theta_{SD}(k)$	$\theta_{RD}(k)$
	Medium	$\theta_{SD}(k)$	$\theta_{RD}(k)$	$\theta_{RD}(k)$
	Large	$\theta_{SD}(k)$	$\theta_{RD}(k)$	$\theta_{RD}(k)$

speed safely. On the other hand, if $\theta_{RD}(k)$ is chosen, it can be concluded that $|\theta_{SD}(k) - \theta_{RD}(k)|$ is large, i.e. the $\theta_{SD}(k)$, presenting the open free space, is in the other direction, and the chosen direction must be a narrow path. The speed controller uses this information and sets a relatively low speed setpoint.

B. The Focus Rule

In order to prevent ION from being distracted by side directions and going off the road, we introduce the "Focus" rules in the steering controller. In the focus rules, we consider the directions pointing forward and leading to the goal point to have higher priorities than others.

In (6.5), $\Delta(\cdot)$ is introduced to reshape the scanning distances curves, $D_R(\cdot)$ and $D_S(\cdot)$. When a direction θ has low priority, the $\Delta(\theta)$ is then set to a large value so that the direction θ is less likely to be in the candidate sets, $\mathcal{I}$ and $\mathcal{J}$. By doing so, the steering controller tends to find the path direction with high priority.

Let $\Delta 1(\theta)$ and $\Delta 2(\theta)$ represent the $|\theta - \theta(k)|$ and $|\theta - \phi(k)|$, respectively, where $\theta(k)$ is the vehicle heading at time k. So, a direction with high priority has a small values of $\Delta 1$ and $\Delta 2$. Let (S, MS, M, ML, L) represent (*Small, Medium Small, Medium, Medium Large* and *Large*), respectively. The fuzzy "Focus" rules that work on the $\Delta(\theta)$ in the equation (6.5) are as follows:

- If ($\Delta 1$ is S) and ($\Delta 2$ is S), then (Δ is S): if θ points forward and leads to the goal direction, then θ has the highest priority to be chosen, i.e. $D_R(\theta)$ or $D_S(\theta)$ has the *smallest* deduction.
- If (($\Delta 1$ is S) and ($\Delta 2$ is M)) or (($\Delta 1$ is M) and ($\Delta 2$ is S)) or (($\Delta 1$ is M) and ($\Delta 2$ is M)), then (Δ is MS).
- If (($\Delta 1$ is S) and ($\Delta 2$ is L)) or (($\Delta 1$ is L) and ($\Delta 2$ is S)), then (Δ is M).
- If (($\Delta 1$ is L) and ($\Delta 2$ is M)) or (($\Delta 1$ is M) and ($\Delta 2$ is L)), then (Δ is ML).
- If ($\Delta 1$ is L) and ($\Delta 2$ is L), then (Δ is L): if the angle θ neither points forward nor leads to the goal direction, it has low priority and the deduction is *large*.

Figure 6.7 is an example of how the "Focus" rules work. The $-\Delta(\theta)$ forms a Λ-shaped curve when $\theta(k)$ is close to $\phi(k)$. Note that the curve can be an M-shaped curve when $\theta(k)$ is separated away from $\phi(k)$. For those θs that $\Delta(\theta) > D_R(\theta)$, we have $\theta \notin \mathcal{I} \cup \mathcal{J}$ according to the focus rules. Therefore, these directions are

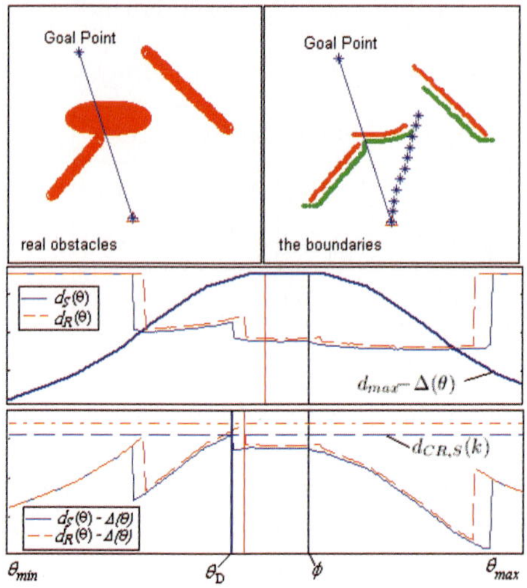

Fig. 6.7. The example of the focus rules: The upper left figure is the "real world" and the upper right figure displays the obstacle boundaries, both real and safe boundaries. The scanning distances to the boundaries are shown in the middle figure. The $d_{max} - \Delta(\theta)$ is a Λ-shaped curve shown in the figure. The adjusted real distance and safe distance curves are plotted in the bottom figure. By selecting the direction θ_D as the steering decision, the steering controller generates a curve of path points shown in the upper right figure, following which the AGV avoids the obstacles and approaches the goal point.

excluded from being selected as the decision direction. By doing so, the steering controller opens a *priority window* for those directions that point forward and lead to the goal point. Thus, the steering controller ignores the distractions from low priority directions. A direction with low priority is selected only when the directions with higher priority are all blocked by obstacles.

In this fuzzy method, triangle membership functions are used to define the fuzzy sets for the $\Delta 1$ and $\Delta 2$, as shown in Figure 6.8. The values assigned to (S, MS, M, ML, L) determine the height of the hump. The larger the values are, the higher the hump is. The values of (a, b, c, d) determine the membership function, consequently determine the width of the hump in the curve of $-\Delta(\theta)$. if the values are small, $-\Delta(\theta)$ is not reduced only in a narrow range of directions. Thus, the "Focus" effort is enhanced. On the other hand, if the values are large, then the "Focus" effort is neutralized and the navigator has a wide range of direction to select from.

C. The Focus vs. Search Rules

The focus rules prevent the distraction by side directions quite well. Nevertheless, the focus rules trade the ability in searching the feasible directions for the

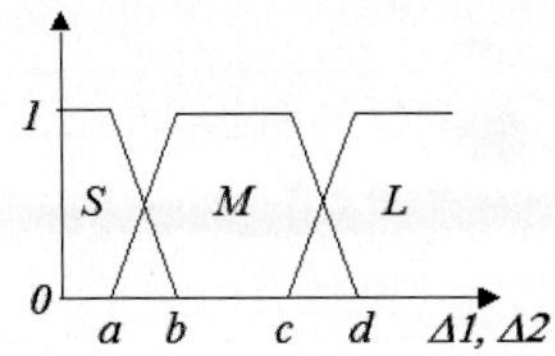

Fig. 6.8. The membership functions for $\Delta 1$ and $\Delta 2$

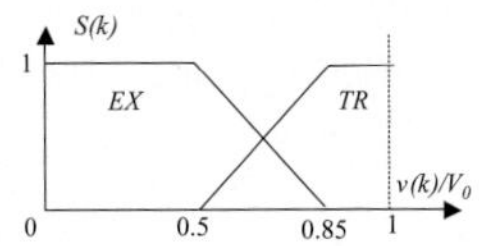

Fig. 6.9. The membership functions for $\Delta 1$ and $\Delta 2$

distraction prevention. It works well when there are few obstacles in the environment. When there are many obstacles in the environment, however, the focus rules impair the search capability of the navigator. Furthermore, as stated previously, the values of (a, b, c, d) in the membership function of the fuzzy sets of $\Delta 1$ and $\Delta 2$ are important parameters of the focus rules. The smaller values of (a, b, c, d) correspond to stronger "Focus" and vice versa.

Motivated by this, the "Focus" rules are improved by considering the fuzzy evaluation of the AGV status. The $v(k)$, the vehicle's speed, is regarded as the index of vehicle's situation: the lower the speed is, the more search capability is required. Therefore, the equations are:

$$\begin{aligned} S(k) &= F_S(v(k), V_0) \\ [a(k)\ b(k)\ c(k)\ d(k)]^T &= F(S(k)) \end{aligned} \tag{6.7}$$

where $S(k)$ is the situation classification, a membership function value of the fuzzy sets: *exploring* and *travelling*.

The membership function is shown in Figure 6.9. The rule $F_S(\cdot)$ states: When the $v(k)/V_0$ is small, $S(k)$ is "*exploring*" (EX). On the other hand, when $v(k)/V_0$ is large, $S(k)$ is "*travelling*" (TR). The AGV speed, $v(k)$, is classified by comparing rather to the base speed than to the absolute values. The base speed, V_0, is provided to the AGV as the current speed limit. In the "Focus vs. Search" rules, the values of (a, b, c, d) are no longer fixed. The fuzzy rules in (6.7) are as follows:

- If $S(k)$ is EX, then the values of (a, b, c, d) increases: When the AGV is nearing obstacles, we assume that the speed controller slows down the AGV. In this case, the $S(k)$ becomes more "EX", and the focus rules are neutralized. In other words, the "Search" capability is enhanced.
- If $S(k)$ is TR, then the values of (a, b, c, d) reduces: vice versa, the focus rules are enhanced in this case.

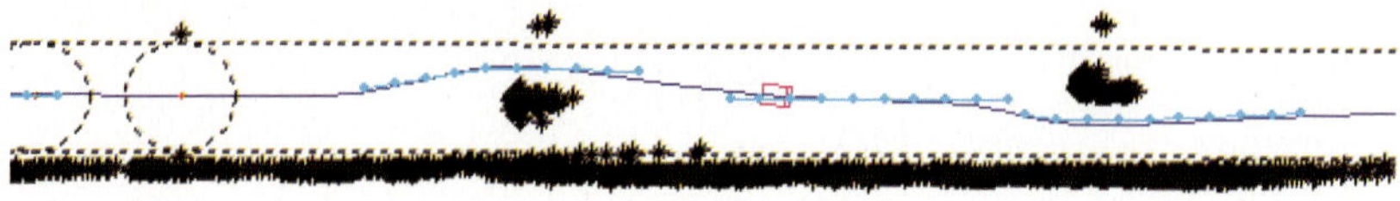

Fig. 6.10. The example of ION's path planning in the two-car-passing section at NQE, GC05. The black stars are the obstacles recovered from ION's sensor logs. Two cars were placed in a corridor of 15ft half-width, as shown in the figure. The distance between the two cars was about 100ft. ION passed the two cars, from left to right in the figure, at the speed of 10 mph, the highest speed permitted in the section.

By applying the rules in (6.7), the AGV is much less likely to be stuck in front of the concave-shaped obstacles. When the AGV approaches the obstacles, the speed is reduced so that the "priority window" is opened up for search and the "focus" effect is neutralized. The "Search" rules help to prevent the AGV from being stuck.

D. The Persistence Rules

Because of the existence of false obstacles, some obstacles may popup or disappear suddenly in the local sensor map. This brings a problem that the *discontinuous* outputs from the steering controller causes the oscillation of the vehicle's heading. In the worst situation, the two consecutive steering decisions may neutralize each other.

To solve this problem, the value of $\theta_D(k-1)$ is introduced to adjust $\Delta(\theta)$ in the focus fuzzy rules. By doing so, the steering controller adjusts the searching window and raise the priority of directions near to the former decision, $\theta_D(k-1)$. Therefore, the steering controller maintains the persistence. The "Persistence" rules weigh on the last decision and give the controller a characteristic of persistence so that the zigzag performance of the AGV is prevented.

When the decision direction is selected, a sequence of pathpoints is generated for the low-level controller. Figure 6.10 shows the path planning result of ION in the two-car-passing section in the NQE, GC05. The dotted lines are the section boundary, the solid line is the trajectory that ION runs and the dark stars are the detected obstacles. Three sections of the 10-dot-curve are the planned path, which avoids the obstacles and stays within the section boundaries. Most planned paths are not shown in order to make the figure clear. ION's trajectory and planned paths almost overlap, which shows that our low level controller follows the planned path very well.

6.4.2.2 Speed Fuzzy Controller Design

The speed controller's main aim is to avoid the collision into the obstacles. Moreover, due to the physical constraint of the vehicle, sharp turning at high

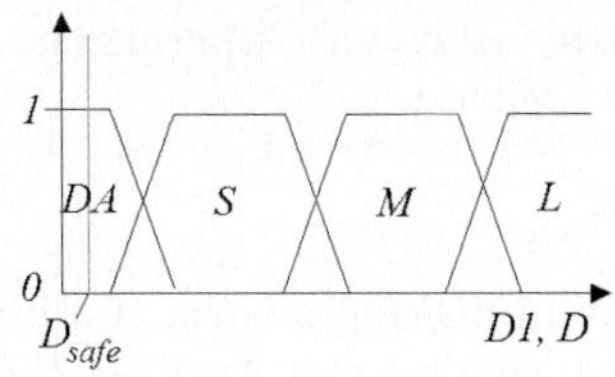

Fig. 6.11. The membership functions for $D1$ and D

speed should be avoided to prevent the AGV from rolling over. The velocity fuzzy controller in ION consists of two sets of rules: the *Anti-collision rules* and the *Safe-steering rules*, shown in the following:

$$\begin{aligned} V_s(k) &= \min[V_0, A(D1, D), T(\Delta\theta)] \\ \Delta\theta &= |\theta_D(k) - \theta(k)| \end{aligned} \tag{6.8}$$

Where $D1$ represents the distance to obstacles in the front of vehicle and D is the value returned from path-planning procedure as in Figure 6.4. $\Delta\theta$ is the angle difference between the decision angle and current heading. V_0 is the speed limit allowed for ION. The final speed set point value is $V_s(k)$.

A. The Anti-collision Rules

The main purpose of the collision rules is to prevent the collision. Let A represent the anti-collision rules. The fuzzy rules, A, have two inputs, $D1$ and D. Let (S, MS, M, ML, L) have the same meaning as stated above, and let (DA) denote the fuzzy set of dangerous distance. The rules are stated as follows:

- If (one of $(D1, D)$ is DA), then A is $STOP$: the vehicle should stop when the distance to the obstacles is too close and it is dangerous to move on.
- If ($D1$ is S), then A is S: the vehicle speed should be reduced when it is close to the obstacles.
- If ($D1$ is M and D is S), then A is MS.
- If ($D1$ is M and D is M), then A is M.
- If (one of $D1$ $(D1, D)$ is L) and (the other is M), then A is ML.
- If (both of $(D1, D)$ are L), then A is L: no obstacle is nearby, the speed is set to a high value.

The boundaries of the fuzzy set is selected according to the dynamic performance of the vehicle and how much risk you want to take. The membership functions used in our system are shown in Figure 6.11. The membership functions for $D1$ and D are the same in our system, although they could be different in the boundaries of the fuzzy sets. Note that the D_{safe}, the minimum safe distance, is marked in the figure, which is totally within the fuzzy set DA. Therefore, the AGV stops before reaching the distance D_{safe} so that collisions are prevented. Should the vehicle be stopped by the rules, for example it is dangerously close to an obstacle, the machine state would switch to the "Rollback" state and wait to be recovered by a sequence of robotic operations. In that case, the vehicle's heading is adjusted for the next try.

B. The Safe-Steering Rules

The other issue is the safety in making a turn. If a high speed set point and the sharp-steering command are sent to the AGV at the same time, it may cause the AGV to roll over. More importantly, it has been observed in our experiments that the low level controller would have some overshoot in path following during sharp turns, thus may cause the collisions with obstacles even if the planned paths are obstacle-free.

To prevent the danger and reduce the overshoot in path following, the safe-steering rules, denoted as T, are introduced. We adopt the value of $|\theta_D(k)-\theta(k)|$, say $\Delta\theta$, to represent the steering command. The safe-steering rule T says: If ($\Delta\theta$ is S) then T is L; If ($\Delta\theta$ is M) then T is M; If ($\Delta\theta$ is L) then S is L. By doing so, the speed set point of the AGV is reduced when the sharp steering command occurs.

6.5 Performance of the Navigation Module

A simulator has been built to test the navigation algorithm designed in this paper. In the simulator environment, the obstacles are already expanded, and the vehicle is regarded as a point mass. The vehicle model in this simulator is commonly known as the Dubins' car model with a minimum turning radius. The kinematic equations are written as:

$$\begin{aligned} \dot{x} &= u\cos\theta \\ \dot{y} &= u\sin\theta \\ \dot{u} &= a \\ \dot{\theta} &= \omega \end{aligned} \tag{6.9}$$

where x and y are the position coordinates. θ is the yaw angle. u stands for the linear velocity, which is assumed to be positive. a is the acceleration. The ω is the angular velocity, a control variable. The model is subject to the constraint:

$$\left|\frac{\omega}{u}\right| \leq \frac{1}{R} \tag{6.10}$$

so that a minimum turning radius R is imposed. In this simulator, we can randomly generate obstacles over a terrain and arbitrarily select a series of check points. Figure 6.12 shows the AGV trajectory over such a terrain. The trajectory of the AGV, the smooth (blue) line in the figure, indicates that the AGV followed the check points, avoided all obstacles and reached the final goal smoothly and safely in the end. The navigation module has been tested with different obstacle densities over 100 times and failed in less than five cases and two of which didn't have feasible path since the obstacles are generated randomly. In the other three

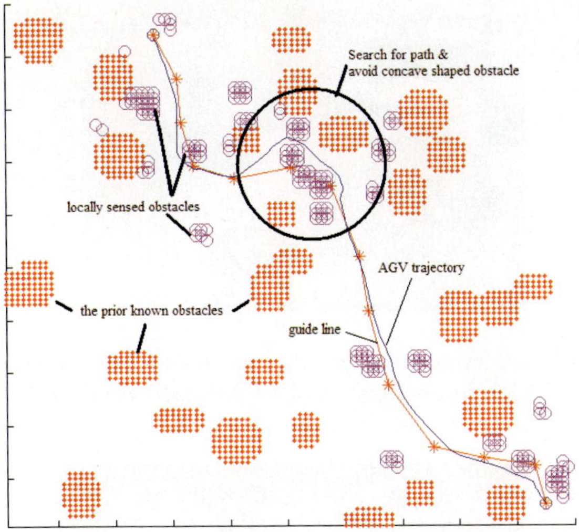

Fig. 6.12. The AGV trajectory over the obstacle-occupied terrain. The AGV is shrunk to a point. The "locally sensed obstacles" are detected by the vehicle and plotted on the map only when they are in the vehicle's sensing range. As a consequence, only those previously unknown obstacles that are near the vehicle trajectory are displayed in the result map as the "locally sensed obstacles".

failed cases, the AGV was stuck because the check points misled the AGV into a wrong fork, which shows that the obstacle avoidance algorithm in this paper is not sufficient in finding a path through a very complicated environment. Nevertheless, the algorithm is sufficient for the GC events.

Figure 6.13 shows the test on small robots. In the indoor robot test, we put an obstacle just behind a gate formed by two obstacles. In this case, even if the obstacles were close to each other, the navigation algorithm was still able to find a path and navigate the robot along the corridor without touching any obstacles. We tested the navigation module for over 20 times on small robots with different obstacles positions and configurations. The small robot failed twice to get to the goal because of the extreme closeness of certain obstacles.

The performance of the navigator designed in this paper was demonstrated by ION in the DARPA GC05 event. ION completely traversed the NQE[1] course successfully four times[2], which fully exhibited the obstacle-avoidance and goal-approaching capabilities of the navigation module and our AGV system. Figure 6.14 shows ION successful passing a car in the NQE.

[1] The National Qualification Event (NQE) in the 2005 GC was the 2.7 mile test track DARPA used as the semifinals. The track provided a series of obstacles and the AGVs are required to go through 51 "gates". It also had a 100ft metal "tunnel" where GPS was lost.

[2] ION tried the NQE course five times in total and only failed once due to a mechanical problem.

Fig. 6.13. Test obstacle avoidance algorithm on small robots. We used an all-terrain-robot-vehicle (ATRV) for outdoor tests and a PIONEER P3-AT robot for indoor tests.

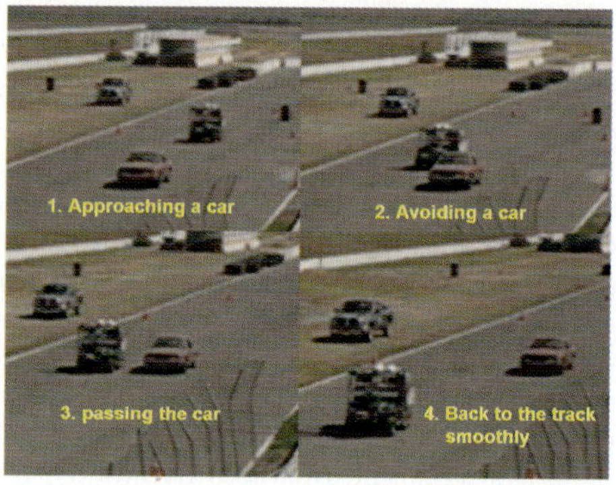

Fig. 6.14. An example: ION's passing a car in NQE, GC05

In GC05, the race route was described by a road definition file which consists of 2935 waypoints in total. Each waypoint was specified by GPS coordinates, the width and the maximum allowed speed of section. The total route distance was about 132 mile. Running with complete autonomous operations, ION covered about 30miles in the Nevada Desert and was terminated at the 576th waypoint from the start. The maximum speed ION reached in some sections was about 25 mph. During the race, ION experienced twice LIDAR failures and both were recovered in two minutes. Without the recovery design, ION could not have reached that far. Also, ION got into the *"robotic operation"* state twice. In the first set of robotic operations, ION tried five times back-and-forth operations before it overcame an obstacle. In the second set of robotic operations, ION was terminated. However, neither DARPA report nor ION's race log indicated that ION had had any collisions or gone off the road in the GC05 race. Also, ION was still totally driveable and stayed in the middle of the road when it was terminated. We guess it was because of the slowness in ION's gear shifting[3]. Since

[3] We were not provided with official reasons of the termination.

it took up to one minute for ION to shift its gear position in some situations, which might be intolerably slow, the second set of robotic operations might present the illusion that ION came to a halt and thus caused the termination. With the watchdog design in the FSM described in Section 6.3, given more time, we expect ION could have reached further in GC05 and even had the potential to finish the whole course.

6.6 Conclusion

ION, the working off-road AGV from team Desert Buckeyes, is a successful integration of a series of technologies, such as sensor fusion, navigation, vehicle control, signal processing, drive-by-wire technology, reliable software and mapping, etc. According to ION's performance in GC05, we realize that the robustness design is of great importance since some modules may not work properly in some situations while being off-road. We have invested extensive effort to improve the robustness of ION, for example, introducing the "Robotic Operation" and "Alarm" states in the FSM design, realizing the recovery design of sensor failures, and implementing the watchdog design for the FSM. Without these, ION could not have reached that far. Nevertheless, ION didn't finish the race because of some unexpected events. The second lesson we learn is that practicing in desert is also important for the race. We have had a lot of off-road practice yet we tested ION in desert for less than one week. Although this one-week testing improves the ION's desert performance greatly, it is still not enough. Practicing in the similar environment to the race can reduce the uncertainties and unexpected events.

This paper briefly introduces the system structure of ION and focuses on ION's navigation module and the obstacle avoidance algorithm utilizing fuzzy logic. The steering fuzzy controller, derived from a basic steering strategy, utilizes several fuzzy rules to deal with complicated situations so that the AGV can reach the goal point and avoid obstacles. The steering controller deals with the goal approaching and obstacle avoidance at the same time, which helps the AGV approach the goal smoothly. The speed controller utilizes fuzzy rules to prevent the AGV from colliding with obstacles and enhance the vehicle path following performance. ION's performance in GC05, both in NQE and the race, justifies the design of the real-time navigation module described in this paper.

Acknowledgement

A large group of researchers and students participated in ION's development. We acknowledge Dr. Keith Redmill, John Martin, Dr. Charles Toth, Matthew Knollman, George McWilliams and Rong Xu for their insightful discussions, valuable suggestions and contributions. We thank the Ohio State University for all their support and assistance. We also acknowledge the contribution of our partners, the team from University of Karlsruhe, which developed the image processing system.

References

Adams, M., & Probert, P. (1990, Jul.). Towards a Real-Tme Navigation Strategy for a Mobile Robot. Paper presented at the IEEE International Workshop on Intelligent Robots and Systems (IROS '90), Tsuchiura, Japan.

Borenstein, J., & Koren, Y. (1989). Real-time Obstacle Avoidance for Fast Mobile Robots. IEEE Transactions on Systems, Man and Cybernetics, 19(5), 1179-1187.

Chen, Q., Özgüner, Ü., & Redmill, K. (2004). The Ohio State University Team at the Grand Challenge 2004: Developing A Completely Autonomous Vehicle. IEEE Intelligent Systems, 19(5), 8-11.

Chen, Q., & Özgüner, Ü. (2005, Jun.). Real-time navigation for autonomous vehicles: a fuzzy obstacle avoidance and goal approach algorithm. Paper presented at the American Control Conference (ACC '05), Portland, OR.

Hatipoglu, C., Özgüner, Ü., & Redmill, K. (2003). Automated lane change controller design. IEEE Transactions on Intelligent Transportation Systems, 4(1), 13-22.

Hodge, N., & Trabia, M. (1999, May). Steering Fuzzy Logic Controller for an Autonomous Vehicle. Paper presented at the IEEE International Conference on Robotics and Automation, Detroit, MI.

Hsiung, C. (1998). First Course in Direrential Geometry. Springer, New York (USA), 1998.

Hummel, B., Kammel, S., Dang, T., Duchow, C., & Stiller, C. (2006, Jun.). Vision-based Path Planning in Unstructured Environments. Paper presented at the IEEE Intelligent Vehicle Symposium (IV '06), Tokyo, Japan.

Hwang, Y., & Ahuja, N. (1988, Apr.). Path planning using a potential field representation. Paper presented at the IEEE International Conference on Robotics and Automation, Philadelphia, PA.

Team Desert Buckeyes. (2005). Team Desert Buckeyes Technical Paper. http://www.darpa.mil/grandchallenge/TechPapers/DesertBuckeyes.pdf.

Khatib, O. (1985, Mar.). Real-Time Obstacle Avoidance for Manipulators and Mobile Robots. Paper presented at the IEEE International Conference on Robotics and Automation, St. Louis, MO.

Kambhampati, S., & Davis, L. (1986). Multiresolution path planning for mobile robots. IEEE Journal on Robotics and Automation, 2(3), 135-145.

Latombe, J.-C. (1991). Robot motion planning. Boston MA: Kluwer Academic Publishers.

Lee, T., Lai, L., & Wu, C. (2005, May). A fuzzy algorithm for navigation of mobile robots in unknown environments. Paper presented at the IEEE International Symposium on Circuits and Systems 2005 (ISCAS '05), Kobe, Japan.

Özgüner, Ü., Hatipoglu, C., & Redmill, K. (1997, Nov.). Autonomy In A Restricted World. Paper presented at the IEEE Conference on Intelligent Transportation System (ITSC '97), Boston, MA.

Özgüner, Ü., Redmill, K., & Broggi, A. (2004, Jun.). Team TerraMax and the DARPA Grand Challenge: A General Overview. Paper presented at the IEEE Intelligent Vehicles Symposium (IVS '04), Parma, Italy.

Redmill, K., Kitajima, K., & Özgüner, Ü. (2001, Aug.). DGPS/INS integrated positioning for control of automated vehicle. Paper presented at the IEEE Conference on Intelligent Transportation System (ITSC '01), Oakland, CA.

Redmill, K., & Özgüner, Ü. (1998, Feb.). The Ohio State University Automated Highway System Demonstration Vehicle. Paper presented at the 1998SAE International Congress and Exposition (SAE Paper 980855), Detroit, MI.

Redmill, K., Martin, J., & Özgüner, Ü. (2006, Jun.). Sensing and Sensor Fusion for the 2005 Desert Buckeyes DARPA Grand Challenge Offroad Autonomous Vehicle. Paper presented at the IEEE Intelligent Vehicle Symposium (IV '06), Tokyo, Japan.

Saffiotti, A. (1997). The uses of fuzzy logic in autonomous robot navigation. Journal of Soft Computing, 1(4), 180-197.

Seraji, H. (2000). Fuzzy Traversability Index: A new concept for terrain-based navigation. Journal of Robotic Systems, 17(2), 75-91.

Stentz, A. (1994, May). Optimal and Efficient Path Planning for Partially-Known Environments. Paper presented at the IEEE International Conference on Robotics and Automation (ICRA'94).

Toth, C., Paska, E., Chen, Q., Zhu, Y., Redmill, K., & Özgüner, Ü. (2006) Mapping Support for the OSU DARPA Grand Challenge Vehicle. Paper submitted to the IEEE Conference on Intelligent Transportation System (ITSC '06), Toronto, Canada.

Warren, C. (1989, May). Global Path Planning Using Artificial Potential Fields. Paper presented at the IEEE International Conference on Robotics and Automation, Scottsdale, AZ.

Yu, H., Chen, Q., & Özgüner, Ü. (2004, Jul.). Control System Architecture for TerraMax - The off-road intelligent navigator. Paper presented at the 5th IFAC/EURON Symposium on Intelligent Autonomous Vehicles, Lisboa, Portugal.

7

The Golem Group / UCLA Autonomous Ground Vehicle in the DARPA Grand Challenge

Richard Mason, Jim Radford, Deepak Kumar, Robert Walters, Brian Fulkerson, Eagle Jones, David Caldwell, Jason Meltzer, Yaniv Alon, Amnon Shashua, Hiroshi Hattori, Emilio Frazzoli, and Stefano Soatto

The Golem Group, 911 Lincoln Boulevard #7, Santa Monica, California 90403

Summary. This paper presents the Golem Group/UCLA entry to the 2005 DARPA Grand Challenge competition. We describe the main design principles behind the development of Golem 2, the race vehicle. The subsystems devoted to obstacle detection, avoidance, and state estimation are discussed in more detail. An overview of the vehicle performance in the field is provided, including successes together with an analysis of the reasons leading to failures.

7.1 Overview

The Golem Group is an independent team of engineers formed to build a vehicle for the 2004 DARPA Grand Challenge (DGC). For the 2005 DARPA Grand Challenge, the Golem Group and UCLA's Henry Samueli School of Engineering and Applied Science joined forces to build a second autonomous vehicle, Golem 2. Performance highlights of this vehicle are summarized in Table 7.1.

The novel aspect of the DGC was to require high-speed autonomous driving in the unstructured or semi-structured environment typical of rough desert trails. GPS waypoint-following was necessary, but not sufficient, to traverse the route, which might be partially obstructed by various obstacles. In order to have a good chance of completing the course, vehicles needed to drive much faster, yet have a lower failure rate, than previously achieved in an off-road environment without predictable cues. High-speed driving over rough ground posed a vibration problem for sensors.

7.1.1 Relation to Previous Work

Prior to the DGC, unmanned ground vehicles had driven at very high speeds in structured, paved environments [Dickmanns, 1997, 2004]. Other vehicles had operated autonomously in unstructured off-road environments, but generally not at very high speed. Autonomous Humvees using ladar to detect obstacles in an off-road environment have been developed at NIST [Coombs et al., 2000, Hong et al., 2000]. The U.S. Army eXperimental Unmanned Vehicle [Bornstein and Shoemaker, 2003] also used ladar to detect obstacles and could navigate unstructured rough ground at somewhat over 6 km/hr. Rasmussen

M. Buehler, K. Iagnemma, and S. Singh (Eds.): DARPA'05, STAR 36, pp. 205–243, 2007.
springerlink.com

Table 7.1. Golem Group / UCLA performance in the 2005 DARPA Grand Challenge

NQE Performance Highlights	
9 min 32 sec NQE run clearing 49/50 gates and 4/4 obstacles	Only Stanford, CMU, and IVST made faster runs clearing all obstacles Only Stanford and CMU made faster runs clearing at least 49 gates
12 min 19 sec NQE run clearing 50/50 gates and 5/5 obstacles	Only Stanford, CMU, Princeton, and Cornell made faster flawless runs
GCE Performance Highlights	
Peak speed (controlled driving) 47 mph Completed 22 race miles in 59 min 28 sec Anecdotally said to be fastest vehicle reaching 16-mile DARPA checkpoint Crash after 22 miles due to memory management failure	

Fig. 7.1. Golem 2 driving autonomously at 30 miles per hour

[2002] used a combination of ladar and vision to sense obstacles and paths in off-road environments. The Carnegie Mellon University Robotics Institute had perhaps the most successful and best-documented effort in the first DGC, building an autonomous Humvee which was guided by ladar [Urmson, 2005, Urmson et al., 2004].

Fig. 7.2. Golem 1 negotiating a gated crossing in the 2004 DARPA Grand Challenge. (Photos courtesy of DARPA).

Golem 1, our own first DGC entry, used a single laser scanner for obstacle avoidance. Golem 1 traveled 5.1 miles in the 2004 Challenge, before stopping on a steep slope because of an excessively conservative safety limit on the throttle control. This was the fourth-greatest distance traveled in the 2004 DGC, a good performance considering Golem 1's small total budget of $35,000.

Our attempts to improve on the performance of previous researchers were centered on simplified, streamlined design—initially in order to conserve costs, but also to enable faster driving by avoiding computational bottlenecks. For example, we relied primarily on ladar for obstacle avoidance, but unlike the majority of ladar users, we did not attempt to build a 3D model of the world *per se*. Instead we only attempted to detect the most important terrain features and track the locations of those on a 2D map. Golem 1, with its single ladar scanner, may have gone too far in the direction of simplicity, and Golem 2 carried multiple ladars to better distinguish slopes and hillsides. Ladar obstacle detection is further discussed in Section 7.3.

As another example of simplification, our path planning process considers possible trajectories of the truck as simple smooth curves in the 2D plane, with curvature of the trajectory as a measure of drivability and distance to the curve as a proxy for danger of collision. We do not consider obstacles in a configuration space of three dimensions, much less six dimensions. Our approach might be inadequate for navigating a wholly general, maze-like enviroment, but more importantly for our purposes, it gives fast results in the semi-structured case of a partially-obstructed dirt road. Trajectory planning is discussed further in Section 7.4.

We did experiment with some vision systems in addition to ladar. Mobileye Vision Technologies Ltd. provided Golem 2 with a monocular roadfinding system, which is discussed in Section 7.6.1 and also in Alon et al. [2006]. This could be considered as extending the cue-based paved-environment work of Dickmanns, and of Mobileye, to an unpaved environment. Experiments with a Toshiba stereo vision system are described in Section 7.6.2.

7.2 Vehicle Design

Each of our vehicles was a commercially-available pickup truck, fitted with electrically-actuated steering and throttle, and pneumatically-actuated brakes.

We felt it was very important that the robot remained fully functional as a human-drivable vehicle. Golem 1 seats two people while Golem 2 seats up to five. A passenger operated the computer and was responsible for testing, while the driver was responsible for keeping the vehicle under control and staying aware of the environment. During testing, it was very convenient to fluidly transition back and forth between autonomous and human control. Individual actuators (brake, accelerator, steering) can be enabled and disabled independently, allowing isolation of a specific problem. Having a human "safety driver" increased the range of testing scenarios we were willing to consider. Finally, having a street-legal vehicle greatly simplified logistics.

Accordingly, a central principle behind actuation was to leave the original controls intact to as great an extent as possible, in order to keep the vehicle street legal. The steering servo had a clutch which was engaged by a pushrod that could be reached from the driver's seat. The brakes were actuated by a pneumatic cylinder that pulled on a cable attached, through the firewall, to the back of the brake pedal. The cable was flexible enough to allow the driver to apply the brakes at any time. The pressure in the cylinder could be continuously controlled via a voltage controlled regulator. A servo was attached directly to the steering column.

A second design aim was to keep the computational architecture as simple as possible. The core tasks of autonomous driving do not require a large amount of computational power. We worked to keep the software running on a single laptop computer. Unburdened by a rack full of computers, we were able to retain working space in the vehicle, but more importantly, any team member could plug in their laptop with a USB cable and run the vehicle. A block diagram of the architecture is shown in Figure 7.3.

7.2.1 Sensors

The sensors mounted on Golem 2 for vehicle state estimation included a Novatel ProPak LBPlus differential GPS receiver with nominal 14-cm accuracy using

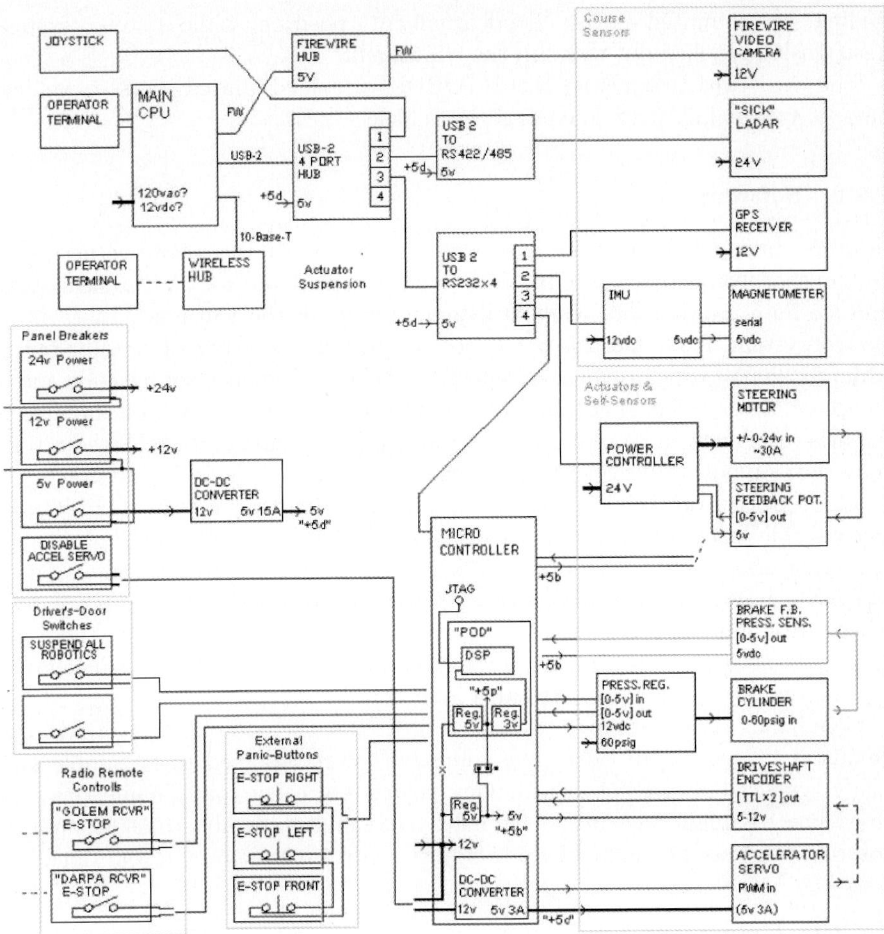

Fig. 7.3. Block diagram of the Golem vehicle architecture

OmniStar HP correction; a BEI C-MIGITS inertial measurement unit (IMU); a custom Hall encoder on the differential for odometry with approximately 10-cm accuracy; and a 12-bit encoder for measuring the steering angle. Vehicle state estimation is discussed further in Section 7.5.

The sensors used for terrain perception included a Sick LMS-221 ladar, which swept a 180-degree arc in front of the vehicle measuring ranges up to 80 meters at 361 samples/sweep, 37.5 sweeps/second, with the samples interleaved 0.5 degrees apart. There were also four Sick LMS-291 ladars, similar in most respects except that they each swept a 90-degree arc while collecting 181 samples/sweep, 75 sweeps/second. The arrangement and function of the ladars is further discussed in Section 7.3. A monocular camera system from Mobileye, used for road finding, was mounted at the "hood ornament" position, while a stereo system from Toshiba looked out through the windshield.

The On-Board Diagnostics Bus II (OBDII) provided limited access to vehicle data; it was mainly used to control the vehicle's horn.

7.2.2 Software

In order for multiple developers to work in parallel, it was plain we needed a system that would give us the freedom to write code in diverse places like deserts and stadium parking lots while still maintaining all the features of a revision control system. We found this in the peer-to-peer revision control system, `darcs` [Roundy, 2005], which maintains repositories containing both source code and a complete history of changes, and allows a developer to push or pull individual patches from one source repository to another. Using `darcs`, we achieved a very tight development-test cycle despite our diverse working environments.

The software ran on a Linux laptop and was divided into two main applications. The program which made decisions based on sensor input, controlled the actuators, and recorded events is known as `golem`. It had no connection to the visualization software `dashboard` other than through the log files it created, which were human-readable plain text. Besides being written to disk, these log files could be piped directly from `golem` to `dashboard` for realtime visualization or replayed offline by `dashboard` at a later time.

Commands could be typed directly to `golem` at the console or received from `dashboard` over a UDP port. The commands were designed to be simple and easily typed while driving in a moving vehicle on a dirt road. While this was convenient, it might have been even more convenient to be able to use an analog control or at least the arrow keys to adjust control parameters in real time.

7.2.2.1 Golem

The main software for data capture, planning, and control, `golem` consisted of two threads. The main thread was completely reactionary and expected to be realtime; it took in data from the sensors, processed it immediately, and sent commands to the actuators. The low level drivers, the state estimators, ladar and obstacle filters, the road/path follower, the velocity profiler, and the controllers all ran in this thread. A second planning thread was allowed to take more time to come up with globally sensible paths for the vehicle, based on snapshots of the accumulated sensor data received at the time the thread was initiated.

`Golem` could be driven by realtime sensor data, by a simple simulator, or from previously recorded log data. The simulator was invaluable for debugging the high level behaviors of the planner, but its models were not accurate enough to tune the low level controllers. The replay mode allowed us to debug the ladar obstacle filters and the state estimators in a repeatable way without having to drive the vehicle over and over.

7.2.2.2 Dashboard

`Dashboard` was created from scratch using C, Gtk+2, and OpenGL, to be an extensible visualization tool which would allow us to play back our log files and learn from their contents. A typical screenshot is shown in Figure 7.4.

The core of `dashboard` consists of a log file parsing library, an OpenGL representation of the world, and a replaying mechanism which controls the flow of time. `Dashboard` takes as input a log file from disk, standard input, or a UDP network connection, and uses rules described in a configuration file to convert human-readable log lines into an internal representation that is easy for other program components to index and access. Log lines are parsed as soon as they are available, by the Perl compatible regular expression (PCRE) library. The internal representation is displayed for the user through the OpenGL window. The user has the ability to play log files, pause them, rewind, fast-forward, measure and move around in the rendered world, and selectively view sensor visualizations they are interested in. This proved indispensable in our development effort.

The data is visualized in a number of different ways, as we found that no one representation was suitable for all debugging situations. The main section is a graphical representation of the world which is viewable from a 2d top-down point of view and also from a 3d point of view. It is rendered using OpenGL. There is also a section that displays inputs as textual data for precise readings. Yet another section contains a set of user programmable graphs. Because of the intentionally generic internal representation, adding a new element to the visualization, such as data from a new sensor, is a very simple process.

7.3 Ladar Obstacle Detection

We considered a non-traversable obstacle to be an object or ground feature which (a) represented a rapid apparent change in ground elevation, with slope magnitude greater than 30 degrees, or an actual discontinuity in ground surface; (b) presented a total change in ground height too large for the vehicle to simply roll over; (c) if discontinuous with the ground, was not high enough for the vehicle to pass under. This was an adequate definition of "obstacle" for the DARPA Grand Challenge.[1] Since obstacles result from changes of ground elevation, the most critical information comes from comparisons of surface measurements adjacent in space.

We did not use sensor data integrated over time to build a map of absolute ground elevation in the world frame, on the hypothesis that this is unnecessarily one step removed from the real information of interest. Instead, we tried to directly perceive, or infer from instantaneous sensor data, regions of rapid change

[1] Ideally one would like to classify some objects, such as plants, as "soft obstacles" that can be driven over even if they appear to have steep sides. Other hazards, such as water or marshy ground, cannot be classified simply as changes in ground elevation. But this increased level of sophistication was not necessary to complete the DGC and remains a topic of future work for us.

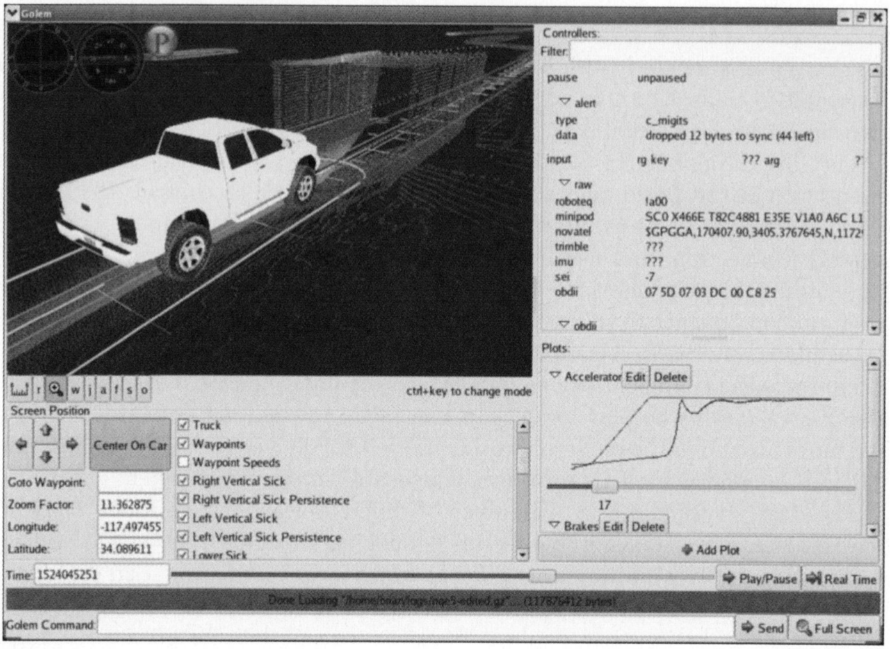

Fig. 7.4. `Dashboard` visualization of Golem 2 entering a tunnel during an NQE run

in ground elevation in the body-fixed frame, and then maintain a map of those regions in the world frame. We did not concern ourselves with any ground slope or surface roughness that did not rise to the threshold of making the ground non-traversable.

7.3.1 Ladar Geometry

We used Sick LMS-291 and LMS-221 laser scanners as our primary means of obstacle detection. It is interesting that the many DGC teams using two-dimensional ladars such as these found a wide variety of ways of arranging them. These ladars sweep a rangefinding laser beam through a sector of a plane, while the plane itself can be rotated or translated, either by vehicle motion or by actuating the ladar mount. The choice of plane, or more generally the choice of scan pattern for any beam-based sensor, represents a choice between scanning rapidly in the azimuthal direction, with slower or sparser sampling at different elevation angles, or the reverse.

A ladar scanning in the vertical plane has the significant advantage that the traversability of the scanned terrain is apparent from a single laser sweep. For example, it is easily determined from the single vertical-plane scan in Figure 7.5 that the ground is continuous, flat, and traversable from a few feet before the truck up to the point where there is a non-traversable vertical surface taller than the vehicle's wheels.

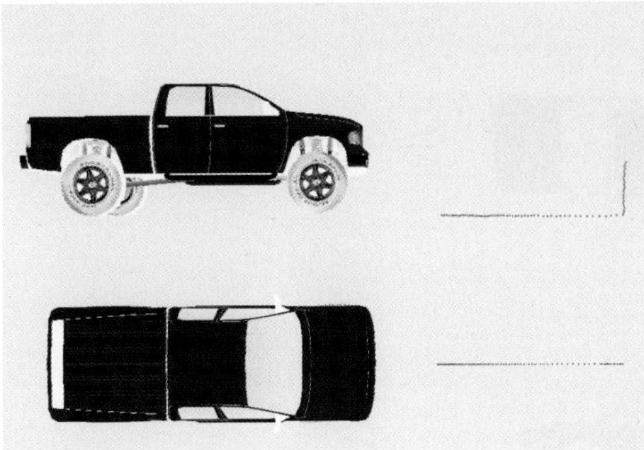

Fig. 7.5. Returns from a ladar oriented in a vertical plane

In our case, a single laser sweep takes 1/75 second and individual sample points are separated by 1/13575 second. On this time scale it is not likely that motion or vibration of the vehicle could distort the vertical scan sufficiently to alter this interpretation (make the traversable ground appear non-traversable or vice versa). Similarly, while small errors in the estimated pitch of the vehicle would induce small errors in the estimated location of the non-traversable obstacle, it is not likely that they could cause a major error or prevent the perception of an obstacle in the right approximate location. Against these advantages is the obvious drawback that a vertical slice only measures the terrain in a single narrow direction. Even if there are multiple vertical scanners and/or the vertical plane can be turned in different directions, the vehicle is likely to have a blinkered view with sparse azimuthal coverage of the terrain, and may miss narrow obstacles.

Conversely, a scan plane which is horizontal or nearly horizontal will provide good azimuthal coverage, and clearly show narrow obstacles such as fenceposts and pedestrians, but the interpretation of any single horizontal scan in isolation is problematic. Lacking measurements at adjacent elevation angles, one cannot determine if a return from a single horizontal scan is from a non-traversable steep surface or from a traversable gently sloping one. Therefore the information from multiple horizontal scans must be combined, but since the crucial comparisons are now between individual measurements taken at least 1/75 seconds apart instead of 1/13575 seconds apart, there is a greater likelihood that imperfectly estimated motion or vibration will distort the data. Small errors in pitch, roll, or altitude could cause the vehicle to misapprehend the height of a ground contour and lead to totally erroneous classification of terrain as traversable or non-traversable.

Our approach was to use a complementary arrangement of both vertical and nearly-horizontal ladars. On the Golem 2 vehicle, there are two nearly-horizontal ladars and three vertically-oriented ladars mounted on the front bumper. The idea is that the vertically-oriented ladars are used to form a profile of the general

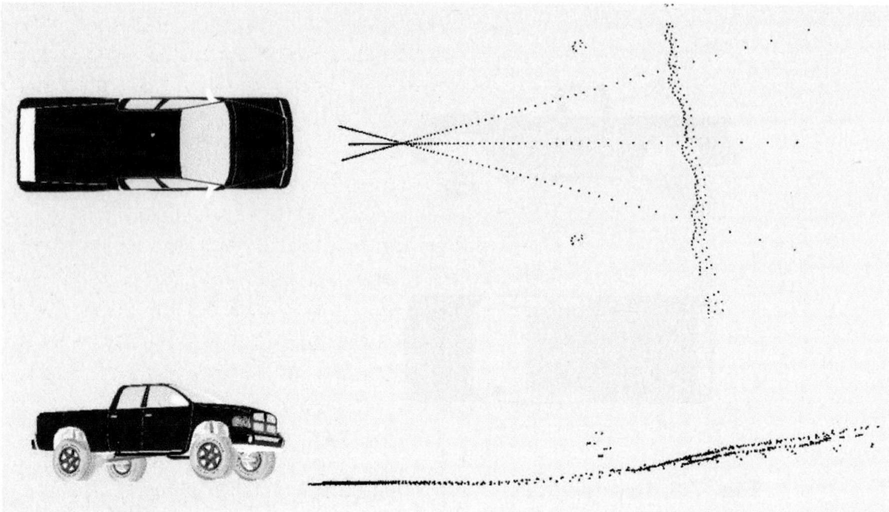

Fig. 7.6. Ladars distinguish obstacles from the ground surface

ground surface in front of the truck, in the truck body-fixed frame (as opposed to a world-fixed frame). The ground model we fit to the data was piecewise linear in the truck's direction of motion and piecewise constant in the sideways direction. The model interpolated the most recent available ladar data. In locations beyond the available data, the ground was assumed to have constant altitude in the body-fixed frame.

The apparent altitude of returns from the nearly-horizontal ladars relative to the ground model was then computed to see if those returns were (a) consistent with a traversable part of the ground model; (b) consistent with a non-traversable part of the ground model; (c) apparently from an elevation moderately higher than the notional ground; or (d) apparently from an object so far above the notional ground that the vehicle should be able to pass under it. In either case (b) or (c), the ladar return is classified as arising from a possible obstacle; after several of these returns are received from the same location at different vantage points, the presence of an obstacle is confirmed. In practice, an object such as a parked car will be represented as a cluster of adjacent obstacles.

For example, Figure 7.6 illustrates ladar data from a portion of the NQE course. The returns from the horizontally-oriented ladars indicated a long straight feature crossing in front of the vehicle, and two smaller features or clusters of returns. The three vertical ladars measure the ground profile in three directions and reveal that the truck is driving towards the foot of an upwards incline. The "long straight feature" is merely a surface contour of the upward-sloping ground, and therefore, since the incline is not too steep, should not be considered an obstacle. The two smaller clusters of returns, however, appear to be at a significant altitude above the notional ground, and are interpreted as non-traversable obstacles. In fact, these are traffic cones.

The entire process is $O(n)$ in the number of ladar measurements n, with a large array providing $O(1)$ access to accumulated obstacle detections in a given location. A many-to-one mapping between latitude-longitude coordinates and cells of the array is needed, with the properties that (a) each connected preimage of an array cell is at least approximately the same physical size and (b) any two distinct regions which are mapped to the same array cell are sufficiently far apart that they will not both need to be considered within the planning horizon of the vehicle. A simple method, which works everywhere except in the immediate vicinity of the poles, is to define a latitude-dependent cell size so that a different integral number of array cells correspond to a fixed extent of longitude at each latitude. It is acceptable that fewer array cells are needed further from the equator and therefore some array cells will have more preimage locations than others. The method could be adjusted to deal with the polar neighborhoods by exception.

Once confirmed, obstacles persist unless and until a period of time (e.g., one second) passes with no ladar returns detected from that location, in which case the obstacle expires. This enables the system to recover from false positive detections and also gives a rudimentary capability to deal with moving obstacles, since obstacles can disappear from one place and be re-detected somewhere else. However we did not attempt to form any velocity estimate of moving objects, and the robot generally operates on the assumption that other objects have zero velocity in the world frame.

In order to save computational time and memory, we did not wish to track obstacles which were well outside the DGC course boundaries. Off-course obstacles should be irrelevant to the vehicle's planning problem, and regularly checking to see whether they have expired should be a waste of time. Therefore non-traversable surfaces off the course were not classified as obstacles as long as the vehicle itself remained on the course. However, sensor measurements indicating off-course obstacles were allowed to accumulate, so that if the vehicle departed from the course boundaries for any reason, and was required to consider off-course obstacles as significant, these obstacles would pass the confirmation threshold very quickly.

7.3.2 Empirical Ladar Detection Results

The NQE obstacle course provided a uniform test environment with well-defined obstacles, of which many different sensor recordings presumably now exist, so it may be useful to report our ladar classification results during traversals of this obstacle course. The ranges at which the intended obstacles (parked cars, tire stacks, tank traps) were recognized as non-traversable by the vehicle are shown in Figure 7.7.

DARPA lined the boundaries of the course with traffic cones, which the vehicle, relying on its own GPS localization measurements, might consider cones to lie either inside or outside the course boundary. Those traffic cones considered to be on the course were classified as non-traversable at ranges indicated in Figure 7.8. Of the four outliers on the low end (cones which were not identified

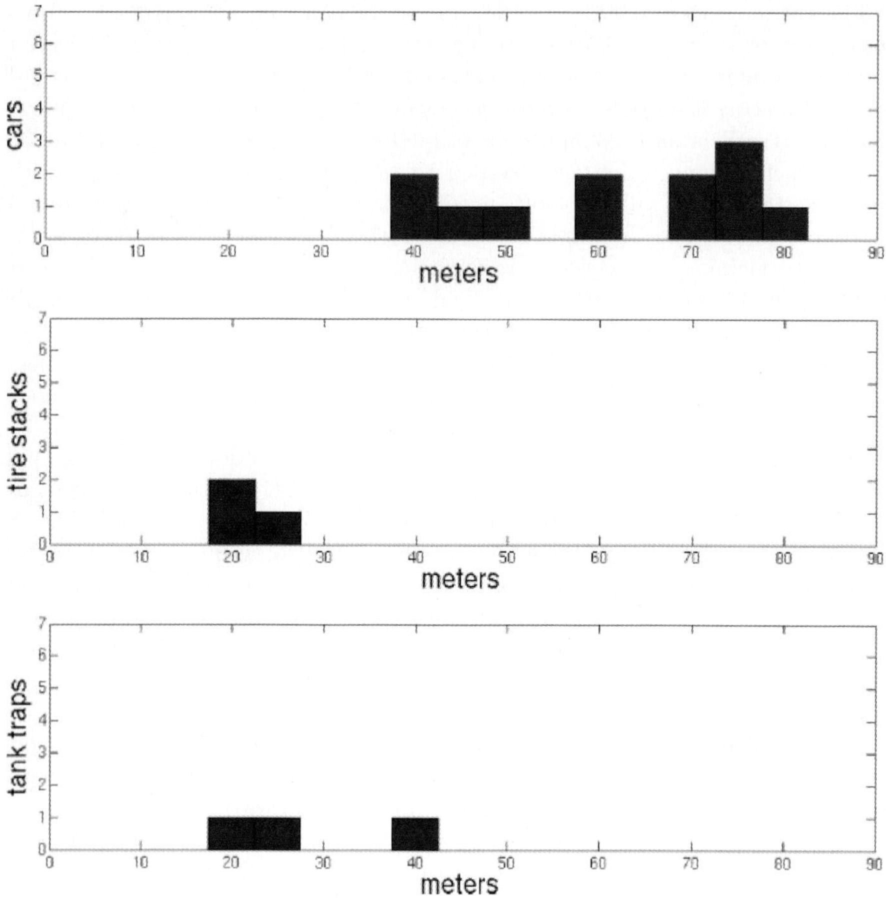

Fig. 7.7. Histograms of NQE obstacles arranged by range of first detection

until they were within less than twenty meters range), two cones were occluded from early view by other objects, and two were at the base of an incline and presumably could not be detected until the vehicle crested the top of the slope and headed down.

There were zero false negatives. Every intentional obstacle and cone on the course was correctly classified, and once classified, was persistently regarded as an obstacle as long as it remained within the 180-degree forward sensor arc of the vehicle.

There were, however, a considerable number of false positives, classifying terrain which was traversable as non-traversable. If these false obstacles appeared in the path of the vehicle, the vehicle would of course try to plan a new trajectory around them, and reduce speed if necessary. However, the false obstacles not did persist as the vehicle approached. Instead all false obstacles eventually expired,

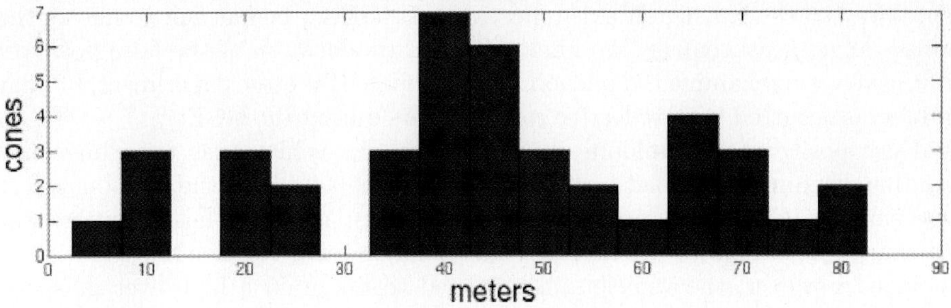

Fig. 7.8. Histogram of NQE traffic cones arranged by range of first detection

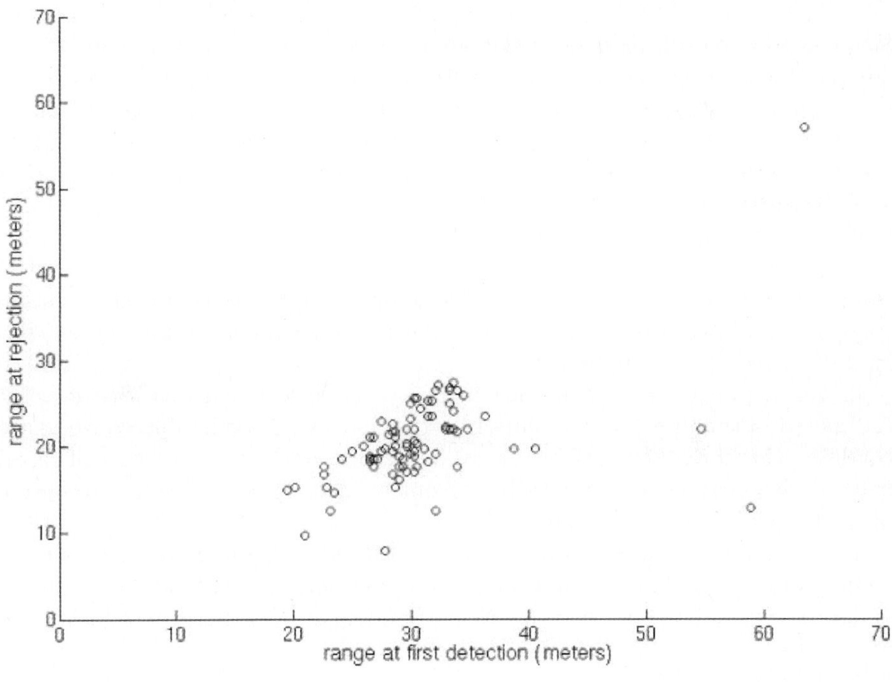

Fig. 7.9. False positive events in which obstacles were "detected" but subsequently rejected

leaving the vehicle free to drive over that terrain. The ranges at which false obstacle detections first appeared and then expired are shown in Figure 7.9. The mean range at first appearance was 30.5 meters; the mean range at expiration was 20.5 meters.

Some of the false positives coincided with irregularities in the ground surface or debris strewn on the course. In hindsight, it appears likely that a large percentage of the false positives were caused by reflective markings on the edges of the speedway track, causing off-axis reflections of the laser beams and confusing the ladar range measurement. We infer this from the location of the false positives but have yet to examine the phenomenon through deliberate experiment. Similar problems occurred with reflective road markers during the GCE.

False positives very seldom occurred when the vehicle was travelling on a continuous, unmarked road surface, e.g., the asphalt track at the NQE, or a dirt road in the GCE. The main NQE performance impact of the false positives was to cause the vehicle to hesitate at transitions crossing the marked edge of the asphalt. However, the vehicle still managed to maintain a high average speed and complete the obstacle course in competitive times. (See Table 7.1.)

7.4 Avoider

We accomplished vehicle trajectory planning using a custom recursive algorithm described in this section. In keeping with the general philosophical approach of the team, we attempted to design for a minimal computation footprint, taking strategic advantage of heuristic rules and knowledge implicit in the DARPA-provided RDDF route. The method bears a resemblance to the generation of a probabilistic roadmap [Kavraki et al., 1996] or a rapidly-exploring random tree [Frazzoli et al., 2002, LaValle, 1998], in that we generate a graph of vehicle configurations connected by smooth trajectories, but instead of generating new configurations on the boundary of an expanding graph, we recursively generate trajectory midpoints in search of the best feasible path connecting a start configuration and end configuration.

In our search for high efficiency in practice, we forfeited any guarantee of the global optimality of our solution paths and even predictable convergence conditions for the routine. However, we found that in the vast majority of cases, traversable paths are found rapidly without burdening the processing resources of the vehicle.

The stage for the planning algorithm is a Cartesian two-dimensional map in the vicinity of the initial global coordinates of the vehicle, populated with a number of discrete obstacles. The obstacles are represented as point hazards that must be avoided by a particular radius or as line segments that can be crossed in only one direction. Point obstacles are centered on locations in the plane that have been identified as non-traversable by the ladar system. The buffer radius surrounding each point obstacle includes the physical width of the vehicle and additionally varies based on the distance from the vehicle to the hazard. A real object such as a wall or car is represented by a cluster of such points, each with its associated radius. Line segment obstacles arise from the Route Definition Data File (RDDF) which specifies corridor boundaries for the vehicle. The vehicle trajectory is a single curve describing the motion of the vehicle frame in the plane.

Our task is to find a drivable path from an initial configuration to a destination configuration that does not encroach on any obstacle.[2] The destination

[2] At this time, we have no classification of obstacles by importance. All obstacles are regarded as equally non-traversable and equally to be avoided.

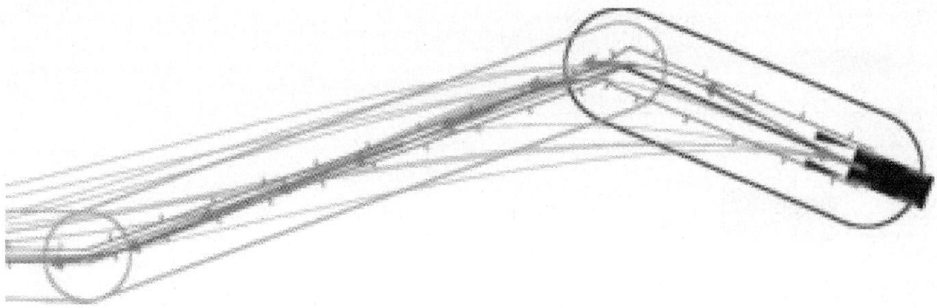

Fig. 7.10. Visualization of Avoider algorithm at GCE start line

is taken to be the center of the RDDF corridor at some distance ahead of our sensor horizon. The simplest and presumably most frequently occurring scenario will require the vehicle to simply drive forward from to without performing any evasive maneuvers. We represent this scenario with a directed acyclic graph containing two nodes and a single connecting edge. This choice of data structure later allows us to take advantage of a topographically sorted ordering that enables the least cost traversal of the graph to be found in linear time (ON edges). Each node in the graph is associated with a possible configuration of the vehicle and each edge is associated with a directed path connecting a starting configuration to a target configuration. At all times, care is taken to maintain the topological correspondence between the nodes in the graph and the locations in the local map.

After initialization we enter the recursive loop of the algorithm, which consists of three stages:

1. Graph Evaluation
2. Path Validation
3. Graph Expansion

This loop is executed repeatedly until a satisfactory path is found from to, or until a watchdog termination condition is met. Typically we would terminate the planner if a path was not found after several hundred milliseconds.

Figure 7.10, Figure 7.11, and Figure 7.12 offer a visualization of the path planning process. Nodes are indicated by bright purple arrows while considered trajectories are drawn as purple lines.

In Graph Evaluation, trajectory segments and cost factors are computed for all new edges in the graph. Each trajectory segment is calculated from the endpoint node configurations that it connects and is afterwards checked for possible obstacle collisions. There may be a large number of different trajectory segments to check and a large number of obstacles currently being tracked, bearing in mind that a single real object of any significant horizontal extent will be tracked as a cloud of smaller obstacles which are close together. In order to efficiently do collision checking, we use a wavefront propagation method to create a map

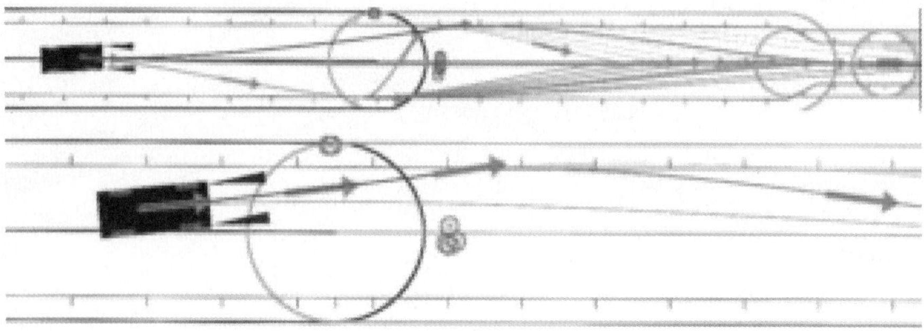

Fig. 7.11. Visualization of Avoider graph as the vehicle navigates around a tank trap obstacle during the NQE

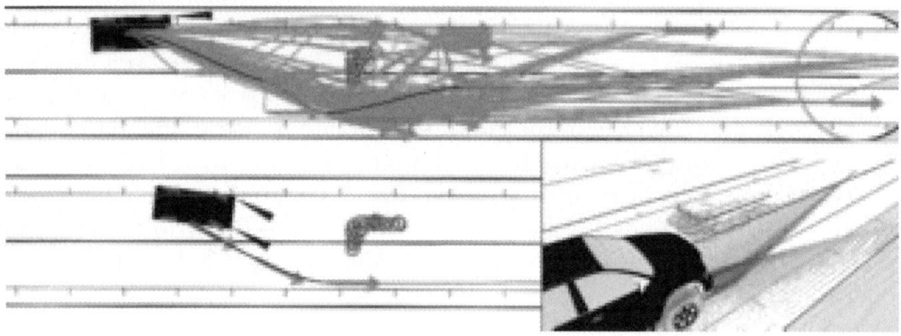

Fig. 7.12. Visualization of Avoider graph as the vehicle navigates around a parked car obstacle during the NQE

indicating approximate distance from each point to the nearest obstacle. Once this map is produced, trajectories can be rapidly checked against it for obstacle collisions. If no obstacles are intersected, the cost of traversal is generated from an integral of the instantaneous curvature along the trajectory. In our experience, the dynamic feasibility of a path can be adequately characterized by this single value.

The loop continues with Path Validation. The optimal path over the entire explored graph is found rapidly by reference to the topological sort of the nodes. If the cost is satisfactory the loop ends and a refinement procedure begins. This refinement process consists of a number of heuristic modifications that add nodes and edges to the graph to "smooth" calculated trajectories. As a final step, a velocity profile is computed along the solution, so that the vehicle's speed will be consistent with the curvature of the path and with any DARPA-imposed speed limits. The planning thread then updates the desired trajectory for the steering and velocity feedback control system.

If the total cost of the best graph traversal is too high, we enter the third phase of the algorithm: Graph Expansion. In this stage, edges that intersect obstacles are split at the point where a collision would occur. A number of new nodes are added to the graph for configurations selected according to heuristic avoidance conditions. Unevaluated edges connect these new nodes to the endpoints of the offending trajectory segment. Trajectory segments for these new edges are then computed when the recursive loop begins again with Graph Evaluation.

In the heuristic techniques applied to select good target configurations for obstacle avoidance, the position component of the target configuration is typically projected perpendicularly out from the obstacle or else interpolated between the obstacle position and the RDDF boundary. We found that vehicle heading was similarly best specified as a function of both the heading at the collision point and the RDDF-defined corridor direction.

7.5 Vehicle State Estimation

The path planning and control subsystems of Golem 2 needed a good estimate of the latitude, longitude, heading and the velocity of the vehicle, at a level of accuracy that was beyond that provided by on-board sensors like the C-MIGITS. Two different state estimators were implemented to carry out this task, as a means to provide analytic redundancy. The first state estimator used a model analogous to a bicycle for the vehicle and relied heavily on the history of state. The second estimator was based on a six-degrees-of-freedom rigid-body model, with the addition of a "soft" non-holonomic constraint on the vehicle's velocity enforced as a virtual heading measurement.

7.5.1 The Bicycle Estimator

In this section we will describe the working of the first estimator, henceforward referred to as the bicycle estimator. The bicycle estimator was a discrete time extended Kalman filter [Gelb, 1974, Kalman, 1960, Kalman and Bucy, 1961, Welch and Bishop, 1995], with the following inputs:

1. Latitude and longitude from a NovAtel GPS sensor at 20 Hz.
2. Rear axle velocity at 30 Hz from a custom Hall sensor system. A set of 16 magnets was installed on the rear axle. Two detectors were attached to the vehicle frame and passed a voltage pulse every time one of the magnets went past them. The two sensors enabled us to have a quadrature encoder, i.e., we were able to distinguish between forward and reverse motion. A discrete time two state Kalman filter used the voltage pulses as input and estimated the rate of rotation, which in turn was scaled by gear ratio and wheel radius to infer the velocity of the vehicle.
3. Steering angle from an absolute encoder at 20 Hz.

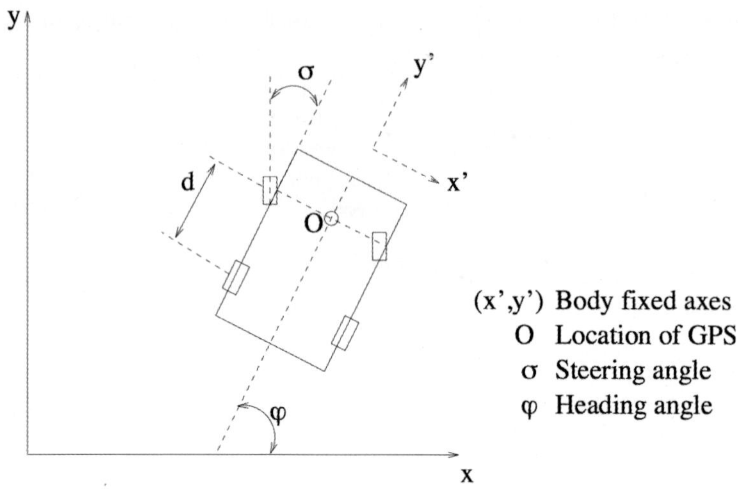

Fig. 7.13. The "bicycle" model

Rear axle velocity and steering encoder were used as inputs to the bicycle estimator. The state propagation equations for the bicycle estimator were:

$$\begin{aligned} \tilde{x}_{k+1} &= \hat{x}_k + \hat{v}_k \Delta t \frac{\cos(\hat{\phi}_k + \hat{\sigma}_k)}{\cos \hat{\sigma}_k} \\ \tilde{y}_{k+1} &= \hat{y}_k + \hat{v}_k \Delta t \frac{\sin(\hat{\phi}_k + \hat{\sigma}_k)}{\cos \hat{\sigma}_k} \\ \tilde{\phi}_{k+1} &= \hat{\phi}_k + \frac{\hat{v}_k \Delta t}{d} \tan \hat{\sigma}_k \end{aligned} \tag{7.1}$$

where as described in Figure 7.13, (x, y) are the local cartesian coordinates of the center of the front axle of the car. The GPS sensor is located at this point. The angle ϕ is the heading of the car with respect to the local x axis. The angle σ is the steering angle of the car as shown in the figure. v is the rear axle velocity of the car. The quantity d is the distance between the front and rear axle of the vehicle. A caret over a variable implies that the variable is an estimate, and a tilde over a variable implies that the variable is predicted or propagated forward. The time elapsed since the last state update is Δt. The subscripts in the above equations refer to successive time indices in state propagation.

The bicycle model worked well when the steering angle was small. For large steering angles however, the model is inaccurate and leads to considerable lag between the updated state and the GPS measurement. The model was also found to be inaccurate for high velocities, when the vehicle slips at turns. Thus, the steering angle, σ, was calculated as:

$$\hat{\sigma}_k = \hat{\gamma}_k(\sigma_{\text{measured}} - \hat{\sigma}_{\text{bias } k}) \tag{7.2}$$

where γ is the steering calibration factor which tries to compensate for model inaccuracies. The steering bias estimate σ_{bias} is used to compensate for bias in the measured steering angle. The rear axle velocity, v, was calculated as:

$$\hat{v}_k = \hat{S}_k(v_{\text{measured}}) \tag{7.3}$$

where v_{measured} is the velocity being input to the bicycle estimator. The slip factor S compensates for variations in the vehicle's apparent wheel radius, and for the fact that the vehicle may be going uphill or downhill, and tries to estimate slip. However, the slip factor was not able to track consistently long periods of slipping.

The state variables were latitude, longitude (translated to local Cartesian coordinates x,y), heading ϕ, slip factor S, and either steering bias σ_{bias} or steering calibration factor γ.

The steering calibration factor γ and the steering bias cannot be estimated simultaneously as the state variables become unobservable. So, the bicycle estimator was operated in the following 2 modes:

1. When the vehicle was going straight, $|\sigma_{\text{measured}}| < 3°$, σ_{bias} was estimated.
2. When the vehicle was turning, $|\sigma_{measured}| > 3°$, γ was estimated. This enabled compensating for model inaccuracies for hard turns.

7.5.2 The Six-Degrees-of-Freedom Estimator

In this section, we describe the design of the six-degree-of-freedom (6DOF) estimator. Like the bicycle estimator discussed previously, the 6DOF estimator is implemented as a discrete-time Extended Kalman Filter. The estimator is designed using fairly standard techniques for strap-down inertial navigation systems. Since a detailed model of the vehicle's dynamics is not available, the filter relies mainly on the rigid-body kinematic equations. However, due to the absence of a magnetometer or other means to measure the vehicle's orientation, and the need to be able to ensure convergence of the non-linear filter without requiring a initial calibration procedures, knowledge of the vehicle's dynamics is exploited in the form of virtual heading measurements from inertial velocity data.

The vehicle is modeled as a rigid body moving in the three-dimensional space; the state of the vehicle can hence be described by a position vector $p \in \mathbb{R}^3$, representing the location with respect to an Earth-fixed reference frame of the on-board Inertial Measurement Unit (IMU), a velocity vector $v = dp/dt$, and a rotation matrix $R \in SO(3)$, where $SO(3)$ is known as the Special Orthogonal group in the three-dimensional space, and includes all orthogonal 3 by 3 matrices with determinant equal to $+1$. The columns of R can be thought of as expressing the coordinates of an orthogonal triad rigidly attached to the vehicle's body (body axes).

The inputs to the estimator are the acceleration and angular rate measurements from the IMU, provided at 100 Hz, and the GPS data, provided at about 20 Hz. In addition, the estimator has access to the velocity data from the Hall sensors mounted on the wheels, and to the steering angle measurement.

In the following, we will indicate the acceleration measurements with $z_{\text{a}} \in \mathbb{R}^3$, and the angular rate measurements with $z_{\text{g}} \in \mathbb{R}^3$. Moreover, we will indicate with

Z_{a} the unique skew-symmetric matrix such that $Z_{\mathrm{a}} v = z_{\mathrm{a}} \times v$, for all $v \in \mathbb{R}^3$. A similar convention will be used for z_g and other three-dimensional vectors throughout this section.

The IMU accelerometers measure the vehicle's inertial acceleration, measured in body-fixed coordinates, minus the gravity acceleration; in other words,

$$z_{\mathrm{a}} = R^T(a - g) + n_{\mathrm{a}},$$

where a and g are, respectively, the vehicle's acceleration, and gravity acceleration in the inertial frame, and n_{a} is an additive, white Gaussian measurement noise. Since the C-MIGITS IMU estimates accelerometer biases, and outputs corrected measurements, we consider n_{a} as a zero-mean noise.

The IMU solid-state rate sensors measure the vehicle's angular velocity, in body axes. In other words,

$$z_{\mathrm{g}} = \omega + n_{\mathrm{g}},$$

where ω is the vehicle's angular velocity (in body axes), and n_{g} is an additive, white Gaussian measurement noise. As in the case of acceleration measurements, n_{g} is assumed to be unbiased.

The kinematics of the vehicle are described by the equations

$$\begin{aligned} \dot{p} &= v, \\ \dot{v} &= a, \\ \dot{R} &= R\Omega, \end{aligned} \tag{7.4}$$

in which Ω is the skew-symmetric matrix corresponding to the angular velocity ω, and we ignored the Coriolis terms for simplicity. As a matter of fact, this is justified in our application due to the low speed, relatively short range, and to the fact that errors induced by vibrations and irregularities in the terrain are dominant with respect to the errors induced by ignoring the Coriolis acceleration terms.

We propagate the estimate of the state of the vehicle using the following continuous-time model, in which the hat indicates estimates:

$$\begin{aligned} \dot{\hat{p}} &= \hat{v}, \\ \dot{\hat{v}} &= \hat{R} z_{\mathrm{a}} + g, \\ \dot{\hat{R}} &= \hat{R} Z_g. \end{aligned} \tag{7.5}$$

An exact time discretization of the above, under the assumption that the (inertial) acceleration and angular velocity are constant during the sampling time is:

$$\begin{aligned} p^+ &= p + v\Delta t + \tfrac{1}{2}\left(\hat{R} z_{\mathrm{a}} + g\right)\Delta t^2, \\ v^+ &= v + \left(\hat{R} z_{\mathrm{a}} + g\right)\Delta t, \\ R^+ &= R\exp(Z_g \Delta t). \end{aligned} \tag{7.6}$$

The matrix exponential appearing in the attitude propagation equation can be computed using Rodrigues' formula. Given a skew-symmetric 3 by 3 matrix M, write it as the product $M = \Omega\theta$, such that Ω is the skew-symmetric matrix corresponding to a unit vector ω; then

$$\exp(M) = \exp(\Omega\theta) = I + \Omega \sin\theta + \Omega^2(1 - \cos\theta). \tag{7.7}$$

The error in the state estimate is modeled as a 9-dimensional vector $\delta x = (\delta p, \delta v, \delta\phi)$, where

$$\begin{aligned} p &= \hat{p} + \delta p, \\ v &= \hat{v} + \delta v, \\ R &= \hat{R}\exp(\delta\Phi). \end{aligned} \tag{7.8}$$

Note that the components of the vector $\delta\phi$ can be understood as the elementary rotation angles about the body-fixed axes that make $\hat{R}$ coincide with R; such a rotation, representing the attitude error, can also be written as $\delta R = \exp(\delta\Phi) = \hat{R}^T R$.

The linearized error dynamics are written as follows:

$$\frac{d}{dt}\delta x = A\delta x + Fn, \tag{7.9}$$

where

$$A := \begin{bmatrix} 0 & I & 0 \\ 0 & 0 & -RZ_{\mathrm{a}} \\ 0 & 0 & -Z_{\mathrm{g}} \end{bmatrix}, \qquad F := \begin{bmatrix} 0 & 0 \\ R & 0 \\ 0 & I \end{bmatrix}. \tag{7.10}$$

When no GPS information is available, the estimation error covariance matrix $P := \mathrm{E}[\delta x(\delta x)^T]$ is propagated through numerical integration of the ODE

$$\frac{d}{dt}P = AP + PA^T + FQF^T.$$

Position data from the GPS are used to update the error covariance matrix and the state estimate. The measurement equation is simply $z_{\mathrm{gps}} = p + n_{\mathrm{gps}}$. In order to avoid numerical instability, we use the UD-factorization method described in Rogers [2003] to update the error covariance matrix and to compute the filter gain K.

Since the vehicle's heading is not observable solely from GPS data, and we wished to reduce calibration and initialization procedure to a minimum (e.g., to allow for seamless resets of the filter during the race), we impose a soft constraint on the heading through a virtual measurement of the inertial velocity, of the form:

$$z_{\mathrm{nhc}} = \arctan\left(\frac{v_{\mathrm{East}}}{v_{\mathrm{North}}}\right) - \lambda\sigma,$$

where σ is the measured steering angle, and λ is a factor accounting for the fact that the origin of the body reference frame is not on the steering axle.

In other words, we effectively impose a non-holonomic constraint on the motion of the vehicle through a limited-sideslip assumption. This assumption is usually satisfied when the vehicle is traveling straight, but may not be satisfied during turns; moreover, when the vehicle is moving very slowly, the direction of the velocity is difficult to estimate, as the magnitude of the velocity vector is dominated by the estimation error. Hence, the virtual measurement is applied only when the steering angle is less than 10 degrees, and the vehicle's speed is at least 2.5 mph (both these values were determined empirically).

The filter described in this section performed satisfactorily in our tests and during the DGC race, providing the on-board control algorithms with position and heading estimates that were nominally within 10cm and 0.1° respectively.

7.5.3 Modeling System Noise

The state was propagated and updated every time a GPS signal arrived. There was no noise associated with the state propagation equations (7.1). It was assumed that there was additive white gaussian noise associated with the inputs (v_{measured} and $\sigma_{\mathrm{measured}}$ for the bicycle model, and z_{a} and z_{g} for the six-degrees-of-freedom model). To efficiently track the constants in the state variables (S, σ_{bias}, γ), it was assumed that there was an additive white gaussian noise term in their propagation. It was also assumed that the GPS measurement (latitude and longitude) had some noise associated with it. The variances assigned to the noise processes described above were tuned to ensure that the innovations[3] were uncorrelated, and that the estimator was stable, and converged reasonably quickly. By tuning the variances, we can alter the "trust" associated with the history of the vehicle state or with the GPS measurement.

7.5.3.1 Determining the Appropriate GPS Measurement Noise Variance

The received data from the GPS consisted of the latitude, longitude and an HDOP (horizontal dilution of precision) value. HDOP is a figure of merit for the GPS measurement which is directly related to the number of satellites visible to the GPS antenna. The HDOP value is related to the noise in each measurement. However, our attempts at mapping the HDOP values to actual variances were futile as we did not observe a monotonic relationship. It was noticed that an HDOP of more than 5 usually corresponded to multipath reception in GPS and consequently, the GPS had an abnormally huge variance in that case. In the absence of any ad hoc relationship between the HDOP and noise variance, we opted to keep a constant variance associated with GPS noise if HDOP was less than 5 and a huge variance otherwise. Also, we noticed that the GPS noise was highly correlated and not white.

7.5.3.2 Projection of the Innovations

The bicycle estimator is based on a very intuitive model of the vehicle, which motivates us to consider the GPS innovations in a physically relevant reference

[3] Innovation of a measured data is the difference between the estimated and the measured data.

frame rather than any arbitrary reference frame. It is insightful to project the innovations into the local frame of the vehicle: parallel to the heading direction and perpendicular to it. The parallel and perpendicular body fixed axes are indicated in Figure 7.13.

For an ideal estimator, the innovations will be uncorrelated with each other. However, we tuned the estimator to just achieve innovations with zero mean. While tuning the estimator it was very useful to consider the parallel and perpendicular innovations. For example, a DC bias in the parallel innovations implied that we were not tracking the slip factor (S) adequately. Thus, to ensure a zero mean parallel innovation the variance associated with the propagation of slip factor should be increased.

7.5.3.3 Adaptive Shaping of the Innovations

The noise in the GPS data was highly correlated and there was very little a priori knowledge of the variance. Very often, especially when the vehicle would drive near a wall, or approach a tunnel, there would be highly erratic jumps in the GPS measurements due to multipath reflections. Without any a priori knowledge of the variance in such cases, the state estimate would bounce around a lot. This was undesirable as it would hamper the path planner and the obstacle detection subroutines in the main program.

To counter these "unphysical" jumps, once the estimator was converged, the innovations were clipped up to a certain maximum absolute value. For example, a GPS measurement corresponding to a perpendicular innovation of 2 meters in 0.05 sec while going straight is unphysical and so perpendicular innovation should be clipped to a nominal value (in our case, six inches). This prevented large jumps in state estimate, but had a grave disadvantage. It was observed that if the innovations were clipped to a fixed range, then in certain situations, the state estimate will lag far behind from "good" set of GPS measurements and take a long time to converge back. To prevent this from happening, the clipping limit of the innovations was determined adaptively as the minimum of either a fixed limit, or the mean of the innovations in the last two seconds scaled by a numerical factor slightly greater than unity. The parallel and perpendicular components of the innovation were clipped separately with different numerical constants used.

7.5.3.4 Countering the Time Delays

There was some finite delay between the time sensor data appeared on the bus and the time it was processed by the control program. Usually this delay was nominal ($\sim$50-200 μs), but it was sporadically very large. A large delay in GPS data manifested in large negative parallel innovation. However, the large innovation was clipped effectively by the method described earlier, and consequently did not effect the state estimate. In future, we plan to implement a routine which synchronizes the time between all the serial/USB data sources.

Performance of the state estimator while going under a bridge is shown in Figure 7.14. "*" represents the GPS data received in a local cartesian cordinate

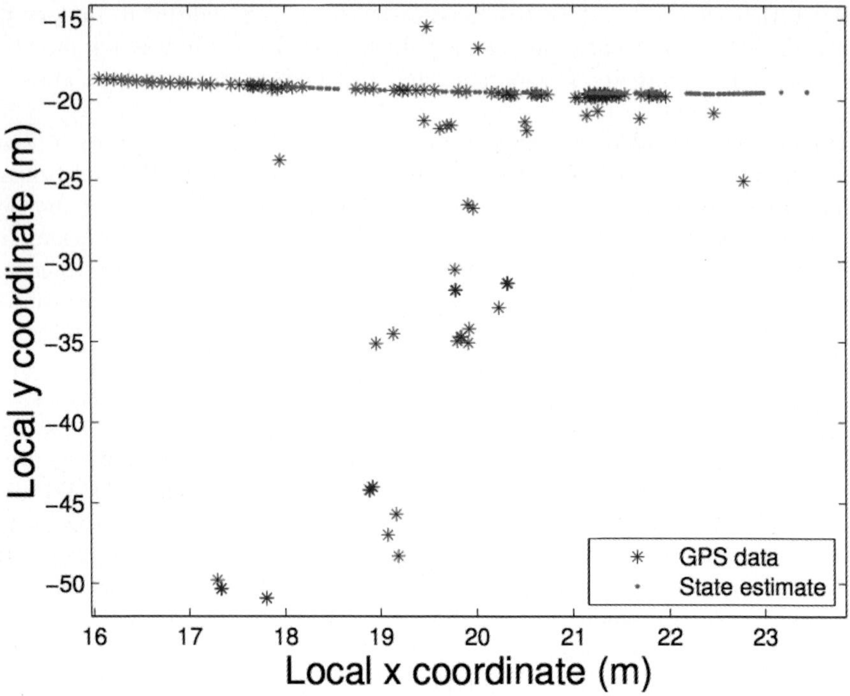

Fig. 7.14. Performance of state estimator while going under a bridge

system. "·" represents the estimated location of the truck. Golem2 was moving from left to right in this figure. Note that the GPS data has a huge variance due to multipath. Also note that the GPS noise is highly correlated, in this case the variance is huge in the direction perpendicular to the motion as compared to the parallel direction. As seen from the figure, the performance of the state estimator is immune to the large "unphysical" jumps in GPS data. Ocassional jumps of length $\sim 0.3m$ are observed in the successive updates of estimated state. These correspond to a delay in received GPS data and large negative parallel innovations as discussed in Section 7.5.3.4.

7.5.3.5 Special Modes of Operation

A couple of situations required special handling instructions:

- GPS signal was observed to drift considerably when the vehicle was at rest. Not only was this unphysical, but it is also detrimental to path planning. Thus when the vehicle was going considerably slowly or was at rest, the variance assigned to the GPS measurements was increased significantly. In such a case, the bicycle estimator essentially worked like an integrator rather than a filter.

- It was observed that the GPS signal would occasionally jump discretely. These jumps usually corresponded to the presence of a power transmission line nearby. The difficulty was that the GPS took a while (>2 s) to re-converge after the jumps. These unphysical jumps were easily detected from the jumps in the parallel and perpendicular innovations. After the detection of such jumps, the GPS variance was increased until it was trustworthy again, i.e., until the innovations were within a certain limit again.

7.5.3.6 Initialization of the IMU

One advantage of this model was that it converged very fast while going straight, usually in approximately 5 s. Once the bicycle model converged, the heading estimate was used to initialize the IMU. While going straight, the performance of the bicycle estimator was extremely good as the heading estimate was within 0.5° of the IMU computed heading. However, on sharp turns, the heading estimate was up to ~3° from the IMU computed heading. Thus, IMU computed heading was given more importance especially when GPS signals were bad. In future, we plan to extend the bicycle model to include the angular rotation and linear displacement data from the IMU.

7.6 Vision Systems

For computer vision, desert paths present a different challenge from paved roads, as the environment is far less structured, and less prior information is available to exploit in constructing algorithms. Our approach was to integrate existing vision technologies which have been proven to work on-road but have the potential to be transferred to the off-road domain. These include learning-based road-finding and binocular stereo reconstruction.

7.6.1 Mobileye Vision System

Golem 2 was equipped with a sophisticated vision system, created by Mobileye Vision Technologies Ltd., consisting of a single camera and a dedicated processing unit. On-road, the Mobileye system can find lane boundaries and detect other vehicles and their positions in real-time. The system was adapted to the off-road environment by Mobileye and the Hebrew University of Jerusalem.

The Mobileye system combines region-based and boundary-based approaches to find path position and orientation relative to the vehicle. The two approaches complement each other, thus allowing reliable path detection under a wide range of circumstances. Specifically, we use a variety of texture filters together with a learning-by-examples Adaboost [Freund and Schapire, 1996] classification engine to form an initial image segmentation into path and non-path image blocks. In parallel, we use the same filters to define candidate texture boundaries and a projection-warp search over the space of possible pitch and yaw parameters in order to select a pair of boundary lines that are consistent with the texture gradients and the geometric model. Both the area-based and boundary-based models are then combined (weighted by their confidence values) to form a final path model for each frame.

7.6.1.1 Region-Based Path Detection

The gray-level image is divided into partially overlapping blocks. A filter bank is applied to all image pixels and a descriptor vector is generated per block. The descriptor contains the mean and standard deviation of the filter response over the block for each of the 16 filters. Each entry in the texture descriptor can be considered a "weak" learner in the sense that it forms class discrimination. The iterative Adaboost algorithm combines the weak learners to form a powerful classification engine that assigns a *path* or *non-path* label to every block according to the training data. The training data was extracted from 200 images that were gathered on various parts of the 2004 Grand Challenge route. The block classification by itself is not sufficient for autonomous vehicle path planning because about 10 percent of the blocks are expected to be misclassified. Filtering methods are used to clear some of the misclassified blocks, followed by detection of path boundaries. The path boundaries are derived via a minimal error separating line on each side of the path. The system confidence is calculated from the separation quality and the sensing range (the distance to the farthest point that we can reliably identify as path). Figure 7.15 shows the results of each of the main parts of the algorithm.

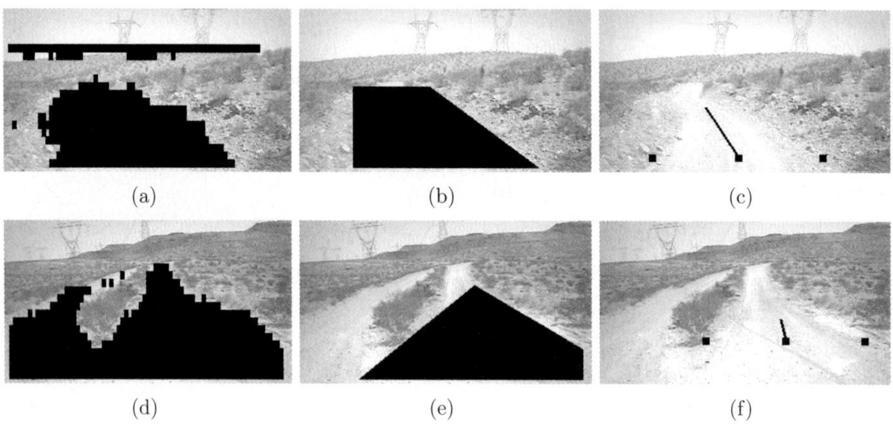

Fig. 7.15. In (a) and (d) blocks are classified by texture into *path* or *non-path*. In (b) and (e), sky blocks are removed and three lines, arranged in a trapezoid, are fitted to the *path* blocks. The trapezoid is considered to represent the path. Finally, in (c) and (f) the path boundaries are calculated at a given distance ahead of the vehicle (5 meters in this example) together with the path center and heading angle.

7.6.1.2 Boundary-Based Path Detection

The boundary-based technique does not rely on prior learned texture information. Instead, it makes the assumption that the path texture properties are different than the surrounding non-drivable areas. For this cue to be reliable, we have to constrain the solution to a strict geometric model where the path boundaries lie on straight parallel edges. This allows us to reduce the problem of finding a drivable path to 4 degrees of freedom: (x, y) position of the vanishing point, and left and right distance to the edge of the path. The geometric constraints resulting from assuming a flat world, perspective camera, and parallel path boundaries in the world suggest the following projection-warp scheme per frame: given a hypothesis of pitch and yaw angles of the camera, the image is warped to form a top view in world coordinates. In the warped image the path boundaries are supposed to be *parallel* vertical lines if indeed the pitch and yaw angles are correct. A projection of the image texture edges onto the horizontal axis will produce a 1D profile whose peaks correspond to vertical texture edges in the warped image. We look for a pair of dominant peaks in the 1D profile and generate a score value which is then maximized by search over the pitch and yaw angles via iterating the projection-warp procedure just described. The search starts with the pitch and yaw angle estimates of the previous frame followed by an incremental pitch and yaw estimation using optic-flow and a small motion model:

$$xw_x + yw_y = yu - xv \tag{7.11}$$

where (u, v) are the flow (displacements) of the point (x, y) and w_x, w_y are the pitch and yaw angles. The warped image is divided into overlapping 10×10 blocks with each pixel forming a block center. Using the same filter bank as in the

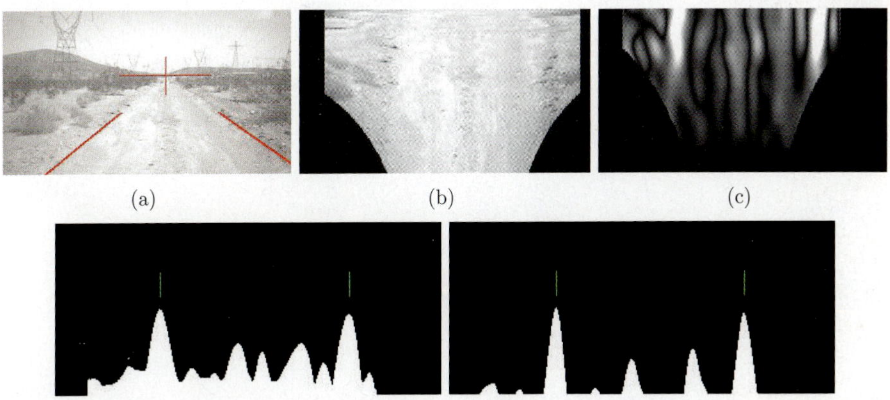

(a) (b) (c) (d) (e)

Fig. 7.16. Projection-Warp Search: (a) original image with the overlaid path boundary and FOE results. (b) the warped image. (c) texture gradients magnitude. (d) projection: vertical sum of gradients. (e) projection profile followed by convolution with a box filter. The two lines on top of the histogram marks the path boundaries.

region-based method, we estimate the the likelihood $e^{-\Delta}$ that the vertical line passing through the block center forms a texture gradient where Δ is the L_1 distance between the texture vector descriptors of the two respective halves of the block. To check a hypothesis (for pitch and yaw), we project the horizontal texture gradients vertically onto the x-axis and look for peaks in this projection. An example result of this projection is shown in Figure 7.16(d). The path boundaries and other vertical elements in the image create high areas in the projection, while low areas are most likely caused by vertical texture gradients that are not continuous and created by bushes, rocks, etc. The peaks in this projection are maximized when the vanishing point hypothesis is correct and the path edges (and possibly other parallel features) line up. By finding the highest peaks for these hypothesis, our system is able to find the lateral position of the left and right boundaries. Figure 7.16(e) shows the "cleaned up" 1D projection profile and the associated pair of peaks corresponding to the path boundary lines.

7.6.1.3 Performance

The system was implemented on a Power-PC PPC7467 1GHZ running at 20 frames per second. The camera was mounted on a pole connected to the front

Fig. 7.17. The camera is mounted inside a housing atop a pole connected to the front bumper. This allows a better field of view than mounting the camera on the windshield.

bumper (Figure 7.17) to allow maximal field of view. We tried both 45-degree and 80-degree field of view lenses, and found the latter to be more suitable for autonomous driving where the vehicle is not necessarily centered over the path. For our applications, the most meaningful overall system performance measure is to count how often (what fraction of frames) the system produced correct path edge positions and, where appropriate, heading angles. Furthermore, it is crucial for the system to know when it cannot determine the path accurately, so that the vehicle can slow down and rely more on information from the other sensors. Our results are broken up by different terrain types. For each, representative challenging clips of 1000 frames were selected and the system performance scored on these sequences by a human observer. The path edge distance accuracy was computed by observing the position of the road edge marks approximately 6 meters in front of the vehicle. A frame was labeled incorrect if the path edge marker at that location appeared to be more then 30 cm ($\approx$ 18 pixels) away from the actual path boundary. For straight paths the perceived vanishing point of the path was also marked, and our algorithm's heading indicator was compared to the lateral position of this point.

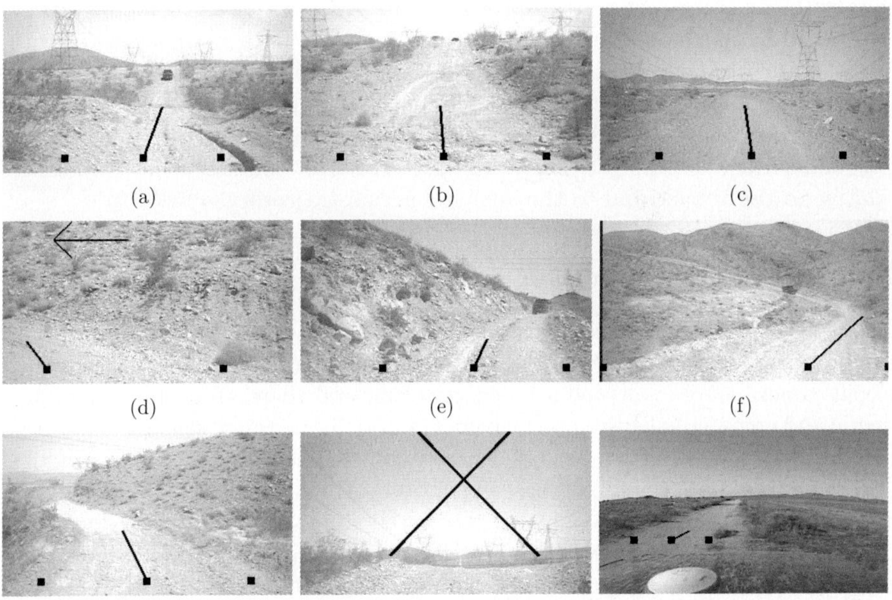

(a) (b) (c)

(d) (e) (f)

(g) (h) (i)

Fig. 7.18. Sample images and system output from 6 hours of driving in the Mojave desert. The path is marked by two left-right boundary points and a center point with heading orientation. The "X" mark in (h) coincides with zero confidence due to short range of visible path. In (i) the path is detected even though the vehicle is not centered on the path (a common situation in autonomous driving).

On relatively straight segments with a comfortably wide path, our system reported availability (high system confidence) **100%** of the time while producing accurate path boundary locations **99.5%** of the time. The mean angular deviation of the heading angle from the human marked vanishing point was 1.7 deg.

The second test clip is an example of more uneven terrains with elevation changes. Here the vehicle passes through a dry river ditch (Figure 7.18(b)) where both the path texture and scene geometry is difficult. When our vehicle is reaching the crest of the hill (Figure 7.18(h)) only a short segment of road is visible. In this case, the system reported unavailability (low confidence) **8%** of the time. When available, however, the accuracy in boundary locations was **98%**.

The final clip contains a winding mountain pass (Figure 7.18(g)), difficult due to path curvature as well as texture variation. Despite these, our system was available throughout the clip and achieved an accuracy of **96%** in detecting the path boundary.

7.6.1.4 Integration

The combination of the learning and geometric approaches yields high quality results with confidence estimates suitable for integration into our control systems. Output from the Mobileye system – which included path boundaries, orientations, and pitch – was integrated into the Golem path planner. Path boundaries were marked as line obstacles, projected onto the world coordinate frame using the known transformation between the Mobileye camera, the GPS antenna, and the road surface. Figure 7.19 shows the detected trail boundary and trail center relative to the vehicle and to the vehicle's actual trajectory.

7.6.2 Toshiba Stereo Vision System

The Golem/UCLA Team experimented with a variety of stereo vision systems for the purposes of obstacle detection, ground-plane registration, and path finding. Fixed-camera stereo is a well studied problem, and there are a number of academic and commercial systems of varying quality. Long range sensing is essential for use in a high-speed automobile since time is of the essence; it is impractical and sometimes impossible to stop to analyze nearby obstacles. This requirement translates into a wide baseline separating the stereo cameras, since the maximum range of the system is determined by this distance. A further performance constraint is processing time, since latency can introduce dangerous errors in path planning and control. Toshiba Research in Kawasaki, Japan, developed a stereo system for on-road driving at high speeds, which we installed on Golem 2. Due to insufficient time, we did not integrate the stereo system into the control system of the vehicle before the GCE, but there is no doubt that it has a high potential for successful autonomous driving. In this section we describe the hardware configuration and implementation details.

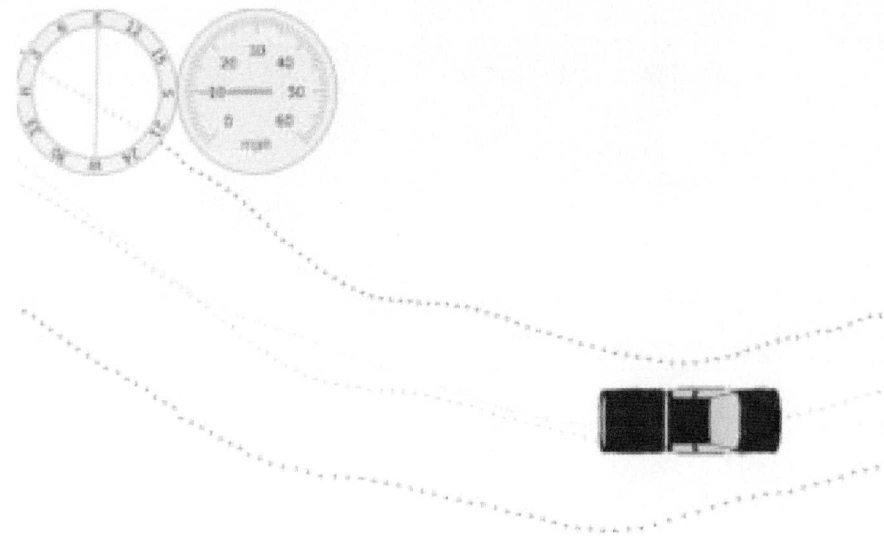

Fig. 7.19. Trail boundaries detected by the Mobileye system

Fig. 7.20. Setup of stereo cameras

7.6.2.1 Hardware Configuration

Figure 7.20 shows the setup of our stereo cameras. Two metal plates sandwich and rigidly fix the cameras so that they can withstand the strong vibrations cased by off-road driving. The distance between the two cameras is 1.2m and each camera is about 1.5m above the ground plane. We use CCD cameras with 7.5mm lenses that have image resolution of 320×240 pixels.

Our stereo system is based on a multi-VLIW processor called *Visconti* [Hattori and Takeda, 2005, Tanabe, 2003]. The processor architecture is designed to ensure efficient performance for general image processing operations while satisfying several requirements for automotive use, e.g., operating temperature range −40 to +85°C, power consumption < 1W@150MHz. Figure 7.21 shows a prototype of an image processing unit using Visconti. It has three video input

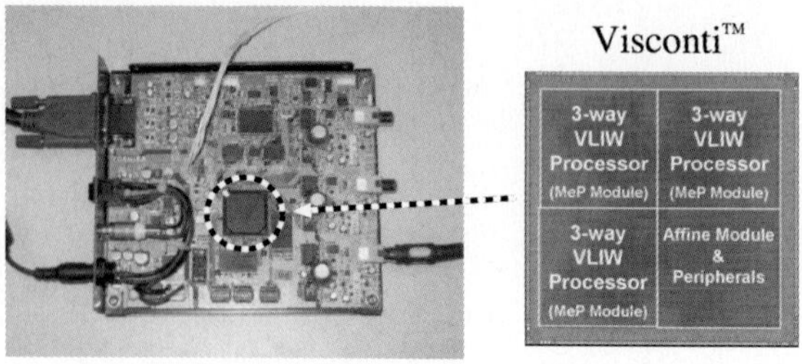

Fig. 7.21. Prototype processing hardware

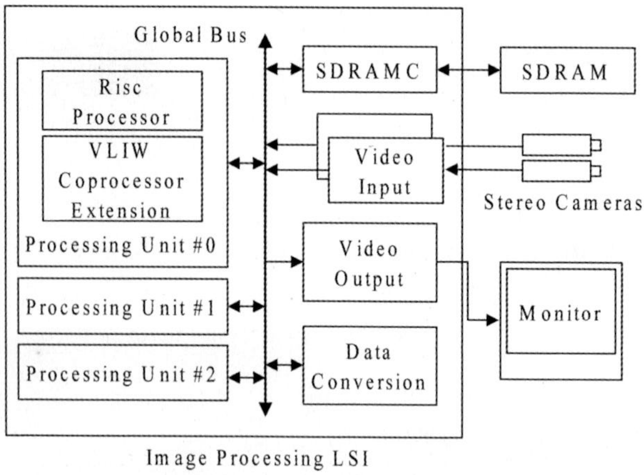

Fig. 7.22. Block diagram of Visconti

channels and a VGA video output to display the processing results. Figure 7.22 shows the block diagram of Visconti. The processor includes one image transformation module and three processing modules operating in parallel. Each of the three processing modules consists of a RISC processor core and a VLIW coprocessor. Several types of SIMD operations required for stereo computation, including convolution, accumulation and pixel shift, are supported in the instruction set of the coprocessor. Each processing module also has a scratch pad memory and a DMA controller so that memory access latency is hidden by double-buffering data translation.

7.6.3 Implementation Details

We adopt sum of absolute differences (SAD) as a matching criterion within a 7×7 window as, SAD is less computationally expensive than other measures such as the sum of squared differences (SSD) and normalized cross correlation (NCC). We also use a recursive technique for efficient estimation of SAD measures

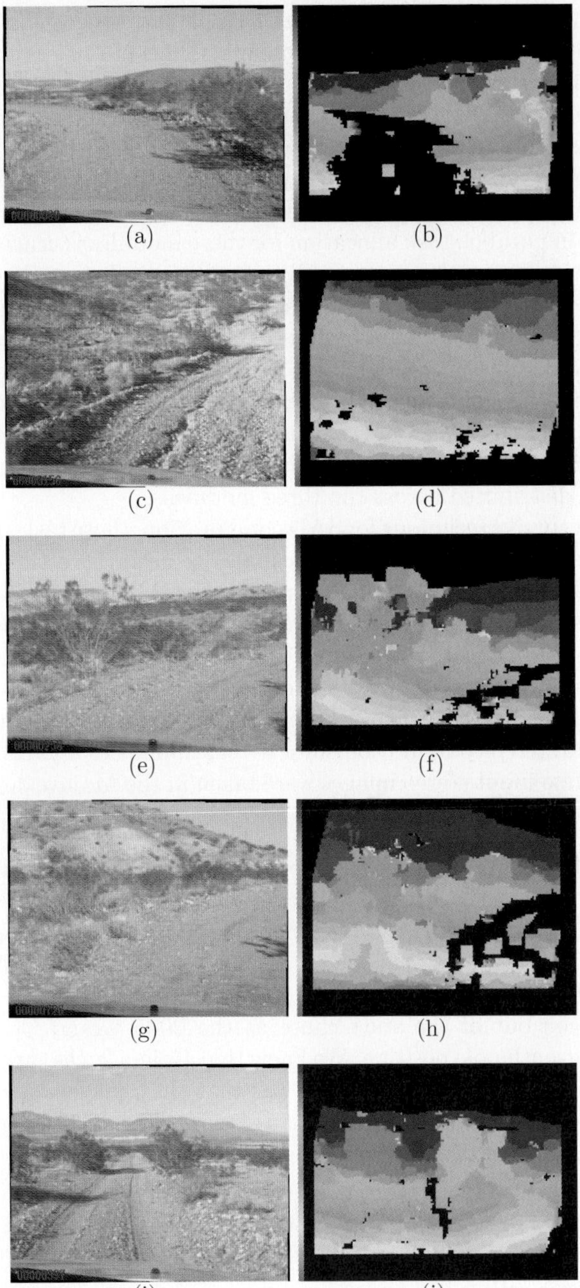

Fig. 7.23. Input images and their disparity maps. More red intensity indicates larger disparity values, i.e., nearer regions, and black indicates texture-less regions. Note that a part of the hood of the vehicle appears on the bottom of input images and these regions are excluded for the disparity estimation.

Faugeras et al. [1993], Hattori and Takeda [2005]. In order to compensate for possible gray-level variations due to different settings of the stereo cameras, the input stereo images are normalized by subtraction of the mean values of the intensities within a matching window at each pixel. Also, the variance of intensities at each point is computed on the reference image to identify those regions which have insufficient intensity variations for establishing reliable correspondences.

As Visconti has one image transformation module and three processing modules operating in parallel, task allocation for these modules is crucial to real-time operation. For instance, the stereo rectification is a indispensable process that transforms input stereo images so that the epipolar lines are aligned with the image scan-lines. The image transformation module carries out the stereo rectification, which is difficult to accelerate by SIMD (single-instruction, multiple-data) operations due to irregular memory access. Also, we divide a pair of images into three horizontal bands which are allocated to those three processing modules. Each of three areas has about the same number of pixels so that the computation cost is equally distributed across the three modules.

Due to the recursive technique for SAD computation, those task allocations and SIMD instructions, our stereo system is capable of disparity estimation at a rate of about 30 frames/sec with up to 30 disparity levels and an image input size of 320×240 pixels (QVGA). This performance is higher than that of a 2.0GHz processor with SIMD instructions. Figure 7.23 shows several examples of disparity images. More red intensity indicates larger disparity values which means closer areas while black indicates texture-less regions. The bottom of input images is excluded from the stereo computation since it corresponds to a part of the hood of the vehicle. These input stereo images were taken in the Mojave desert.

7.7 Results

Golem 2's qualifying runs on the National Qualification Event (NQE) obstacle course were among the best of the field, as shown in Table 7.1, although we also failed on two runs for reasons discussed in Section 7.7.1.

Golem 2 raced out of the start chute at the 2005 Grand Challenge Event (GCE) in the seventh pole position. We knew that Golem 2 was capable of driving well at high speeds. Our speed strategy was that the vehicle would drive at the maximum allowed speed whenever this was below 25 mph. If the recommended speed was greater than 25 mph (implying that the maximum allowed speed was 50 mph) then Golem 2 would exceed the recommended speed, by small amounts at first, but more and more aggressively as the race continued, until it was always driving at the maximum legal speed, except, of course, when modulating its speed during a turn.

As expected, Golem 2 made rapid time on dirt roads and over a dry lakebed. On a paved bridge, Golem 2's laser sensors misperceived the reflective "Botts' dots" in the center of the road as obstacles, which seemed to vanish like a mirage as the vehicle got closer. (See Figure 7.24.) This caused the vehicle to weave back and forth on the bridge, alarming the DARPA observers in the chase vehicle.

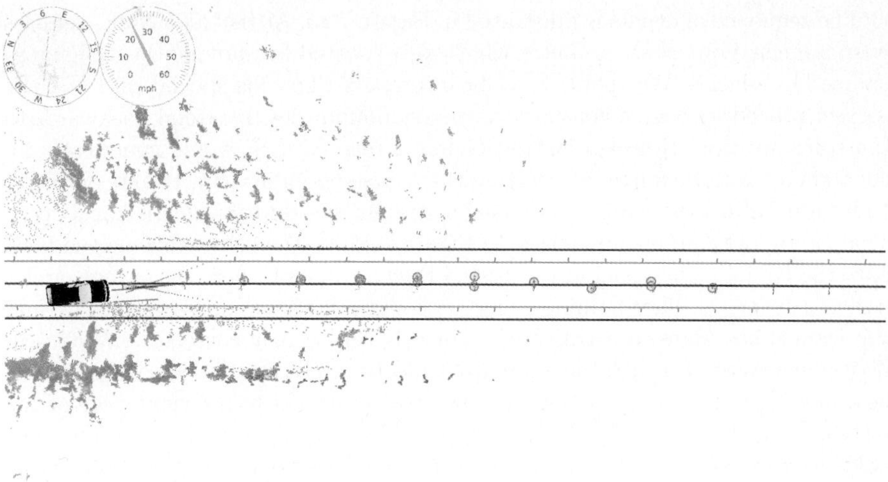

Fig. 7.24. Golem 2 misperceives reflective dots on a paved road surface as obstacles

But our vehicle kept going and once it reached dirt road again, it straightened out and resumed progress at over thirty miles per hour.

The DARPA observers characterized Golem 2 as initially "skittish" and compared it to a teenage driver, but stated that once it left the paved road and entered the desert, they were impressed by its performance and believed they had a winner on their hands.

Unfortunately, after driving twenty-two miles in just under one hour, the computer crashed due to faulty memory management. The uncontrolled vehicle departed from the course boundaries at high speed, crashing through vegetation. The DARPA "pause" button was no longer functional, since no software was running, and the DARPA observers did not press the "disable" button in case the vehicle might recover. Golem 2 hurtled more than half a mile off the course before pounding from the rough terrain finally shook connectors free from its fusebox, killing the engine.

7.7.1 Causes of Failure

Golem 2 crashed on three significant occasions: twice during NQE trials and once during the GCE. We think that all of these failures should be considered mere "bugs" rather than fundamental flaws in the design. Nevertheless it may be interesting to review the causes of these failures.

On its first attempt at the NQE course, Golem 2 immediately veered off to the right and crashed into one side of a gate intended to simulate a cattle crossing. It knocked down the fence beside the gate and came to a stop. The primary cause of this failure was that one of the vertical ladars had been repositioned and miscalibrated (due to a missing decimal point). Mishandling of the course start conditions was a contributing factor.

The sequence of events is illustrated in Figure 7.25. At the 2005 NQE, vehicles were launched out of start chutes which were located far outside the designated course boundaries. We should have been prepared for this special case and the correct procedure was to consider the course boundaries to extend backwards to the start position. However instead Golem 2 reacted as it would generically react to being far off course, by relaxing the course boundary constraints outward. In Figure 7.25(a) the vehicle is moving a straight trajectory which is hardly constrained by the course boundaries. In Figure 7.25(b), the vehicle has moved back onto the RDDF course and also detected the gate ahead where the orange circles indicate obstacles. The planning boundary constraints have contracted inward and the vehicle has planned a trajectory which is very nearly correct, i.e., a trajectory which passes through the first gate. Unfortunately, the boundary constraints have not tightened quite enough and the trajectory skirts the right edge of the RDDF course. In Figure 7.25(c), because of the miscalibrated ladar, the vehicle has misperceived a cloud of phantom obstacles where no obstacles actually exist, and planned to swerve around them to the right. In Figure 7.25(d), the vehicle has perceived that its new trajectory collides with a real obstacle, the fence, but it cannot find a plan that does not appear to result in collision. In Figure 7.25(e), collision is imminent and the vehicle belatedly attempts to brake. In Figure 7.25(f), the fence has been knocked down and the vehicle has come to a stop. Although the vehicle was physically capable of proceding forward over the crushed fence, there was no "restart" logic enabling the vehicle to begin moving again.

Our reaction to this failure was, of course, to fix the calibration of the vertical ladar, correctly handle the special case of starting outside the course boundaries, improve the reaction of the velocity controller to obstacles, and prevent the vehicle from coming to a permanent stop if not paused. We were able to carry out these fixes only because of the very useful `dashboard` visualization tool, shown in Figure 7.25 and elsewhere, that enabled us to closely examine the results of the failed run and simulate what would result from changes to the software.

After two very successful completions of the NQE course, Golem 2 crashed again on the fourth attempt. This time the bug was in the path planner, which failed to properly validate all the possible candidate trajectories and ended up selecting a degenerate trajectory containing two sharp 180-degree turns. The impossibly tight loop in the desired trajectory caused the vehicle to jerk suddenly into a concrete barrier. This event motivated increased evaluation of candidate trajectories, and repair and reinforcement of Golem 2's steering actuator.

Golem 2's final and most distressing failure occurred during the GCE due to static memory overallocation. Large amounts of random access memory were set aside at start time for use in recording obstacles, trajectories, and sensor history. In fact, the memory was overallocated, but this did not become apparent until a large sensor history had accumulated, which, because of the mapping between geographic coordinates and elements of the sensor observation array, only occurred when a large amount of terrain had been covered. Golem 2 had made experimental autonomous runs of 10 miles or so, but had never made a continuous overland journey on the scale of the GCE. Furthermore, an endurance

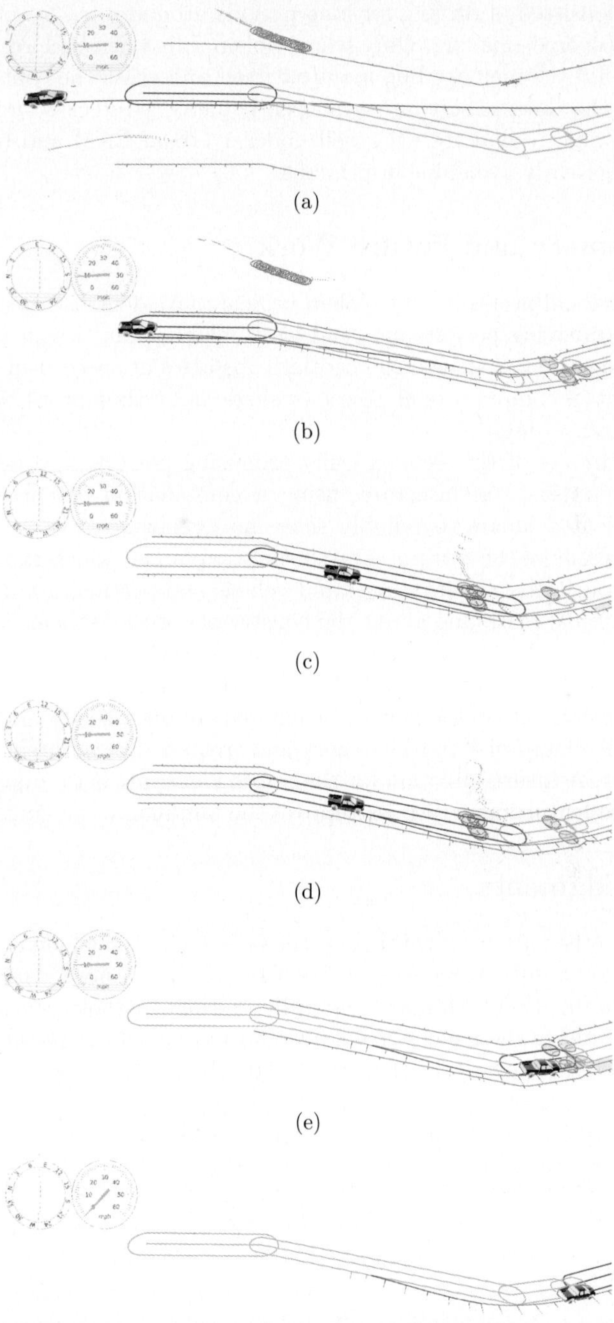

Fig. 7.25. An initial failure of Golem 2

trial which consisted of driving for long periods around a track would probably not have uncovered this bug. Only when Golem 2 had driven across 22 miles of new terrain did the memory bug manifest itself and crash the control program.

Although the Golem 2 software is inherently memory-intensive in its approach, it should be able to operate with well under 1 GB of RAM and therefore this failure was perfectly avoidable in principle.

7.8 Summary and Future Work

Despite occasional problems, the Golem vehicles have demonstrated a high level of high-speed driving performance and we think that our design approach has promise. The system managed to negotiate obstacles at speed using a relatively small amount of computational power (a single 2.2 GHz laptop) and relatively sparse laser range data.

The key drivers of this economically-promising performance include a simplified computational architecture; using a combination of horizontally- and vertically-oriented ladars to reliably sense major obstacles while disregarding inessential details of the terrain; a fast heuristic planner which rapidly finds solutions in typical driving situations; and vehicle state estimation using both an IMU and physical reasoning about the constraints of the vehicle.

The "false" obstacles sometimes announced by the ladar system were not entirely spurious, but generally represented road markings or minor, traversable obstacles. Ideally these would neither be ignored nor mistaken for non-traversable obstacles, but placed in a third category and treated appropriately by the planning software. A future direction for the sensing software is to improve the handling of these features and also to improve the handling of moving obstacles.

Acknowledgments

In addition to the authors of this paper, the Golem Group included Jim Swenson, Jerry K. Fuller, Josh Arensberg, Kerry Connor, Jeff Elings, Izaak Giberson, Maribeth Mason, Brent Morgan, and Bill Caldwell, without whose invaluable technical assistance the Golem vehicles would not have been possible.

The financial support of the Henry Samueli School of Engineering and Applied Sciences at the University of California, Los Angeles, is gratefully acknowledged. In addition, we received financial support and in-kind donations from a number of other organizations, including BEI Technologies, Mobileye, NovAtel, Sick, and OmniStar.

References

Alon, Y., Ferencz, A., and Shashua, A. (2006). Off-road path following using region classification and geometric projection constraints. International Conference on Computer Vision and Pattern Recognition.

Bornstein, J. A. and Shoemaker, C. M. (2003). Army ground robotics research program. In *Unmanned Ground Vehicle Technology V*, pages 303–310.

Coombs, D., Murphy, K., Lacaze, A., and Legowik, S. (2000). Driving autonomously offroad up to 35 km/h. In *Proceedings of the IEEE Intelligent Vehicles Symposium.*

Dickmanns, E. (1997). Vehicles capable of dynamic vision. In *Proceedings of the International Joint Conference on Artificial Intelligence*, pages 1577–1592.

Dickmanns, E. D. (2004). Dynamic vision-based intelligence. *AI Magazine*, 25(2): 10–30.

Faugeras, O., Hots, B., Mathieu, H., Vieville, T., Zhang, Z., Fua, P., Theron, E., Moll, L., Berry, G., Vuillemin, J., Bertin, P., and Proy, C. (1993). Real time correlation-based stereo: algorithm, implementations, and applications. Technical Report Technical report 2013, INRIA, Sophia-Antipolis, France.

Frazzoli, E., Dahleh, M. A., and Feron, E. (2002). Real-time motion planning for agile autonomous vehicles. *AIAA Journal of Guidance, Control, and Dynamics*, 25(1):116–129.

Freund, Y. and Schapire, R. E. (1996). Experiments with a new boosting algorithm. In *International Conference on Machine Learning*, pages 148–156.

Gelb, A. (1974). *Applied Optimal Estimation.* MIT Press, Cambridge, MA.

Hattori, H. and Takeda, N. (2005). Dense stereo matching in restricted disparity space. In *Proceedings of the IEEE Intelligent Vehicles Symposium*, pages 118–123.

Hong, T. H., Abrams, M., Chang, T., and Shneier, M. (2000). An intelligent world model for autonomous off-road driving. *Computer Vision and Image Understanding.*

Kalman, R. E. (1960). A new approach to linear filtering and prediction problems. *Transactions of the ASME*, 82:35–45.

Kalman, R. E. and Bucy, R. S. (1961). New results in linear filtering and prediction theory. *Transactions of the ASME*, 83:95–107.

Kavraki, L., Svestka, P., Latombe, J., and Overmars, M. (1996). Probabilistic roadmaps for path planning in high-dimensional configuration spaces. In *IEEE Transactions on Robotics and Automation*, volume 12, pages 566–580.

LaValle, S. (1998). Rapidly-exploring random trees: A new tool for path planning. Technical Report TR 98-11, Ames, IA: Iowa State University.

Rasmussen, C. (2002). Combining laser range, color, and texture cues for autonomous road following. In *Proceedings of the IEEE International Conference on Robotics and Automation.*

Rogers, R. M. (2003). *Applied Mathematics in Integrated Navigation Systems.* AIAA Education Series. AIAA, Reston, VA, second edition.

Roundy, D. (2005). *David's Advanced Revision Control System.* `http://darcs.net/`.

Tanabe, J. (2003). Visconti: multi-vliw image recognition processor based on configurable processor. In *Custom Integrated Circuits Conference*, pages 185–188.

Urmson, C. (2005). Navigation regimes for off-road driving. Technical Report CMU-RI-TR-05-23, Carnegie Mellon University Robotics Institute.

Urmson, C., Anhalt, J., Clark, M., Galatali, T., Gonzalez, J., Gowdy, J., Gutierrez, A., Harbaugh, S., Johnson-Roberson, M., Kato, H., Koon, P., Peterson, K., Peterson, B., Smith, B., Spiker, S., Tryzelaar, E., and Whittaker, W. (2004). High speed navigation of unrehearsed terrain: Red team technology for grand challenge. Technical Report Technical Report TR-04-37, Carnegie Mellon University Robotics Institute.

Welch, G. and Bishop, G. (1995). An introduction to the kalman filter. Technical Report TR 95-041, University of North Carolina at Chapel Hill, Chapel Hill, NC.

Fig. 8.1. CajunBot

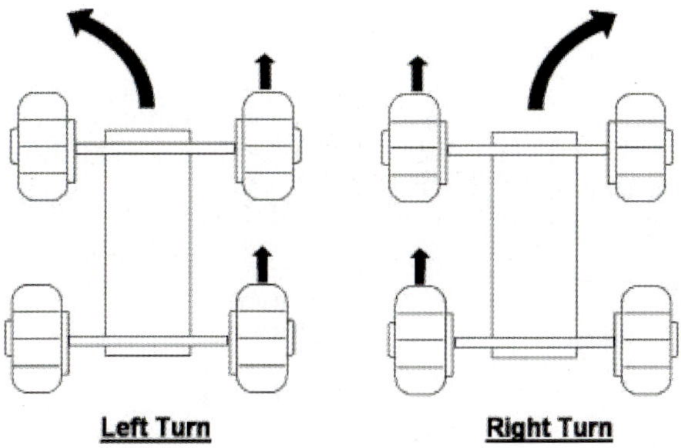

Fig. 8.2. Skid Steered Vehicle

route. The next two properties meant that we had more room to play in the software. The last property made it easy for us to build a drive-by-wire system.

In regards to the drive-by-wire system, it is instructive to know the mechanical points of contact for throttle, braking, and turning. The throttle on the vehicle is pulled by a cable, similar to that in a lawn mower or a motor cycle. Being a skid-steered vehicle, its braking and turning operations are interconnected. The vehicle turns by braking the wheels on one side, see Figure 8.2. The vehicle has two levers, with each lever controlling the transmission and brakes of the set of wheels on one side. Pulling a lever engages the brakes. Releasing the lever engages the gear. Lastly there is a region in between that represents neutral.

The choice of a MAX IV had its downsides. The top speed of the vehicle, about 45kph, meant it could not be a serious contender for a speed oriented track. The vehicle's five gallon tank was clearly insufficient for a 10+ hours run, and had to be retrofitted with a larger tank. The vehicle did not have any enclosure or roof, thus, a frame had to be built to house the electronics and for mounting sensors. Its power generation capacity was woefully inadequate for our needs requiring us to add a generator. Finally, absence of air conditioning implied we had to improvise the cooling system for computers and electronics.

The most significant drawback of a MAX IV is that it lacks any active suspension. Its wheels are its suspension. Any bump or shock not absorbed by the wheel is transferred to the rest of the body. We used LORD's center-bonded shock mounts (CBA 20-300) to provide vibration and shock isolation to the new frame, and therefore to the sensors mounted on the frame. In addition we used a MIL-spec Hardigg Case with rack mount to further isolate the computers from shocks.

The shock mounts, however, did not change the fact that the vehicle moves like a brick on wheels. Any movements felt by its six axles, as the vehicle travels over bumps, are transferred to the frame. This led to interesting challenges when processing sensors data, as elaborated later.

8.2.2 Electrical (Power) System

CajunBot requires around 1670W of power for simultaneous peak performance of all the equipment on board. When operating in the field the necessary power is generated by a Honda EU2000i generator, fed to a 2200VA UPS unit, which

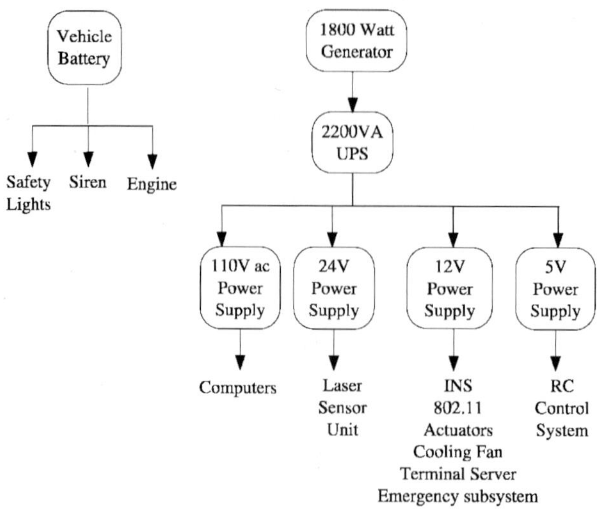

Fig. 8.3. CajunBot E-Power Subsytem

then conditions the power, and provides four power supplies - 5VDC, 12VDC, 24VDC and 110VAC, as shown in Figure 8.3. In the lab environment, the power may be switched to a wall outlet.

8.2.3 Electronics

Figure 8.4 shows a schematic diagram of CajunBot's electronics, which may be viewed as being composed of three major systems:

- Sensor Systems
- Drive-By-Wire System
- Computing System

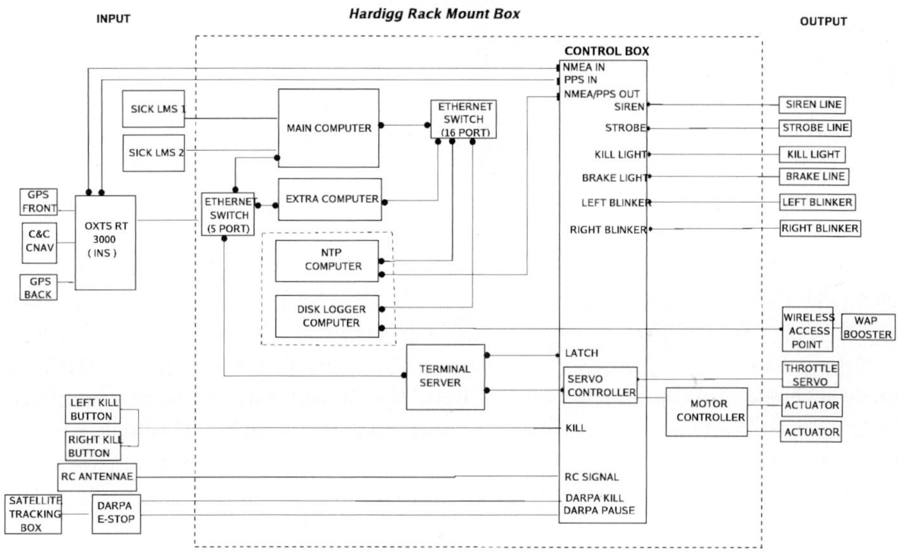

Fig. 8.4. CajunBot Electronic System

8.2.3.1 Sensor Systems

The components shown in the left column of Figure 8.4, identified as INPUT, constitute CajunBot's sensor systems. The sensors may be further classified as those needed for autonomous operation and those for monitoring and emergency control.

CajunBot uses an INS (Oxford Technology Solutions RT3102) and two LIDAR scanners (SICK LMS 291) for autonomous operation. The accuracy of the INS is enhanced by Starfire differential GPS correction signals provided by a C&C Technologies C-Nav receiver. The LIDARs are mounted to look at 16m and

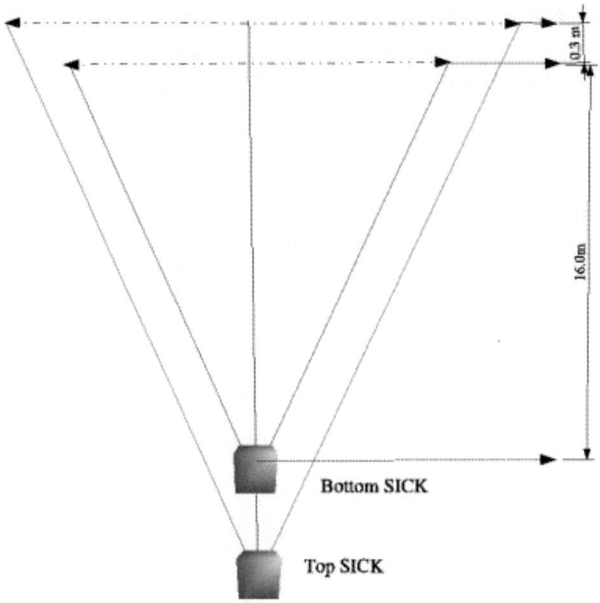

Fig. 8.5. The Top View of the Top and Bottom LIDAR's

16.3m in the front of the vehicle, see Figure 8.5. CajunBot also performs well, albeit at a reduced speed, with only one LIDAR.

The sensors for monitoring and emergency control include the DARPA E-Stop, when performing in the Grand Challenge; an RC Receiver to communicate with an RC Controller, used during testing and for moving the vehicle around when not in autonomous mode; a wireless access point to broadcast data for real-time monitoring in a chase vehicle; and two kill switches on each side of the vehicle.

8.2.3.2 Drive-by-Wire System

The box annotated as 'Control Box' and the components listed on the right column, annotated as OUTPUT, in Figure 8.4, constitute the Drive-By-Wire system.

The Drive-By-Wire System provides (1) an interface for computer control over the vehicle's mechanics, i.e., throttle and levers, and signals, i.e., siren, safety lights, and brake indicators; (2) emergency control of the vehicle in autonomous mode; and (3) manual control of the vehicle when not in autonomous mode.

Computer control. The drive-by-wire system provides serial communication interfaces for controlling the vehicle through software. There are five serial interfaces, one each for servo, left lever, right lever, signals, and controls.

The servo interface receives a single byte value representing the position of the servo connected to the throttle cable of the vehicle. The control box translates the value into appropriate servo signals. The interfaces for the left and right lever also behave similarly, except that a motor controller is used to communicate the commands to two actuators, one connected to each lever.

The serial interface for signals uses one bit for each of: left and right turn signals, brake lights, kill lights, strobe lights, and siren. The control box translates the bits received into an appropriate electrical signal to activate/deactivate a relay for each device.

While the other interfaces principally receive commands from the computers, the control interface provides input to the computers. The control interface currently provides single bit status indicators for pause and kill signals.

Emergency control. The system supports three emergency control operations: disable, manual override, and pause. The first two operations are implemented entirely in the Control Box, and override the computers. The vehicle may be disabled by the kill buttons on the vehicle, DARPA Kill-switch, or kill button on the RC Controller. When a kill signal is received, the drive-by-wire system pulls the left and right levers to the brake position, cuts the throttle, kills the engine, turns on a flashing light, and also turns on the kill signal on the control interface to inform the control software. For safety reasons when the vehicle goes outside the RC controller's range, a kill signal is issued internally, unless the RC kill is disabled.

The kill signal brings the vehicle to an abrupt halt, which could be detrimental when a vehicle is traveling at a high speed. Depending on the dexterity of the operator in the chase vehicle, it may sometimes be preferred to take manual control of the vehicle, and bring it to a safe state. This is achieved by toggling a button on the RC controller. The drive-by-wire system then ignores the computer commands and takes commands from the RC controller.

Finally, there is the pause signal, which is simply passed on to the software, which then stops the vehicle with maximum safe deceleration. In the pause mode, both the software and the hardware continue to operate. When the pause mode is removed, the vehicle resumes autonomous operation.

Manual control. In non-autonomous (or manual) mode the vehicle is operated using an RC controller. This operation is completely at the hardware level, and does not require the computers to be turned on.

8.2.3.3 Computing System

The computing system, the collection of four computers as shown in Figure 8.4, provides the computational power of CajunBot. The computers labeled "Main Machine" and "Extra Machine" are Dell PowerEdge 750s. CajunBot performed in the GC with only the main machine. The extra machine was in place if there was a need to distribute the computation. The other two computers "NTP Machine" and "Disk Logger Machine" are mini-ITX boards, used because of their low cost and small physical size. Both the boards are mounted on the same

1-U case. The computers were mounted on a shock proof MIL-spec Hardigg cases to dampen the shocks

The NTP Machine provides Network Time Protocol service, a service necessary to synchronize data from multiple sensors and computers. Though NTP is a light process a separate machine is dedicated to it for pragmatic reasons. To setup a Linux machine as an NTP server requires applying a PPS patch. The patch was available for Linux 2.4 Kernel, not for Linux 2.6 Kernel, the base of Fedora Core 2 OS used on our main computing machines. The Disk Logger Machine is used for logging data during a run, typically for post analysis. The logging operation is moved to a separate machine so that a disk failure, a very likely possibility in a 10 hour run, does not interfere with the autonomous operation. Using flash media for logging data would have circumvented the need to have a separate machine for this purpose. However, we were unable to boot the Dell PowerEdge 750s from flash media, and had to use machines with disks.

Though all the devices could potentially be connected on the same network, the system is configured with three networks, once again for pragmatic reasons. The first network, connects all the computers and the control box (via a Digi Terminal Server) through the 16-port gigabit Ethernet switch. This network carries data required for distributed processing. The INS is not connected on the same network because it required a certain network configuration for optimal performance. Since the INS data is used for all the phases of the processing, a second network consisting of the INS and all the computational machines is configured. A third network is configured to support real-time monitoring of the system from a chase vehicle. This network consists of the Disk Logger Machine connected to a wireless access point. The access point is not connected to the first network because the amount of data flowing on it saturates the wireless device, thus disrupting real-time monitoring. To overcome this problem the Disk Logger Machine samples the data before broadcasting.

8.3 Software Architecture

Figure 8.6 depicts CajunBot's software architecture. The system is decomposed into several components along functional boundaries. Each component, except the Middleware, runs as an independent process (program). The modules "Obstacle Detection", "Planner", and "Navigator" implement the *Core Algorithms* for autonomous behavior, and are discussed in the next section. All other modules are considered *Support Modules*, and are described in this section along with the design criteria that influence the software architecture.

The Drivers, Simulator, and Playback modules are mutually exclusive. While the Drivers module provides an interface to a physical device, the Simulator and the Playback modules provide virtual devices, as described below. Only one of the three modules can be active at any time. The mutual exclusion of the three modules is annotated in the architecture diagram by the walls between these modules, akin to the '|' symbol used in regular expressions.

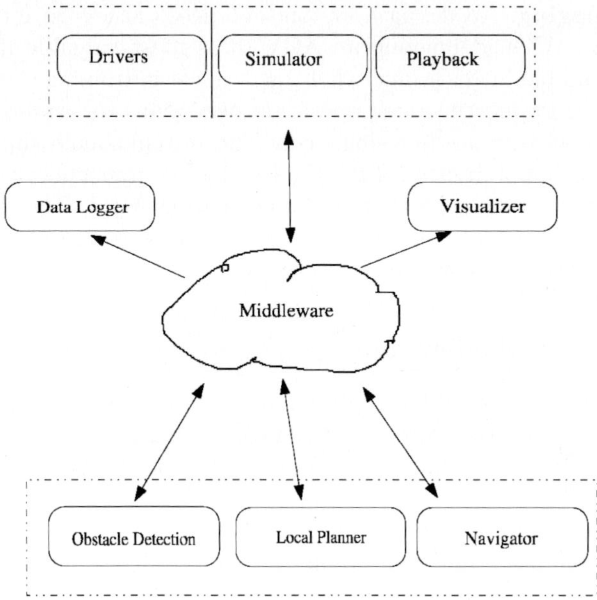

Fig. 8.6. Software Architecture

8.3.1 Design Criteria

The following design criteria influence the software architecture:

Device independence. There are multiple vendors for sensors, such as, GPS, INS, IMU, LIDARs, and so forth. The core algorithms of the system should not depend on the specific device. It should be possible to replace an existing device with another make/model or to introduce a new device while making only localized changes to the system.

Algorithm independence. Development of the system is an iterative process, which involves choosing between competing algorithms for the same task. It should be possible to develop each algorithm in isolation, that is, in a separate program, and switch the algorithm being used by selecting some configuration values.

Scalability. The computational requirements of the system may vary as the system's design evolves. For instance, if CajunBot did not perform adequately with two LIDARs and there was a need to add a third, it would require more computational power. It should be easy to add additional sensors and also seamlessly distribute the application on multiple computers.

Off-line Testability. The definitive way to test an AGV is to run it in the field. However, it is expensive to test each submodule of the system and every change by running the vehicle in the field. The system should enable off-line (in the lab) testing of various components and compositions of components.

Ease of debugging. To debug a system one needs access to internal data of the system. When debugging an AGV, it is most valuable if the internal data is available in real-time, when the vehicle is running. Debugging also requires performing the same operation over and over again, for one may not observe all the cues in a single run. The system should support (1) real-time monitoring of internal state of its various components and also (2) the ability to replay the internal states time-synchronized. The system should also support (3) presenting the data, which is expected to be voluminous, in a graphical form to enable ease of analysis.

8.3.2 CajunBot Middleware

The Middleware module, CBWare, provides the infrastructure for communication between distributed processes (CajunBot programs), such that the producers and consumers of data are independent of each other. Except for the properties of the data written to or read from CBWare, a module in the system does not need to know anything else about the module that has generated or will consume the data. This decoupling of modules is central to achieving the design criteria listed above.

CBWare provides two types of interfaces. A typed queue interface, CBQueues, for reading and writing messages. And a typed message packet interface, CB-Packets, for only writing messages. A typed message packet may also find its way into a queue, from where it may be read.

CBQueues provides distributed queues using a combination of POSIX Shared Memory [Marshall, 1999] and UDP communication. On an individual machine the queues are maintained as circular lists in the shared memory. The data written to a queue is distributed to other computers using a UDP broadcast. Figure 8.7 depicts how distributed interprocess communication is achieved by replicating shared memory queues across machines. This feature of CBWare to

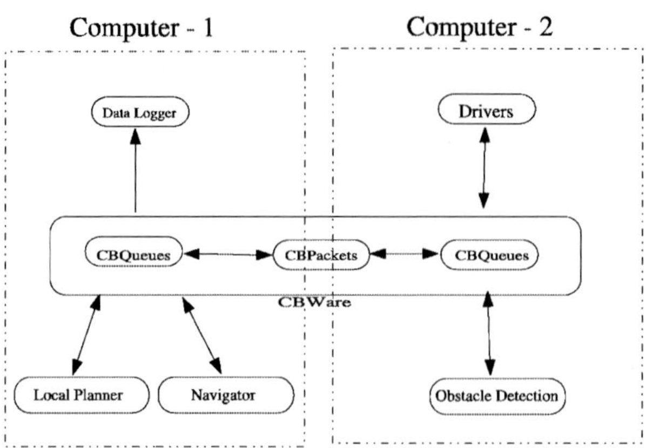

Fig. 8.7. Distributed Onboard Software Architecture

distribute queues over other machines allowed easy porting of programs over to multiple machines, achieving easy scalability of computational power, one of the design criteria.

CBQueues imposes an important constraint: each queue can have only one writer, but there is no limit on the number of readers. The single writer (producer) restriction ensures that the data in each distributed queue can be temporally ordered on the time the data was produced. If multiple producers of similar type of data exist, such as multiple LIDARs, a separate queue is maintained for each producer.

Besides providing the usual interfaces to access a queue, CBQueues also provides an interface to find in a queue two data items produced around a particular time. This capability, made possible due to temporal ordering of data in the queues, provides support for fusion of data from multiple sources based on the time of production. When two sources generate data at different frequencies, it may not always be appropriate to use the most recent data from both sources. Doing so may lead to the fusion of mutually inconsistent data. For instance, when a LIDAR scan is mapped to global coordinates using the INS data, the resulting coordinates would have significant error if the vehicle experienced a sharp bump immediately after the scan. In such cases it is better to fuse data in close temporal proximity. Along the same lines, instead of using the data generated directly by a source, sometimes it is preferred to interpolate the data for the specific time when data from another source is produced. In our LIDAR and INS example, it may be preferred to interpolate the position of the vehicle to match the time of LIDAR scan.

The CBPackets interface provides support for multiple writers and multiple readers. However, in so doing it cannot support temporal fusion of data. This interface is most useful for distributing status, warning, and error messages. Such messages are used in isolation, that is, they are not fused with other messages, and are mostly used for monitoring, not control.

CBWare serves the same purpose as NIST's Neutral Message Language (NML) [Shackleford et al., 2000], Simmons & Dale's CMU-IPC [Simmons and James, 2001], or RTI's NDDS [Pardo-Castellote and Hamilton, 1999], to cite a few middleware frameworks for real-time, distributed systems. While CBWare shares several similarities with each of these systems, such as publish-subscribe communication, the fundamental, and most crucial, difference is that CBWare supports fusion of sensor (and other data) based on the time of production. This support has been an important contributor in CajunBot's ability to leverage rough terrain to increase its sensor performance.

8.3.3 Data Logger

The Data Logger is run on the Disk Logger machine. Besides logging the data on disk, the Data Logger broadcasts data on the wireless network for real-time monitoring in a chase vehicle. CajunBot communicates with the chase vehicle on an 802.11G wireless network. The variety of wireless communication equipment we have tried tends to crash when all data produced in the queues is put in the air. This is particularly true when, for the purpose of debugging, the internal states of Obstacle Detection and Local Planner modules are broadcast. To accomodate for the shortcomings of the wireless network, the data logger has provision to sample the data at some prescribed interval. If disk space is an issue, it also provides support to save only a sample of the data.

8.3.4 Drivers

Device independence is achieved by having a separate program, referred to as a Driver, to interact with a particular device. The Drivers are divided into two classes. 1) Sensor drivers, which read input data from sensors, such as the INS and LIDARs. 2) Control drivers, which control devices, such as throttle, left and right levers, safety lights, siren, kill lights, brake lights, and indicator lights.

Besides hiding the details of communicating with the device, a driver also transforms the data to match units and conventions used by the rest of the system. For instance, the CajunBot system measures angles in the anti-clockwise direction, with East as zero. If an IMU or INS uses any other convention for measuring angles, its corresponding device driver transforms angles from the device's convention to CajunBot's convention. Similarly, most GPS and INS equipment tends to provide vehicle position in latitude/longitude, which is translated by the driver to UTM coordinates, as used by the CajunBot system.

8.3.5 Simulator

Off-line testability is a direct outcome of device independence. Since the core algorithms are unaware of the source/destination of the data, the data does not have to come from or go to an actual device. The algorithms may as well interact with virtual devices and a virtual world. The Simulator module (and the Playback module, discussed later) creates a virtual world in which the core algorithms can be tested in the laboratory.

CajunBot's Simulator, CBSim, is a physics-based simulator developed using the Open Dynamics Engine (ODE) physics engine. Along with simulating the vehicle dynamics and terrain, CBSim also simulates all the onboard sensors. It populates the same CBWare queues with data in the same format as the sensor drivers. It also reads vehicle control commands from CBWare queues and interprets them to have the desired effect on the simulated vehicle.

While CBSim is a physics-based simulator like Stage [Gerkey et al., 2003] and Gazebo [Vaughan, 2000], it has two interesting differences. First, CBSim does not provide any visual/graphical interface. The visualization of the world and the vehicle state is provided by the Visualizer module, discussed later. Second, CBSim also generates a clock, albeit a simulated one, using the CBWare queues.

The simulated clock helps in synchronizing the distributed programs when running in a virtual world. The distributed programs of CajunBot have a read-process-output-sleep loop, as elaborated later. The frequency at which a program is scheduled is controlled by choosing the duration for which it sleeps. When operating in the real-world, the duration of 'sleep' is measured in elapsed real-time. However, it is not beneficial to use elapsed real-time to control the sleep duration of a process when operating in a virtual environment. In particular, using the real-time for controlling sleep forces the simulation to execute at the same pace as the real program, even when one may be using faster and better computers. Thus, when operating in virtual environment we use the elapsed simulated time to determine how long a processs sleeps.

By maintaining a system wide simulated time, the CajunBot system is able to create a higher fidelity simulation than that provided by Stage and Gazebo. The computation in the entire system can be stopped by stopping the clock; and its speed can be altered by slowing down or speeding up the clock. This also makes it feasible to run the application in a single step mode, executing one cycle of all programs at a time, thereby significantly improving testing and debugging.

8.3.6 Playback

Offline-testing and debugging is further aided by the Playback module. This module reads data logged from the disk and populates CBWare queues associated with the data. The order in which data is placed in different queues is determined by the time stamp of the data. This ensures that the queues are populated in the same relative order. In addition, the Playback module, like the Simulator module, generates the simulator time queue representing the system-wide clock.

This simple act of playing back the logged data has several benefits. In the simplest use, the data can be visualized (using the Visualizer module) over and over again, to replay a scenario that may have occured in the field or the simulator. It offers the ability to replay a run after a certain milestone, such as a certain amount of elapsed time or a waypoint is crossed. In a more significant use, the playback module can also be used to test the core algorithms with archived data. This capability has been instrumental in helping us refine and tune our obstacle detection algorithm. It is our common operating procedure to drive the vehicle over some terrain (such as during the DARPA National Qualifying Event), playback the INS and LIDAR data, apply the obstacle detection algorithm on the data, and then tune the parameters to improve the obstacle detection accuracy.

8.3.7 Visualizer

Real-time and off-line debugging is supported by CBViz, the Visualizer module. CBViz is an OpenGL graphical program that presents visual/graphical views of the world seen by the system. It accesses the data to be viewed from the CBWare queues. Thus, CBViz may be used to visualize data live during field tests and simulated tests, as well as visualizing logged data using the Playback module.

Since communication between processes occurs using CBWare, CBViz can visualize data flow between processes. We have also found it beneficial to create special queues in CBWare to provide CBViz with data that is otherwise internal to a process. This capability has been very effective, and almost essential, in achieving the Ease of Debugging design criterion as listed above.

8.4 Algorithms

This section presents the algorithms underlying CajunBot's autonomous behavior.

Figure 8.8 presents the system level data flow diagram. Although CajunBot's system uses the blackboard architecture, see Figure 8.6, the system level DFD of Figure 8.8 shows the actual data flows. Each step in the system level DFD is implemented by one or more independent programs. Each program has the following pattern:

```
while (true) {
   read inputs;
   perform processing;
   generate outputs;
   sleep for a specified time
}
```

The overall steps are depicted in Figure 8.8. The INS data read in Step 1 gives the vehicle's state, which includes position in UTM coordinate space, orientations along the three dimensions (i.e., heading, roll, and pitch), speed over ground, and

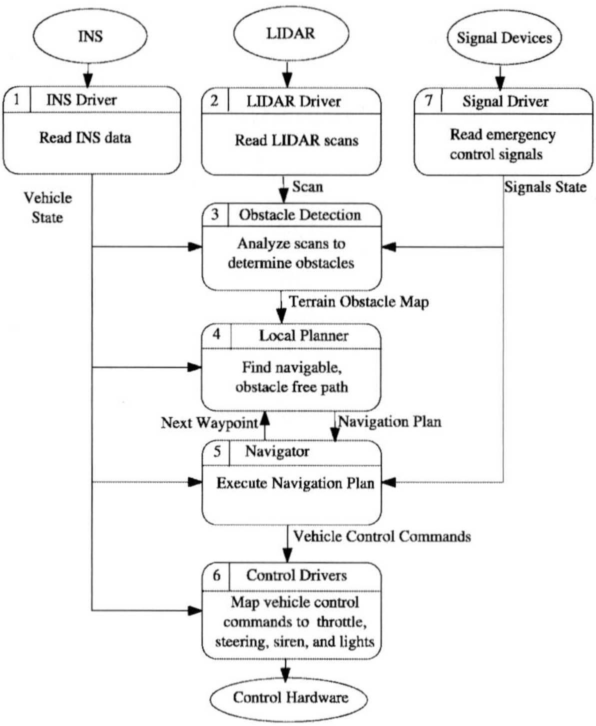

Fig. 8.8. System Level Data Flow Diagram

accelerations along the three dimensions. The LIDAR generates a sequence of scans, with each scan containing a collection of beams. Each beam is a value in polar coordinate space. The LIDAR scans are read in Step 2. In Step 3, the LIDAR scans and vehicle state (at the time of the scan) are used by the Obstacle Detection Module to create a Terrain Obstacle Map (TOM). The TOM is a map of obstacles in the vicinity of the vehicle. The Local Path Planner Module (Step 4) uses the TOM and the vehicle's state to generate a Navigation Plan. This consists of a sequence of waypoints and the recommended speed along each segment. The Navigation Plan is used by the Navigator Module, Step 5, to generate steering and throttle commands. In executing the plan the Navigator takes into account safety of the vehicle. Finally, the Control Drivers (Step 6) map the steering and throttle commands to physical devices. This includes the throttle servo position and actuator positions for brake control. The Pause and Kill Signals are read by the Signal Driver (Step 7). These signals are used by the Obstacle Detection and Navigator Modules, as described later. The Navigator Module also controls the emergency signals, siren and flashing lights.

The system does not use any explicit real-time primitives. Each program runs in an endless loop, reading from its input CBWare queues and writing to its output CBWare queues. When explicit syncronization is needed between the producer and consumer of some data, the consumer uses CBWare primitives that block it if no new data is available from the respective producer. Further, the sleep step of each program is tuned to have the program run at a certain frequency. The rate at which various programs of Figure 8.8 operate are provided in Table 8.1.

Table 8.1. Operating frequencies of programs of Figure 8.8

Program	Frequency (Hz)
INS Driver	100
LIDAR Driver	75
Obstacle Detection	15
Local Planner	5
Navigator	20
Control Drivers	20

The algorithms for the Obstacle Detection, Local Path Planner, and Motion Controller Modules are key to providing the autonomous behavior and are described below.

8.4.1 Obstacle Detection Module

This section summarizes CajunBot's obstacle detection algorithm and highlights the specific features that enable it to take advantage of vibrations along the height axis, i.e., bumps, to improve its ability to detect obstacles.

8.4.1.1 The Algorithm

The data flow diagram in Figure 8.9 enumerates the major steps of the obstacle detection algorithm. The algorithm takes as input the vehicle's state and LIDAR scans. The vehicle state data is filtered to attend to spikes in data due to sensor errors (Step 3.1), and then used to compute the global coordinates for the locations from which the beams in a LIDAR scan were reflected (Step 3.2). The global coordinates form a 3-D space with the X and Y axes corresponding to the Easting and Northing axes of UTM Coordinates, and the Z axis giving the height above sea level. Virtual triangular surfaces with sides of length 0.20m to 0.40m are created with the global points as the vertices. The slope of each such surface is computed and associated with the centroid of the triangle (Step 3.3). A vector product of the sides of the triangle yields the slope. The height and slope information is maintained in a digital terrain map, which is an infinite grid of 0.32m $\times$ 0.32m cells. A small part of this grid within the vicinity of the vehicle is analyzed to determine whether each cell contains obstacles (Step 3.4). This data is then extracted as a Terrain Obstacle Map.

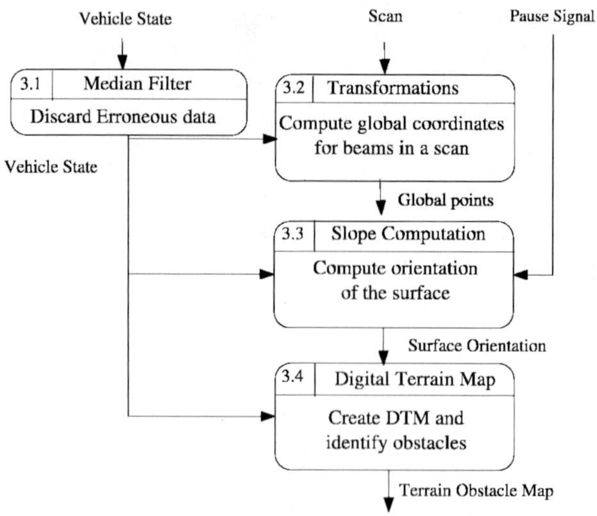

Fig. 8.9. Data Flow Diagram for the Obstacle Detection Module

Figure 8.10 graphically depicts data from the steps discussed above. The figure presents pertinent data at a particular instant of time. The grey region represents the path between two waypoints. The radial lines emanating from the lower part of the figure show the LIDAR beams. There are two sets of LIDAR beams, one for each LIDAR. Only beams that are reflected from some object or surface are shown. The scattering of black dots represent the global points, the points where

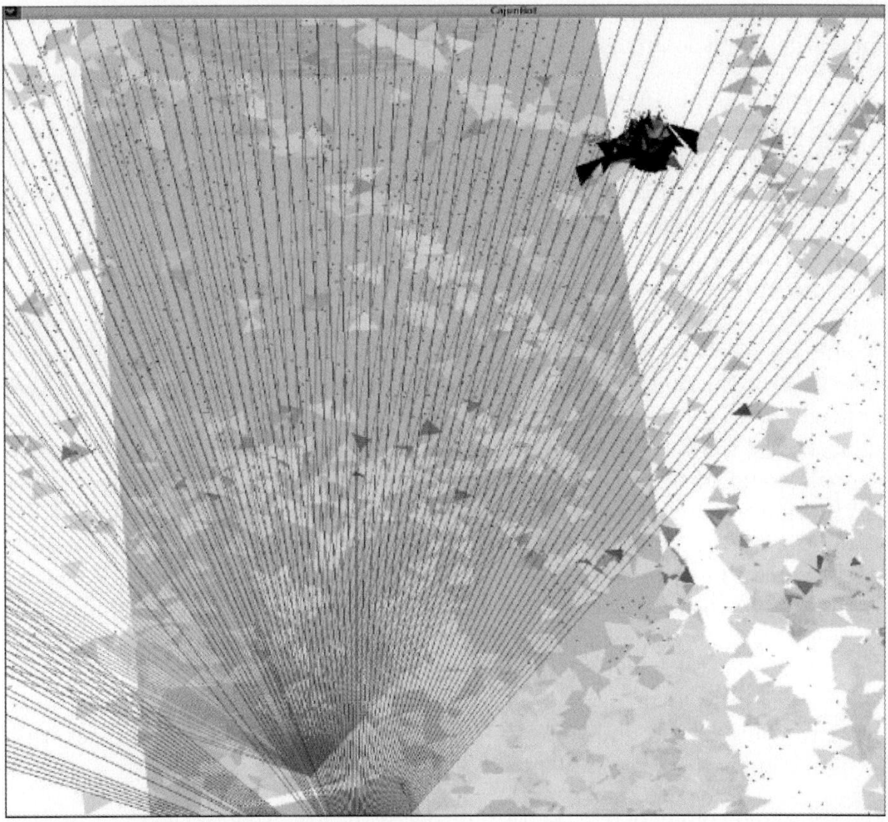

Fig. 8.10. Virtual Triangle Visualization

LIDAR beams from some previous iteration had reflected. The figure is scattered with triangles created from the global points. Only global points that satisfy the spatio-temporal constraints, discussed later, are part of triangles. There is a lag in the data being displayed. The triangles shown, the global points, and the LIDAR beam are not from the same instant. Hence, some points that can make suitable triangles are not shown to form triangles. The shade of the triangles in Figure 8.10 represents the magnitude of slopes. The black triangles have high slope, +/- 90 degrees, and the ones with lighter shades have much smaller slopes. In the figure, a trash can is detected as an obstacle, as shown by the heap of black triangles. The data was collected in UL's Horse Farm, a farm with ungraded surface. The scattering of dark triangles is a result of the uneven surface.

An obstacle is classified as an obstacle using the following steps. First, a cell is tagged as a 'potential' obstacle if it satisfies one of three criteria. The number of times a cell is categorized as a potential obstacle by a criterion is counted. If this count exceeds a threshold–a separate threshold for each criterion–it is deemed an obstacle. The criteria used to determine the classification of a cell as a potential obstacle are as follows:

High absolute slope. A cell is deemed as a potential obstacle if the absolute maximum slope is greater than 40 degrees. Large objects, such as, cars, fences, and walls, for which all three vertices of a triangle can fall on the object, are identified as potential obstacles by this criterion. The threshold angle of 40 degrees is chosen because CajunBot cannot physically climb such a slope. Thus, this criterion also helps in keeping CajunBot away from unnavigable surfaces.

High relative slope. A cell is deemed as a potential obstacle if (1) the maximum difference between the slope of a cell and a neighbor is greater than 40 degrees and (2) if the maximum difference between the heights of the cell and that neighbor is greater than 23cm. This criterion helps in detecting rocks as obstacles, when the rock is not large enough to register three LIDAR beams that would form a triangle satisfying the spatio-temporal constraint. The criterion also helps in detecting large obstacles when traveling on a slope, for the relative slope of the obstacle may be 90 degrees, but the absolute slope may be less than 40 degrees. The test for height difference ensures that small rocks and bushes are not deemed as a potential obstacle. The height 23cm is 2cm more than the ground clearance of CajunBot.

High relative height. A cell is deemed as a potential obstacle if the difference between its height and the height of any of its neighbor is greater than 23cm. This criterion aids in detecting narrow obstacles, such as poles, that may register very few LIDAR hits.

The threshold counts of 5, 5, and 12, respectively, are used for the three criteria to confirm a potential obstacle as an obstacle.

As a matter of caution, Step 3.3 disables any processing when the Pause Signal is activated. This prevents the system from being corrupted if someone walks in front of the vehicle when the vehicle is paused, as may be expected since the Pause Signal is activated during startup and in an emergency.

8.4.1.2 Utilizing Bumps to Enhance Obstacle Detection Distance

Figure 8.11 presents evidence that the algorithm's obstacle detection distance improves with roughness of the terrain (bumps). The figure plots data logged by CajunBot traveling at 7m/s through a distance of about 640m of a bumpy section during the 2005 GC Final. The X-axis of the plot represents the absolute acceleration along the height (Z) axis at a particular time. Greater acceleration implies greater bumps. The Y-axis represents the largest distance from the vehicle at which an obstacle is recorded in the Terrain Obstacle Map. The plot is the result of pairing, at a particular instance, the vehicle's Z acceleration with the furthest recorded obstacle in the Terrain Obstacle Map (which need not always be the furthest point where the LIDAR beams hit). The plot shows that the obstacle detection distance increases almost linearly with the severity of bumps

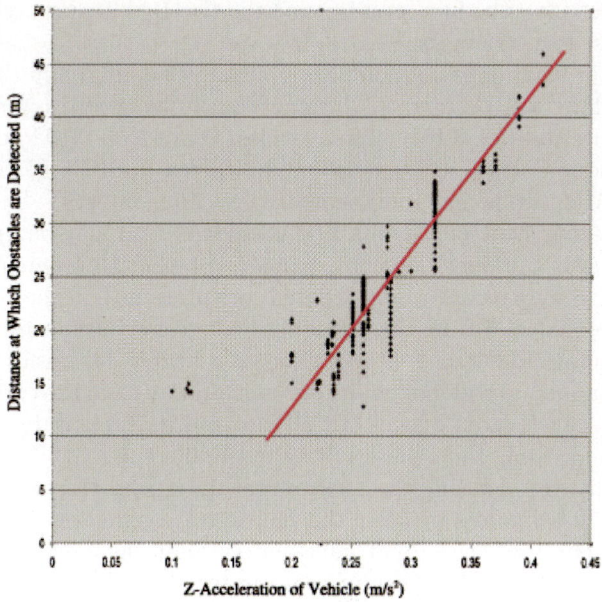

Fig. 8.11. Graph Showing Distance to Detected Obstacles Versus Z-Acceleration

experienced by the vehicle. The absolute vertical acceleration was never less than 0.1 m/s^2 because the vehicle travelled at a high speed of 10 m/s on a rough terrain. That the onboard video did not show any obstacles on the track and that the obstacle detector also did not place any obstacles on the track leads us to believe that the method did not detect any false obstacles.

Bumps along the road have impact on two steps of the algorithm, Step 3.2, where data from the INS and LIDAR is fused and, Step 3.3, when data from beams from multiple LIDAR scans are collected to create a triangular surface. The issues and solutions for each of these steps are elaborated below.

In order to meaningfully fuse INS and LIDAR data it is important that the INS data give orientation of the LIDARs at the time a scan is read. Since it is not feasible to mount an INS on top of a LIDAR, due to the bulk and cost of an INS, the next logical solution is to mount the two such that they are mutually rigid, that is, the two units experience the same movements. There are three general strategies to ensure mutual rigidity between sensors: (1) Using a vehicle with a very good suspension so as to dampen sudden rotational movements of the whole body and mounting the sensors anywhere in the body. (2) Mounting the sensors on a platform stabilized by a Gimbal or other stabilizers. (3) Mounting all sensors on a single platform and ensuring that the entire platform is rigid (i.e., does not have tuning fork effects). Of course, it is also possible to combine the three methods.

CajunBot uses the third strategy. The sensor mounting areas of the metal frame we created is rigid, strengthened by trusses and beams. In contrast, most other GC teams used the first strategy and the two Red Teams used a combination of the first two strategies.

Strategy 3 in itself does not completely ensure that mutually consistent INS and LIDAR data will be used for fusion. The problem still remains that the sensors generate data at different frequencies. Oxford RT 3102 generates data at 100Hz, producing data at 10ms intervals, whereas a SICK LMS 291 LIDAR operates at 75Hz, producing scans separated by 13ms intervals. Thus, the most recent INS reading available when a LIDAR scan is read may be up to 9ms old. Since a rigid sensor mount does not dampen rotational movements, it is also possible the INS may record a very different orientation than the time when the LMS data is recorded. Fusing these readings can give erroneous results, more so because an angular difference of a fraction of a degree can result in a LIDAR beam being mapped to a global point several feet away from the correct location.

The temporally ordered queues of CBWare and its support for interpolating data help in addressing the issue resulting from differences in the throughput of the sensors. Instead of fusing the most recent data from the two sensors, Step 3.2 computes global points by using the vehicle state generated by interpolating the state immediately before and immediately after the time when a LIDAR scan was read. Robots with some mechanism for stabilizing sensors can fuse a LIDAR scan with the most recent INS data because the stabilizing mechanism dampens rotatonal movements, thus ensuring that the sensors will not experience significantly different orientations in any 10ms period.

Fig. 8.12. LIDAR Beams Scattered Due to Bumps

Absence of a sensor stabilizer also influences Step 3.3, wherein triangular surfaces are created by collecting global points corresponding to LIDAR beams. Since CajunBot's sensors are not stabilized, its successive scans do not incrementally sweep the surface. Instead, the scans are scattered over the surface as shown in Figure 8.12. This makes it impossible to create a sufficient number of triangular surfaces of sides 0.20m to 0.40m using points from successive scans (or even ten successive scans). It is always possible to create very large triangles, but then the slope of such a triangle is not always a good approximation for the actual slope of its centroid.

If the GPS/INS data were very precise then triangles of desired dimensions could be created by saving the global points from Step 3.2 in a terrain matrix, and finding groups of three points at a desired spatial distance. This is not practical because of Z-drift, the drift in Z values reported by a GPS (and therefore by the INS) over time. When stationary, a drift of 10 cm - 25 cm in Z values can make even a flat surface appear uneven.

The Z-drift issue can be addressed by taking into account the time when a particular global point was observed. In other words, a global point is a 4-D value (x, y, z, and time-of-measurement). Besides requiring that the spatial distance between the points of a triangular surface be within 0.20m and 0.40m, Step 3.3 also requires that their temporal distance be under three seconds.

To recap, the following features of the Obstacle Detection Module enables it to utilize bumps to improve obstactle detection distance.

- A rigid frame for mounting all sensors.
- Fusing mutually consistent LIDAR scan with INS data based on the time of production of data.
- Using 4-D space and spatio-temporal constraints for creating triangles to compute the slope of locations in the 3-D world.

8.4.1.3 Performance Evaluation on 2005 GC Data

We now turn to the question of efficiency and scalability of the algorithm. Tables 8.2 and 8.3 present the following data for a single LIDAR and LIDAR pair configuration.

CPU Utilization. The average 'percentage CPU utilization', as reported by the Linux utility `top`, sampled every second.

Increase in CPU. The percentage increase in CPU utilization going from one LIDAR configuration to a configuration of two LIDARs.

Table 8.2 gives the data when the terrain was not very bumpy, whereas Table 8.3 presents data for bumpy terrain in the actual Grand Challenge Final Run. In both the situations, adding another LIDAR reduces the obstacle detection time at a higher rate (38-48%) than the increase in the CPU utilization (22-28%). This implies our algorithm scales well with additional LIDARs, since the benefits of adding a LIDAR exceeds the costs.

Table 8.2. Effect of number of LIDARs, with low average bumps: $0.11m/s^2$

# LIDARs	1	2
CPU Utilization	12.4%	15.9%
Increase in CPU		28.23%

Table 8.3. Effect of number of LIDARs, with high average bumps: $0.24m/s^2$

# LIDARs	1	2
CPU Utilization	11.2%	13.7%
Increase in CPU		22.32%

Comparing data across the Table 8.2 and Table 8.3 further substantiates that our algorithm takes advantage of bumps. Compare the data for the single LIDAR configurations in the two tables. The CPU utilization is lower when the terrain is bumpy. The same is true for the dual LIDAR configuration. The more interesting point is that adding another LIDAR does not lead to the same increase in CPU utilization for the two forms of terrain. For the bumpy terrain the CPU utilization increased by 22.32%, which is significantly less than the 28.23% increase for the smoother terrain. The efficient and scalable implementation is due to two factors. First, in Step 3.3 it is not necessary that the triangles be created using global points observed by the same LIDAR. The data may be from multiple LIDARs. The only requirement is that the triangles created satisfy the spatio-temporal constraints. The second factor is that we utilize an efficient data structure for maintaining the 4-D space. Though the 4-D space is infinite, an efficient representation is achieved from the observation that only the most recent three seconds of the space need to be represented. This follows from the temporal constraint and that one point of each triangle created in Step 3.3 is always from the most recent scan.

8.4.2 Local Path Planner Module

The RDDF (route data description format) file provided as input to the vehicle may be considered a global plan, a plan computed in response to some global mission. The Local Path Planner Module's role is to create a navigation plan to advance the vehicle towards its mission taking into account the ground realities observed by the sensors. The navigation plan is a sequence of *local* waypoints that directs the vehicle towards an intermediate goal, while staying within the lateral boundary and avoiding obstacles. The intermediate goal–a point on the RDDF route at a fixed distance away from the vehicle–advances as the vehicle makes progress. Also associated to each local waypoint is a speed that is within the speed limits prescribed in the global plan and one that can be safely achieved by the vehicle.

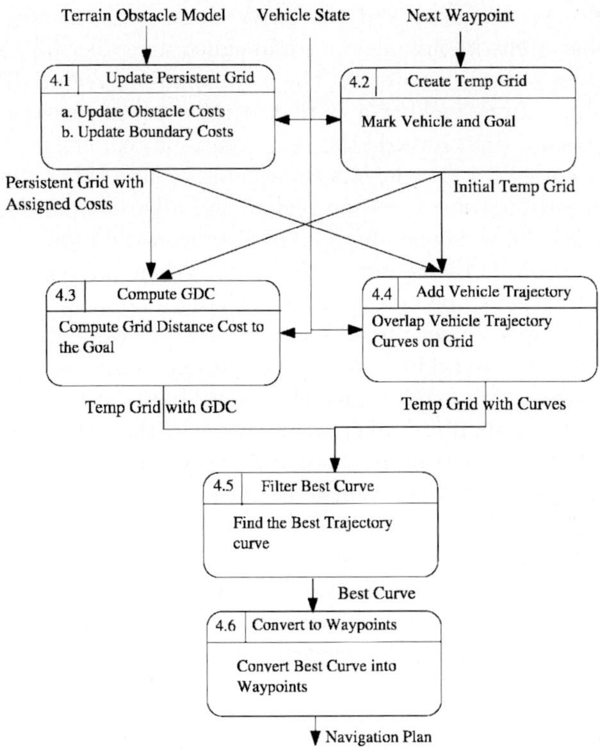

Fig. 8.13. Data Flow Diagram for the Navigator Module

Figure 8.13 presents the logical data flow diagram of the local path planner[1]. The algorithm maintains two grid representations of the world, a persistent grid and a temporary grid. The persistent grid uses Easting and Northing coordinates whereas the temporary grid uses the vehicle's own coordinate space. As the name suggests, the persistent grid retains data between iterations of the algorithm, whereas the temporary grid is created afresh for each iteration. Following is the explanation of the steps followed by local path planner as shown in Figure 8.13.

Step 4.1 updates 'assigned costs' in the Persistent Grid. A cell in the persistent grid is assigned two types of costs, *obstacle cost* and *boundary cost.* The obstacle cost is based on the proximity of a cell to an obstacle. The boundary cost is based on its proximity to the boundary. The locations of obstacles are obtained from the Terrain Obstacle Model. To enable treating the vehicle as a point object, each obstacle is expanded in two concentric circles, as shown in Figure 8.14. First the region within the inner circle is called the hard obstacle expansion region. Each cell inside this circle is assigned an infinite obstacle cost. That the point vehicle is in this region implies the vehicle has crashed into some object. The hard expansion indicates the area the vehicle must avoid at any cost. Second,

[1] Note that branches in a DFD represent branching of data flow, not of control.

the region between the inner and the outer circle is called the soft obstacle expansion region. Cells in this region are assigned a smaller obstacle cost, and the cost decreases radially outwards. The soft expansion region discourages the local path planner from picking a path too close to an obstacle, unless it is absolutely necessary. Cells outside the soft expansion region are assigned a zero obstacle cost. Figure 8.14 also depicts three types of lateral boundary expansion region: warning region, soft expansion region, and a hard expansion region. The warning region is 0.3m wide and about 0.7m distance within the lateral boundary in the direction towards the center of the track. When the (point) vehicle is in this region, its wheels will be barely inside the lateral boundary. The cells in the warning region are given a very small boundary cost, a cost sufficient to keep the vehicle away from the lateral boundary. The soft expansion region starts outside of the warning region and extends past the lateral boundary for about 2m. Cells in this region have a slightly higher boundary cost than those in the warning region. This cost is intended to 'aggressively' prevent the vehicle out from cross the lateral boundary, and if the vehicle happens to be in the lateral boundary it aggressively pushes it back. The region outside of the track, further past the soft expansion region is called the hard expansion region. Cells in this region are assigned an infinite boundary cost. Like the hard obstacle expansion region, this region is considered to be unsafe for the vehicle and should be avoided at any cost. Cells inside the track, between the warning region, are assigned a zero boundary cost.

Step 4.2 uses the Vehicle State and the Next Global Waypoint information to create the temporary grid, which involves allocating memory, marking the cells for the intermediate goal, and marking the position of the vehicle.

Step 4.3 computes the Grid Distance Cost (GDC) [Maida et al., 2006, Barraquand et al., 1992] for the cells in-between the vehicle and the intermediate goal on the temporary grid. Each cell of the temporary grid has three costs associated with it: obstacle cost, lateral boundary cost and cell cost. Obstacle costs and lateral boundary costs are taken from the corresponding cell of the persistent grid, computed in Step 4.1. The cell cost represents the cost of traveling from that cell to the goal cell. It includes distance to the goal, penalty for traveling close to obstacles and lateral boundary.

The following recursive equations describe the relation between the cell cost $C(i)$ of a cell i and the cell costs of its neighbors $nbs(i)$.

$C(i) = 0$, i is a goal cell.
$C(i) = min\{C(j) + T(j,i) \mid j \in nbs(i)\}$, i is not a goal cell

where $T(j,i) = CF(j,i) \times (1 + B(j) + O(j))$, is the cost of traveling from cell j to cell i, $CF(j,i)$ is the Chamfer Factor [de Smith, 2004], $B(j)$ is the boundary cost of cell j, and $O(j)$ is the obstacle cost of cell j.

The fixed point of the above system of equations gives the desired cost. We use the A* algorithm to efficiently direct the propagation of costs from the goal towards the vehicle.

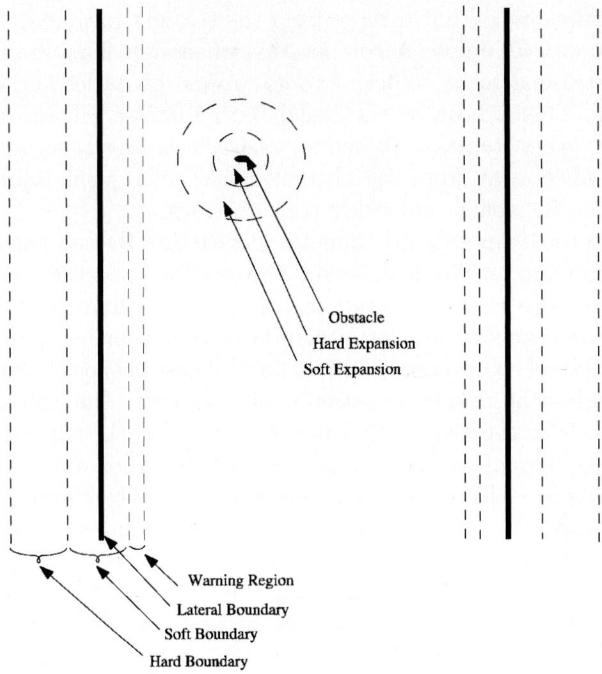

Fig. 8.14. Obstacle and boundary costs

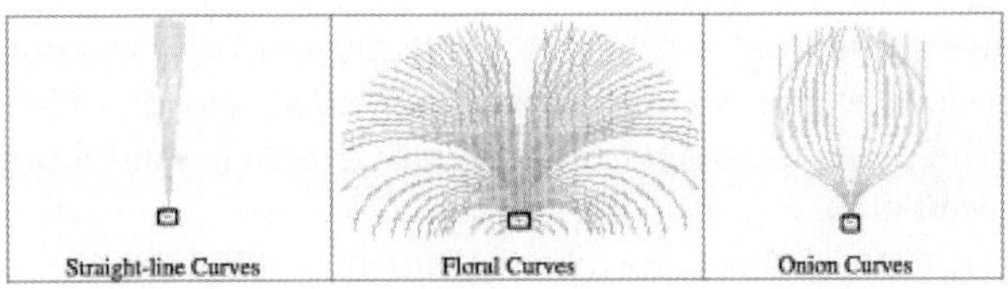

Fig. 8.15. Maneuverable Curves of CajunBot

Step 4.4 overlays on the temporary grid a set of pre-computed curves representing acceptable movements of the vehicle. The curves are truncated at the point where they enter an obstacle cell, or a lateral boundary cell. Figure 8.15 shows the maneuverable curves in an obstacle free world. The curves originate at the vehicle and spread out along the orientation of the vehicle. These curves are pre-computed only once by simulating the vehicle's steering control [Scheuer and Xie, 1999, Scheuer and Laugier, 1999]. Unlike curve computations based on Dubin's car [Dubins, 1957], our collection of curves does not represent all possible maneuvers that can be made by the vehicle. Instead, as evident from Figure 8.15, our collection contains *straight-line curves*, those curves emanating in the direction of the vehicle's heading; *floral curves*, curves that diverge from the current heading of the vehicle like a floral arrangement; and *onion curves*, the curves that diverge from the straight path and then converge back, like layers in an onion. A combination of straight-line, floral curves, and onion-curves is found to be sufficient to go around obstacles and make various types of turns. This is because the Local Path Planner generates a new navigation plan in every iteration. When an obstacle is first seen, an onion curve diverts the vehicle away from the obstacle. In a subsequent iteration, a floral curve brings the the vehicle smoothly back on track.

Step 4.5 uses the temporary grid annotated with the GDC and projected curves to select the Best Curve. This is done in two steps. First, a set of candidate curves are selected from the projected curves. Second, a Best Curve is selected from the candidate curves. Each curve is evaluated on three properties, namely, a) length of the curve, b) final cost to the goal, i.e., the cell cost of the cell where the curve terminates, and c) the rate of reduction in the cell cost. The latter is the reduction in cell cost from the start of the curve to the end of the curve divided by the cost of traveling through each cell–$T(j, i)$ described earlier–along the curve. The candidate curves are selected from the projected curves by eliminating curves as follows. First, all curves that are smaller than a minimum acceptable length are removed. Next, of the remaining curves the smallest final cost to the goal is found. All curves whose final cost to the goal is greater than some threshold more than the smallest final cost are removed. Finally, the highest rate of reduction in cell cost of the remaining curves is found. Any curve whose rate of reduction is less than some threshold of the highest is removed. Now, the Best Curve of the set of candidate curves is the curve with the end point closest to the centroid of the all the candidate curves. If the set of candidate curves is empty, the path planner reports that it cannot generate any path.

Step 4.6 converts the best curve into a sequence of local waypoints. This step also annotates each waypoint with a recommended speed, which is computed using the RDDF speed, distance to the nearest obstacle, and amount of turn to make at that waypoint.

The key innovation of our algorithm is that it is a hybrid of discrete and differential algorithms (see [LaValle, 2006] for a comprehensive survey on planning algorithms). Step 4.3, the discrete part of the algorithm, computes GDC without taking into account the vehicle's state. Steps 4.4 and 4.5, the differential components of the algorithm, take the vehicle's state and its maneuverability into account to pick the Best Curve. By combining discrete and differential algorithms, we are able to get the best of both worlds. The discrete algorithm is fast and efficient, and hence can be used for planning over a large area. The differential component computes a path only in the vicinity of the vehicle. But, since the path selected is based on GDC, the curve chosen may be influenced by terrain conditions much further away. Thus, the navigation plan created is maneuverable within the vicinity of the vehicle and also brings the vehicle to a position close to an optimal path.

There are other non-trivial and interesting aspects of the algorithm that are worthy of explanation: 1) representation of the goal; and 2) representation of robot and obstacles to aid in creation of navigable paths.

Representation of goal. The size and shape of the goal has very significant effect on the paths created by the algorithm. If the goal is a point object on the center line, and there is an obstacle near the goal, the path generated will hug the boundaries of the obstacle even if there is a large amount of free space on the track. If the goal is a straight line encompassing the whole segment, perpendicular to the direction of motion, then the path may hug the corners after a turn, since that will be the shortest path.

We use a V-shaped goal, with an angle of about 150 degrees at the vertex. When there are no obstacles on the track, the tip of the V-goal smoothly brings the vehicle to the center of the track. However if there is an obstacle on the center lane of the track, the path generated will smoothly deflect away from the obstacle aiming for some other point on the V-shaped goal.

Representation of robot and obstacles. The classical approach to path planning for a circular robot with unlimited mobility is to model the robot as a point object and to expand an obstacle cell in a circle the same dimension as the robot [Feng et al., 1990, Stentz and Hebert, 1995]. Whether the robot placed in a particular cell will collide with an obstacle can be determined by checking if the cell falls in the expansion region of an obstacle. This model of robot and obstacle leads to a very efficient test for collision.

We extend the above approach for the AGV domain. The Local Path Planner represents the vehicle as a point object and expands an obstacle cell in two concentric circles. The inner circle is called the *hard expansion* and is considered a hard obstacle. Presence of the (point) vehicle in a hard expansion cell implies imminent collision. This is encoded by associating an extremely high cost for being in that cell. The ring between the inner and outer circle is called the soft expansion. It is an area that is not very desirable for the vehicle to be in, unless there is no other option. The inner cells of the soft expansion are considered as far less desirable than the outer cells. This is encoded by using a higher cost for the inner cells than the outer cells.

CajunBot is 1.5m wide, that is, 0.75 m wide from center of the body to a side. Hence the hard expansion radius is set to 1m, giving an extra distance of 0.25m for sensor and vehicle control inaccuracies. The soft expansion radius is set to 2.5m. The combination of hard and soft expansion ensures that the path chosen using the discrete algorithm is navigable by CajunBot even though the vehicle is rectangular, and not circular.

8.4.3 Navigator

The Navigation Module is responsible for executing the plan generated by the Local Path Planner taking into account the vehicle's dynamics. As shown in Figure 8.8, the module takes as input the Navigation Plan, the Vehicle State and Safety Signals, and generates Vehicle Control Commands for the Control Drivers Module and the Next Waypoint (index to next global path waypoint) for the Local Path Planner Module.

It is the Navigation Module's responsibility to make the vehicle drive smoothly even when successive straight-line segments in the input plan are at sharp angles and have different speed limits. To provide such a driving experience, the Navigation Module performs four tasks: instantaneous steering guide, instantaneous speed guide, steering controller, and speed controller, as elaborated below.

Instantaneous Steering Guide. The instantaneous steering guide determines the desired heading for the vehicle in order to follow the Navigation Plan provided by the Local Path Planner. If the vehicle is not on the path suggested in the Navigation Plan, the instantaneous steering guide computes the instantaneous desired heading to smoothly bring the vehicle back on to the path. It is the task of the steering controller to achieve the instantaneous desired heading so computed.

The instantaneous desired heading is computed using a variation of the *follow-the-carrot* method [Hebert et al., 1997]. The vehicle's position is projected on the segment of the Navigation Plan being executed (see Figure 8.16). A carrot point is marked along this segment at a *look-ahead-distance* away from the vehicle's projected position. The orientation of the line joining the vehicle and the carrot point gives the instantaneous desired heading. Our method is different from follow-the-carrot in that carrot point is marked at a *look-ahead-distance* on the current segment, not at a distance on the path. If the length of the segment is shorter than the *look-ahead-distance* we extend the segment for the purpose of placing the carrot point. In contrast, the follow-the-carrot method, travels along the path to find the carrot point.

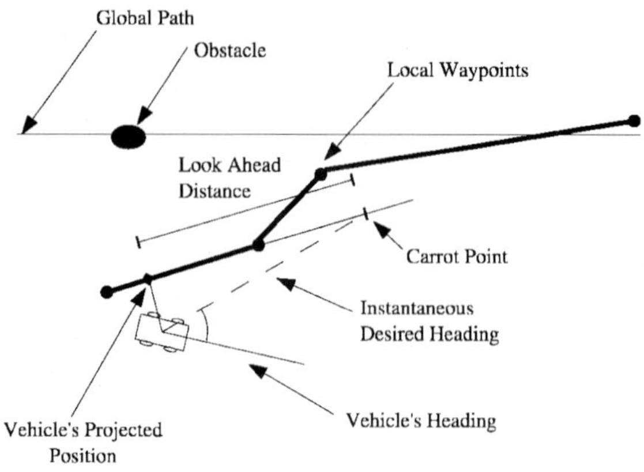

Fig. 8.16. Steering Guide

Instantaneous Speed Guide. The instantaneous speed guide computes the desired speed that vehicle should try to achieve based on the present speed, the vehicle's kinematic limits, and the safe speed limit. The safe speed limit depends on the track speed given in the Navigation Plan, the safe speed for making turns ahead, and, if the pause signal is activated then, the safe speed to bring the vehicle to a stop.

Steering Controller. The responsibility of the steering controller is to ensure that the vehicle maintains the desired heading. It generates low level steering commands for the left and right brakes in order maintain this desired heading. The low level steering commands are generated as a floating point value between -1 and 1. A negative value refers to amount of the left brake to apply and positive value refers to the amount of the right brake to apply.

The controller uses an incremental PID to generate steering commands by using the difference between the desired heading and the present vehicle heading as error input. The controller uses different PID constants at different speeds, as CajunBot, a skid steered vehicle shows varying steering responsiveness for same steering commands at different speeds.

Speed Controller. The speed controller is responsible for achieving the desired speed suggested by the instantaneous speed guide. The speed controller emits a floating point value in the range of -1 to +1, where a negative value represents amount of brakes to apply and a positive value represents the amount of throttle to give. Thus, the controller will never try to perform both active braking and throttling at the same time. The speed controller is also a classic incremental PD controller that uses the difference between desired speed and present vehicle speed as error input.

8.5 Field Experience

This section summarizes Team CajunBot's experience during the NQE (National Qualifying Event) and the 2005 GC Final.

The primary goal of the team was to be amongst the GC finalists and travel over 11.8km (over 7.3 miles), the distance traveled by Red Team's Sand Storm in 2004 GC. Due to its inherent low maximum speed 10.35m/s (23 miles/hr), CajunBot was never a contender to win the race based on the speed against the faster competitors like Sandstorm and Stanley. Further, the Max IV ATV, the underlying vehicle, is not normally used to transport for very long distances, running continuously for hours. So we were unsure if the vehicle's mechanics would hold together for the entire run. However, we were assured by the manufacturer and the dealer that the vehicle was rugged enough to last that far. In any case, the team felt that the core problems to be solved were in software, and that our ability to solve them could be amply demonstrated by successfully completing the NQE and traveling a significant distance in the final.

Indeed during the NQE and GC, CajunBot demonstrated the ability to navigate the course, control speed, and detect and avoid obstacles at reasonable speeds. In the GC final CajunBot started at the 21st position and covered about 28.324km (17.6miles) before it was killed due to a mechanical failure. CajunBot was placed 15th on the basis of the distance covered. Along the way it overtook two vehicles that were still in the running, achieved its maximum speed of 10.35 m/s, and covered the 28.324km distance at an average speed of 4.95m/s (11 miles/hr).

8.5.1 National Qualifying Event (NQE)

Of the six NQE runs, CajunBot completed two runs (Runs 4 and 5), ran into the last car on the final stretch in Run 6, did not compete in Run 3 because of a mechanical failure, suddenly stopped after going through the tunnel due to a GPS/INS related failure in Run 2, and climbed the hay bails when approaching the tunnel due to a rounding error in the path planner in Run 1.

Run 1: In the first run of the qualification round, CajunBot started well passing through the gates and cones, going up the slope, and, through the speed section. The next part of the course was a narrow section bounded by hay bails which lead to the tunnel. While going through this section CajunBot made a sudden left turn to avoid the hay bails on the right, and in so doing climbed up on the bails on the left. All three wheels on left side of CajunBot were off-ground as the vehicle tried to climb the hay bail. Ironically, bringing the vehicle back on the track required spinning the wheels which were in the air and thus without traction. This was the first time we experienced the limitation of skid-steering. CajunBot had to be taken off the track.

Of course, the incident would not have happened if CajunBot had chosen a straight path along the center of the route. We were puzzled why it did not do so. The reason turned out to be a rounding error in the mapping of a curve path to the grid cell. The effect of the rounding error was magnified because the location that was rounded off was very close to the vehicle, leading to a very high change in instantaneous heading. Ironically, again, the rounding error was not a complete oversight. The single line of faulty code was annotated as 'FIX ME' for later. The approximate calculation was put in place, to be corrected when other pieces were completed. Addressing the 'FIX ME' never became a priority because it never led to any failure until that NQE Run.

Run 2: Very much like the first run, for the second NQE run, CajunBot started confidently, avoiding the initial set of obstacles, passing through the gates and the slope. In the speed section the vehicle reached its top speed of 23 miles/hr. As expected from our testing in the simulator, CajunBot followed a straight-line path through the narrow section bounded by hay bails and the tunnel that followed. It avoided the tires, stationary cars and went swiftly through the gravel section. The navigation system was impeccable as it made a smooth curve by the wall, passing through the narrow section between the cones and the wall. And then CajunBot simply stopped. After waiting for about 10 minutes, the vehicle was manually driven off the track.

The analysis of the logged data revealed a sudden change in the GPS height value in the region where the vehicle stopped. The height value (Z) had jumped by 30 meters in a fraction of a second causing the software to detect a wall-like obstacle in front of the robot. The entire region was filled with the obstacles, forcing the navigation software to stop giving paths.

The reason for the spike in Z value turned out to be due to loss of GPS signal in the tunnel. After the GPS signal was lost the INS started using ded-reckoning to estimate change in its position. Even after passing through the tunnel and regaining the GPS signal, the INS continued ded-reckoning, or so we understand. At some point it switched from ded-reckoning to using the Z-value reading from the GPS. But by then ded-reckoning errors had accumulated and we experienced a sudden spike in the data.

A median filter was added to the software, which monitors the data from the GPS/INS. The filter discards the data if any unreasonable rate of change in the height and position information of the GPS/INS data was detected. Also, upon detecting a spike in data, the obstacle detection module also flushes its spatio-temporal data to ensure that the data before and after the spike are not mixed during analysis.

Run 3: As CajunBot was being brought for the third run, it was observed that the vehicle steering incorrectly while been driven using the RC control. The reason turned out to be a broken transmission. We had to forgo this run as the vehicle was not in operational condition.

This was a good lesson about the consequences of using an uncommon vehicle. The nearest dealer of Max ATVs was about 320km (200miles) away. We were extremely lucky that "2 The Max ATV" dealer was very magnanimous. She was willing to bring a brand new Max IV ATV to us, and allow us to scavenge it for spares, at no additional cost but to replace the parts when we were done. The team changed the entire transmission of the vehicle in less than seven hours and was ready for the fourth run the next morning.

Run 4: The fourth run was the first successful run for CajunBot at the NQE. The vehicle started smoothly, detected and avoided every obstacle, and successfully completed the run.

For the fourth run, the speed section was shifted to a more bumpy stretch as compared to the earlier runs. The logged data showed that CajunBot saw a line of false obstacles a couple of times when the vehicle was at its maximum speed on a considerably bumpy track. The sensors were pointing way far apart. The top LIDAR was aimed at 25m in front of the robot whereas the bottom LIDAR was aimed at 7m in front of the robot. During testing a few days before the NQE the configuration of the sensors was changed from the one described in Figure 8.5. The bottom LIDAR was reoriented to reduce the blind spots at turns. This change had the unintended consequence of increasing false obstacles in rough conditions. The configuration of Figure 8.5 was chosen to ensure rapid creation of triangles that satisfied the spatio-temporal constraints. With only 0.3m separation in their range, triangles of Figure 8.10 could be created using the global points resulting from the most recent scans of the two LIDARs. Increasing the separation in their range to 18m significantly decreased the chances of grouping beams from the most recent scans from the two LIDARs. When the global points from the two LIDARs were grouped, they would be expected to be temporally a part, and therefore the triangles were prone to errors. By increasing the number of such triangles, we increased the chances of detecting false obstacles. We could address the problem by reverting the sensor configuration. But doing so implied more risk since we did not have the time to test the effect of the change. Hence, the team decided not to make any changes and live with the consequences.

Run 5: The fifth run was also a successful run, with the only difference that the vehicle could not detect the lower leg of the tank trap, placed in the extreme end of the run. It brushed the tank trap on its way to the end of the course.

Low data density on the lower leg of the tank trap caused the software not to detect it as an obstacle. A solution for this was to increase the time for which the LIDAR data is used to from the triangles or to point the LIDAR's more closer such that there is more data density. The first option was ruled out as the GPS signal was not reliable after the vehicle passes through the tunnel. Increasing the time might have led to an increase in false obstacles. The second option was not tested in the recent past, hence, the team decided to keep the same configuration for the rest of the runs.

Run 6: The final run of the NQE, the run 6, was a near perfect run until the very last part. The last 200m of the track had two cars and a tank trap as obstacles in the path. CajunBot detected and avoided the first car perfectly. However, while trying to avoid the second car it hit the car in the corner. Like the first run, CajunBot's left wheels were in the air and it needed to spin those wheels to turn right. CajunBot had to be taken off the track.

Based on the analysis of the logged data, as CajunBot turned to avoid the first car, the second car was in the blind spot of the top LIDAR. By the time the bottom LIDAR could detect the car as an obstacle, CajunBot was dangerously close to it. The delay in detecting the obstacle did not give enough time for the local path planner to steer the vehicle around the obstacle. The vehicle hit the right rear end of the car while trying to steer away from it.

Having completed two runs successfully and one near successful run, CajunBot was selected as one of the 23 finalists to compete in the DARPA Grand Challenge 2005. The team and the bot moved to Primm, Nevada.

In the time between the NQE and the GC final, we were afforded a window of opportunity to fix the configuration of our LIDARs. We reverted the LIDARs to the configuration given in Figure 8.5 and tested it near Slash X, the starting point of the 2004 Grand Challenge.

8.5.2 DARPA Grand Challenge Final Run

On the grand finale, CajunBot was the 21st robot to start. She was flagged off at 8:30am. Just about that time the weather took a turn. The winds picked up, blowing through the dry lake bed and causing a big sand storm. In no time, CajunBot was out of sight, in the thick of the storm. We knew that the LIDARs could not see through the sand, and were pretty nervous. To our relief the DARPA scoreboard showed CajunBot was still moving. It took about an hour for CajunBot to complete the 12.874km (8 mile) loop and pass the spectator stand. By then the weather had settled and she was moving really well. For the next hour, CajunBot passed several disabled vehicles and two vehicles still in the run. We were pleased that she was going strong. It was now poised to enter a region that was also part of the tail end of the course. The leading bots, Stanley, Sandstorm, and Highlander, were about to enter this region. To give precedence to the leading bots, CajunBot was paused. It took about 45 minutes for the lead bots to clear that path. When CajunBot was un-paused, she simply failed to start. After attempting to restart for 20 minutes, the vehicle was disabled.

The reason why CajunBot failed to restart turned out to be very mundane, and related to the transmission failure before Run 3. The transmission has two plungers that connect to two levers. One lever is for braking the left wheels or engaging the left transmission. The second one is for the right side. We control each lever using a lead-screw linear actuator. The actuators consume current when the lead screw is tightened to pull a load. However, if power is cut-off it stays locked in a position. We map the thrust of the actuator to a range from 0 to 1, to correspond to the maximum desired movement of the levers. After the transmission failure, we did not calibrate the actuators correctly. Its '1' was mapped to a position that the transmission could not physically reach. When the vehicle was put in to pause mode, to engage the brakes the levers had to be moved to '1'. However, this position could not be reached and the motor controller continued to attempt to move it. In the process, for 45 minutes the motor was fed its peak current, a current it can withstand only for short duration. That caused the motors on the actuators to burn out.

Further analysis of the logged data revealed how CajunBot weathered the sand storm. The on-board video show CajunBot completely engulfed in the sand. That led it to see 'false obstacles', forcing it to go out of the track to avoid them. When the vehicle strays too far outside the lateral boundary we disable path planning and force it back to the track. Once it was back on track it would repeat the same behavior, thus appearing as though it is wandering aimlessly. However, after the sand storm cleared, the video shows CajunBot running very much along the middle of the track, and passing stalled or stopped vehicles. The logged data shows absolutely zero false obstacles throughout the run after the storm, even in areas where the vehicle experienced severe bumps.

Regardless of the outcome of the race, Team CajunBot came away with a much better understanding of autonomous navigation. We are very confident that, but for the mechanical failures, the vehicle would have completed the track, especially since there was no weather related disturbance in the later part of the day. The team is looking forward for the next challenge and is working on the new proposed entry, the RaginBot - a 2004 Jeep Rubicon.

8.6 Future Work

Our experience suggests that field testing is one of the most expensive parts of developing an AGV. To field test, one must have a fully operational vehicle, a field for testing it, correct weather conditions, and a significant size staff. Unless the procedures for bringing the vehicle to the field are very well-defined, small issues, such as insufficient gas in the generator, can consume significant time.

Having a fully operational vehicle is no small requirement, given that an AGV has linear dependencies between the automotive, the electromechanical components, the electrical, electronics, sensors, and the software. Failure in any one of the components can hold back the testing.

Yet testing in the current generation of simulation environments, such as CB-Sim, Stage [Gerkey et al., 2003] and Gazebo [Vaughan, 2000] are quite limited. While these environments are good for doing integration testing, their simulation abilities are quite limited providing information about how the vehicle may perform in the real-world, such as, in different terrain and weather conditions.

We are working on developing a higher fidelity simulation and visualization of the real-world and the vehicle. The environment will utilize a cluster of computers and a 6-surface cave to create an immersive visualization of real-time simulation.

There is another aspect of field testing that can be improved. How does one evaluate the performance of a vehicle in the field? A report of observations by the testing team while useful leaves room for subjectivity and human error. A 10 hour run can be pretty long for someone to correctly recount and to pay attention for taking notes.

There is ongoing work in our lab to develop automated, objective methods to analyze logged data from a field run and to evaluate a vehicle's performance.

In addition we are also working on improving the system further, such as, to introduce camera vision capability; improve reliability by introducing the ability to restart individual software components or the whole system; and porting the system to a completely new vehicle, Ragin'Bot, a 4x4 Jeep Wrangler.

8.7 Conclusion

Participating in the DARPA Grand Challenge has been an intense experience, unlike anything we have experienced during normal research. The most significant difference was to have an end-to-end system that would perform in a real-world situations, and in the course of the development solve some significant research problems.

The requirement to develop an end-to-end system implied we could not have tunnel vision and get carried away improving only one component of the system. That the vehicle would perform in real-world situations required us to consider solutions for a multitude of scenarios, rather than be content with developing solutions for some simplified scenarios.

This paper presents the overall hardware and software architecture of the CajunBot. More importantly, it highlights the specific innovations resulting from the effort. It is hoped that these innovations in some small way help in advancing the overall field of AGV.

Acknowledgements

Special thanks to following Team CajunBot members for their contribution: Adam Lewis, Adrian Aucoin, Christopher Mire, Daro Eghagha, Joshua Brideveaux, Muralidar Chakravarthi, Patrick Landry, Santhosh Padmanabhan, Scott Wilson, Vidhyalakshmi Venkitakrishnan.

The project was supported in part by the Louisiana Governor's Information Technology Initiative, and benefited from the sponsorship of C&C Technologies, MedExpress Ambulance Service, Lafayette Motors, Firefly Digital, Oxford Technical Solutions, and SICK USA, Inc.

References

Barraquand et al., 1992. Barraquand, J., Langlois, B., and Latombe, J. C. (1992). Numerical potential field techniques for robot path plannning. *IEEE Transactions on Systems, Man, and Cybernetics*, 22:224–241.

de Smith, 2004. de Smith, M. J. (2004). Distance tranforms as a new tool in spatial analysis, urban planning, and gis. *Environment and Planning B: Planning and Design*, 31:85–104.

Dubins, 1957. Dubins, L. E. (1957). On curves of minimum length with a constraint on average curvature, and with prescribed initial and terminal positions and tangents. *American Journal of Mathematics*, 79:497–516.

Feng et al., 1990. Feng, D., Singh, S., and Krogh, B. H. (1990). Implementation of dynamic obstacle avoidance on the CMU NavLab. In *Proceedings of the 1990 IEEE Conference on Systems Engineering*, pages 208–211.

Gerkey et al., 2003. Gerkey, B., Vaughan, R. T., and Howard, A. (2003). The Player/Stage project: Tools for multi-robot and distributed sensor systems. In *Proceedings of the 11th International Conference on Advanced Robotics (ICAR'03)*, pages 317–323, Coimbra, Portugal.

Hebert et al., 1997. Hebert, M., Thorpe, C., and Stentz, A., editors (1997). *Intelligent Unmanned Ground Vehicles: Autonomous Navigation Research at Carnegie Mellon University*. Kluwer Academic Publishers.

LaValle, 2006. LaValle, S. M. (to appear in 2006). *Planning Algorithms*, chapter 1, page 3. Cambridge University Press. `http://msl.cs.uiuc.edu/planning/`.

Maida et al., 2006. Maida, A. S., Golconda, S., Mejia, P., and Lakhotia, A. (2006). Subgoal-based local-navigation and obstacle-avoidance using a grid-distance field. *International Journal of Vehicle Autonomous Systems*.

Marshall, 1999. Marshall, A. D. (1999). *Programming in C, Unix System Calls and Subroutines using C*, chapter IPC: Shared Memory. `http://www.cs.cf.ac.uk/Dave/C/node27.html`.

Pardo-Castellote and Hamilton, 1999. Pardo-Castellote, S. S. G. and Hamilton, M. (1999). NDDS: The real-time publish-subscribe middleware. Technical report, Real-Time Innovations, Inc.

Scheuer and Laugier, 1999. Scheuer, A. and Laugier, C. (1999). Planning sub-optimal and continuous-curvature paths for car-like robots. In *Proceedings of the IEEE International Conference on Intelligent Robots and Systems*, volume 1, pages 25–31, Victoria, Canada.

Scheuer and Xie, 1999. Scheuer, A. and Xie, M. (1999). Continuous-curvature trajectory planning for manoeuvrable non-holomorphic robots. In *Proceeding of IEEE-RSJ International Conference on Intelligent Robots and Systems*, volume 3, pages 1675–1680.

Shackleford et al., 2000. Shackleford, W. P., Proctor, F. M., and Michaloski, J. L. (2000). The neutral message language: A model and method for message passing in heterogeneous environments. In *Proceedings of the World Automation Conference*, Maui, Hawaii. `http://www.isd.mel.nist.gov/documents/shackleford/Neutral_Message_Language.pdf`.

Simmons and James, 2001. Simmons, R. and James, D. (2001). *IPC - A Reference Manual, version 3.6.* Robotics Institute, Carnegie Mellon University. `http://www.cs.cmu.edu/afs/cs/project/TCA/ftp/IPC_Manual.pdf`.

Stentz and Hebert, 1995. Stentz, A. and Hebert, M. (1995). A complete navigation system for goal acquisition in unknown environments. *Autonomous Robots*, 2(2): 127–145.

Vaughan, 2000. Vaughan, R. T. (2000). Gazebo: A multiple robot simulator. Technical Report IRIS-00-394, Institute for Robotics and Intelligent Systems, School of Engineering, University of Southern California.

9

SciAutonics-Auburn Engineering's Low Cost High Speed ATV for the 2005 DARPA Grand Challenge

Robert Daily[1], William Travis[1], David M. Bevly[1], Kevin Knoedler[2], Reinhold Behringer[2], Hannes Hemetsberger[3], Jürgen Kogler[3], Wilfried Kubinger[3], and Bram Alefs[4]

[1] Auburn University, Auburn, Alabama 36849
[2] SciAutonics, Thousand Oaks, California 91360
[3] ARC Seibersdorf Research GmbH, A-1220 Vienna, Austria
[4] Advanced Computer Vision, A-1220 Vienna, Austria

Summary. This paper presents a summary of SciAutonics-Auburn Engineering's efforts in the 2005 DARPA Grand Challenge. The areas discussed in detail include the team makeup and strategy, vehicle choice, software architecture, vehicle control, navigation, path planning, and obstacle detection. In particular, the advantages and complications involved in fielding a low budget all-terrain vehicle are presented. Emphasis is placed on detailing the methods used for high-speed control, customized navigation, and a novel stereo vision system. The platform chosen required a highly accurate model and a well-tuned navigation system in order to meet the demands of the Grand Challenge. Overall, the vehicle completed three out of four runs at the National Qualification Event and traveled 16 miles in the Grand Challenge before a hardware failure disabled operation. The performance in the events is described, along with a success and failure analysis.

9.1 Introduction

The 2005 DARPA Grand Challenge was a competition to spur the development of autonomous ground vehicle capabilities. It consisted of a 132 mile course that had to be completed in less than 10 h by vehicles with no human intervention. The course was mostly desert terrain including dry lake beds, rough roads, long tunnels and underpasses, and numerous obstacles. Initially, 195 teams entered the Challenge; 43 were invited to the National Qualification Event (NQE). SciAutonics-Auburn Engineering was one of 23 teams chosen from these semifinalists to compete in the final 132 mile course.

SciAutonics formed to compete in the initial DARPA Grand Challenge in 2004. The core technical team was initially comprised mainly of engineers at Rockwell Scientific Corporation (RSC) and received a large portion of its funding from RSC. ATV Corporation donated the vehicle platform and a second test vehicle, and provided continual technical support throughout the project. Auburn University joined the team to develop the vehicle control and navigation

M. Buehler, K. Iagnemma, and S. Singh (Eds.): DARPA'05, STAR 36, pp. 281–309, 2007.
springerlink.com

aspects of the system (Behringer, Gregory et al., 2004). In 2005, the team name changed to SciAutonics- Auburn Engineering, and more collaborators joined to complement the existing expertise. Seibersdorf Research provided a stereo vision system for object detection and road segmentation. The City of Thousand Oaks was also a partner, providing an area of land for performing vehicle tests and the required DARPA site visit.

This paper discusses the SciAutonics-Auburn Engineering effort in the DARPA Grand Challenge 2005. In particular, it covers both the components that comprised the entry vehicle and the strategies that allowed the team to compete successfully in the Challenge. Particular emphasis is placed on the localization, obstacle detection, and vehicle control algorithms.

9.1.1 System Development Strategy

A system that exhibits autonomous driving capability is, by its very nature, quite complex and consists of subsystems with a high degree of interdependence. Since the team was mostly a volunteer effort comprised of people working on this project in their spare time, it was a challenge to map the system structure onto the various team members in a way that was efficient and could allow the team members to work on subtasks independently.

The vehicle team initially set up the hardware: Vehicle components and the lowest level actuation controllers. This could be done relatively independently as other subteams addressed different technical challenges. As the hardware implementation progressed, the work on the control system became more relevant. Two students from Auburn University worked remotely on identifying the vehicle model for throttle and steering control, as well as the navigation algorithm. The sensor team addressed the feasibility of sensor systems and performed independent sensor tests and characterization in desert conditions. ATV Corporation gave the team access to a second vehicle, which had the same driving characteristics but was not equipped with any means for automatic driving. This vehicle served as a platform for mounting and testing sensors while not removing the capability to manually drive. The software team built the framework for communication, sensor data acquisition, and processing. The intelligent behavior subteam addressed the issues related to the design of the autonomous concept, such as path planning and obstacle avoidance.

The team met regularly on evenings and weekends to test, implement, and integrate the different modules in the complete system context. Trials were performed at the Lang Ranch area, where the City of Thousand Oaks had given permission to conduct these tests.

9.1.2 Entry Vehicle

A small All Terrain Vehicle (ATV) platform (Figure 9.1) was chosen for the entry vehicle due to its size, agility, and ruggedness (Behringer, Sundareswaran et al., 2004). The Prowler by ATV Corp. is an ATV modified for military use

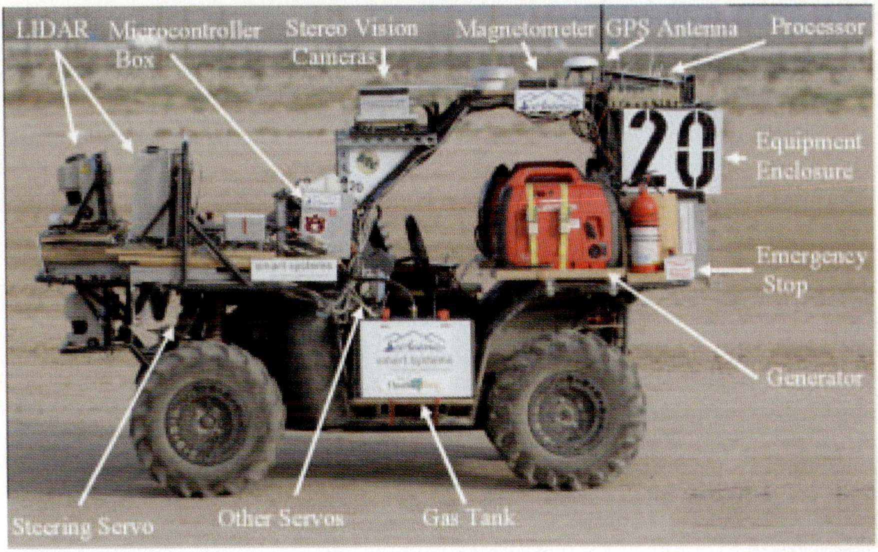

Fig. 9.1. RASCAL

that is equipped with a 660 cc Yamaha engine, enhanced suspension, full roll cage, run-flat tires, and cargo rack. This combination of power, ruggedness, and useable space proved to be an excellent foundation for an off-road autonomous vehicle. The 1,000 lb payload capacity and heavy duty suspension handled the multitude of motors, sensors, and computers that were mounted on the cargo rack and in the roll cage. The independent suspension with 8-9 in. of travel, and high 12.5 in. ground clearance, allowed the vehicle to traverse difficult terrain with relative ease.

Modifications were made to the ATV, dubbed RASCAL (Robust Autonomous Sensor Controlled All terrain Land vehicle), for automation. A servomotor was installed in the engine bay and attached to the steering system to actuate the front wheels. The motor output, 6.5 ft lb of torque, was fed into a 14:1 gearbox, placing a total of 90 ft lb of torque on the steering rack. The throttle, brake, and gear were controlled with smaller servos, which output 27 ft lb. All of the servos were directed by microcontrollers, which communicated with a computer via serial ports. An emergency stop mechanism, with ultimate authority over the microcontrollers, was wired in series to the vehicle's power and the brake and throttle servos to eliminate the possibility of losing control of the vehicle in the event of a software or hardware failure. Two 15 gallon gas tanks were added to the side of the ATV and gave RASCAL more than enough fuel to finish the course. Two 2,000 W generators provided additional power needed to operate the on-board electronics. An enclosure mounted in the rear contained the delicate hardware.

9.2 Software Architecture

9.2.1 Autonomous System Concept

One of the main ideas in the autonomous system was the modularity of the architecture (Behringer et al., 2005). The vehicle control and GPS/INS navigation were the core modules providing the basic autonomous functions. The other modules were "optional add-ons" (Figure 9.2). They provided information about the environment, as well as objects to be avoided. If these modules were disconnected or failed, the core vehicle control still continued to operate using solely GPS and inertial input for computing the control output. Of course, in this mode, the vehicle operated blindly; therefore, the maximum speed was reduced to 2 m/s, a compromise between avoiding heavy damage in a collision and being able to continue driving to fulfill the given mission.

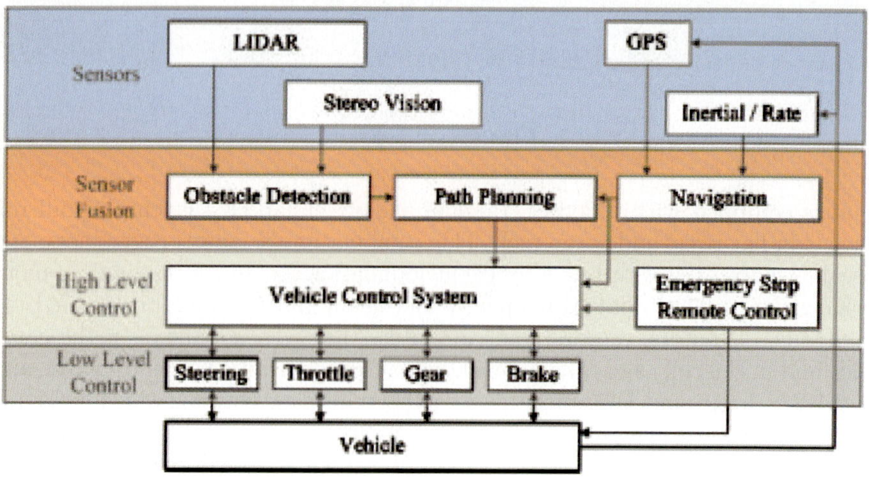

Fig. 9.2. System architecture for autonomous driving on RASCAL

9.2.2 Software Structure

The software was structured to allow easy partitioning among physical central processing units (CPUs) and offline debugging. The overall structure of the software was a collection of modules running as independent processes in the Linux operating system. The modular design allowed fault isolation and parallel development of the modules. A network was set up to allow communication between the modules, via a user datagram protocol (UDP) (Postel, 1980) with timeouts for detecting fault conditions, but no acknowledgements. Running each module as a process provided CPU and memory usage isolation, as well as easy partitioning among the physical processing units. This was important as the CPU utilization of each module was not known in advance. The easy partitioning also allowed all modules to be run on a single system for debugging, or any one module to be run on a debug system on or off the vehicle, as long as it was on the vehicle network. For the events, two laptops were used to run the vehicle.

Linux was chosen as the operating system for a number of reasons. It provided acceptable real-time behavior; while the default 2.6 kernel is certainly not a hard real-time kernel, it provided consistent cycle times for control loops and good isolation of the processes from one another. Linux could also be run on developer desktops, as well as the actual vehicle, which improved development efficiency. With the use of the Gazebo simulator (Koenig & Howard, 2004) and tools for playing back recorded vehicle data, much of the debugging and development could be carried out on individual laptops; so development work could continue when the vehicle was not available. During vehicle test sessions, results from one run could be analyzed and changes could be made, while a different set of tests were running on the vehicle.

The real-time nature of the vehicle did result in some challenges, and highlighted some of the limitations of Linux for this application. In some cases, code modules would go into tight loops due to bugs. This would cause other modules running on the same CPU to get insufficient processor time. In other cases, too many modules or overly complex algorithms were run, again using too many CPU cycles. The control loops, starved of cycles, would either timeout or in borderline cases not behave as desired. A more interesting case was when one of the CPUs would reduce its frequency based on temperature limits. The behavior would be correct in most cases; but for this application, the system would not have enough CPU cycles to complete all tasks. A hard real-time operating system would limit the impact of these issues to only one process. However, if that process was a key process (as it often was) the vehicle would still be disabled. Debugging these issues required looking at the timestamps of messages passed between the modules.

9.3 Localization

9.3.1 Hardware

RASCAL contained a variety of sensors to determine states critical to the vehicle controller, such as position, heading, and speed. A strategy of redundancy was employed to provide measurements from some sensors when others were not available or in the event of the failure of a particular sensor.

The cornerstone of vehicle localization was a single antenna Navcom differential global positioning system (DGPS) receiver with Starfire corrections broadcast by Navcom. It generated unbiased measurements of position (north and east), velocity (north, east, and up), and course at 5 Hz. It is important to note that vehicle course is the angle from north created by the vehicle's velocity vector, not the vehicle's pointing vector. With the corrections broadcast by Navcom, this GPS receiver is capable of producing position measurements accurate to less than 10 cm. However, the output rate of the receiver was too low to adequately control the vehicle.

An inertial measurement unit (IMU) was used to obtain high update rate measurements. A Rockwell Collins GIC-100 tactical-grade six degrees of freedom IMU measured translational accelerations and angular rates at 50 Hz. These measurements, however, were inherently corrupted with biases and noise. Dead reckoning with the unit provided acceptable results for a short period of time if the initial biases were eliminated. However, the biases did not remain constant and therefore had to be continually updated and removed from the measurement to provide a reliable navigation solution.

The ATV's onboard speedometer was used as an additional speed sensor. The output rate of this sensor was dependent upon vehicle speed, so compensation was needed to provide a more consistent measurement to the controller. In addition, the measurement was corrupted by wheel slip, which appeared as a sudden large change in the bias. The sensor also contained a calibration error that would corrupt the speed estimate during a GPS outage.

Magnetometers are often used in aerial applications to provide orientation information. They sense the earth's magnetic field, thus all measurements are referenced from magnetic north and not true north. This difference can easily be calibrated and remains fairly constant if the sensor remains in a region near its calibration point. RASCAL utilized two magnetometers to measure the vehicle's heading, roll, and pitch angles. A TCM2 magnetometer provided 16 Hz measurements that contained high noise, but a slow bias drift rate. A Microstrain 3DM-GX1 IMU and magnetometer provided 50 Hz measurements that had very little noise, but the bias drifted quickly (the IMU was not used). It was discovered that the magnetometers could help initialize the navigation algorithm, but once the vehicle started moving they were of little use because the magnitude and drift rate of their biases were greater than that of the other sensors.

9.3.2 Algorithms

Kalman filtering is a proven method for blending measurements to eliminate various sensor deficiencies while utilizing the strengths of each sensor by statistically weighting each measurement. The localization algorithm used was an extended Kalman- Bucy filter (EKF), outlined in detail by Stengal (1994). This algorithm handled the system nonlinearities by continuously propagating the system model to calculate the time update, and discretely propagating the measurement update. The EKF combined the bias-free low update GPS measurements with the other measurements to produce a bias-free high update (50 Hz) navigation solution. An EKF is as accurate and less computationally intensive

Table 9.1. Variable definitions

V – Velocity	$b_{[r,\dot{\phi},\dot{\theta}]}$ – Rate gyro bias (yaw, roll, pitch)
ψ – Heading	$b_{M[\psi,\phi,\theta][1,2]}$ – Magnetometer bias (heading, roll, pitch) (TCM2,Microstrain)
ϕ – Roll	g – Gravity
θ – Pitch and longitudinal accelerometer bias	a_x – Measured longitudinal acceleration
N – North	r – Measured yaw rate
E – East	$\dot{\phi}_{meas}$ – Measured roll rate
θ_g – Road Grade	$\dot{\theta}_{meas}$ – Measured pitch rate

as some higher-order filters if the sample rate is high enough (St. Pierre & Gingras, 2004). This efficiency was an advantage with the algorithm because the 16 element state vector already imposed a moderate computational burden. GPS and inertial measurements were loosely coupled, meaning the inertial errors were corrected with a computed GPS solution. A tightly coupled system, where the GPS pseudo-range measurements amend the IMU errors (Farrell & Barth, 1999), was considered, but due to time constraints and the overall satisfaction with the loosely coupled system, it was not constructed.

The estimated states were chosen to provide the vehicle controller with the necessary position, velocity, and heading data; to fully orient the vehicle; and to calculate sensor biases. The nonlinear differential equations of the states are listed below [Eq. (9.1)], and the variable definitions are described in Table 9.1:

$$
\begin{array}{lll}
 & & \dot{b}_{M\psi 1} = 0; \\
\dot{V} = a_x - g\theta - g\theta_g; & \dot{\phi} = \dot{\phi}_{meas} - b_{\dot{\phi}}; & \dot{b}_{M\phi 1} = 0; \\
\dot{\psi} = r - b_r; & \dot{b}_{\dot{\phi}} = 0; & \dot{b}_{M\theta 1} = 0; \\
\dot{b}_r = 0; & \dot{\theta} = \dot{\theta}_{meas} - b_{\dot{\theta}}; & \dot{b}_{M\psi 2} = 0; \\
\dot{N} = V\cos(\psi); & \dot{b}_{\dot{\theta}} = 0; & \dot{b}_{M\phi 2} = 0; \\
\dot{E} = V\sin(\psi); & \dot{\theta}_g = 0; & \dot{b}_{M\theta 2} = 0;
\end{array}
\quad (9.1)
$$

These equations were continuously integrated in the time update of the EKF. Noise terms do not appear in Eq. (9.1) because they are included in the time update of the EKF by utilizing the process noise covariance matrix. Methods defined by Bevly (2004) were used to derive these equations of motion. The coordinate frame used is depicted in Figure 9.3.

The bias states have no dynamic response according to the equations of motion, but in actuality, biases randomly drift over time. The process noise covariance

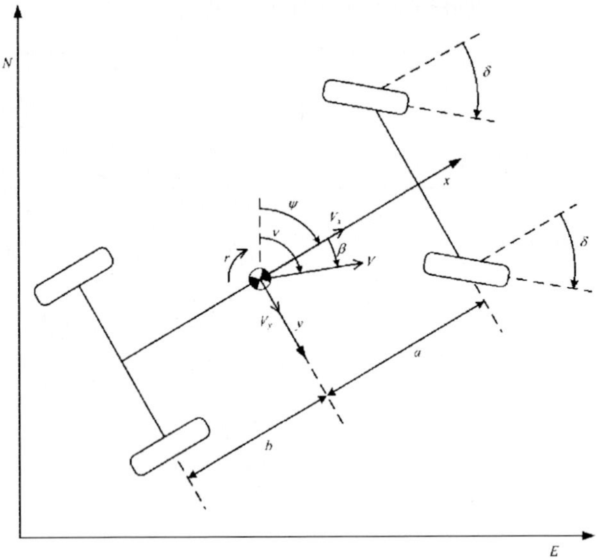

Fig. 9.3. Vehicle body coordinates and reference frame

Table 9.2. Process noise statistics

$\sigma_{az}^2 = 0.65^2 dt$	$\sigma_r^2 = 0.038129^2 dt$	$\sigma_{br}^2 = 10^{-10}$	$\sigma_{\dot{\phi}}^2 = 0.17851^2 dt$
$\sigma_{b\phi}^2 = 10^{-10}$	$\sigma_{\dot{\theta}}^2 = 0.088147^2 dt$	$\sigma_{b\theta}^2 = 10^{-8}$	$\sigma_N^2 = 0.000001^2 dt$
$\sigma_E^2 = 0.000001^2 dt$	$\sigma_{b\theta_g}^2 = 10^{-7}$	$\sigma_{bM\psi 1}^2 = 10^{-4}$	$\sigma_{bM\phi 1}^2 = 10^{-6}$
$\sigma_{bM\theta 1}^2 = 10^{-8}$	$\sigma_{bM\psi 2}^2 = 0.01$	$\sigma_{bM\phi 2}^2 = 0.1$	$\sigma_{bM\theta 2}^2 = 0.1$

matrix included values for the bias states to account for this random bias walk. These were used as tuning parameters for the EKF because they directly influenced the amount of filtering on the estimated states. The other entries in the matrix captured the system noise, which was determined during static tests. The process noise covariance matrix is a diagonal matrix with the covariances in Table 9.2. The time update in the EKF requires this matrix to be continuous; therefore, the measured discretized values are multiplied by the sample rate (dt) as presented by Stengal (1994). This discrete to continuous conversion is only valid for very small sample rates.

The discrete measurement update in the EKF utilized statistically weighted measurements, from the sensors listed in Section 9.3.1, to overcome integration errors. The measurement matrix was adjusted accordingly depending upon the availability of the different measurements. Two calculations were included in addition to the raw sensor measurements. GPS forward and vertical velocities were used to solve for road grade [Eq. (9.2)], and linear equations of vehicle roll

Table 9.3. Measurement noise statistics

$\sigma_V^2 = 0.05^2$	$\sigma_{VWS}^2 = 0.3^2$	$\sigma_\psi^2 = \left(\sigma_V / V\right)^2$	$\sigma_N^2 = 0.1^2$	$\sigma_E^2 = 0.1^2$
$\sigma_{\theta_g}^2 = \left(\sigma_V / V\right)^2$	$\sigma_{\theta_{calc}}^2 = 0.08135^2$	$\sigma_{M\psi 1}^2 = 0.19995^2$	$\sigma_{M\phi 1}^2 = 0.024445^2$	$\sigma_{M\theta 1}^2 = 0.036297^2$
	$\sigma_{M\psi 2}^2 = 0.003^2$	$\sigma_{M\phi 2}^2 = 0.003^2$	$\sigma_{M\theta 2}^2 = 0.003^2$	

as a function of lateral acceleration were derived using knowledge of the vehicle's dynamics [Eq. (9.3)]:

$$\theta_g = \tan^{-1}\left(\frac{V_{GPS\,up}}{V_{GPS\,x}}\right) \approx \frac{V_{GPS\,up}}{V_{GPS\,x}} \tag{9.2}$$

$$\phi = \frac{1}{g}\left(a_y - V\left(r - b_r\right)\right) \tag{9.3}$$

Using these two measurements, the magnetometer roll and pitch biases were observable. It should be noted that the roll estimate contained the lateral accelerometer bias because of the method defined in Eq. (9.3). With the existing sensor suite, there was no measurement available to observe and remove the lateral accelerometer bias.

Noise statistics were found by recording long periods of static data. The covariance values shown in Table 9.3 were loaded into the diagonal measurement noise covariance matrix for use in the measurement update.

9.3.3 Experimental Validation

The algorithm's performance was evaluated based on the amount of error during a simulated GPS outage. Two reasons for this evaluation method are as follows: First, when enough satellites are in view and the receiver is outputting valid messages, the EKF successfully tracks the GPS measurements; and second, a real GPS outage would eliminate the truth measurement, degrading the accuracy of the error analysis. Figure 9.4 is a plot of GPS and estimated position during a test run. The circles signify the beginning and end of the outage, starting before the first turn and concluding at the end of the straight section.

Clearly, error growth occured at the onset of the outage. Figure 9.5 is a plot displaying the magnitude of the error. Over the 25 s outage, the vehicle was traveling 3.2 m/s. The maximum error for this period of time at this speed was slightly over 1.5 m. Since the vehicle was required to remain in a 10 m corridor, this level of error was acceptable.

The IMU and speedometer biases had the most negative impact on the navigation solution. As stated earlier, the magnetometers were not of much use during a run, and were statistically weighted out of the EKF after initialization.

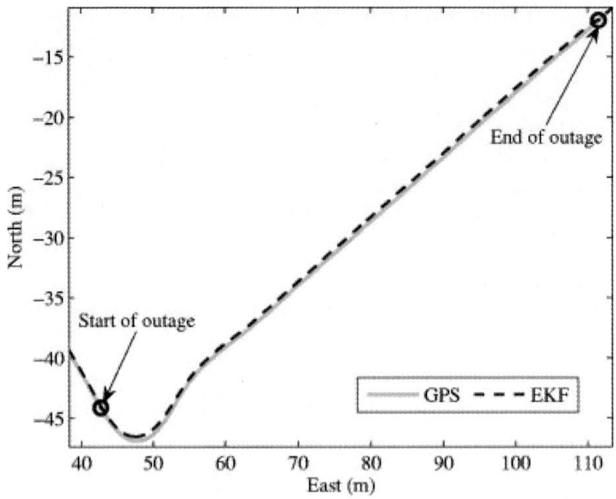

Fig. 9.4. Position estimation during an artificial GPS outage

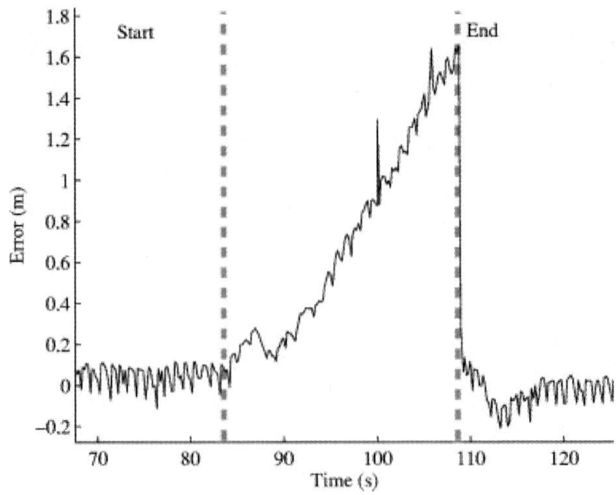

Fig. 9.5. Position error growth during a GPS outage

The tactical-grade IMU contained mechanical rate gyros with a bias drift rate of less than 1° per square-root-h, so the bias error over this period of time is 1 min. The speedometer contained a nonlinear scale factor that could be estimated as a bias. However, when GPS was lost, bias states were held constant because they were nonobservable. If the vehicle slightly changed speeds during an outage, this bias estimate would have been incorrect. This specific algorithm without the

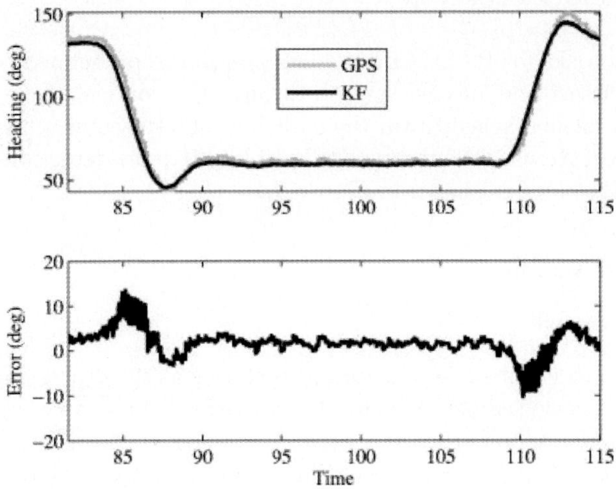

Fig. 9.6. Vehicle heading and discrepancies due to sideslip

estimated bias state was compared with another that included this bias state. It was determined there was no benefit to including the bias state, because the errors in both algorithms grew similarly. The leading error source during this run was due to the incorrect scale factor on the speedometer.

Another source of navigation error is vehicle slip. Vehicle slip can disrupt a navigation algorithm even in the presence of GPS. Longitudinal and lateral slip occur on moving ground vehicle's. Longitudinal slip is generated by a difference in the vehicle's velocity and the wheel's velocity, and lateral slip is created when the vehicle translates laterally. Wheel slip can corrupt the speedometer measurement by causing a sudden jump in the estimated bias (if the modeled bias dynamics have a high enough bandwidth). In addition, it reports a false speed value to the Kalman filter, which can directly inject error into the speed and position estimates. Sideslip also leads to a less accurate estimate, because GPS and integrated IMU measurements differ. The GPS course measurement and an integrated yaw rate gyroscope are typically used to estimate the vehicle's heading. In reality, the sensors are providing two different measurements; the measurements just happen to be similar for the majority of the time on ground vehicles. GPS course (ν) is the direction of the vehicle's velocity vector, while an integrated yaw gyro is the direction of the vehicle's pointing vector (ψ) when roll and pitch angles are absent. Evidence of sideslip-induced error can be seen in Figure 9.5 when the vehicle enters a turn. Figure 9.6 is a plot of the estimated heading when GPS is available. A discrepancy due to sideslip can be seen between estimated heading and GPS course. The error caused by vehicle slip is seen on multiple estimated states and is influenced by the initial tuning of the EKF. This phenomenon is discussed more in depth by Travis (2006).

9.4 Obstacle Detection

One of the key tasks for RASCALwas to remain in the predefined corridor while choosing the fastest and/or easiest path through the corridor. Five sensors were used to search for obstacles within the corridor: The stereo vision system (SVS) from Seibersdorf Research and four SICK LMS-221 light detection and ranging device (LIDAR) units as shown in Figure 9.1.

9.4.1 Stereo Vision

9.4.1.1 Hardware

A high-performance embedded platform was chosen as the processing unit for the vision system. A sealed camera box, hardware platform, and pair of cameras with a 300 mm fixed baseline were the three main components comprising the embedded vision system. Figure 9.7 shows the system mounted on RASCAL during test runs in the desert. Two communication types were necessary for operation. One was the communication with RASCAL, especially with the path planner software module, and the other was communication with the cameras. The communication of the stereo sensor system with RASCAL was carried out using 100 mbits Ethernet; the messages were sent via UDP packets. The cameras were controlled with the DCAM standard 1.31 using the IEEE1394a FireWire bus. The images acquired from the cameras were also transferred using the FireWire bus.

Fig. 9.7. The stereo vision system mounted on RASCAL

9.4.1.2 Algorithms

The obstacle detection algorithm detected relevant objects within the field of view. It used a predictive variant of the V-disparity algorithm (Labayrade, Aubert & Tarel, 2002), which provided a coarse grid for searching the disparity space efficiently based on prior knowledge about vehicle state and road geometry. This met real-time constraints with a processing time of about 20 ms for two images of 640 X480 pixels. The algorithm consisted of two modules: Pitch determination and obstacle detection. First, the pitch was determined from the vertical movements of far objects. Second, the ground plane was determined using this pitch information. Third, obstacles were detected that were above the ground plane and met constraints of width and height. Hence, this method combined dynamic pitch determination based on image features and the detection of obstacles near the ground plane.

9.4.1.2.1 Pitch Determination

Vehicle pitch was defined with respect to world coordinates. For obstacle detection, the relative pitch with respect to the road surface was relevant, not the absolute pitch. Since the relative pitch depends on the scene in front of the vehicle, it cannot be determined from vehicle dynamics or GPS data only. An algorithm was developed to determine relative pitch from the image sensor by tracking vertical changes of image features for a specified region of interest (ROI).

The relative pitch was defined by the vertical angular difference between the optical axis, i.e., the axes perpendicular to the image plane and the track to be followed on the terrain in front of the vehicle. As the relative pitch was defined with respect to the elevation of the track to be followed, it could be shown that the long-term average relative pitch for reaching a point is zero. In fact, for RASCAL, changes in pitch (absolute or relative) were dominated by angular vibrations of the vehicle itself. These changes included the camera system with respect to the wheels due to a vibration isolation system. Such system vibrations were typically periodic and corresponded primarily to changes in velocity. Figure 9.8 shows an example of the change of the estimated pitch, which was dominated by the vehicle dynamics. The solid line indicates the change

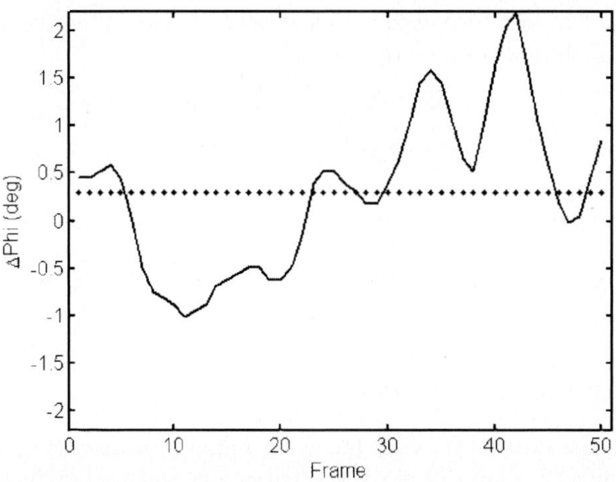

Fig. 9.8. Pitch determination with stereo vision cameras

of pitch during a typical test that included velocity variations; the dashed line indicates the average. As seen in Figure 9.8, pitch varied around a constant equilibrium position, depending only on the camera mount.

Assuming planar terrain geometry and change in pitch due to vehicle vibrations, the relative pitch could be determined by summing changes in the relative pitch and correcting it with its long-term average. The pitch determination module was implemented by choosing a ROI in the center of the field of view, which predominantly consisted of objects near the horizon or the horizon itself. In the case of high obstacles and terrain elevations at about camera height, the change in vertical position due to the vehicle movement in the forward direction was given by the following:

$$\Delta z_{vehicle} = f \frac{\Delta z}{x} \tag{9.4}$$

where $\Delta z_{vehicle}$ was the vertical change of objects in image coordinates and f was the focal length in pixels. z and x were the vehicle coordinates as shown in Figure 9.3. $\Delta z_{vehicle}$ could be calculated as follows:

$$\begin{aligned} \Delta z_{vehicle} &= \left(z_2 - z_1\right) = f \cdot H \left(\frac{1}{x - \Delta x} - \frac{1}{x} \right) \\ &= f \cdot H \frac{\Delta x}{x\left(x - \Delta x\right)} \end{aligned} \tag{9.5}$$

where H was the camera height above the ground plane and Δx was the forward change in position of the vehicle; z_1 and z_2 were positions of an obstacle at t=1 and t=2, respectively. For this region, the horizontal edges were determined using a gradient filter, and for three horizontally divided subregions, a vector was determined with the edge density as a function of the vertical image position. For each region, these vectors were matched between two subsequent frames and the total vertical shift was given by

$$\begin{aligned} \Delta z &= \Delta z_{vehicle} + \Delta z_{pitch} \\ \Delta z &\approx \Delta z_{pitch} \end{aligned} \tag{9.6}$$

where Δz_{pitch} was the vertical shift of image features due to the pitch change between two subsequent frames. The relative pitch was determined by integrating Δz over time and each 10 frames correcting with the average value of the previous 30 frames.

9.4.1.2.2 Obstacle Recognition

It was assumed the path trajectory basically followed a drivable track or road. Obstacles that would naturally block the trajectory, such as ravines, rock faces, or landslides, were not to be expected. Expected obstacles included objects, natural or not, which were put there by man. Such obstacles, consisting of boxes, vehicle tires, traffic cones, trunks, etc., were of limited size and shape, i.e., artificial to their surroundings.

Compactness, together with positioning, described the size and position of the evaluation window. Typical object sizes were 75×75 cm, positioned on the ground plane. Using the epipolar geometry and assuming a minimal detection range of 10 m and a minimal window size of 10×10 pixels in the image, relevant disparities ranged from a 26 down to 4 pixel shift between the left and right images.

The pronouncement of objects was given by the strengths of its edges. For the matching of images with a horizontal baseline, only vertical edges were relevant; the horizontal edge strength was omitted. Edges were determined using a gradient filter, as with the pitch determination. For each time instance, t, and evaluation window, the correlation between the left and right edge image was determined, providing a matrix $C_t(y, d)$ consisting of 114×23 elements, for each horizontal image position (y) and disparity value (d).

The object stability was determined by tracking different modes in $C_{t-2}(y, d)$, $C_{t-1}(y, d)$ and $C_t(y, d)$, considering a constant velocity, and the reciprocal relation between the distance and disparity value. Objects, whose summed correlation

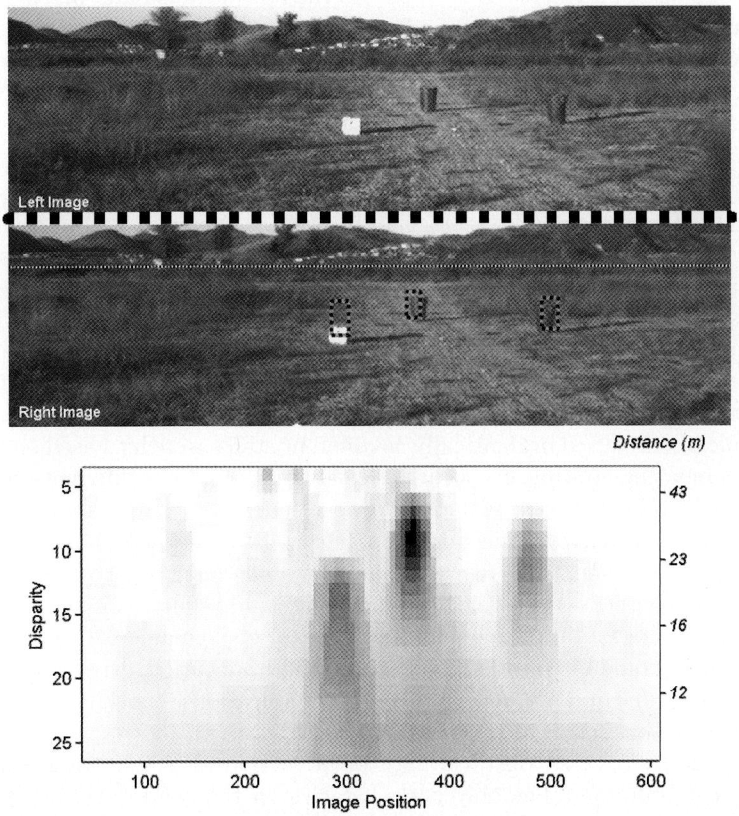

Fig. 9.9. Left and right camera images (top) with correlation magnitude (bottom)

exceeded an adaptive threshold, were included as obstacles. Figure 9.9 shows the left and right images with three obstacles, and the projected horizon from pitch determination (horizontal line). Each of the three obstacles within range was detected, as indicated with dashed rectangles in the right image of Figure 9.9. The lower plot is an example of the correlation magnitude for 2,622 image regions at different disparity values; between 4 (far end) and 26 (near end). A dark color indicates a high correlation result for the specific window.

9.4.2 LIDAR

9.4.2.1 Hardware

A LIDAR sends pulses of infrared light and records the reflection time and intensity to measure the distance and angle to an object. The scans range from 0° to 180° at 75 Hz to produce a two-dimensional image of the environment in polar coordinates. The range of the scan is adjustable in software, and can reach 80 m if desired.

Two of the LIDARs were mounted above the front tires to scan vertically, one was mounted on RASCAL's "hood" so that the scanned line was parallel to the ground (horizontal LIDAR), and one was mounted below the hood so that the scanned line intersected the ground at approximately 5 m in front of the vehicle. Each sensor data set was processed by an algorithm specialized to that sensor/orientation, and the processed data were sent to the path planner.

9.4.2.2 Algorithms

The horizontal LIDAR was processed to filter out grass and weeds while still detecting actual obstacles. This was done based on three principles. First, weeds did not present a consistent surface as an obstacle or berm. Within a LIDAR scan line, the points could be checked if the distances for adjacent points were within a threshold of one another. Second, the LIDAR image of the weeds varied as the vehicle moved. A solid object provided a consistent LIDAR return from scan to scan while weeds varied dramatically as different stalks were hit based on the angle. By comparing multiple consecutive scans, weeds could be differentiated from real obstacles based on the correlation between scans. Third, weeds (at least small ones) tended to be narrow while obstacles (fence posts, trees, telephone polls, cars, berms, etc.) were wide. LIDAR scans that showed very narrow obstacles ($<$ 5cm) were considered weeds. While the techniques were quite effective, it was important to be aware of the limitations. A thin steel pole (0.7 cm in diameter), as used for an electric fence, would be considered a weed. In this application, driving over a steel pole would not harm RASCAL; but in other applications, knocking down fences that may contain livestock would be discouraged.

Figure 9.10 shows an obstacle map on a section of a test track with and without weed filtering. The only real obstacles on the course are the cylinders seen in the photo.The arrow on the graphs indicates the direction of the vehicle; the circles represent the cylinders.

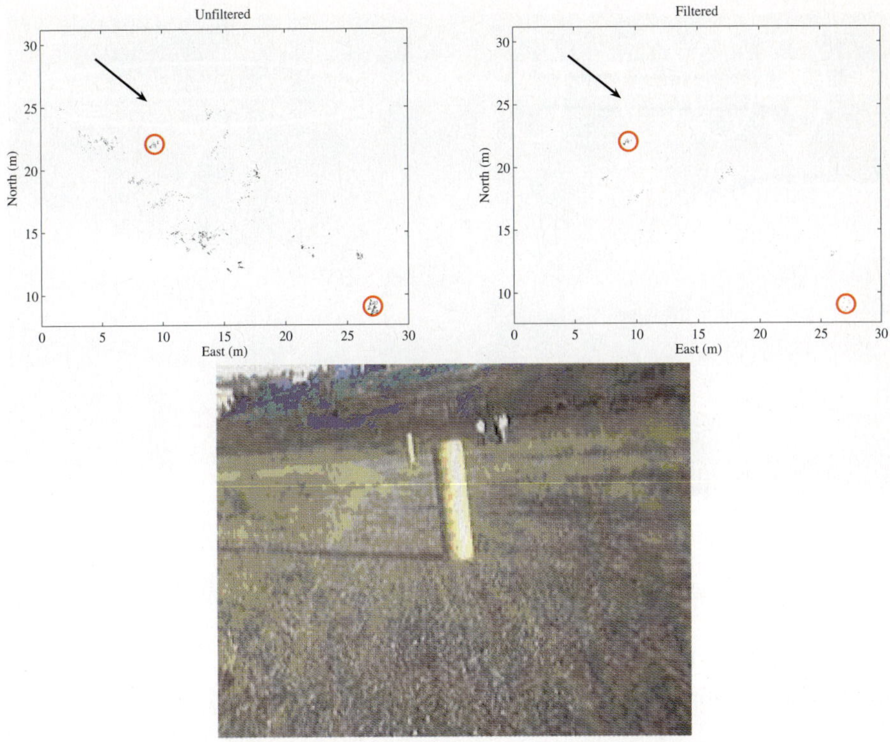

Fig. 9.10. Filtered and unfiltered LIDAR data (top) of obstacles (bottom)

When RASCAL pitched up, the downwardfacing LIDAR was processed as if it were a horizontal LIDAR. Normally, the downward-facing LIDAR saw the ground approximately 5 m in front of RASCAL, and was processed to reject debris thrown up by the front wheels and to detect berms or dropoffs on the edges of the road. The vertical LIDARs were similarly processed to detect obstacles and dropoffs in front of the vehicle.

9.4.3 Experimental Validation

The obstacle detection systems were evaluated in many stand-alone tests; however, the real testing and experimental verification of robust and correctly working algorithms could only be carried out during realistic trials. The NQE (described below) offered the opportunity to verify the obstacle detection in real-world scenarios. Under these conditions, there were several differences between the detection capabilities of the LIDAR and stereo vision systems.

Figure 9.11 shows the tunnel section and the detected obstacles of the NQE. Obstacles that were detected by the stereo vision sensor are represented by light dashes, and the obstacles detected by the LIDAR are shown with dark lines.

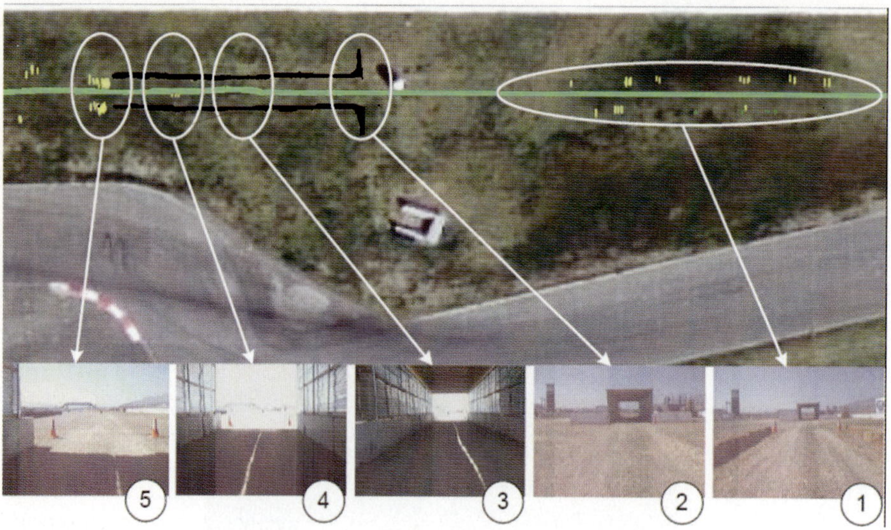

Fig. 9.11. Section with the tunnel from NQE Run 3

Pictures captured by a webcam mounted in front of RASCAL, are also depicted. Picture No. 1 shows the hay bales; these obstacles were only seen by the SVS because they were too low for the LIDAR to perceive. The SVS was able to detect the hay bales if they had any vertically edges in their textures. The left (lower) hay bale border had a long consistent shadow; therefore, detection was not consistent. However, on the right side, more visible transitions between the hay bales existed, and thus more obstacles were detected. Picture No. 2 shows the tunnel entry, which was only seen by the LIDARs. The stereo vision sensor was not able to detect this as an obstacle, because the algorithm was configured to detect obstacles with a width less than 1 m. Picture No. 3 shows the inside of the tunnel, and picture No. 4 displays the exit point of the tunnel. In the tunnel, the stereo vision sensor detected a "ghost obstacle." This happened because of the sunbeams that shone through a crack in the tunnel. In picture No. 5, the two detected cones that were placed on the exit of the tunnel can be seen. The repeated detection of cones (more than one line per cone) was due to the vibrating sensor head. The vibrating differences between the left and the right cameras created differences in distance calculations.

9.5 Path Planning

9.5.1 Initial Path

The initial route definition data file (RDDF) given to the team by DARPA, had points spaced variably from a few meters out to hundreds of meters. The control algorithms needed more consistent information than this offered. Two approaches were used to generate the nominal path that the vehicle would follow through DARPA's corridor: Speed optimization and map tuning. The speed optimization worked on the principle of maximizing the attainable speed, while staying within the corridors defined by DARPA. To this end, a path made up of a series of lines and arcs was generated. This path took into account the turning capability, acceleration, deceleration, and maximum speed of RASCAL to minimize the time through the course. Each line/arc segment included information about allowable speed, as well as position. The mapping team then adjusted the path based on map data, such as location of roads, cliffs, etc. The map tuning also included modifications of allowable speed to account for such factors as passing through and exiting tunnels, etc. As the vehicle traversed the course, this initial path was modified to take into account previously unknown obstacles.

9.5.2 Path Regeneration

The processed output of the five obstacle detection sensors was used to modify the initial path as RASCAL drove. The processed output of each sensor was a set of obstacles and a weight associated with each obstacle. That weight was an indication of how likely it was that RASCAL would need to avoid the obstacle. For example, in the LIDAR weed filtering, items that passed the filter were given a high weight, while items categorized as weeds were given a very low weight. The intent was to always avoid real obstacles, but not to drive through weeds if a perfectly clear path was also available.

The currently planned path was then compared against the list of obstacles. This list contained a global map of the obstacles currently in sight, as well as past obstacles that had moved out of view. If RASCAL was on a path that would intersect an obstacle, a new path was produced. A number of alternative paths were generated using several different methods. One method generated the sharpest possible turns based on the current velocity and deceleration capability of the vehicle. This allowed the vehicle to avoid near obstacles in the shortest distance possible. Other alternative paths were generated that gradually shifted the current path to the left and right, again taking into account the vehicle velocity.

The alternative paths and the original path were then scored. Paths that would certainly take RASCAL out of bounds were eliminated. Since the position is not known precisely, paths that might have taken RASCAL out of bounds, or nearly out of bounds, were penalized based on an estimated position error, i.e., paths that came closer to the corridor boundary were penalized. Paths that would have taken longer to traverse the same distance were penalized. Paths that went through low probability obstacles were penalized. Paths that intersected high probability obstacles were rejected. All of the paths were then ranked, and the best was chosen. If no path was found that was above a threshold, the current path was held and velocity was reduced to 1 m/s. RASCAL would then continue to look for alternative paths, while traveling at a safe speed. This fall-back state was incorporated to handle false obstacles, so that RASCAL could continue even with uncertain sensor data. Further work was planned to integrate ultrasonic and contact sensors to discriminate between false and real obstacles, but was not implemented.

9.6 Vehicle Control

The vehicle control module consisted of three parts: Path interface, speed control, and heading control. The path interface took the path segment from the path planner module, and determined where the vehicle should drive based on the vehicle's position, orientation, and speed. The speed and heading control then took this information and determined what throttle and steering inputs to apply to the vehicle.

9.6.1 Path Interface

The path planner module sent a series of waypoints to the vehicle controller. Each waypoint contained a position and desired speed. The vehicle's current position was known from the navigation solution. Using the current position and path segment, a waypoint to drive toward was chosen. The waypoint chosen had to meet two conditions, illustrated in Figure 9.12. First, it had to be in front of the vehicle, and second it had to be at least a given distance from the vehicle. If no waypoints satisfied these two criteria, RASCAL slowly circled to find a valid point. This distance increased with speed and made the steering smoother by effectively adding damping to the heading controller. Humans have a similar response: At low speeds, focus is near the front of the vehicle; as the speed increases, however, it becomes necessary to look further ahead.

Once the point was chosen, a heading error was computed as the angle between the vehicle heading and the line from the vehicle to the chosen waypoint. Speed error was also computed as the difference in the desired and actual speed. These errors were fed to the speed and heading controllers.

This method of path following (waypoint tracking) does have limitations. In particular, because of the necessity to look ahead to upcoming points, it can turn too tightly on corners, particularly tight ones, causing oscillations as the error is reduced. The fact that the look ahead distance is reduced with speed mitigates this problem to a large degree; for tight corners, the car is already moving slowly, thus not looking ahead much, and not cutting the corner. Another problem may arise if the vehicle is significantly off the path. In this situation, the controller tends to drive straight toward the nearest point on the path, instead of smoothly merging onto the path, possibly leading to oscillation about the desired path. However, this method of designating the path allows the steering controller to be much more robust to uncertainties or changing parameters in the vehicle model.

9.6.2 Speed Controller

The speed controller set the throttle position to drive the speed error to zero. Braking was also used to slow the vehicle; however, because the brakes were either on or off (not progressive), they were only used in more extreme situations. The dynamics, from throttle position to vehicle speed, are generally modeled as a first-order lag described by a DC gain and a time constant. The time constant was identified by performing step inputs into the system; from a stop, the throttle was set to a known position, and the time to reach a percentage of the steady-state speed was recorded. In general, this method should also identify the DC gain of the system, but the drive train of RASCAL contained a continuously variable transmission (CVT) making the DC gain variable. The CVT, along with other engine/vehicle dynamics, make the time constant also somewhat dependent on speed.

Once a rough time constant and DC gain were identified, the controller was designed to obtain a smooth yet responsive reaction to disturbances or changes in desired speed. The controller was a PI type. The integrator was needed to counteract the uncertainty in the DC gain. Special care was taken to ensure the integrator did not wind up and harm the engine or drive train. In particular, a limit was placed on the integrated error, and the error did not accumulate unless the actual vehicle speed was within a bound of the desired speed. This bound was tuned based on the uncertainty in the DC gain.

9.6.3 Heading Controller

The heading controller was more complicated than the speed controller. The complication was due to the dynamics between steer angle (δ) and yaw rate (r); heading (ψ) simply added an integrator to the system (Figure 9.3). The

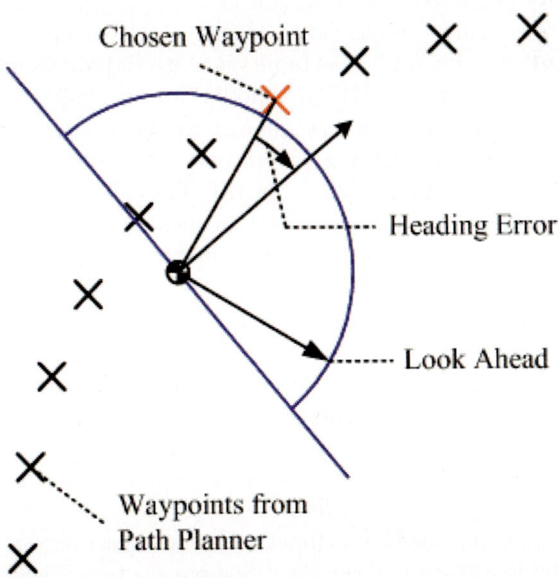

Fig. 9.12. Vehicle control waypoint selection geometry

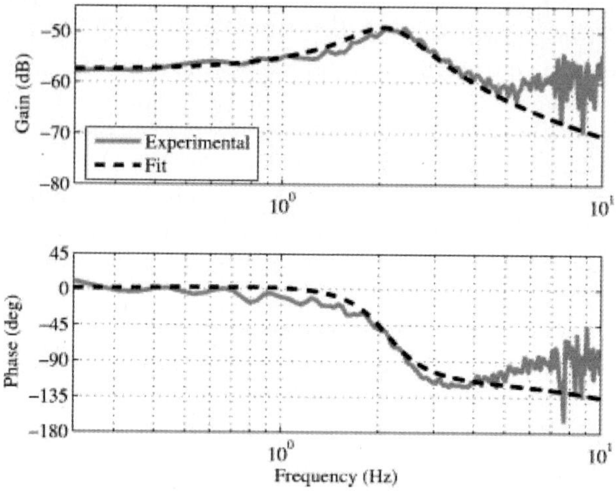

Fig. 9.13. RASCAL open loop steering response

simplest model that captures all of the important lateral vehicle dynamics is a second-order model with one zero as shown below:

$$\frac{r}{\delta} = \frac{k(s+n)}{s^2 + 2\zeta\omega_n s + \omega_n^2} \tag{9.7}$$

The coefficients in the transfer function needed to be identified through dynamic testing. To capture the full range of the vehicle's response, a wide range of frequencies needed to be applied as inputs; a chirp signal was used to meet this requirement. To fit the data, an ARMAX model (Ljung, 1999) produced the least residual errors. One frequency response, along with its fit, is shown in Figure 9.13. The small gain is due to the steer angle input being in counts applied to the servo; 32,000 counts were approximately 30° at the tire. The chirp signal stopped at 7 Hz, meaning the experimental data above this frequency are not valid. A higher-order model could fit the data better; particularly, the higher frequencies. However, in order to keep the controller simple, the fit was left as second order.

Terrain and other varying parameters could also create inaccuracies in the identified model. In general, the parameters that capture vehicle models, tire cornering stiffness, in particular, are assumed to be terrain independent. The tire saturation force is variable, but this does not become a factor, except in extreme cases (Gillespie, 1992).

One characteristic of vehicles is that the lateral dynamic characteristics change with speed, so the chirp input and fit were performed over a range of constant speeds. The identified parameters as functions of speed are shown in Figure 9.14. A curve fit was applied to each of the identified parameters. These curve fits were then used to design the steering controller. It is interesting to note that the trend of these parameters with velocity did not follow that of typical models for ground

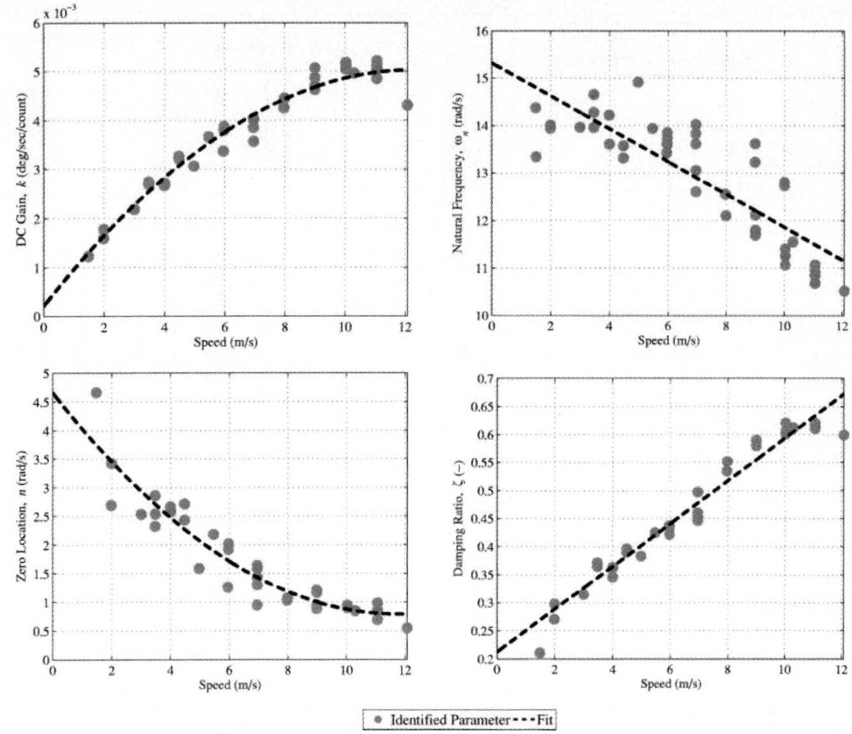

Fig. 9.14. Steering parameters behavior with speed

vehicles, in particular the bicycle model (Gillespie, 1992). The most likely cause was the size of the vehicle. The natural frequency was higher than that of most ground vehicles (2 Hz as opposed to 1 Hz); this meant that other typically ignored factors, such as tire dynamics, were affecting the system response.

A PD controller was used to drive the heading error to zero. As was previously mentioned, the transfer function from steer angle to heading is given by Eq. (9.7) with an additional integrator, making it third order. Two states were fed back; heading error and yaw rate error; therefore, two closed-loop poles could be chosen for the system. Over the identified range, the PD gains were scheduled with velocity, based on the parameter curve fits, to keep these closed-loop poles constant. The third closedloop pole was floating; it was checked to guarantee stability for the entire range of speeds. To account for any model inaccuracies, the controller gains were conservatively designed. This also necessitated the simpler path structure discussed earlier.

9.6.4 Experimental Validation

Figure 9.15 shows the response of the vehicle and throttle controller to steps in the reference speed. In general, the controller performed well; the error is

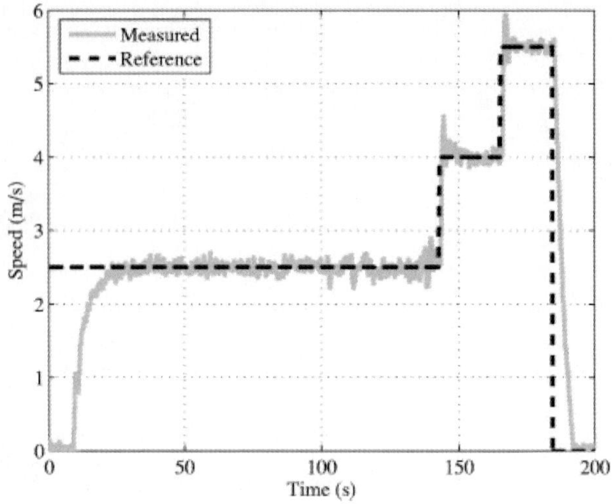

Fig. 9.15. Controlled speed response

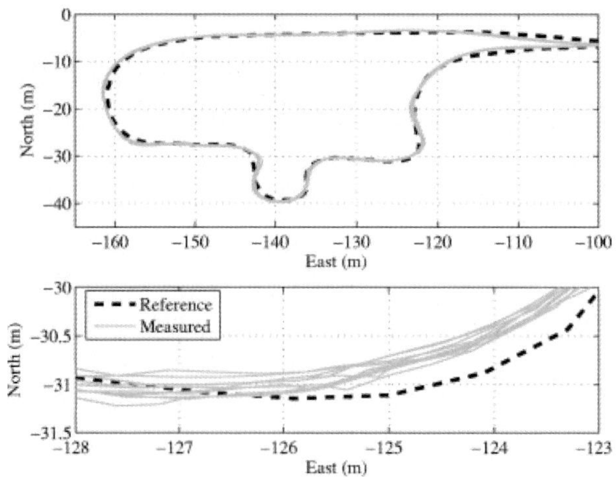

Fig. 9.16. Waypoint tracking response

typically around 0.1 m/s. It is interesting to note that the initial step (to 2.5 m/s) is overdamped, while the other two (to 4 and 5.5 m/s) are underdamped. The difference is due to the unmodeled dynamics moving the effective closed-loop pole and the fact that the integrator is switching on and off based on the magnitude of the speed error. The controller was tested at much higher speeds (up to 18 m/s) and still performed with no noticeable degradation.

Figure 9.16 shows a portion of the path created for the application video required by DARPA, along with laps run by RASCAL around this path. There are some areas where RASCAL does not exactly follow the desired path; after one particularly tight high-speed corner, the error was 1 m. This error was mostly due to look ahead in the steering controller as discussed above. It could have been reduced with a higher fidelity controller/vehicle model at the expense of robustness; however, with the parameters defined by DARPA (i.e., typical corridor width), the error was deemed acceptable. The pass-to-pass repeatability was quite accurate; typically around 20 cm. RASCAL's response to a given situation was very predictable.

While the vehicle was at Auburn University, the controllers were also stress tested for over 100 miles at high speeds; between 25 and 40 mph. For a vehicle the size of an ATV, these speeds were significant, as they corresponded to a sports utility vehicle traveling at highway speeds. These tests were conducted without the obstacle detection algorithms in order to concentrate on the behavior of the navigation and control modules alone. The error grew as the speed increased, reaching 2 m at 40 mph; some of this was due to an incorrect calibration in the steering servo. However, at these speeds, RASCAL would have been on a wide road during the Grand Challenge, and that amount of error was acceptable.

9.7 Results

9.7.1 The National Qualification Event

The NQE was the final measure for entry in the Grand Challenge. Of the 195 initial entrants, 43 teams successfully completed a preselection qualification and

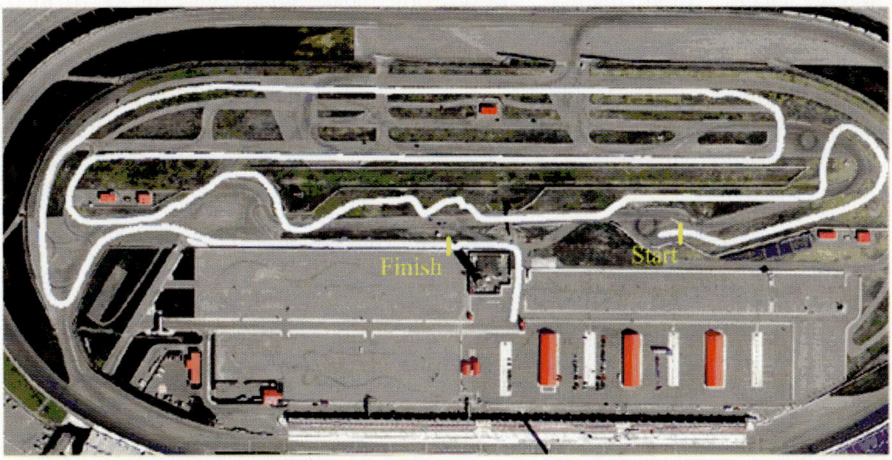

Fig. 9.17. Aerial picture of an NQE obstacle course

Table 9.4. RASCAL NQE results

	NQE1	NQE2	NQE3	NQE4
Time	Did not finish	16min	14min	14min
Passed Gates	22/50	46/50	48/50	48/50
Passed Obstacles	1/4	4/4	4/5	5/5

participated at the NQE. After the NQE, the best 23 teams got the opportunity to start the Grand Challenge. The main goal of the NQE was to complete four different tracks, each approximately 2.5 miles long. Figure 9.17 shows an aerial picture of an NQE track. In these tracks, DARPA officials composed courses with different sets of obstacles that could be found in the desert. Each track consisted of a small hill, a high-speed section, a tunnel where the GPS signal was blocked, gates, and several manmade obstacles. The four tracks differed in types and location of obstacles. A run consisted of 50 gates, a tunnel section, and four or five obstacles. As presented in Table 9.4, RASCAL completed three of the four runs in the NQE. This success rate allowed RASCAL to be one of ten teams selected for early qualification.

The initial run ended soon after the tunnel; Figure 9.18 shows the cause of the failure. The light thin line indicates the estimated position from the localization algorithm, and the light dots represent the recorded GPS position. As expected, RASCAL lost GPS immediately upon entering the tunnel and began to dead reckon to estimate position. The dead reckoning worked exceptionally well for 2 min, during which time RASCAL avoided several obstacles. GPS measurements were reacquired as RASCAL was initiating its obstacle avoidance procedure to avoid a parked car. The GPS receiver incorrectly reported its measurements valid to within 10 cm. Instead, the position measurement was off by 10 m to the north and east, and the velocities were reported as pure zeros causing the localization algorithm to crash. The addition of a few simple lines of code ignored these false messages, at the expense of having to dead reckon for approximately 4 min after a GPS outage. Once this fix was applied, the remaining NQE runs were successfully completed.

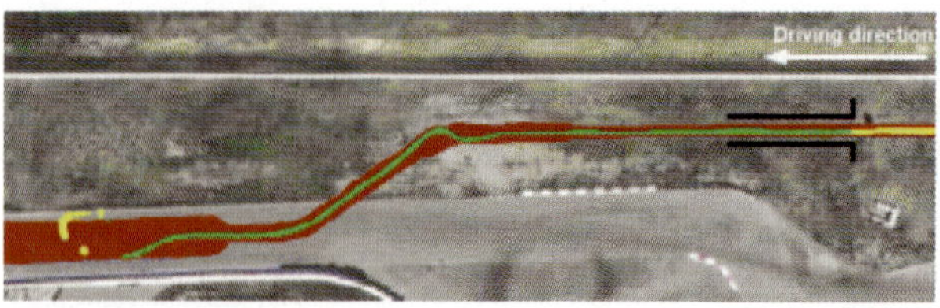

Fig. 9.18. Dead reckoning through NQE Run 1

9.7.2 DARPA Grand Challenge

RASCAL was the tenth vehicle to start the 2005 Grand Challenge. It efficiently negotiated the first segments of the course, passing two competitor vehicles within the first few miles. The initial portion of the course was smoothly traversed with no difficulty. However, problems developed and officials had to stop RASCAL because of severe performance degradation 16 miles into the course. For the duration of the run, the vehicle traveled an average of 4.6 m/s, and reached maximum speeds of 11.5 m/s. Although the vehicle was capable of operating at higher speeds, the obstacle detection system could not process the data reliably beyond 11 m/s. Figure 9.19 displays the whole course, along with a closeup of the portion RASCAL completed.

After an analysis of the recorded data, the cause of failure is fairly certain. Shortly before RASCAL went off road, it lost all LIDAR data. Soon thereafter, it lost all vehicle state data. The LIDAR and the internal sensors were connected via USB hubs to the processing computers. Speculation is that one of the following faults occurred and ended RASCAL's day: USB hubs lost power terminating the connection between the computer and sensors, or the USB hubs overheated and ceased to function.

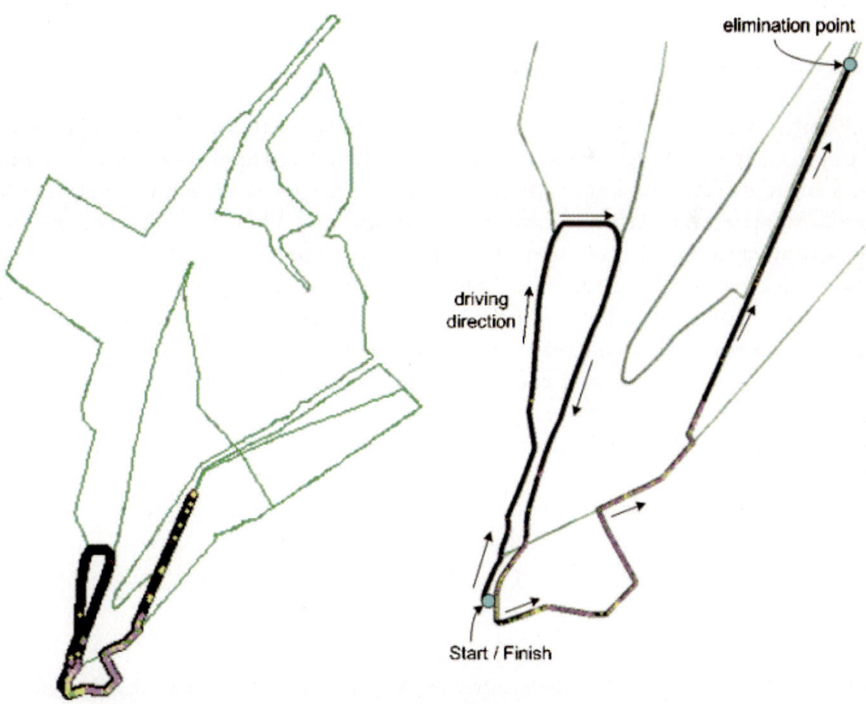

Fig. 9.19. Grand Challenge race course

9.8 Conclusion

SciAutonics-Auburn Engineering, along with Seibersdorf Research, managed to build a strong contender for the DARPA Grand Challenge 2005. A team of volunteer engineers with limited resources managed to stay competitive among teams with more time, money, and resources. Out of 195 initial entrants, 23 teams started at the DARPA Grand Challenge, and five teams finished the race. The 16 mile effort by RASCAL was the 16th longest run among the remaining entrants. The vehicle successfully demonstrated the cohesive hardware and software integration at the NQE, but minor events harmed the vehicle's endurance and ended its run in the 2005 Grand Challenge.

The key to the Grand Challenge was not necessarily the incredibly accurate sensing technology or immense amounts of computing power. RASCAL was able to detect and avoid the same obstacles as teams with twice the number of sensors and considerably more than twice the computing power. With intelligent yet efficient algorithms and a few key sensors, these hurdles could be overcome. The failure that finally disabled RASCAL was a simple hardware connection malfunction. More time needed to be spent by the team to harden the vehicle. RASCAL's concept was validated by its showing in the Grand Challenge; with more time, the realization of its potential would also have been reached.

Acknowledgments

The authors acknowledge the financial contributions to the SciAutonics, LLC team, from Rockwell Scientific (RSC), issued by CEO Derek Cheung. They also acknowledge the provision of SICK LIDAR sensors by Rockwell Collins, thanks to the support by Patrick Hwang and his group. They further acknowledge the strong support of their team by ATV Corporation through Amos Deakon. Thanks are extended to the stockholders for the monetary support and enthusiasm they provided. The Auburn University College of Engineering and Mechanical Engineering Department must also be thanked for their support of this project. In addition, without the help of all the volunteers, this project and the achievements would not have been possible. Special thanks go to the team members who could not contribute to this specific paper, but without whom RASCAL would not have driven, particularly: Team leader— John Porter, Wayne Guthmiller, Dick Elsley, Bob Addison, Gary Bostrup, Monte Khoshnevisan, Don Harris, Ron Dilley, James Whiteley, and Alon Yaari.

References

Behringer, R., Gregory, B., Sundareswaran, V., Addison, B., Elsley, R., Guthmiller, W., deMarchi, J., Daily, R., Bevly, D., Reinhart C. (2004, July). Development of an Autonomous Road Vehicle for the DARPA Grand Challenge. IFAC Symposium on Intelligent Autonomous Vehicles 2004. Lisbon, Portugal.

Behringer, R., Sundareswaran, V., Gregory, B., Elsley, R., Addison, B., Guthmiller, W., Daily, R., Bevly D. (2004, June). The DARPA Grand Challenge – Development of an Autonomous Vehicle. IEEE Symposium on Intelligent Vehicles 2004. Parma, Italy.

Behringer R., Travis, W., Daily, R., Bevly, D., Kubinger, W., Herzner, W., Fehlberg, V. (2005, September). RASCAL – An Autonomous Ground Vehicle for Desert Driving in the DARPA Grand Challenge 2005. IEEE ITSC 2005. Vienna, Austria.

Bevly, D. (2004). Global Positioning System (GPS): A Low-Cost Velocity Sensor for Correcting Inertial Sensor Errors on Ground Vehicles. Journal of System Dynamics, Measurement, and Control, Vol. 126. pp 255–264.

Farrell, J., Barth, M. (1999). The Global Positioning System and Inertial Navigation. New York, NY: McGraw-Hill Companies, Inc.

Gillespie, T. (1992). Fundamentals of Vehicle Dynamics. Warrendale, PA: Society of Automotive Engineers.

Koenig, N., Howard, A. (2004, September). Design and Use Paradigms for Gazebo, An Open-Source Multi-Robot Simulator. IEEE International Conference on Intelligent Robots and Systems. Sendai, Japan.

Labayrade, R., Aubert, D., Tarel, J. (2002, June). Real Time Obstacle Detection in Stereo Vision on Non-Flat Road Geometry Through "V-Disparity" Representation. IEEE International Vehicle Symposium. Versailles, France.

Ljung, L. (1999). System Identification: Theory for the User, 2nd ed. Upper Saddle River, NJ: Prentice Hall PTR.

Postel, J. (1980). RFC 768 – User Datagram Protocol. St. Pierre, M., Gingras, D. (2004, June). Comparison Between the Unscented Kalman Filter and the Extended Kalman Filter for the Position Estimation Module of an Integrated Navigation Information System. IEEE Symposium on Intelligent Vehicles 2004. Parma, Italy.

Stengal, R. (1994). Optimal Control and Estimation. Mineola, NY: Dover Publications. Travis, W. (2006). Methods for Minimizing Navigation Errors Induced by Ground Vehicle Dynamics. M.S. thesis, Auburn University, Auburn, Alabama.

10

Team CIMAR's NaviGATOR: An Unmanned Ground Vehicle for the 2005 DARPA Grand Challenge

Carl D. Crane III[1], David G. Armstrong II[1], Robert Touchton[1], Tom Galluzzo[1], Sanjay Solanki[1], Jaesang Lee[1], Daniel Kent[1], Maryum Ahmed[1], Roberto Montane[1], Shannon Ridgeway[1], Steve Velat[1], Greg Garcia[1], Michael Griffis[2], Sarah Gray[3], John Washburn[4], and Gregory Routson[4]

[1] University of Florida, Center for Intelligent Machines and Robotics, Gainesville, Florida
[2] The Eigenpoint Company, High Springs, Florida
[3] Autonomous Solutions, Inc. Young Ward, Utah
[4] Smiths Aerospace, LLC Grand Rapids, Michigan

Summary. This paper describes the development of an autonomous vehicle system that participated in the 2005 DARPA Grand Challenge event. After a brief description of the event, the architecture, based on version 3.2 of the Department of Defense Joint Architecture for Unmanned Systems (JAUS), and the design of the system are presented in detail. In particular, the "smart sensor" concept is introduced which provided a standardized means for each sensor to present data for rapid integration and arbitration. Information about the vehicle design, system localization, perception sensors, and the dynamic planning algorithms that were used is then presented in detail. Subsequently, testing results and performance results are presented.

Keywords: DARPA Grand Challenge, autonomous navigation, path planning, sensor fusion, world modeling, localization, JAUS.

10.1 Introduction

The DARPA Grand Challenge is widely recognized as the largest and most cutting-edge robotics event in the world, offering groups of highly motivated scientists and engineers across the U.S. an opportunity to innovate in developing state-of-the-art autonomous vehicle technologies with significant military and commercial applications. The U.S. Congress has tasked the military with making nearly one-third of all operational ground vehicles unmanned by 2015 and The DARPA Grand Challenge is one in a number of efforts to accelerate this effort. The intent of the event is to spur participation in robotics by groups of engineers and scientists outside the normal military procurement channels including leaders in collegiate research, military development, and industry research.

M. Buehler, K. Iagnemma, and S. Singh (Eds.): DARPA'05, STAR 36, pp. 311–347, 2007.
springerlink.com

(a) Team CIMAR's 2004 DARPA Grand Challenge Entry

(b) Team CIMAR's 2005 DARPA Grand Challenge Entry

Fig. 10.1. The NaviGATOR

Team CIMAR is a collaborative effort of the University of Florida Center for Intelligent Machines and Robotics (CIMAR), The Eigenpoint Company of High Springs, Florida, and Autonomous Solutions of Young Ward, Utah. The goal of Team CIMAR is to develop cutting-edge autonomous vehicle systems and solutions with wide ranging market applications, such as intelligent transportation systems and autonomous systems for force protection. Team CIMAR focused on proving their solutions on an international level by participating in both the 2004 and the 2005 DARPA Grand Challenges.

In 2003, Team CIMAR was one of 25 teams selected from over 100 applicants nationwide to participate in the inaugural event. Team CIMAR was also one of the 15 teams that successfully qualified for and participated in the inaugural event in March 2004; and finished in eighth place. Team CIMAR was accepted into the inaugural DARPA Grand Challenge in late December 2003 and fielded a top-10 vehicle less than three months later. The team learned a tremendous amount from the initial event and used that experience to develop a highly advanced new system to qualify for the second Grand Challenge in 2005 (see Figure 10.1).

10.2 System Architecture and Design

The system architecture that was implemented was based on the Joint Architecture for Unmanned Systems (JAUS) Reference Architecture, version 3.2 (JAUS, 2005). JAUS defines a set of reusable components and their interfaces. The system architecture was formulated using existing JAUS-specified components wherever possible along with a JAUS-compliant intercomponent messaging infrastructure. Tasks for which there are no components specified in JAUS required the creation of so-called "Experimental" components using "User-defined" messages. This approach is endorsed by the JAUS Working Group as the best way to extend and evolve the JAUS specifications.

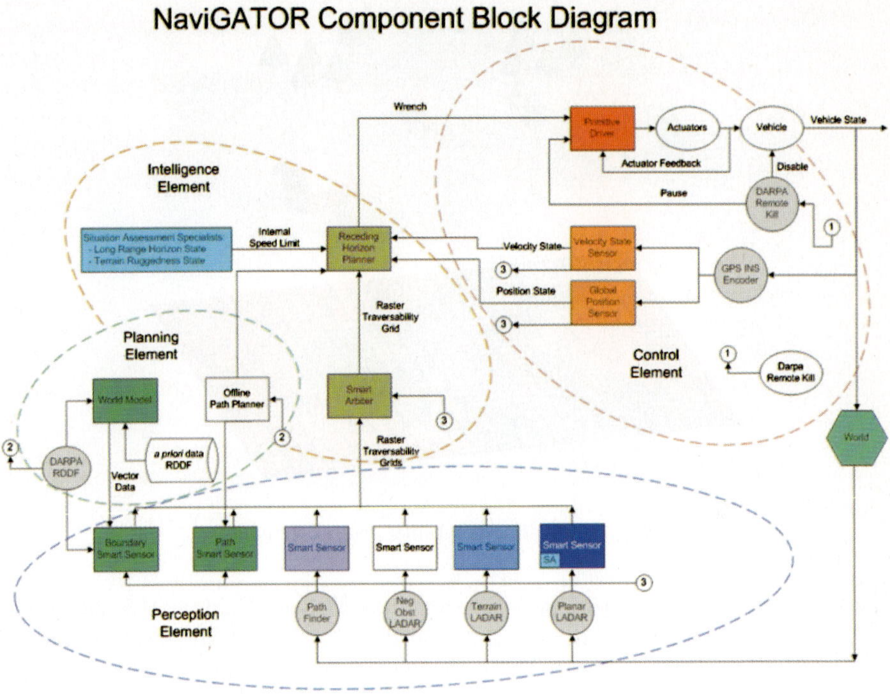

Fig. 10.2. The NaviGATOR's JAUS-compliant architecture

10.2.1 High-Level Architecture

At the highest level, the architecture consists of four fundamental elements, which are depicted in Figure 10.2:

- Planning Element: The components that act as a repository for *a priori* data, including known roads, trails, or obstacles, as well as acceptable vehicle workspace boundaries. Additionally, these components perform offline planning based on that data.
- Control Element: The components that perform closed-loop control in order to keep the vehicle on a specified path.
- Perception Element: The components that perform the sensing tasks required to locate obstacles and to evaluate the smoothness of terrain.
- Intelligence Element: The components that act to determine the "best" path segment to be driven based on the sensed information.

10.2.2 Smart Sensor Concept

The Smart Sensor concept unifies the formatting and distribution of perception data among the components that produce and/or consume it. First, a common data structure, dubbed the Traversability Grid, was devised for use by all Smart

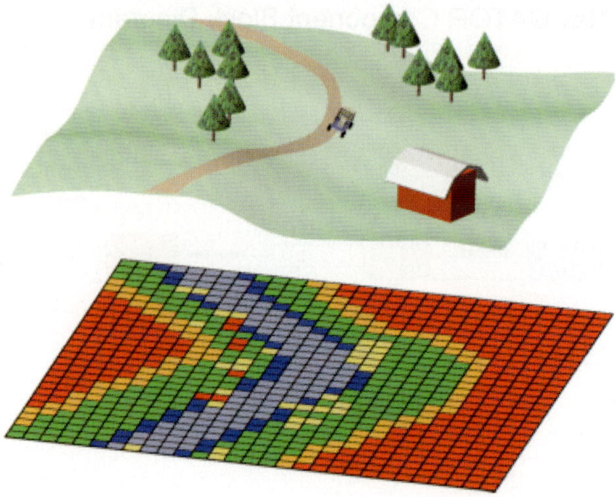

Fig. 10.3. Traversability grid portrayal

Sensors, the Smart Arbiter, and the Reactive Driver. Figure 10.3 shows the world as a human sees it in the upper level, while the lower level shows the Grid representation based on the fusion of sensor information. This grid was sufficiently specified to enable developers to work independently and for the Smart Arbiter to use the same approach for processing input grids, no matter how many there were at any instant in time.

The basis of the Smart Sensor architecture is the idea that each sensor processes its data independently of the system and provides a logically redundant interface to the other components within the system. This allows developers to create their technologies independent of one another and process their data as best fits their system. The sensor can then be integrated into the system with minimal effort to create a robust perception system. The primary benefit of this approach is its flexibility, in effect, decoupling the development and integration efforts of the various component researchers. Its primary drawback is that it prevents the ability of one sensor component to take advantage of the results of another sensor when translating its raw input data into traversability findings.

The Traversability Grid concept is based on the well-understood notion of an Occupancy Grid, which is often attributed to Alberto Elfes of Carnegie Mellon University (Elfes, 1989). His work defines an Occupancy Grid as "a probabilistic tesselated representation of spatial information." Sebastian Thrun provides an excellent treatise on how this paradigm has matured over the past 20 years (Thrun, 2003). The expansion of the Occupancy Grid into a Traversability Grid has emerged in recent years in an attempt to expand the applicability and utility of this fundamental concept (Seraji, 2003), (Ye & Borenstein, 2004). The primary contribution of the Traversability Grid implementation devised for the NaviGATOR is its focus on representing degrees of traversability including terrain conditions and obstacles (from absolutely blocked to unobstructed level pavement) while preserving real-time performance of 20 Hz.

The Traversability Grid design is 121 rows (0 – 120) by 121 columns (0 – 120), with each grid cell representing a half-meter by half-meter area. The center cell, at location (60, 60), represents the vehicle's reported position. The sensor results are oriented in the global frame of reference so that true north is always aligned vertically in the grid. In this fashion, a 60m by 60m grid is produced that is able to accept data at least 30m ahead of the vehicle and store data at least 30m behind it. To support proper treatment of the vehicle's position and orientation, every Smart Sensor component is responsible for establishing a near-real-time latitude/longitude and heading (yaw) feed from the GPOS component.

The scoring of each cell is based on mapping the sensor's assessment of the traversability of that cell into a range from 2 to 12, where 2 means that the cell is absolutely impassable, 12 means the cell represents an absolutely desirable, easily traversed surface, and 7 means that the sensor has no evidence that the traversability of that cell is particularly good or bad. Certain other values are reserved for use as follows: 0 → "out-of-bounds," 1→ "value unchanged," 13→ "failed/error," 14→"unknown," and 15→ "vehicle location." These discrete values have been color-coded to help humans visualize the contents of a given Traversability Grid, from red (2) to gray (7) to green (12).

All of these characteristics are the same for every Smart Sensor, making seamless integration possible, with no predetermined number of sensors. The grids are sent to the Smart Arbiter, which is responsible for fusing the data. The arbiter then sends a grid with all the same characteristics to the Reactive Driver, which uses it to dynamically compute the desired vehicle speed and steering.

The messaging concept for marshalling grid cell data from sensors to the arbiter and from the arbiter to the reactive driver is to send an entire traversability grid as often as the downstream component has requested it (typically at 20 Hz). In order to properly align a given sensor's output with that of the other sensors, the message must also provide the latitude and longitude of the center cell (i.e., vehicle position at the instant the message and its cell values were determined). An alternative approach for data marshalling was considered in which only those cells that had changed since the last message were packaged into the message. Thus, for each scan or iteration, the sending component would determine which cells in the grid have new values and pack the row, column, and value of that cell into the current message. This technique greatly reduces the network traffic and message-handling load for nominal cases (i.e., cases in which most cells remain the same from one iteration to the next). However, after much experimentation in both the lab and the field, this technique was not used due to concerns that a failure to receive and apply a changed cells message would corrupt the grid and potentially lead to inappropriate decisions, while the performance achieved when sending the entire grid in each message never became an issue (our concern about the ability of the Smart Sensor computers, or the onboard network, to process hundreds of full-grid messages per second did not manifest itself in the field).

In order to aid in the understanding, tuning, and validation of the Traversability Grids being produced, a Smart Sensor Visualizer component was developed. Used primarily for testing, the SSV can be pointed at any of the Smart Sensors, the Smart Arbiter, or the Reactive Driver and it will display the color-coded Traversability Grid, along with the associated vehicle position, heading, and speed. The refresh rate of the images is adjustable from real time (e.g., 20 Hz) down to periodic snapshots (e.g., 1 s interval).

10.2.3 Concept of Operation

The most daunting task of all was integrating these components such that an overall mission could be accomplished. Figure 10.4 portrays schematically how the key components work together to control the vehicle. Figure 10.4 also shows how the Traversability Grid concept enables the various Smart Sensors to deliver grids to the Smart Arbiter, which fuses them and delivers a single grid to the Reactive Driver. Prior to beginning a given mission, the *a priori* Planner builds the initial path, which it stores in a Path File as a series of global positioning system (GPS) waypoints. Once the mission is begun, the Reactive Driver sequentially guides the vehicle to each waypoint in the Path File via the Primitive Driver. Meanwhile, the various Smart Sensors begin their search for obstacles and/or

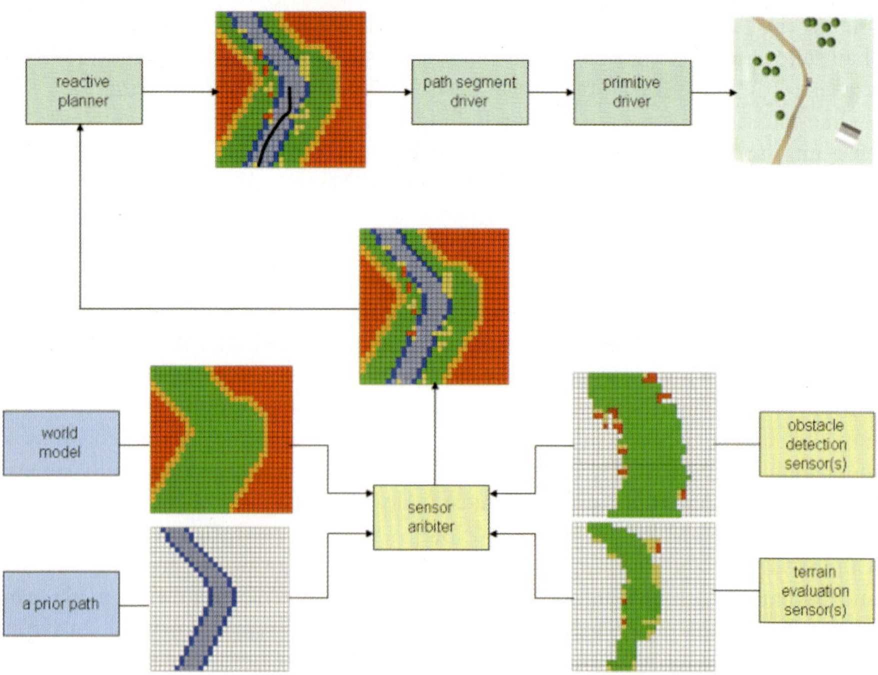

Fig. 10.4. Operational schematic (including traversability grid propagation)

smooth surfaces and feed their findings to the Smart Arbiter. The Smart Arbiter performs its data fusion task and sends the results to the Reactive Driver. The Reactive Driver looks for interferences or opportunities based on the feed from the Smart Arbiter and alters its command to the Primitive Driver accordingly. Finally, the goal is to perform this sequence iteratively on a subsecond cycle time (10 to 60 Hz), depending on the component, with 20 Hz as the default operational rate.

10.3 Vehicle Design

The NaviGATOR's base platform is an all terrain vehicle custom built to Team CIMAR's specifications. The frame is made of mild steel roll bar with an open design. It has 9" Currie axles, Bilstein Shocks, hydraulic steering, and front and rear disk brakes with an emergency brake to the rear. It has a 150 HP Transverse Honda engine/transaxle mounted longitudinally, with locked transaxle that drives front and rear Detroit Locker differentials (4 wheel drive, guaranteed to get power to the road). The vehicle was chosen for its versatility, mobility, openness, and ease of development (see Figure 10.5).

(a)

(b)

Fig. 10.5. Base mobility platform

The power system consists of two, independent 140A, 28V alternator systems (Figure 10.5a). Each alternator drives a 2400W continuous, 4800W peak inverter and is backed up by 4 deep cell batteries. Each alternator feeds one of two automatic transfer switches (ATS). The output of one ATS drives the computers and electronics while the other drives the actuators and a 3/4 Ton (approx. 1kW cooling) air conditioner. Should either alternator/battery system fail the entire load automatically switches to the other alternator/battery system. Total system power requirement is approximately 2200W, so the power system is totally redundant.

The system sensors are mounted on a rack that is specifically designed for their configuration and placement on the front of the vehicle (see Figure 10.6). These

Fig. 10.6. View of sensor cage

sensors include a camera that finds the smoothest path in a scene. Equipped with an automatic iris and housed in a waterproof and dust proof protective enclosure, the camera looks through a face that is made of lexan and covered with a polarizing scratch-resistant film. Also mounted on the sensor cage are two SICK LADARs that scan the ground ahead of the vehicle for terrain slope estimation; one tuned for negative obstacle detection and the other for smooth terrain detection. Also, an additional SICK LADAR aimed parallel to the ground plane is mounted on the front of the vehicle at bumper level for planar obstacle detection. Additional sensors were mounted on the vehicle for experimental purposes, but were not activated for the Darpa Grand Challenge (DGC) event. Each sensor system is described in detail later in this paper.

The computing system requirements consists of high-level computation needs, system command implementation, and system sensing with health and fault monitoring. The high level computational needs are met in the deployed system via the utilization of eight single-processor computing nodes targeted at individual computational needs. The decision to compartmentalize individual processes is driven by the developmental nature of the system. A communications protocol is implemented to allow interprocess communication.

The individual computing node hardware architecture was selected based on the subjective evaluation of commercial off-the-shelf hardware. Evaluation criteria were centered on performance and power consumption. The deployed system maintains a homogenous hardware solution with respect to the motherboard, random access memory (RAM), enclosure, and system storage. The AMD K8 64

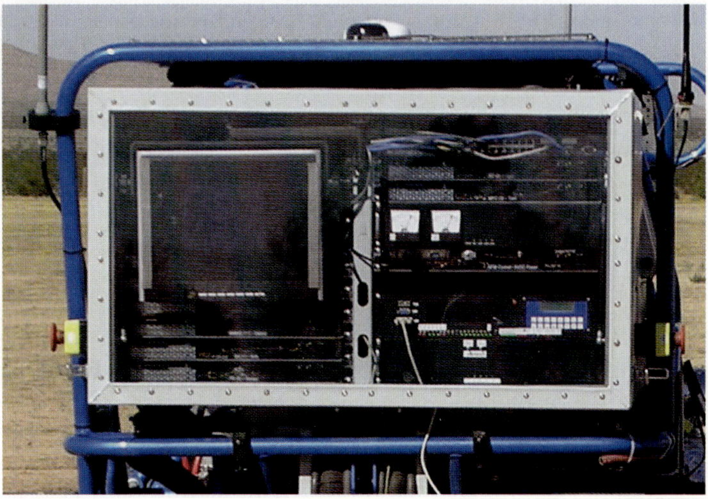

Fig. 10.7. Computer and electronics housing

bit microprocessor family was selected based on power consumption measurement and performance to allow tailoring based on performance requirements with the objective of power requirement reduction. Currently, three processor speeds are deployed: 2.0 GHz, 2.2 GHz, and 2.4 GHz. The processors are hosted in off-the-shelf motherboards and utilize solid-state flash cards for booting and long-term storage. Each processing node is equipped with 512 to 1028 MB of RAM. JAUS communication is effected through the built-in Ethernet controller located on the motherboard. Several nodes host PCI cards for data i/o. Each node is housed in a standard 1-U enclosure. The operating system deployed is based on the 2.6 Linux kernel. System maintenance and reliability are expected to be adequate due to the homogenous and modular nature of the compute nodes. Back-up computational nodes are on hand for additional requirements and replacement. All computing systems and electronics are housed in a NEMA 4 enclosure mounted in the rear of the vehicle (see Figure 10.7).

10.4 Route Pre-planning

The DARPA Grand Challenge posed an interesting planning problem given that the route could be up to 175 miles in length and run anywhere between Barstow, California and Las Vegas, Nevada. On the day of the event, DARPA supplied a Route Data Definition File (RDDF) with waypoint coordinates, corridor segment width, and velocity data. In order to process the *a priori* environment data and generate a route through DARPA's waypoint file, Team CIMAR used Mobius, an easy to use graphical user interface developed by Autonomous Solutions Inc. for controlling and monitoring unmanned vehicles. Mobius was used to plan the initial path for the NaviGATOR in both the National Qualification Event and the final Grand Challenge Event.

The route pre-planning is done in three steps: Generate corridor data, import and optimize the DARPA path, and modify path speeds. A World Model component generates the corridor data by parsing DARPA's RDDF and clipping all other environment information with the corridor such that only elevation data inside the corridor are used in the planning process (see Figure 10.8). The RDDF corridor (now transformed into an ESRI shapefile) is then imported into Mobius and displayed to the operator for verification.

In the next step, Mobius imports the original RDDF file for use in path generation. Maximum velocities are assigned to each path segment based on the DARPA assigned velocities at each waypoint. From here, the path is optimized using the NaviGATOR's kinematics constraints and a desired maximum deviation from the initial path. The resultant path is a smooth drivable path, from the start node to the finish node, that stays inside the RDDF corridor generated specifically for the NaviGATOR (see Figure 10.9). Mobius is then used to make minor path modifications where necessary to create a more desirable path.

The final step of the pre-planning process is to modify path velocities based on *a priori* environment data and velocity constraints of the NaviGATOR itself. Sections of the path are selected and reassigned velocities. Mobius assigns the minimum of the newly desired velocity and the RDDF-assigned velocity to the sections in order to ensure that the RDDF-assigned velocities are never exceeded. During the DARPA events, the maximum controlled velocity of the NaviGATOR

Fig. 10.8. RDDF corridor (parsed with elevation data)

was 25 miles per hour so, in the first pass, the entire path was set to a conservative 18 mph except in path segments where the RDDF speed limit was lower. From there, the path is inspected from start to finish and velocities are increased or decreased based on changes in curvature of the path, open environment (dry lake beds), elevation changes, and known hazards in the path (e.g., over/under passes). After all velocity changes are made, the time required to complete the entire path can be calculated. For the race, it was estimated that it would take the NaviGATOR approximately 8 hours and 45 minutes to complete the course. Finally, the path is saved as a comma-separated Path File and transferred to the NaviGATOR for autonomous navigation.

10.5 Localization

The NaviGATOR determines its geolocation by filtering and fusing a combination of sensor data. The processing of all navigation data is done by a Smiths Aerospace North-finding Module (NFM), which is an inertial navigation system. This module maintains Kalman filter estimates of the vehicle's global position and orientation, as well as linear and angular velocities. It fuses internal accelerometer and gyroscope data, with data from an external NMEA GPS, and external odometer. The GPS signal provided to the NFM comes from one of the two onboard sensors. These include a NavCom Technologies Starfire 2050, and a Garmin WAAS Enabled GPS 16. An onboard computer simultaneously

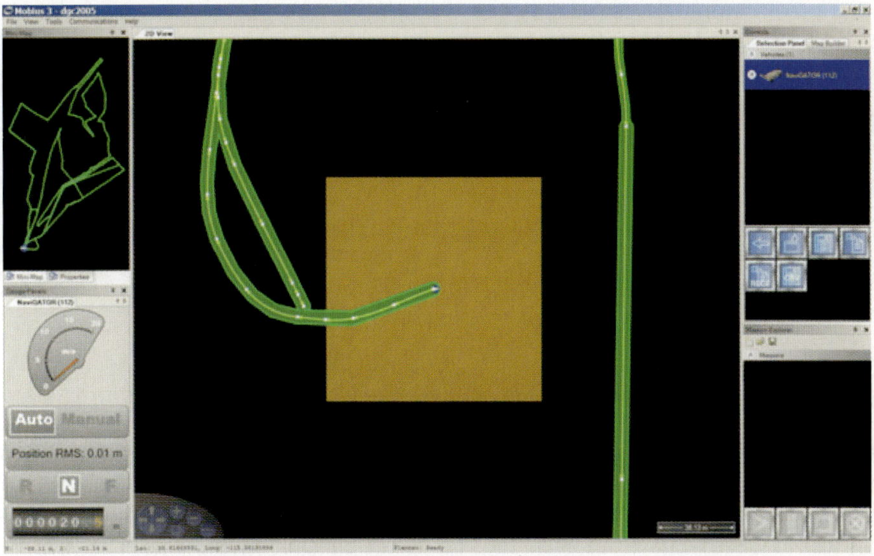

Fig. 10.9. Mobius screen shot with path optimized for the NaviGATOR. The race RDDF is shown in the upper left corner and the start/finish area is centered on the screen.

parses data from the two GPS units and routes the best-determined signal to the NFM. This is done to maintain valid information to the NFM at times when only one sensor is tracking GPS satellites. During valid tracking, the precision of the NavCom data is better than the Garmin, and thus the system is biased to always use the NavCom when possible.

In the event that both units lose track of satellites, as seen during GPS outages, which occurs when the vehicle is in a tunnel, the NFM will maintain localization estimates based on inertial and odometry data. This allows the vehicle to continue on course for a period of time; however, the solution will gradually drift and the accuracy of the position system will steadily decrease as long as the GPS outage continues. After a distance of a few hundred meters, the error in the system will build up to the point where the vehicle can no longer continue on course with confidence. At this point, the vehicle will stop and wait for a GPS reacquisition. Once the GPS units begin tracking satellites and provide a valid solution, the system corrects for any off-course drift and continues autonomous operation.

The Smith's NFM is programmed to robustly detect and respond to a wide range of sensor errors or faults. The known faults of both GPS systems, which generate invalid data, are automatically rejected by the module, and do not impact the performance of the system, as long as the faults do not persist for an extended period of time. If they do persist, then the NFM will indicate to a control computer what the problem is, and the system can correct it accordingly. The same is true for any odometer encoder error, or inertial sensor errors. The NFM will automatically respond to the faults and relay the relevant information to control computers, so the system can decide the best course of action to correct the problem.

10.6 Perception

This section of the paper discusses how the NaviGATOR collects, processes and combines sensor data. Each of the sensor components is presented, organized by type: LADAR, camera, or "pseudo" (a component that produces an output as if it were a sensor, but based on data from a file or database). Finally, the Smart Arbiter sensor fusion component is discussed.

10.6.1 LADAR-Based Smart Sensors

There are three Smart Sensors that rely on LADAR range data to produce their results: the Terrain Smart Sensor (TSS), the Negative Obstacle Smart Sensor (NOSS) and the Planar LADAR Smart Sensor (PLSS). All three components use the LMS291-S05 from Sick Inc. for range measurement. The TSS will be described in detail and then the remaining two will be discussed only in terms of how they are different than the TSS.

A laser range finder operates on the principle of time of flight. The sensor emits an eye-safe infrared laser beam in a single-line sweep of either 180° or 100°, detects the returns at each point of resolution, and then computes single line range image. Although three range resolutions are possible (1°, 0.5°, or 0.25°) the resolution of 0.25° can only be achieved with a 100° range scan. The accuracy of the laser measurement is +/- 50 mm for a range of 1 to 20 m, while its maximum range is ~80 m. A high-speed serial interface card is used to achieve the needed high-speed baud rate of 500 kB.

10.6.1.1 Terrain Smart Sensor

The sensor is mounted facing forward at an angle of 6° toward the ground. For the implementation of the TSS, the 100° range with a 0.25° resolution is used. With this configuration and for nominal conditions (flat ground surface, vehicle level), the laser scans at a distance of ~20 m ahead of the vehicle and ~32 m wide. The TSS converts the range data reported by the laser in polar coordinates into Cartesian coordinates local to the sensor, with the Z axis vertically downward and the X axis in the direction of vehicle travel. The height for each data point (Z component) is computed based on the known geometry of the system and the range distance being reported by the sensor. The data is then transformed into the global coordinate system required by the Traversability Grid, where the origin is the centerline of the vehicle at ground level below the rear axle (i.e., the projection of the GPS antenna onto the ground), based on the instantaneous roll, pitch, and yaw of the vehicle.

Each cell in the Traversability Grid is evaluated individually and classified for its traversability value. The criteria used for classification are:

1. The mean elevation (height) of the data point(s) within the cell.
2. The slope of the best fitting plane through the data points in each cell.
3. The variance of the elevation of the data points within the cell.

The first criterion is a measure of the mean height of a given cell with respect to the vehicle plane. Keep in mind that positive obstacles are reported as negative elevations since the Z-axis points down. The mean height is given as

$$\mu = \frac{\Sigma Z_i}{n}, \tag{10.1}$$

where μ is the mean height, ΣZ_i is the sum of the elevation of the data points within the cell, and n is the number of data points.

The second criterion is a measure of the slope of the data points. The equation for the best fitting plane, derived using the least squares solution technique, is given as

$$S_{\text{optimum}} = (G^T G)^{-1} G^T b, \tag{10.2}$$

where:

S_{optimum} is the vector perpendicular to the best fitting plane,

G is an $n \times 3$ matrix given by

$$G = \begin{bmatrix} x_1 & y_1 & z_1 \\ x_2 & y_2 & z_2 \\ - & - & - \\ x_n & y_n & z_n \end{bmatrix}, \tag{10.3}$$

b is a vector of length "n" given by

$$b = \begin{bmatrix} -D_{01} \\ -D_{02} \\ - \\ -D_{0n} \end{bmatrix}. \tag{10.4}$$

Assuming that D_{0i} is equal to 1, the above equation is used to find $S_{optimum}$ for the data points within each cell. Once the vector perpendicular to the best-fitting plane is known, the slope of this plane in the "x" and "y" directions can be computed. Chapter 5 of Solanki (2003) provides a thorough proof of this technique for finding the perpendicular to a plane.

The variance of the data points within each cell is computed as

$$\text{Variance} = \frac{\Sigma(Z_i - \mu)^2}{n}. \tag{10.5}$$

A traversability value between 2 and 12 is assigned to each cell, depending on the severity values of the mean height, slope, and variance information. A cell must contain a minimum of three data points or else that cell is flagged as unknown. This also helps in eliminating noise. Each of the parameters is individually mapped to a corresponding traversability value for a given cell. This mapping is entirely empirical and non-linear. A weighted average of these three resulting traversability values is used to assign the final traversability value.

10.6.1.2 Negative Obstacle Smart Sensor

The NOSS was specifically implemented to detect negative obstacles (although it can also provide information on positive obstacles and surface smoothness like the TSS). The sensor is configured like the TSS, but at an angle of 12° toward the ground. With this configuration and for nominal conditions, the laser scans the ground at a distance of ~10 m ahead of the vehicle. To detect a negative obstacle, the component analyzes the cases where it receives a range value greater than would be expected for level ground. In such cases, the cell where one would expect to receive a hit is found by assuming a perfectly horizontal imaginary plane. As shown in Figure 10.10, this cell is found by solving for the intersection of the imaginary horizontal plane and the line formed by the laser beam. A

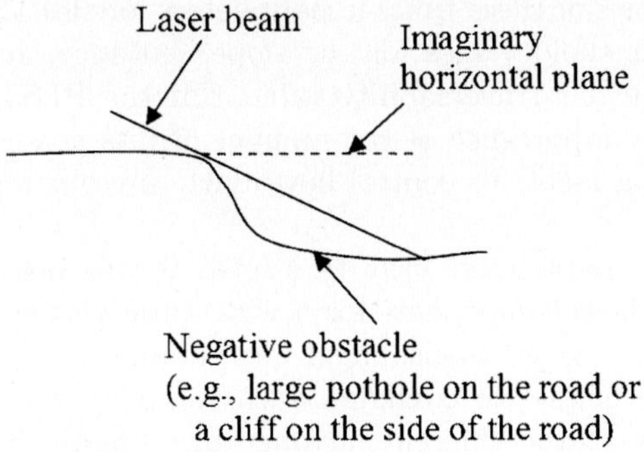

Fig. 10.10. NOSS implementation (side view)

traversability value is assigned to that cell based on the value of the range distance and other configurable parameters. Thus, a negative obstacle is reported for any cell whose associated range data are greater than that expected for an assumed horizontal surface. The remaining cells for which range value data is received are evaluated on a basis similar to the TSS.

10.6.1.3 Planar LADAR Smart Sensor

The sensor is mounted 0.6 m above the ground, scanning in a plane horizontal to the ground. Accordingly, the PLSS only identifies positive obstacles and renders no opinion regarding the smoothness or traversability of areas where no positive obstacle is reported. For the PLSS, the 180° range with a 0.5° resolution is used. The range data from the laser is converted into the Global coordinate system and the cell from which each hit is received is identified. Accordingly, the "number of hits" in that cell is incremented by one and then, for all the cells between the hit cell and the sensor, the "number of missed hits" is incremented by one. Bresenham's line algorithm is used to efficiently determine the indices of the intervening cells.

A traversability value between 2 and 7 is assigned to each cell based on the total number of hits and misses accumulated for that cell. The mapping algorithm first computes a score, which is the difference between the total number of hits and a discounted number of misses in a cell (a discount weight of 1/6 was used for the event). This score is then mapped to a traversability value using an exponential scale of 2. For example, a score of 2 or below is mapped to a traversability value of "7," a score of 4 and below is mapped to a "6," and so on, with a score greater than 32 mapped to a "2." The discounting of missed hits provides conservatism in identifying obstacles, but does allow gradual recovery from false positives (e.g., ground effects) and moving obstacles.

10.6.1.4 Field Testing

The parameters of the algorithm that affect the output of the component are placed in a configuration file so as to enable rapid testing and tuning of those parameters. Examples of these tunable parameters for the TSS and NOSS components are the threshold values for the slope, variance, and mean height for mapping to a particular traversability value. For the PLSS, these parameters include the relative importance of the number of hits versus misses in a given cell, and a weighting factor to control how much any one scan affects the final output.

By making these parameters easy to adjust, it was possible to empirically tune and validate these components for a wide variety of surroundings such as a steep slopes, cliffs, rugged roads, small bushes, and large obstacles, like cars. This approach also helped to configure each component to work in the most optimum way across all the different surroundings. Finally, it helped in deciding on the amount of data/time the component required to build confidence about an obstacle or when an obstacle that was detected earlier has now disappeared from view (e.g., a moving obstacle).

10.6.2 Camera-Based Smart Sensor

The Pathfinder Smart Sensor (PFSS) consists of a single color camera mounted in the sensor cage and aimed at the terrain in front of the vehicle. Its purpose is to assess the area in the cameras scene for terrain that is similar to that on which the vehicle is currently traveling, and then translate that scene information into traversability information. The PFSS component uses a high-speed frame-grabber to store camera images at 30 Hertz.

Note that the primary feature used for analytical processing is the red, green and blue (RGB) color space. This is the standard representation in the world of computers and digital cameras, and is therefore often a natural choice for color representation. Also, RGB is the standard output from a CCD-camera. Since roads typically have a different color than nondrivable terrain, color is a highly relevant feature for segmentation. The following paragraphs describe the scene assessment procedure applied to each image for rendering the Traversability Grid that is sent to the Smart Arbiter.

10.6.2.1 Preprocess Image

To reduce the computational expense of processing large images, the dimensions of the scene are reduced from the original digital input of 720×480 pixels to a 320×240 reduced image. Then, the image is further preprocessed to eliminate the portion of the scene that most likely corresponds to the sky. The segmentation of the image is based simply on the physical location within the scene (tuned based on field testing), adjusted by the instantaneous vehicle pitch. This very simplistic approach is viable because the consequences of inadvertently eliminating ground are minimal due to the fact that ground areas near the horizon will likely be beyond the 30 m planning distance of the system. The motivation for this step in the procedure is that the sky portion of the image hinders the classification procedure in two ways. First, we considered that the sky portion slows down the image processing speed by spending resources evaluating pixels that could never be drivable by a ground vehicle. Second, there could be situations where parts of the sky image could be misclassified as road.

10.6.2.2 Produce Training and Background Data Sets

Next, a 100 × 80 sub-image is used to define the drivable area, and two 35 × 50 sub-images are used to define the background. The drivable sub-image is placed in the bottom-center of the image, while the background sub-images are placed at the middle-right and middle-left of the image, which is normally where the background area will be found, based on experience (Lee, 2004) (see Figure 10.11). When the vehicle turns, the background area that is in the direction of the turn will be reclassified as a drivable area. In this case, that background area information is treated as road area by the classification algorithm.

Fig. 10.11. Scene segmentation scheme

10.6.2.3 Apply Classification Algorithm

A Bayesian decision theory approach was selected for use, as this is a fundamental statistical approach to the problem of pattern classification associated with applications such as this. It makes the assumption that the decision problem is posed in probabilistic terms, and that all of the relevant probability values are known. The basic idea underlying Bayesian decision theory is very simple. However, this is the optimal decision theory under Gaussian distribution assumption (Morris, 1997).

The decision boundary that was used is given by

$$\frac{1}{(2\pi)^{d/2}|\Sigma_1|^{1/2}}\exp\left[-\frac{1}{2}(\mathbf{x}-\boldsymbol{\mu}_1)^T\Sigma_1^{-1}(\mathbf{x}-\boldsymbol{\mu}_1)\right]$$
$$=\frac{1}{(2\pi)^{d/2}|\Sigma_2|^{1/2}}\exp\left[-\frac{1}{2}(\mathbf{x}-\boldsymbol{\mu}_2)^T\Sigma_2^{-1}(\mathbf{x}-\boldsymbol{\mu}_2)\right], \tag{10.6}$$

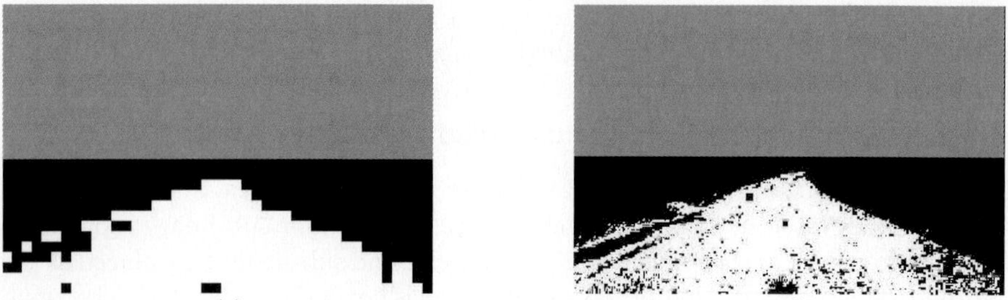

(a) (b)

Fig. 10.12. Processed images

where μ_1 and $\sum_1$ are the mean vector and covariance matrix of the drivable-area RGB pixels in the training data, μ_2 and $\sum_2$ are those of the background pixels, and $\mathbf{x}$ contains RGB values of the entire image.

The decision boundary formula can be simplified as

$$(\mathbf{x} - \boldsymbol{\mu}_1)^T \Sigma_1^{-1} (\mathbf{x} - \boldsymbol{\mu}_1) = (\mathbf{x} - \boldsymbol{\mu}_2)^T \Sigma_2^{-1} (\mathbf{x} - \boldsymbol{\mu}_2). \tag{10.7}$$

A block-based segmentation method is used to reduce the segmentation processing time. 4×4 pixel regions are clustered together and replaced by their RGB mean value, as follows:

$$\mu^L_{(x,y)} = \frac{1}{N^2} \sum_{i=1}^{N} \sum_{j=1}^{N} P^L_{(i,j)}, \tag{10.8}$$

where $\mu(x, y)$ is the new pixel mean value for the 4×4 block, P is the raw pixel data, (i, j) is the raw pixel orientation, (x, y) is the new pixel orientation, L$\epsilon\{1, 2, 3\}$ for RGB, and N is the block size.

The clusters, or blocks, are then segmented, and the result, as shown in Figure 10.12(a), has less noise compared with pixel-based approaches, Figure 10.12(b). Also, the segmentation process is accomplished faster than pixel-based classification. A disadvantage, however, is that edges are jagged and not as distinct.

10.6.2.4 Transform to Global Coordinate System

After processing the image, the areas classified as drivable road are converted by perspective transformation estimation into the global coordinates used for the Traversability Grid (Criminisi, 1997). The perspective transformation matrix is calculated based on camera calibration parameters and the instantaneous vehicle heading. Finally, the PFSS assigns a value of 12 (highly traversability) to those cells that correspond to an area that has been classified as drivable. All other cells are given a value of 7 (neutral). Figure 10.13 depicts the PFSS Traversability Grid data after transformation into global coordinates.

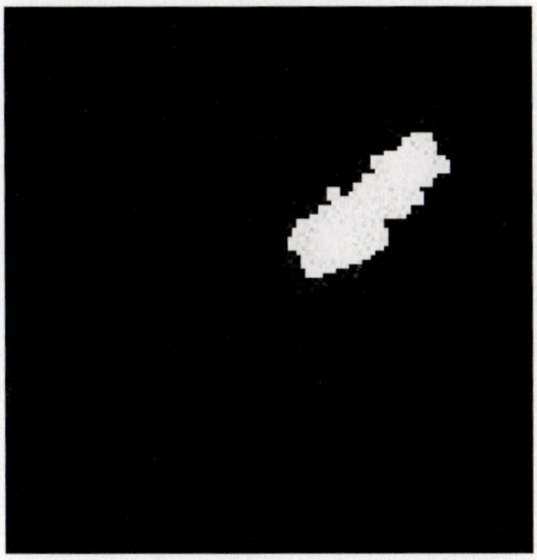

Fig. 10.13. Transformed image

10.6.3 Pseudo Smart Sensors

There are two Smart Sensors that produce Traversability Grids based on stored data: The Boundary Smart Sensor (BSS) and the Path Smart Sensor (PSS).

The BSS translates boundary knowledge, defined as boundary polygons prior to mission start, into real-time Traversability Grids, which assures that the vehicle does not travel outside the given bounds. The BSS is responsible for obtaining the boundary information from a local spatial database. The BSS uses these data to determine the in-bounds and out-of-bounds portions of the traversability grid for the instantaneous location of the vehicle. The BSS also has a configurable "feathering" capability that allows the edge of the boundary to be softened, creating a buffer area along the edges. This feature provides resilience to uncertainties in the position data reported by the GPOS component. Figure 10.14

Fig. 10.14. Traversability Grid showing boundary data

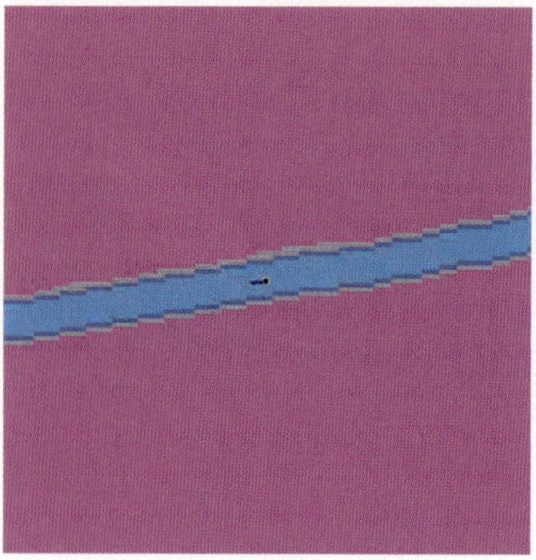

Fig. 10.15. Traversability Grid showing *a priori* path data

shows a typical grid output from the BSS indicating the vehicle's location within the grid, and the drivable region around it. By clearly demarking areas of the grid as out-of-bounds, the BSS allows the Smart Arbiter to summarily dismiss computation of out-ofbounds grid cells and the Reactive Driver to prune its search tree of potential plans.

The PSS translates the *a priori* path plan, stored as a "path file" prior to mission start, into real-time Traversability Grids. The PSS uses these path data to superimpose the originally planned path onto the traversability grid based on the instantaneous location of the vehicle. The PSS has a configurable "feathering" capability that allows the width of the path to be adjusted and the edges of the path to be softened. This feature also allows the engineer to select how strongly the originally planned path should be weighted by setting the grid value for the centerline. A 12 would cause the Arbiter and Planner to lean toward following the original plan even if the sensors were detecting a better path, while a 10, which is what was used at runtime, would make the original plan more like a suggestion that could be more easily overridden by sensor findings. Figure 10.15 shows a typical grid output from the PSS indicating the vehicle's location within the grid and the feathered *a priori* planned path flowing through the inbounds corridor.

10.6.4 Sensor Fusion

With the Traversability Grid concept in place to normalize the outputs of a wide variety of sensors, the data fusion task becomes one of arbitrating the matching cells into a single-output finding for that cell for every in-bounds cell location in the grid.

10.6.4.1 Grid Alignment

First, the Smart Arbiter must receive and unpack the newest message from a given sensor and then adjust its center-point to match that of the Arbiter (assuming that the vehicle has moved between the instant in time when the sensor's message was built and the instant in time when the arbitrated output message is being built). This step must be repeated for each sensor that has sent a message. The pseudo-code for this process is:

Determine current location of vehicle,
Adjust centerpoint of Smart Arbiter Grid to match current location
For each active Smart Sensor:
  Adjust centerpoint of Smart Sensor Grid to match current location.

At this point, all input grids are aligned and contain the latest findings from its source sensor. To support the alignment of Traversability Grids with current vehicle position, a so-called "torus buffer" object was introduced. This allows the system to use pointer arithmetic to "roll the grid" (i.e., change the row and column values of its center-point) without copying data.

10.6.4.2 Data Arbitration

Now the Smart Arbiter must simultaneously traverse the input grids, cell-by-cell, and merge the data from each corresponding cell into a singleoutput value for that row/column location. Once all cells have been treated in this fashion, the Smart Arbiter packs up its output grid message and sends it on the Reactive Driver.

For early testing, a simple average of the input cell values was used as the output cell value. Later work investigated other algorithms, including heuristic ones, to perform the data fusion task. The Smart Arbiter component was designed to make it easy to experiment with varying fusion algorithms in support of on-going research. The algorithm that was used for the DGC event entailed a two-stage heuristic approach. Stage 1 is an "auction" for severe obstacles for the cell position under consideration. Stage 2 then depends on the results of the "auction". If no sensor "wins" the auction, then all of the input cells at that position are averaged, including the arbiter's previous output value for that cell. The pseudo-code for this algorithm is:

For each cell location:

IF *any* sensor reported a "2,"
  THEN decrement the Arbiter's output cell by *decr* (min=2),
ELSE, IF *any* sensor reported a "3,"
  THEN decrement the Arbiter's output cell by *decr/2* (min=3),
ELSE
  Arbiter's output cell= Average(input cells+ Arbiter's prior output cell),

where *decr* is a configurable parameter (= 2 for the Grand Challenge Event).

Thus, a sensor must report a severe obstacle for several iterations in order for the arbiter to lower its output value, thus providing a dampening effect to help circumvent thrashing due to a sensor's output values. The averaging of input values along with the arbiter's previous output value also provides a dampening effect.

The Smart Arbiter, using the algorithm described here, was able to achieve its specified processing cycle speed of 20 Hz. The premise used for all algorithms that were explored was to keep the arithmetic very simple and in-place since the data fusion task demands can reach 2 million operations per second just to process the algorithm (7 grids/cycle * 14,641 cells/grid * 20 cycles/second). Thus, complex probabilistic-based and belief-based approaches were not explored. However, adding highly traversable cells to the auction (i.e., 11's and 12's) and post-processing the output grid to provide proximity smoothing and/or obstacle dilation were explored, but none of these alternatives provided any better performance (in the sense of speed or accuracy) than the one used for the event.

10.7 Real-Time Planning and Vehicle Control

The purpose of online planning and control is to autonomously drive the NaviGATOR through its sensed environment along a path that will yield the greatest chance of successful traversal. This functionality is compartmentalized into the Reactive Driver (RD) component of the NaviGATOR. The data input to this component include the sensed cumulative traversability grid, assembled by the Smart Arbiter component, vehicle state information, such as position and velocity, and finally the *a priori* path plan, which expresses the desired path for the vehicle to follow sans sensor input. Given this information, the online real-time planning and control component, seeks to generate low-level actuator commands, which will guide the vehicle along the best available path, while avoiding any areas sensed as poorly traversable.

10.7.1 Receding Horizon Controller

The objective of the RD component is to generate an optimized set of the actuator commands (referred to as a "wrench" in JAUS), which drives the vehicle through the traversability grid and brings the vehicle to a desired goal state. The NaviGATOR accomplishes this real-time planning and control simultaneously, with the application of an innovative heuristic-based receding horizon controller.

Receding horizon is a form of model predictive control (MPC), an advanced control technique, used to solve complex and constrained optimization problems. In this case, the problem is to optimize a trajectory through the localized traversability space, while adhering to the nonholonomic dynamics constraints of the NaviGATOR. An in-depth explanation and analysis of the technique is provided in (Mayne, 2000), and the application of suboptimal MPC to nonlinear systems is given in (Scokaert, 1999). This method was selected because it unifies the higher-level planning problem with the lower-level steering control of the vehicle. Separate components are not needed to plan the geometry of a desired path, and then regulate the vehicle onto that path.

The controller attempts to optimize the cost of the trajectory by employing an A* search algorithm (Hart, 1968). The goal of the search is to find a set of open-loop actuator commands that minimizes the cost of the trajectory through the traversability space, and also bring the vehicle to within a given proximity of a desired goal state. The goal state is estimated as the intersection of the *a priori* path with the boundary of the traversability grid. As the vehicle nears the end of the path, and there is no longer an intersection with the grid boundary, the desired goal state is simply the endpoint of the last *a priori* path segment.

Special consideration was given to formulating the units of cost (c) for the search. An exponential transformation of the traversability grid value (t), multiplied by distance traveled (d) was found to work best. The cost equation is given here, where the exponent base is represented by (b):

$$c = db^t. \tag{10.9}$$

Thus, the cost of traversing a grid cell scales nonlinearly with its corresponding traversability value. An intuitive comparison is best to describe the effect of this transformation and why it works well: With a linear transformation, the cost of a path traveling through a traversable value of two is only twice as high as the same path through a value of one. (Note, these values are just given for the purpose of an example and are not actually encountered in the NaviGATOR system.) Therefore, the search would possibly choose a path driving through up to twice as much distance in the value of one, rather than a much shorter path driving through a value of two. Whereas, an exponential transformation ensures that there is always a fixed ratio between neighboring integer traversability values. Thus, this ratio can be used as a tuning parameter to allow the algorithm to achieve the desired tradeoff between the length and cumulative traversability cost of a selected path. Conveniently, the ratio used for tuning is equal to the base of the exponent given in the cost equation.

Closed-loop control with the receding horizon controller is achieved by repeating the optimization algorithm as new traversability data are sensed and vehicle state information is updated. Thus, disturbances, such as unanticipated changes in traversability or vehicle state, are rejected by continually reproducing a set of near optimal open-loop commands at 20 Hz, or higher.

The search calculates different trajectories by generating input commands and extrapolating them through a vehicle kinematics model. The cost of the resulting trajectory is then calculated by integrating the transformed traversability value along the geometric path that is produced through the grid. The search continues until a solution set of open-loop commands is found that produce a near-optimal trajectory. The first command in the set is then sent to the actuators, and the process is repeated. A typical result of the planning optimization is shown in (Figure 10.16, where the dark line is the final instantaneous solution).

Rather than plan through the multidimensional vector of inputs, i.e., steering, throttle, and brake actuators, the search attempts to optimize a one dimensional set of steering commands at a fixed travel speed; the control of the desired speed is handled separately by a simple proportional integral differential (PID) controller. Since it may be necessary to change the vehicle's desired speed in order to optimize the planned trajectory through the search space, extra logic is

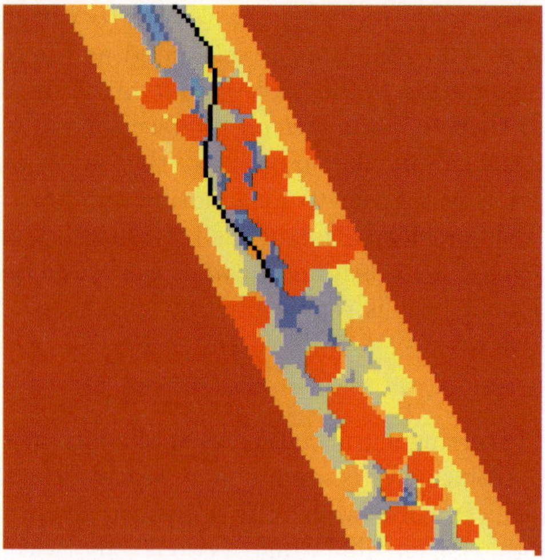

Fig. 10.16. Sample planning result through traversability grid

included in the search to either speed up or slow down the vehicle according to the encountered data.

10.7.2 Vehicle Model

The kinematics model used for response prediction of an input command to the system is that of a front-wheel steered rear-wheel-drive vehicle. The input signals to this model are the desired steering rate (u), and linear vehicle velocity (v). The model states include the vehicle Cartesian position and orientation (x, y, θ) and the angle of the front steering wheels (φ) with respect to the vehicle local coordinate frame. The kinematics equation is given here, where (b) represents the wheel base of the vehicle:

$$\begin{bmatrix} \dot{x} \\ \dot{y} \\ \dot{\theta} \\ \dot{\varphi} \end{bmatrix} = \begin{bmatrix} \cos\theta & 0 \\ \sin\theta & 0 \\ \frac{1}{b}\tan\varphi & 0 \\ 0 & 1 \end{bmatrix} \begin{bmatrix} v \\ u \end{bmatrix}. \tag{10.10}$$

In the algorithm implementation, additional constraints were added to the model to limit the maximum steering rate, and also the maximum steering angle. These values were determined experimentally and then incorporated into the software as configurable parameters. Also, due to the complex nature these system dynamics, obtaining a solution to the differential equations is not feasible; therefore, a series of simplifications and assumptions were made to allow for fast computation of future state prediction. The underlying assumption is that, since the resolution of the traversability grid is relatively low (0.5 m), very accurate estimates for the vehicle's predicted motion are not necessary. The assumptions made were that for a short period of time or traveled distance, the curvature of the path that the vehicle followed was near constant, and this constant curvature could be taken as an average over the predicted interval. Thus, the predicted path trajectory was simply a piece-wise set of short circular arcs. An implementation of Bresenham's algorithm (Bresenham, 1977) was used to integrate the traversability grid value along the determined arcs.

As an additional measure for vehicle stability, a steering constraint was added to limit the maximum steering angle as a function of speed (v) and roll angle (Φ) (due to uneven terrain). The goal of this constraint was to limit the maximum lateral acceleration (n_y) incurred by the vehicle due to centripetal acceleration and acceleration due to gravity (g). Thus, if the vehicle were traveling on a gradient that caused it to roll toward any one direction, the steering wheels would be limited in how much they could turn in the opposite direction. Additionally, as the vehicle increased in speed, this constraint would restrict turns that could potentially cause the NaviGATOR to roll over. This constraint is given by the following equation:

$$k_{\max} = \frac{\pm n_{y\max} + g \sin(\Phi)}{v^2}. \tag{10.11}$$

The value for maximum lateral acceleration was determined experimentally with the following procedure. A person driving the NaviGATOR would turn the wheels completely to one direction, and then proceed to drive the vehicle in a tight circle slowly increasing in speed. The speed in which the driver felt a lateral acceleration that was reasonably safe or borderline comfortable was recorded, and the acceleration value was calculated. This was done for both left and right turns, and the minimum of the two values were taken for conservatism. The value found to be a reasonable maximum was, 4 mps^2, and was hard coded as a constraint to the vehicle model.

10.7.3 Desired Speed Logic

The determination of the commanded and planned vehicle speed is derived from many factors. The RD receives several sources of requested speed, calculates its own maximum speed, and then chooses the minimum of these to compute the throttle and brake commands to the vehicle. Each of the input speed commands was calculated or originated from a unique factor of the vehicles desired behavior. At the highest level, vehicle speed was limited to a maximum value that was determined experimentally based on the physical constraints of the NaviGATOR platform. The next level of speed limiting came from the *a priori* path data, which itself was limited to the speeds provided by the RDDF corridor. Additionally, the speed provided by the path file can be assessed to see if a slower desired speed is approaching ahead of the vehicle. Therefore, if it is about to become necessary to slow down the vehicle, the RD can allow for natural deceleration time. Also, desired speed as a function of pitch was added to slow down the vehicle on steep ascents or descents. This ensures that the NaviGATOR does not drive too fast while encountering hilly terrain.

Another important speed consideration resides with the planning search. Embedded in the search itself is the ability to slow the vehicle down in the cases where the desired trajectory comes within proximity to poorly traversable grid cells. The planner uses a look-up table that enumerates all of the maximum speeds that the vehicle is allowed to drive while traveling through areas of low traversability. Thus, if the vehicle attempts to avoid an obstacle or travel down a narrow road with hazardous terrain to either side, it is commanded to slow down, thus providing a lower risk while allowing for a more comprehensive search to find the best course of action. Also, if the search was unable to find a reasonable solution (i.e., only a solution that goes through very poor areas was found), then the desired speed is lowered. In its next iteration, the RD attempts to find a better solution at that slower speed. This approach is reasonable because the vehicle has greater maneuverability at low speed, and therefore the planner has a better chance of finding a less costly route to its goal.

Additional speed control is provided by a Situation Assessment component consisting of a Long Range Obstacle Specialist and a Terrain Ruggedness Specialist. The Long Range Obstacle Specialist uses a data feed from the PLSS LADAR to determine whether the space directly in front of the vehicle is free of obstacles beyond the 30 m planning horizon (i.e., 30 m out to the 80 m range limit of the LADAR). The Terrain Ruggedness Specialist uses the instantaneous pitch rate and roll rate of the vehicle (provided by the Velocity State Sensor) to classify the current terrain as "Smooth," "Rugged," or "Very Rugged." Based on the Long Range Obstacle State and Terrain Ruggedness State, with appropriate hysteresis control and dampening, the permitted speed of the vehicle is selected and sent to the RD. For example, if the terrain is Smooth and no Long Range Obstacles are detected, then the RD is permitted to drive the vehicle up to its highest allowable speed and thus faster than an empirically derived Obstacle Avoidance speed of 7.2 mps (16 mph).

10.7.4 Controller Fault Detection

There are four faults that the RD is capable of detecting during normal operation of the vehicle. They are the cases where the NaviGATOR has: become blocked, become stuck, collided with an obstacle, or gone out of the bounds of the RDDF. In each of these scenarios, it is possible for the system to take corrective action. The most commonly found of these errors is the blocked condition. In this case, there is no viable path planning solution, even when the search is attempting to plan a trajectory at the vehicle's most maneuverable speeds. It was determined through analysis of the collected data that this case was most often occurring due to sensor misclassifications. The corrective action in these scenarios is to simply wait a short period of time for the sensor data to correct itself, allowing the planner to find a solution. Sometimes, the data will not correct without the vehicle changing its position, therefore an active correction is taken to automatically "nudge" the vehicle forward after a brief wait, and continue with the planned path once the blockage is clear.

10.8 Testing and Performance

This section of the paper summarizes the testing and performance that occurred at each of several key venues. This section is supplemented by a video depicting the NaviGATOR operating in each of these venues (see multimedia).

10.8.1 The CIMAR Lab

Testing began with the JAUS messaging system on the ten computers that would drive the NaviGATOR. The JAUS messaging would need to be capable of sending up to 500 messages per second per node for over 14 hours. On race day, over 20 million JAUS messages were actually sent and received. Next, initial testing of the individual JAUS components, discussed in this paper, took place in the spring of 2005 primarily in the CIMAR lab at the University of Florida. The goal was to get each component working by itself, "on the bench" in a controlled laboratory environment. To support bench testing, a simple vehicle simulator component was devised that sends out position- and velocity-related JAUS messages as if the vehicle were moving through an RDDF corridor. Once each individual component had been successfully tested and declared operational, then various combinations of components were integrated and tested together as the system began to take shape. The base vehicle platform had been assembled during the same period of time as the various JAUS components were being bench tested. With both the vehicle assembled and the JAUS components operational, the various JAUS components were then mounted in the NaviGATOR.

10.8.2 The Citra Test Site

Next, it was time to take the system to the field. On 20 April 2005, a test site was designed and constructed at the University of Florida's Plant Science Unit located in Citra, Florida. The course was laid out in an open field and consisted mainly of a figure eight, an oval, and several left and right sharp turns (see Figure 10.17). Various segments were added to this course to replicate terrain that was expected in the desert. While this course had a few tough obstacles, it was basically the "safest" place to test. This was Team CIMAR's main test site and was used for extensive development of the system as well as the location where the DARPA site visit took place on 6 May 2005. On 20 May 2005, the NaviGATOR was put into a 1/2 mile loop, and it ran for 12 miles before stopping due to a minor problem. This was the furthest it would run prior to heading west in September, as it spent the next three months undergoing major upgrades to both hardware and software.

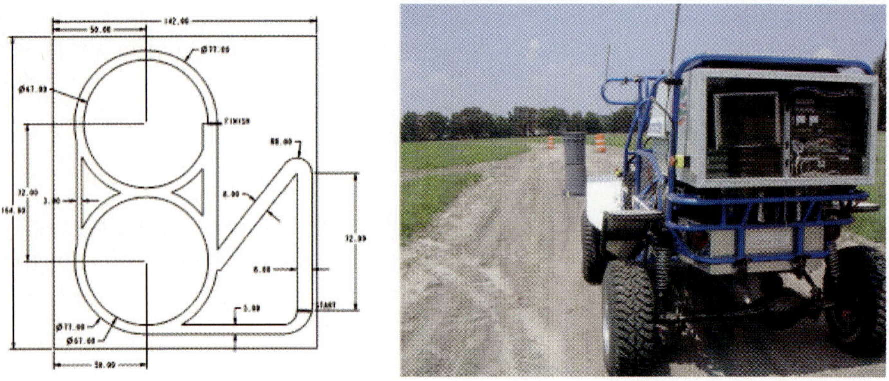

Fig. 10.17. The Citra test site

(a) (b)

Fig. 10.18. PLSS testing images

Part of the Citra testing effort was devoted to the initial tuning of the sensors. Figures 10.18 and 10.19 depict scenes of the terrain at Citra and the accompanying Smart Sensor output, as captured during the sensor tuning process. Figure 10.18(a) shows evenly spaced orchard poles, while Figure 10.18(b) shows a snapshot of the PLSS Traversability Grid while traveling on the graded road in which the poles have been clearly detected and scored as impassable obstacles. This area was specifically chosen to assure that the output of the PLSS accurately maps obstacles onto the grid. Note that the PLSS algorithm has been tuned to accurately locate the poles, even though most of them are occluded for periods of time as the vehicle moves past them. Figure 10.19(a) shows a roadway with rough terrain appearing to the left of the vehicle when traveling in the indicated direction. Figure 10.19(b) shows a snapshot of the TSS Traversability

(a) (b)

Fig. 10.19. TSS testing images

Grid for that same section of road, having scored the rough area as somewhat undesirable, but not absolutely blocked (i.e., 4's, 5's and 6's).

By 20 August 2005, the major hardware and software upgrades were complete and the system was ready for one last round of testing at the Citra site prior to heading west. On 25 August 2005, however, while performing a high-speed radar test, the vehicle suffered a serious failure. One of the rear shocks snapped and the engine and frame dropped onto the rear drive shaft and odometer gear. The sudden stop also caused the front sensor cage struts to snap and the sensor cage collapsed forward. The causes of the failures were determined and the system was re-designed and re-built in approximately one week. With the damage repaired, the NaviGATOR returned to Citra for several days to verify that the system was operational and ready to graduate to the desert for a more serious round of testing.

10.8.3 The Stoddard Valley OHV Area

On 11 September 2005, the NaviGATOR headed west to the Stoddard Valley Off Highway Vehicle (OHV) Area near Barstow, California (see Figure 10.20). The team first attempted some short test runs to ensure system operation. This also marked the first times the team had run the NaviGATOR with a chase vehicle setup (see Figure 10.21(a)). These system tests were done in the OHV area of Stoddard Valley (marked 1 in Figure 10.20). This test route is approximately 4 miles long and included the first serious autonomous uphill and downhill climbs, allowing the team to evaluate the performance of the system during both accent and decent maneuvers. Speeds during these tests stayed in the 10 mph range. Overall, the system showed an almost surprising ability to handle the terrain, prompting the team to accelerate their efforts in finding more challenging test paths. Following these successful tests, the team moved the vehicle on to Slash X.

Fig. 10.20. Stoddard Valley OHV test sites

Fig. 10.21. Testing in the Stoddard Valley OHV area

Slash X was the site of the start of the DARPA Grand Challenge 2004 (DGC04) event and during their time there, Team CIMAR shared the area with several other DARPA Grand Challenge 2005 teams. First, the team ran the start and first mile of the DGC04 route (site 2 in Figure 10.20). This allowed the team to test and tune the sensors specifically against the barbed wire fence that was the downfall of the 2004 NaviGATOR. This path also provided a good place to test higher speed navigation. Between path 2 and an open area behind Slash X, the team was able to test and tune the NaviGATOR up to 30 mph, with an empirically determined obstacle avoidance speed of 16 mph.

Sunday 18 September 2005 turned out to be a historic day for Team CIMAR. Team TerraMax graciously gave us one of their RDDFs through the desert (labeled site 3 in Figure 10.20). The team took the file and after several false starts finally launched the vehicle at 4 pm. Path 3 is approximately 20 miles each way (with a built-in turnaround). The speed testing had not yet been completed and the first test was done at a cap of 10 mph. The team had never seen nor traversed this path prior to this first test. Not knowing exactly where they were going, the NaviGATOR led the way (see Figure 10.21(b)). Surprising even the team members, the NaviGATOR successfully navigated the entire 20 mile distance on the first try, stopping only to give its human handlers time to drink and rest.

Over the following week, the team tested the NaviGATOR several more times on this course, reaching speeds of 25 mph and completing the entire 40-mile course several times. The path included long straight roads, a mountain climb, and areas covered by power lines; all terrain the team expected to encounter during the DARPA Grand Challenge event.

The last significant area of testing in Stoddard Valley (marked 4 in Figure 10.20) was another portion of the DGC04 event. Known as Daggett Ridge, this was the area that the farthest teams had reached during the previous event and consisted of very dangerous mountain switchbacks and drop-offs of hundreds of feet. The sensor team made several trips with the vehicle to tune and test the sensor suite on the path during manual drive, especially focusing on detecting negative obstacles (in the form of cliffs and washouts).

During two weeks of dawn-to-dusk testing in the Stoddard OHV area, the NaviGATOR went from a personal best of 12 miles in a 1/2-mile circuit to 40-mile runs across miles of desert terrain. The team was able to scale the system quickly, going from 10 mph runs to 25 mph with reliable obstacle avoidance at speeds up to 16 mph, along with tuning and validating the software that dynamically determines which speed should be used. That time in the desert was perhaps the best time spent testing during the entire DARPA Grand Challenge project, both in progress for the vehicle and the team members. While more testing time would have been very useful, on 27 September 2005 the team left for the California Speedway and the National Qualification Event.

10.8.4 The National Qualification Event

Immediately following the opening ceremony, the NaviGATOR was the fourth team in line for the first qualification run. The qualification course is shown in Figure 10.22. It consisted of a 2.3 mile long path with three parked cars, a rough

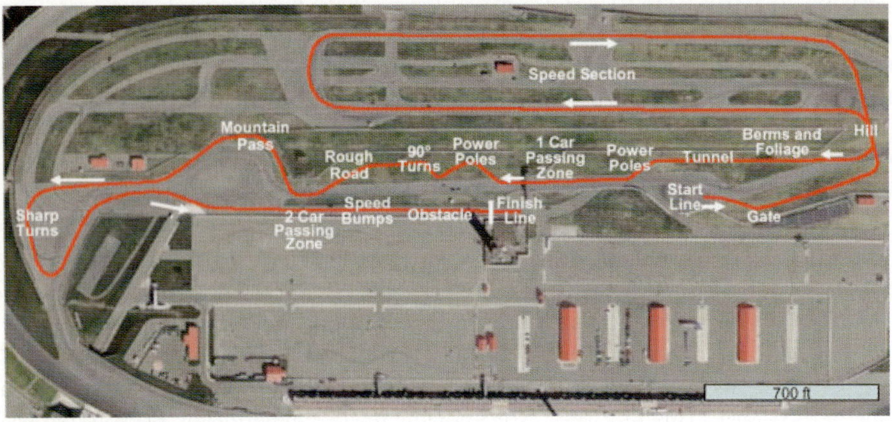

Fig. 10.22. Qualification course at the California Speedway

Fig. 10.23. NaviGATOR at NQE

terrain section, a simulated mountain pass, a tunnel, and finally a wooden "tank trap" obstacle.

The NaviGATOR completed the entire course on the first attempt. Figure 10.23 depicts the NaviGATOR on the NQE course. However, three lane-marking cones had been hit and the tank trap obstacle at the end of the course had been slightly brushed. Two changes were made to the NaviGATOR for the second run. The desired speed on the high-speed section of the course was increased from 16 mph to 20 mph and the dilated size of the perceived obstacles was increased in an attempt to completely miss the tank trap obstacle. During the second run, the NaviGATOR began oscillating and became unstable on the high-speed section and the run was aborted. The problem was that the high-speed section of the qualification course was on pavement whereas all high-speed testing had been conducted off-road. The disturbances caused by the constant four-wheel drive on pavement were responsible for the oscillation.

For the third run on the qualification course, all parameters were reset to those used during the first run. All went well until the vehicle scraped the concrete wall in the mountain pass section of the course, snapping the front steering linkage. The vehicle was quickly repaired. For future runs, the path centerline as reported by the PSS was shifted 12 in. away from the wall in the mountain pass section. After this, the qualification course was successfully completed two more times. In summary, the NaviGATOR completed the entire qualification course three out of five times, and the team was selected by DARPA to compete in the desert race.

10.8.5 The Race

The team received the RDDF containing the course waypoints in the early morning of 8 October 2005. Two hours were allocated for processing the data, which primarily consisted of setting desired speeds for each section of the course. The path file was then uploaded to the vehicle and by 9:30 a.m. the NaviGATOR was off. After leaving the start gate, the NaviGATOR headed off into the desert and then circled around past the crowd at about the eight-mile mark. The NaviGATOR headed past the spectators at approximately 24 mph, performing very well at this point in the race (see Figure 10.24). After following the dirt road a bit further, the NaviGATOR encountered a paved section of the course and started to oscillate. It stopped and did a couple of turns, criss-crossed the road, and then regained its composure

Fig. 10.24. NaviGATOR passing the stands at the 2005 DARPA Grand Challenge Event

and headed back in the right direction. As during the second qualification run, the desired speed was set too high for operation on pavement.

The NaviGATOR next flawlessly traversed a bridge over a railroad track and disappeared into the brown desert haze. Shortly before 11 a.m., the team received word from the chase truck that was following NaviGATOR that the vehicle had inexplicably run off the road and stopped. NaviGATOR appeared reluctant to move forward into and out of low brush in front of it, although its off-road capabilities would have easily carried it through. After several attempts to pause and restart the NaviGATOR, the driver called back to say the vehicle was moving, but slowly and still off the road. After about one-half of a mile of starting, stopping, and driving very slowly over brush, it regained the road and took off again at high speed following the road perfectly. However, after about another mile, the vehicle again went off the road and this time stopped in front of a bush. This time, DARPA officials quickly declared the NaviGATOR dead. The time was shortly before noon, and NaviGATOR had traveled past the 24-mile marker. NaviGATOR placed 18th among the 23 finalists. A total of five teams actually completed the entire course, with Stanford's Stanley taking the $2 million prize for the shortest time of six hours, 53 minutes and 58 seconds.

10.8.6 What Stopped the NaviGATOR?

Team members went out on the course the day after the race and found the NaviGATOR tire tracks at the two locations where the vehicle went off the right side of the road. From this information and data that were logged on the vehicle, it appears that the calculated GPS position drifted by approximately twenty feet causing the vehicle to want to move to the right of the actual road. From the tire tracks and from the traversability grid (see Figure 10.25), it was apparent that the vehicle wanted to move to the right, but the obstacle avoidance sensors were

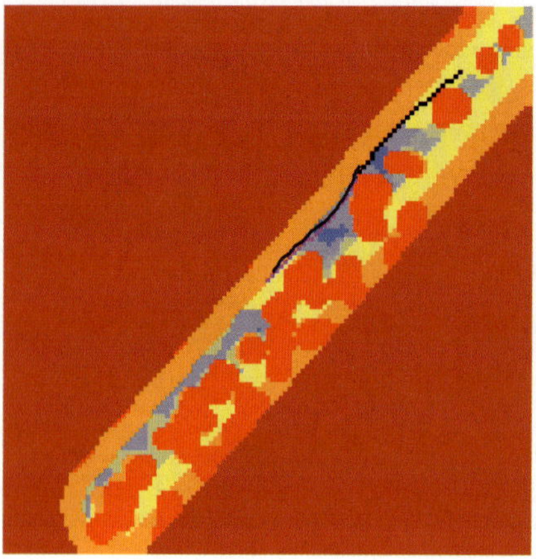

Fig. 10.25. Traversability Grid (during time of position system drift)

Fig. 10.26. Location where NaviGATOR veered off the course and was stopped

detecting the bushes and berms on the right side of the road. From the vehicle's perspective (see Figure 10.25) it appeared that the corridor was littered with objects and the best it could do was to travel along the left side of the corridor on the verge of going out of bounds on the left. In reality, the vehicle was hugging the right side of a very navigable dirt road, however most of the open road was being classified as out of bounds.

Both times that the vehicle went off course were due to the fact that the right side became free of obstacles and the vehicle attempted to move to the center of its incorrect corridor. Figure 10.26 shows the location where the NaviGATOR moved off the course for the second time whereupon DARPA officials stopped it. In summary, a twenty-foot position error caused a corresponding shift of the boundary smart sensor that eliminated the actual sensed road as an option to the planner.

10.9 Conclusion

Overall the team was very pleased with the NaviGATOR system. The base vehicle is very capable and has excellent mobility in very rough terrain. The obstacle and terrain detection sensors, and sensor integration approach, worked very well as did the reactive planner module. Overall, the control loop (from sensed objects to determination of vehicle actuation parameters) operated at a rate of over 20 Hz. Also, a significant contribution of the effort was to show that JAUS could be used successfully in a situation such as this, and that the standardized messaging system defined by JAUS could greatly simplify the overall integration effort.

There are four key areas that are currently being pursued by the team. The first two of these focus on resolving specific issues encountered while competing at the Grand Challenge event. The other two are improvements that will make the NAVIGATOR system more resilient to such problems when they occur.

1. **Stability.** The stability of the controller can be improved simply by putting additional time into getting the control parameters properly tuned. The goal is to achieve stable control at 25 mph on pavement and 30 mph on dirt in the near future.
2. **Position System.** We are currently improving the accuracy of the position system's estimate of error so that when the output of the system is degraded, it can inform the rest of the system appropriately. A better version of the GPS switching code is being implemented that will allow the system to decide which GPS to use as the input to the NFM, the NavCom or the Garmin, based on which is better at the time. At the same time, NavCom and Smiths Aerospace are working together to further improve the overall accuracy of the system.
3. **Dynamic BSS and PSS.** As discussed earlier, the reason the NaviGATOR got stuck off the road in the race was due to a position error causing the Boundary Smart Sensor to shift the drivable corridor off the road. To prevent this from happening in the future, the width of the corridor created by the BSS will be made a function of the position system root mean square error (RMS). For example, if the position RMS is good then the BSS corridor in the grid will be correspondingly tight; but when the position RMS degrades, then the BSS will stroke a correspondingly wide corridor through its traversability grid. In this way, the BSS will no longer eliminate the road as an option, thus allowing the sensors to find the road off to the side. Similarly, the weight of the PSS can be adjusted such that its recommended path is painted with tens when the position RMS is good, but only sevens or eights when the position RMS degrades, thus reducing its influence accordingly.
4. **Adaptive Planning Framework.** A more extensive implementation of the situation assessment specialists and high-level decision-making capabilities is currently underway. This will allow the NaviGATOR to do such things as determine when it has become blocked and decide how to best fix the problem, such as backing up and re-planning. Other examples include altering the aggressiveness of the plan (risk) based on mission parameters and altering the contribution of a given sensor based on the environmental situation.

The first three items on this list are relatively short term and should be completed before this paper is published. With a tuned controller, the position system upgraded, and the BSS and PSS dynamically adjusting to the position RMS, the NaviGATOR should be capable of completing the 2005 DARPA course in under 10 hours. The maturation of the Adaptive Planning Framework will likely continue into the future for some time.

In retrospect, the team would have benefited from more testing time in the California desert. The issues associated with the positioning system and the high-speed control on pavement could have been resolved. However, the project was very successful in that an entirely new vehicle system was designed, fabricated, and automated in a nine-month period, ready to compete in the 2005 DARPA Grand Challenge. This was a monumental effort put on an aggressive time and resource schedule.

Acknowledgments

The authors would like to acknowledge the work of all the DARPA staff members who organized and operated a very successful event. Also, thanks are given to the members of the JAUS Working Group whose efforts are leading to an open architecture that will promote interoperability of military robotic systems. Finally, the authors gratefully acknowledge the technical and financial support of the team's Title Sponsor, Smiths Aerospace.

References

Bresenham, J. (1977). A Linear Algorithm for Incremental Digital Display of Circular Arcs. *Communications of the ACM*, 22(2), 100–106.

Criminisi, A., Reid, I., and Zisserman, A. (1997). A Plane Measuring Device [Electronic Version]. Retrieved April, 2006 from `http://www.robots.ox.ac.uk/~vgg/presentations/bmvc97/criminispaper/`

Elfes, A. (1989). Using Occupancy Grids for Mobile Robot Perception and Navigation. *IEEEComputer Magazine*, pp 46–57.

Hart, P. E., Nilsson, N.J., Raphael, B. (1968). A Formal Basis for the Heuristic Determination of Minimum Cost Paths. *IEEE Transactions on Systems Science and Cybernetics*, SSC4(2), 100–107.

JAUS. (2005). *Joint Archtiecture for Unmanned Systems Reference Architecture, version 3.2*: JAUS Working Group (`http://www.jauswg.org/`).

Lee, J., Crane, C., Kim, S., and Kim J. (2004). *Road Following in an Unstructured Desert Environment using Monocular Color Vision as Applied to the DARPA Grand Challenge.* Paper presented at the ICCAS2005, Seoul.

Mayne, D., Rawlings, J., Rao, C., Scokaert, P. (2000). Constrained Model Predictive Control: Stability and Optimallity. *Automatica*, 36(6), 789–814.

Morris, R. D., Descombes, X., and Zerubia, J. (1997). Fully Bayesian image segmentation - an engineering perspective. *Proceedings of IEEE International Conference on Image Processing.*

Scokaert, P., Mayne, D., Rawlings, J. (1999). Suboptimal Model Predictive Control (Feasibility Implies Stability). *IEEE Trans. Automatic Control*, 44, 648–652.

Seraji, H. (2003). New Traversability Indices and Traversability Grid for Integrated Sensor/Map- Based Navigation. *Journal of Robotic Systems*, 20(3), 121–134.

Solanki, S. (2003). *Implementation of laser range sensor and dielectric laser mirrors for 3D scanning of glove box environment.* Unpublished Master's Thesis, University of Florida, Gainesville.

Thrun, S. (2003). Learning Occupancy Grid Maps with Forward Sensor Models. *Autonomous Robots*, 15, 111–127.

Ye, C., and Borenstein, J. (2004). T-transformation: Traversability Analysis for Navigation on Rugged Terrain. *Proceedings of the Defense and Security Symposium, Unmanned Ground Vehicle Technology VI (OR54).*

11

Prospect Eleven: Princeton University's Entry in the 2005 DARPA Grand Challenge

Anand R. Atreya, Bryan C. Cattle, Brendan M. Collins, Benjamin Essenburg, Gordon H. Franken, Andrew M. Saxe, Scott N. Schiffres, and Alain L. Kornhauser

Princeton University, E-407 Engineering Quadrangle,
Princeton, New Jersey 08544

Summary. This paper describes Princeton University's approach to the 2005 DARPA Grand Challenge, an off-road race for fully autonomous ground vehicles. The system, Prospect Eleven, takes a simple approach to address the problems posed by the Grand Challenge including obstacle detection, path planning, and extended operation in harsh environments. Obstacles are detected using stereo vision, and tracked in the time domain to improve accuracy in localization and reduce false positives. The navigation system processes a geometric representation of the world to identify passable regions in the terrain ahead, and the vehicle is controlled to drive through these regions. Performance of the system is evaluated both during the Grand Challenge and in subsequent desert testing. The vehicle completed 9.3 miles of the course on race day, and extensive portions of the 2004 and 2005 Grand Challenge courses in later tests.

11.1 Background

Prospect Eleven was Princeton University's entry in the 2005 DARPA Grand Challenge, a competition for autonomous vehicles held on October 8, 2005 on an off-road course in the vicinity of Primm, Nevada. The race was organized by the Defense Advanced Research Projects Agency (DARPA) to promote research in autonomous ground vehicles. This was the second instance of the race, the first having been held a year earlier. Princeton University did not participate in the first race, and no entrants completed the course that year.

The Princeton team consisted entirely of undergraduates under the direction of Prof. Alain Kornhauser. Among Prospect Eleven's unique features are its inexpensive and simple design and its reliance on stereo vision as its only means of obstacle detection. What follows is a high-level description of Prospect Eleven's main systems, the lessons learned while developing these systems, and a discussion of the vehicle's accomplishments to date.

M. Buehler, K. Iagnemma, and S. Singh (Eds.): DARPA'05, STAR 36, pp. 349–361, 2007.
springerlink.com

11.2 Mechanical Systems

The Princeton University team received a stock 2005 GMC Canyon truck to serve as a development platform. The throttle, braking, and steering systems were modified to allow for drive-by-wire operation.

Like many modern cars, the gas pedal in the Canyon is fully electronic so it was not necessary to establish a mechanical linkage to the engine throttle. Instead, a computer-generated voltage simulates the behavior of the physical pedal. The brake pedal is mechanically controlled by two independent systems: a custom-built linear ball-screw actuator used under normal operation, and a pneumatic piston capable of applying 670 N of force for emergency use. These are connected to the brake pedal with sheathed steel cable, and either system may be in operation without preventing the operation of the other. An inline tension sensor is used to measure the degree of brake application. Steering control is accomplished via a DC motor mounted under the steering column and attached to the steering wheel with a set of gears. An optical rotary encoder, also attached to the steering wheel, provides precise position feedback.

11.3 Vehicle Control

The vehicle's steering wheel angle and vehicle speed are maintained with modified Proportional, Integral, Derivative (PID) control loops. During normal autonomous operation, each of the PID controllers runs at approximately 20 Hz.

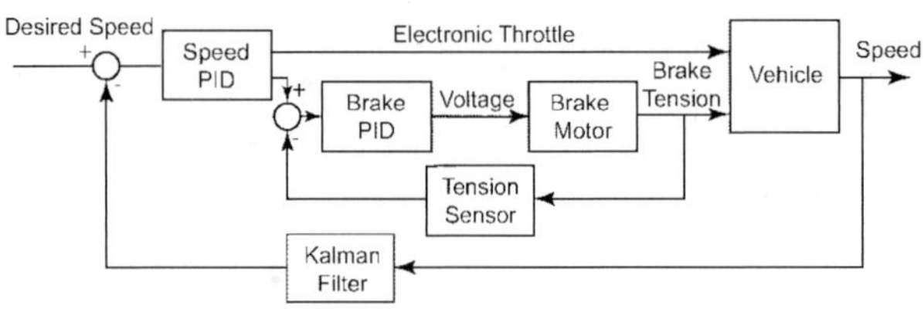

Fig. 11.1. Speed Control Functional Block Diagram

Steering control uses the optical encoder as input and controls the steering motor as needed. A two-layer system of PID controllers regulates Prospect Eleven's speed. In the first PID control layer, the reference input is the car's velocity and the output is a throttle voltage if the output is positive, or a desired brake tension if negative. The desired brake tension is then monitored by the second PID controller. This controller takes the current brake tension as input and controls the braking motor. Figure 11.1 depicts the speed control block diagram. Figure 11.2 shows desired and actual velocity during Prospect Eleven's first run at the National Qualification Event (NQE).

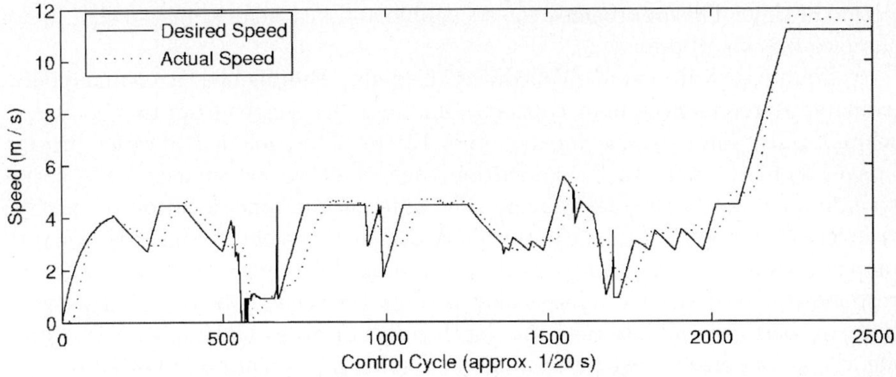

Fig. 11.2. Speed vs. Time at start of NQE Run 1

11.4 Computing

All computing is performed by two standard desktop computers. These are mounted in a shock-isolated rackmount case behind the passenger seat. The rack sits on four fluidic shock mounts designed to attenuate the high-frequency vibrations which can cause hard drive failure. To provide further protection, each computer has a RAID array which mirrors the contents of the primary hard drive.

Vision processing and obstacle detection algorithms were written in C++. All other car control, data acquisition, and decision making control systems were implemented in C# ("C-sharp") on the other computer. Both systems run Microsoft Windows. The C# language proved particularly effective as a development platform; the extensive Microsoft .NET libraries and intuitive object-oriented structure allowed rapid debugging while permitting the implementation of advanced functionality such as multithreading and low-level I/O.

11.5 Stereo Vision

Prospect Eleven relied solely on stereo vision to detect and range obstacles. In doing so, it was unique among the contestants present at the Grand Challenge finals.

Obstacle detection using stereo vision can be broken down into three problems: (1) obtaining an accurate depth map of the scene ahead from pixel disparities between the two cameras, (2) identifying obstacles in the scene, and (3) calculating the range of detected obstacles so they can be avoided.

This process is susceptible to many environmental sources of error, including unfavorable lighting conditions and irregular terrain. Central to the approach of this paper is the assumption that most of this error is random, and hence can be averaged out by filtering many measurements of obstacle position over a period of time. This approach suggests simple and fast algorithms, so that many samples may be obtained.

A Point Grey Research (Vancouver, Canada) Bumblebee, a commercially-available stereo camera pair, captures simultaneous images from two black and white CCDs. The baseline separation is 12 cm. The included libraries process images from the camera by comparing each pixel in one image to the corresponding feature in the other image. The difference in pixel location (disparity) is inversely proportional to the pixel's depth in the scene. From this, a depth map is calculated. A depth map is simply an image in which each pixel value corresponds to the disparity value of that pixel in the scene. The included software libraries perform validation of the depth map. In order to simplify the overall algorithm and ensure fast performance, subpixel interpolation is not used. As a result, disparities may only take on integer values. For example, at a range of 20 m, the difference in range between two adjacent disparity values is 3 m.

Several strategies were found to be effective in improving the number of accurate and validated matches—particularly in poor lighting conditions. Red photographic filters mounted in front of each lens were used to increase contrast by blocking blue light and reducing UV haze, mitigating problems such as CCD "bleeding" on bright days and boosting the brightness of the ground—the area of interest in each frame. Results were further improved by custom camera gain control which was designed to optimize the exposure in the ground plane at the expense of the upper half of the frame.

11.5.1 Obstacle Detection

As mentioned, it is important that Prospect Eleven range an obstacle many times such that many measurements may be filtered to increase accuracy. This is only possible with a fast obstacle detection algorithm. This section presents an algorithm which is very fast and well-suited for heavily quantized data. When faced with conditions like those encountered on the Grand Challenge course, the system performs sufficiently well for obstacle avoidance.

Several authors have also examined the problem of fast obstacle detection. One approach, adopted by Matthies and Grandjean (1994), is to consider the slope of a pixel relative to a ground plane. It is supposed that a significant obstacle will have slope greater than some threshold. Similarly, Broggi, Caraffi, Fedriga, and Grisleri (2005) simply search each column in the depth map for large intervals at similar disparities. Indeed, for a camera mounted nearly parallel to the ground plane, as was the case for Prospect Eleven and Broggi et al., this approach approximates thresholding vertical slope over some window. The algorithm in this paper parallels that of Broggi et al.

Were disparity values not so heavily quantized, a logical measure of similarity might be variance. However, as a result of quantization, the algorithm simply looks for a contiguous span within the column for which the disparity is the same value throughout the span. Such spans occur in flat scenes in the image, so it must be required that they be of at least a certain length l to be classified as an obstacle. In contrast to Broggi et al., this paper's approach is to make

l dependent on the extent to which the current interval is above the row-wise median disparity. The justification for this is simple: for a relatively flat scene, the disparity of a pixel which belongs to an obstacle should be above the median disparity in its row. Setting l to be lower for pixels above the median enables information from the entire image to be considered while detecting obstacles in a single column. We set l to be 12 for pixels which are above the median by 2, 20 for pixels above the median by 1, and 35 for pixels at the median disparity value. Use of the median may not be effective in unstructured environments, but since the Grand Challenge course was graded, it was assumed that most rows would be roughly homogeneous over the traversable region.

The algorithm can be structured as a finite state machine which scans each column from top to bottom and has states `IN OBSTACLE` and `NOT IN OBSTACLE`. When the state is `NOT IN OBSTACLE`, the code simply looks for an interval satisfying the above criteria. In state `IN OBSTACLE`, it grows the interval downward until it first fails to meet a slightly weaker form of the above criteria. The weaker form differs only in the constants used, and is employed because the base of an obstacle is more similar to the background than the top. Figure 11.3 shows a scene image, with obstacle pixels highlighted, as well as the corresponding depth map. One column in Figure 11.3a is highlighted. Figure 11.4 shows depth values and the algorithm's output on this highlighted column.

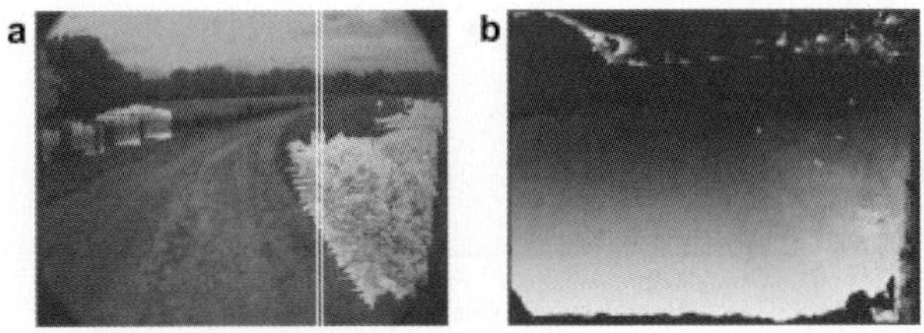

Fig. 11.3. (a) Sample scene image, with detected obstacle pixels highlighted and sample column outlined, as analyzed in Figure 11.4. (b) The corresponding disparity map.

The time complexity of the algorithm is linear. Each pixel need only be examined once, and median calculation can be performed efficiently using a radix sort. Figure 11.5 shows the computation times of 193 images at 640x480 resolution versus the proportion of pixels in the image which were identified as obstacles.

Once each pixel is classified as being an obstacle or not an obstacle, bounding boxes are constructed around each connected obstacle region. In doing so, a central point and width are computed for each box. A confidence measure is computed based on the detected size of the obstacle and the similarity of disparity values within the obstacle. A box is classified as an obstacle if its confidence measure exceeds a threshold. Table 11.1 gives the performance of this algorithm on several obstacles at various distances. A ✓ indicates successful detection, and an ✗ indicates that the obstacle was not detected.

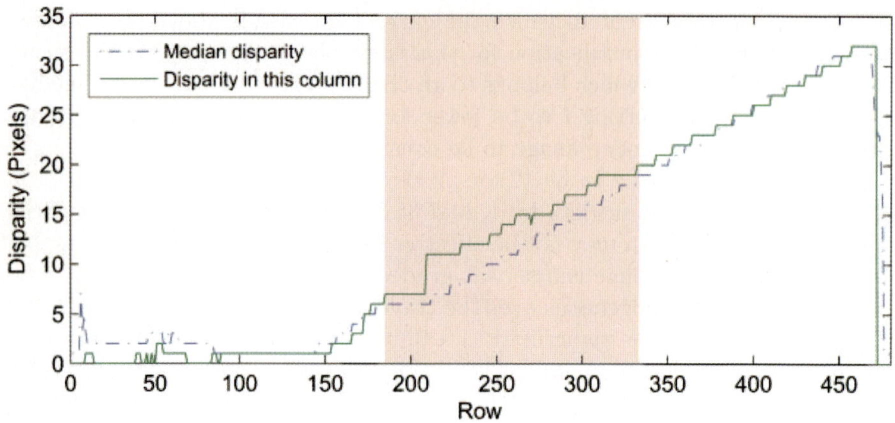

Fig. 11.4. Disparity values and obstacle detection in the highlighted column of Figure 11.3a. The shaded region is detected as an obstacle.

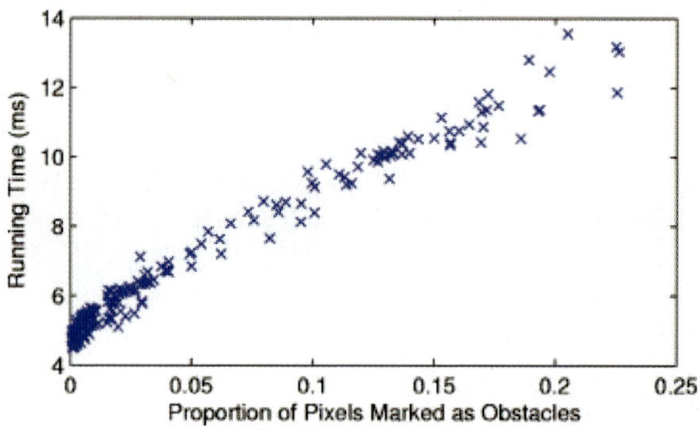

Fig. 11.5. Computation time versus proportion of obstacle pixels over 193 640x480 images

Table 11.1. Detection of objects at various ranges (in m)

Object	4.5	6	7.5	9	10.5	12	13.5	15	16.5
Downturned cinderblock (19.5 cm)	✓	✓	✓	×	×	×	×	×	×
Upright cinderblock (40 cm)	✓	✓	✓	✓	✓	×	×	×	×
Shelves (65 cm)	✓	✓	✓	✓	✓	✓	✓	✓	×
Trashcan (69 cm)	✓	✓	✓	✓	✓	✓	✓	✓	×

Note: ✓=Successful detection; ×=Obstacle not detected.

As can be seen, the range at which short obstacles can be reliably detected is quite limited. Fortunately the primary function of obstacle detection during the Grand Challenge was to detect graded berms on either side of the course. Data recorded during the Grand Challenge indicates that Prospect Eleven was able to do so at ranges of approximately 8 m, which was adequate for navigation during the race. The limited detection distance for small but dangerous obstacles capped the vehicle's maximum speed.

11.5.2 Tracking in the Time Domain

Detected obstacles are tracked in the time domain to improve accuracy in positioning and limit false positives. When a new image is processed, the list of obstacles from that image is compared to the list of currently tracked obstacles. Each new obstacle is matched to the closest existing one, or declared a new obstacle if no suitable match exists. A confidence measure is maintained for each obstacle, based on the aforementioned confidence of each detection, number of frames in which it was detected, and the number of frames in which it was *not* detected despite being in detection range. Only obstacles whose confidence measure exceeds a threshold are used in path planning. A Kalman filter maintains the estimate of the obstacle's location.

Matching is effective in improving localization, particularly for obstacles at ranges above 8 m. The measurement error for localization decreases quadratically as a function of range. Though ground-truth data is not available, a direct approach to obstacles at randomized positions is simulated with a wide variety of parameters. Figure 11.6 gives the mean precision of localization at various ranges over 10,000 simulations. Values simulated include vehicle speeds in the range 6 m/s to 13 m/s, minimum detection ranges between 1.5 m and 5 m, maximum detection ranges between 15 m and 20 m, detection frequencies between 6 Hz and 10 Hz, and disparity calculations with unbiased Gaussian error with σ^2 between 0.05 pixels and 0.5 pixels. The error model assumes that error is caused entirely by miscalculation and quantization of disparity values. The periodic behavior of the measurement error is a result of quantization: for ranges which correspond to a nearly integer disparity value, quantization causes very little error. Though there are certainly many more sources of error than those modeled, and actual error is much greater, the simulation demonstrates that tracking is effective at reducing error in localization.

11.6 Navigation

The choice of navigation scheme was governed by the specific structure of the competition. DARPA's rules and a review of the 2004 competition suggested the following:

1. DARPA would provide a high level, high detail GPS path from start to finish
2. A traversable path would exist within course boundaries
3. The course would be narrow and lie on desert roads

An algorithm which chooses a steering angle at each instant without preplanning a path ahead is sufficient for the Grand Challenge. Global convergence problems are largely mitigated due to the provided GPS path. This path performs the function of a high-level planned path. For these reasons, the navigation

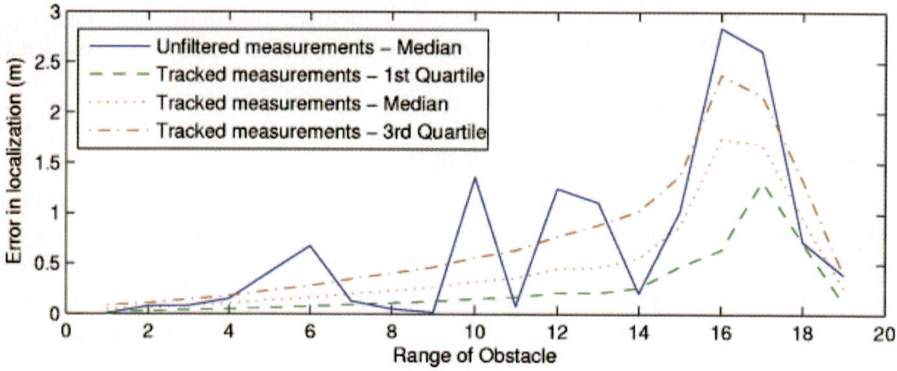

Fig. 11.6. Error in localization versus range in meters

software implements a reactive algorithm that processes a region limited to the detection range of the stereo vision system.

The algorithm implemented is a Nearness Diagram (ND) approach (Minguez and Montano, 2004) modified to suit the structure of the competition. The ND navigation scheme as implemented by Minguez and Montano begins by forming a polar plot of the distance from the vehicle to the nearest obstacle at each angle. This plot is manipulated to identify available gaps in the obstacles surrounding a robot, and a control vector is selected from among five different control strategies based on the particular structure of available openings.

Our implementation differs by (1) changing the gap calculation to accommodate road boundaries, (2) changing the gap representation from angular width to physical end points and width in meters, and (3) utilizing the physical dimensions of the gap to collapse ND's five control schemes into one. These modifications adapt the ND approach to travel along roads.

11.6.1 World Representation

The navigation system maintains an internal model of the world composed of the DARPA defined GPS course and detected obstacle locations. The model is a Cartesian plane with the origin located at the first course waypoint. Terrain elevation is neglected. Approximating the globe as a plane was found to be sufficient throughout the race. The course representation was geometric, with course segments represented as rectangles capped with semicircles and obstacles represented as circles of varying diameter and location. Approximating obstacles as circles leads to inaccuracies in representing planar obstacles, such as when walls appear as a string of small circles. However, the nature of the competition lessened the impact of this shortcoming as frequently-encountered objects like gate posts, tank traps, and bushes are all well-approximated by circles.

This internal representation stands in contrast to cost map approaches. The geometric model requires less memory than large cost maps, and in general can calculate intersections and other quantities analytically. However, a drawback of this method is that its obstacle representation is binary—space must either be fully traversable or blocked. Given that the stereo vision system cannot distinguish between obstacles of different severity, for example rocks one would prefer to avoid versus a parked car one must avoid, the binary representation was not a limitation.

11.6.2 Gap Calculation

The central task of the navigation system is to extract the location of "gaps" in the terrain in front of the vehicle, where gaps are defined as physical openings wide enough for the vehicle to travel through. The first step in locating these is the construction of a polar tube plot, which closely parallels the nearness diagram. Whereas the nearness diagram graphs distance to the nearest obstacle in an angular sector, the tube plot graphs the distance to the nearest obstacle in a rectangular region (a tube) extending at a given angle off of the car's heading (Figure 11.7). Using tubes instead of sectors preserves the actual width of gaps at different distances from the vehicle. Next, discontinuities in the plot

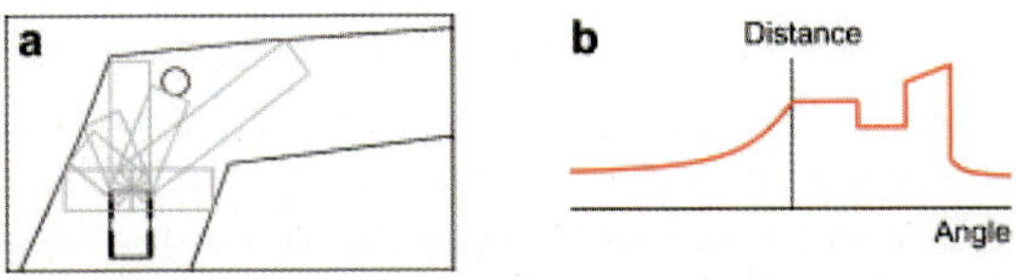

Fig. 11.7. (a) A sample world configuration consisting of GPS course boundaries, a circular obstacle, and the vehicle. Projected tubes are shown in gray. (b) The resulting polar tube plot.

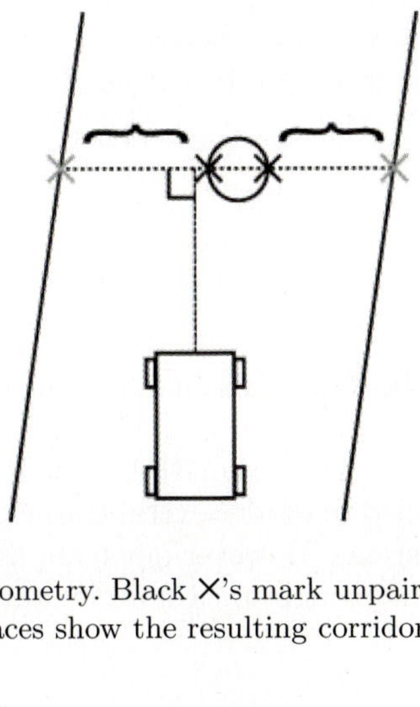

Fig. 11.8. Corridor gap geometry. Black X's mark unpaired endpoints, gray X's mark corridor endpoints, and braces show the resulting corridor gaps.

are identified, and gaps are formed from a pair of adjacent left and right endpoints. Because road borders are continuous lines, no discontinuities are present to form appropriate gaps between road edges and obstacles. Another mechanism was developed to achieve suitable gap formation on roads or in corridors. Additional "corridor" gaps are formed between an unpaired endpoint and a point along the adjacent corridor boundary. The appropriate contact point along the boundary is taken to be the point of intersection of the boundary with a line passing through the unpaired endpoint perpendicular to the vehicle's direction of travel (Figure 11.8).

11.6.3 Control Action Selection

A control action comprising desired heading and speed is calculated based on the available gaps. First, a target gap is selected based upon width and angle from the vehicle's heading. Wider gaps are more desirable, as are those which require the least deviation from current heading. Once chosen, a gap is biased to be targeted in subsequent time steps to prevent alternating between gaps. This is a serious problem, because such indecision effectively takes the average between the two gaps, where an obstacle lies. Next, a pursuit point along the course centerline is selected. In the absence of obstacles, aiming at this pursuit point will guide the vehicle along the GPS course. To cause the vehicle to dodge obstacles by a margin of d where possible, the target gap endpoints are each moved toward each other by $d + \frac{w}{2}$, where w is the width of the vehicle. If the gap is too narrow to move each endpoint the required distance, both endpoints are placed at the middle of the gap, so the vehicle will still pass through narrower openings if d is unavailable. Desired heading is calculated to be the heading within the narrowed target gap closest to the heading of the pursuit point. This method allows the vehicle to dodge obstacles by precise distances.

Desired speed is computed as the minimum of three terms: (1) the DARPA mandated speed limit, (2) a safety speed limit based on the average curvature of the track ahead of the car, and (3) a reactive term proportional to the length of the tube projected from the front of the vehicle. This last term slows the vehicle while dodging obstacles.

11.7 Results

11.7.1 Site Visits

To earn an invitation to the National Qualification Event (NQE) in Fontana, California, teams had to demonstrate basic GPS-following and obstacle avoidance capabilities during a site visit by DARPA officials. During the first site visit in May, Prospect Eleven failed to evade several trashcans, and was not originally placed in the top 40 contestants. However our team did earn a second site visit in June, and performed well enough to earn an invitation to the NQE as one of three alternates.

11.7.2 National Qualification Event

The NQE consisted of 5 runs over a 3.5 km course. The course included a 100-ft tunnel under which robots lost GPS fix, rumble strips, four parked cars, a tank trap, hay bales, a simulated mountain pass, and tire stacks. Each run was judged on time, evasion of obstacles, and completion of course gates. Though Prospect Eleven performed admirably on three runs, its poor performance on the other two demonstrated serious software reliability issues. Table 11.2 shows the results of the five NQE runs. During runs 1-3, the vehicle's GPS system was misaligned. Despite this, during runs 1 and 3 the vision system was able to keep Prospect Eleven within boundaries by detecting physical course markings. In run 2, the misalignment caused Prospect Eleven to collide with the first obstacle. During run 4, slow software performance resulted in unstable steering control. This was a result of several extraneous processes left running on the vehicle, as well a systemic software bug, discussed in the next section. The fifth run was essentially perfect and within two minutes of the course record.

Table 11.2. Results at the National Qualification Event. "DNF" denotes "did not finish."

Result	Run 1	Run 2	Run 3	Run 4	Run 5
Obstacles avoided (of 5)	5	0	5	2	5
Gates passed (of 50)	48	0	50	8	50
Time	13:03	DNF	12:34	DNF	12:11

Note: DNF=did not finish.

11.7.3 Grand Challenge Event

The top twenty-three robots from the NQE were invited toparticipate in the Grand Challenge Event on October 8, 2005 in Primm, Nevada. Prospect Eleven was seeded tenth. The race started smoothly, and Prospect Eleven appeared on schedule at the eight mile mark. Shortly thereafter, at approximately 9.4 miles, steering control became unstable. The DARPA chase vehicle disabled Prospect Eleven, and Prospect Eleven's race was over.

Analysis of the logs made it clear that the culprit was again slow software performance. Normally, Prospect Eleven's control loops run at 16-20 Hz, however at the time the vehicle was disabled they were only running at 0.3 Hz. Further analysis revealed that this was caused by a bug in the obstacle tracking code, as obstacles were never entirely cleared from the list of tracked obstacles when passed. Tracking the position of thousands of irrelevant obstacles overwhelmed the processor, and starved critical code.

11.7.4 Post-grand Challenge

In order to evaluate Prospect Eleven's performance without this software bug, the course was attempted once more on October 31, 2005. By this point course conditions had changed considerably from the day of the Grand Challenge: Several formerly dry lake beds had been filled with rain in the intervening weeks. Also, the heavy rains cut a number of deep washouts across the path. In addition, some parts of the course had been altered following the Grand Challenge, with ramps bulldozed and a short stretch of track deliberately rendered impassable. In all of the above cases, Prospect Eleven was removed from autonomous operation and manually driven around the impasse. In addition, Prospect Eleven suffered a communications failure between the GPS unit and the guidance computer just before Beer Bottle Pass, a mountain pass near the end of the course, that would have ended a fully autonomous attempt. This was fixed en route. Nevertheless, Prospect Eleven drove the rest of the course autonomously, successfully navigating two tunnels, multiple gates, and descending winding Beer Bottle Pass at night without any intervention.

To assess repeatability, Prospect Eleven ascended and descended Beer Bottle Pass again the following day. Although traversal of the pass itself was uneventful, the vehicle blew out its left front tire at the base of the pass on the descent, following a collision with a small, sharp rock. The front wheels were also jarred out of alignment. This failure demonstrates a limit of the vehicle's stereo vision technology, which could not detect small but crucial features of this size. The descent occurred at night. The vehicle headlights provided sufficient illumination for robust obstacle detection.

As a final test, Prospect Eleven attempted the 2004 Grand Challenge course backward, from Primm, NV to Barstow, CA. Again, the vehicle was unable to complete a fully autonomous traversal of the course due to environmental factors. Manual control had to be taken back a number of times to guide the vehicle around washouts, and in one case to divert the vehicle around an underpass that had filled with silt. Also, three hardware failures would have ended a fully autonomous attempt of the course: a communications cable came loose, the steering position encoder became jammed with sand, and the vehicle's spare tire, installed to replace the old left front tire, was eventually destroyed by the terrain. The failure of the spare was expected due to the misalignment of the wheels after the previous collision. Despite these issues, Prospect Eleven navigated a substantial distance autonomously. The vehicle traversed the three mountain passes of the course without incident, including a descent of Daggett Pass in total darkness.

11.8 Conclusion

Over the course of the Grand Challenge qualification, competition, and subsequent testing, Prospect Eleven performed well. Though the system was designed to exploit the specific structure of the Grand Challenge race, it showed the capability to perform in some environments more complicated than those originally envisioned. This demonstrates the promise of a simple approach to autonomous systems. Though it is useful to consider implementation as a system of interconnected modules, each of which may be individually optimized, Prospect Eleven shows that overall performance relies on carefully considering the integration of components in the system as a whole. For instance, the use of a binary obstacle representation, though in itself suboptimal, contains all information the stereo vision system can provide and more accurately reflects the limitations of the detector. Prospect Eleven shows the feasibility of a simple autonomous design that can operate effectively in difficult environments.

Acknowledgments

This work was supported by the CSX Transportation Research Endowment Fund, Eric Huber, Katherine Kornhauser, General Motors, Applanix Corporation, Tom Haines, and ALK Technologies, Inc.

The students would particularly like to thank Prof. Alain Kornhauser for his steadfast support of this project.

References

1. Broggi, A., Caraffi, C., Fedriga, R.I., & Grisleri, P. (2005). Obstacle Detection with Stereo Vision for Off-Road Vehicle Navigation. Procs. of the IEEE Wks. on Machine Vision for Intelligent Vehicles, 65–72.
2. Matthies, L. & Grandjean, P. (1994). Stochastic Performance Modeling and Evaluation of Obstacle Detectability with Imaging Range Sensors. IEEE Transactions on Robotics and Automation, Special Issue on Perception-based Real World Navigation, 10(6), 783–792.
3. Minguez, J. & Montano, J. (2004). Nearness Diagram (ND) Navigation: Collision Avoidance in Troublesome Scenarios. IEEE Transactions on Robotics and Automation. 20(1), 45–59.

12

Cornell University's 2005 DARPA Grand Challenge Entry

Isaac Miller*, Sergei Lupashin, Noah Zych, Pete Moran,
Brian Schimpf, Aaron Nathan, and Ephrahim Garcia**

Sibley School of Mechanical and Aerospace Engineering
Cornell University, Ithaca NY, 14853

Summary. The 2005 DARPA Grand Challenge required teams to design and build autonomous off-road vehicles capable of handling harsh terrain at high speeds while following a loosely-defined path. This paper discusses the critical subsystems of Cornell University's entry, an autonomous Spider Light Strike Vehicle. An attitude and position estimator is presented with modifications for specific problems associated with high-speed autonomous ground vehicles, including GPS signal loss and reacquisition. A novel terrain estimation algorithm is presented to combine attitude and position estimates with terrain sensors to generate a detailed elevation model. The elevation model is combined with a spline-based path planner in a sensing / action feedback loop to generate smooth, human-like paths that are consistent with vehicle dynamics. The performance of these subsystems is validated in a series of demonstrative experiments, along with an evaluation of the full system at the Grand Challenge.

12.1 Introduction

Field robotics has grown substantially in the past few decades, but a divide still remains between theoretical systems realized in ideal environments and those that must contend with the real world. The DARPA Grand Challenge sought to bridge this gap with a number of demanding technical and physical tasks, where only platforms robust both in theory and construction could hope to compete. The difficulty of the Grand Challenge spurred many unique approaches to the problem, each largely shaped by its team's technical, intellectual, and financial resources.

Cornell University entered the Grand Challenge in the summer of 2004 as a new team composed largely of undergraduate students and a few faculty advisors from the Cornell schools of Mechanical and Aerospace Engineering, Electrical Engineering, and Computer Science. The addition of Singapore Technologies Kinetics as a primary sponsor in the fall provided the necessary financial support to make Cornell's entry a serious competitor. In the beginning, Team Cornell divided its work

* Graduate Research Fellow, Sibley School of Mechanical and Aerospace Engineering.
** Associate Professor, Laboratory for Intelligent Machine Systems.

M. Buehler, K. Iagnemma, and S. Singh (Eds.): DARPA'05, STAR 36, pp. 363–405, 2007.
springerlink.com

among three main sub teams: a vehicle sub team, a sensing sub team, and an artificial intelligence sub team. As integration problems emerged between the three sub teams, however, a more goal-oriented approach was necessary. The final solution divided the team according to four general goals: building a mobile hardware platform, determining the platform's orientation and location, sensing its surrounding environment, and planning a path through that environment.

This paper presents Cornell University's approach to accomplish each of these four goals, expanding on the high-level overview presented in Team Cornell's DARPA Grand Challenge technical report (Team Cornell, 2005). First, section 12.2 presents a block diagram of Cornell's approach that outlines the critical subsystem as well as information flow between them. The subsequent sections present each subsystem in turn. Section 12.3 discusses the capabilities of Cornell's hardware platform, the Spider Light Strike Vehicle, and its actuation for drive-by-wire capabilities. Section 12.4 discusses the approach used to fuse inertial and navigation information to estimate the Spider's position, velocity, and orientation, with considerations for improving the robustness of the system in autonomous ground vehicles. Section 12.5 presents a novel terrain estimation algorithm that utilizes attitude and position estimates and terrain sensors to generate a dense elevation model, along with a representation of uncertainty, for path planning without explicit obstacle identification. Section 12.6 presents a method for controlling the terrain sensors to improve the utility of data entering the terrain estimator. Section 12.7 presents a cubic spline-based path planner that utilizes terrain estimates and physical constraints to determine the final path traversed by the vehicle. Section 12.8 concludes with a performance analysis of Cornell's entry in the 2005 Grand Challenge.

12.2 System Subdivision and Information Flow

Cornell's approach to the Grand Challenge divides the problem into four separate goals: developing a robust platform, estimating the orientation and position of that platform, sensing its surroundings, and planning a path through its environment. The advantage of dividing the problem in this particular manner is that sub teams can work on these goals independently using either simulated or logged data from other sub teams. This structure allowed the entire team to work under a common architecture without integration bottlenecks.

Figure 12.1 shows the relation between the four components of Cornell's solution. The heart of the solution is the attitude and position estimator, which provides the other three subsystems with the Spider's current position, velocity, and orientation via direct serial connection, as well as a serial timing pulse for synchronization. The terrain estimator combines that information with terrain sensors to form a map of the world, which it sends to the path planner over a dedicated gigabit Ethernet connection. The path planner combines the environment map with the vehicle's state to generate a path, which it converts to a desired steering angle and speed. It then sends the desired steering angle and speed to the Spider's microcontroller directly over a 500 kbps controller area

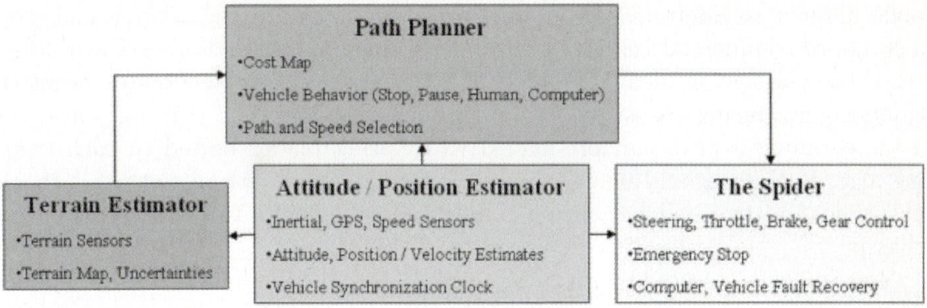

Fig. 12.1. System division in Cornell's Spider. Each block uses separate computational resources and communicates to other blocks via serial, Ethernet, or controller area network (CAN) connections.

network (CAN) bus. This low level microcontroller then generates the steering, brake, throttle, and transmission commands necessary to control the Spider.

12.3 Hardware and Actuation

12.3.1 The Spider Light Strike Vehicle

Cornell's Grand Challenge entry is based on a Spider Light Strike Vehicle, manufactured by Singapore Technologies Kinetics (STK), shown in Figure 12.2. The Spider, nominally an off-road 6-passenger military vehicle, is better suited to carry computers across harsh environments than a commercial sports utility vehicle or truck. Its fully-independent suspension and 35-inch Goodyear Maximum Traction radial tires give it more than 16 inches of ground clearance, and its heavy-gauge steel tube-frame chassis, roll cage, and full underbody skid

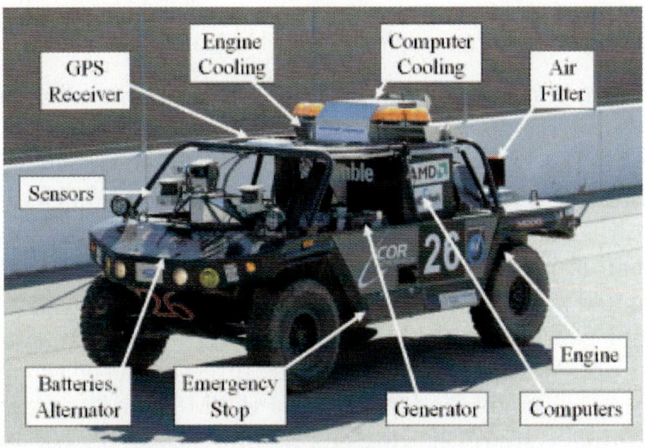

Fig. 12.2. Cornell's Modified Spider Light Strike Vehicle

plate allow it to safely traverse desert brush and small ditches that would stop a standard commercial vehicle. The Spider's unequal-length double A-arm front suspension design is ideal for protecting its payload, as it generates negative camber gain during chassis roll to aid vehicle stability. The rear suspension is a semi-trailing arm design pre-loaded with spring blades bolted to each trailing arm. Both front and rear suspensions are equipped with nitrogen gas shock absorbers and two boilover spring / shock absorber pairs to further isolate the chassis.

The Spider is powered by a VM Motori 2.8 Liter, 4 cylinder common-rail diesel engine with a maximum output of 163 horsepower and 295 ft-lbs of torque. The engine is coupled to an Audi 3-speed semi-automatic transmission and configured for 4-wheel drive, with a locked 1 : 1 front to rear output ratio. With a top speed of 50 miles per hour off-road and the ability to traverse grades of nearly 60%, the Spider's factory configuration is well-equipped to drive the nominal Grand Challenge route and also to tolerate mistakes made by its autonomous components (Singapore Technologies Kinetics, 2006).

The Spider's electric components are run from three separate power buses: 12 and 24 V DC sources, and a 110 V AC source, for design flexibility and compatibility with off-the-shelf automotive and computer components. The pre-installed 24 V source, which powers the Spider's actuators, sensors, gimbal, and emergency stop, is reinforced with two Optima deep-cycle car batteries charged by an engine-driven alternator. The 12 V source, powered from the 24 V source via DC-to-DC converter, is used to accommodate the standard automotive siren, engine control unit, and gauges, which are only available in 12 V configurations. A fuel-injected 5.5 kW Onan Commercial Mobile series generator provides 110 V AC power for the Spider's computers, which require clean, uninterrupted power even during sharp turns or when the Spider's engine is off. Five APC Smart-UPS 1000 VA units provide approximately 10 minutes of reserve power to the AC source, allowing time for the generator to be automatically restarted if it stalls.

12.3.2 Design for Drive-by-Wire Operation

The Spider's steering, throttle, brake, and transmission are all actuated for computer control and drive-by-wire operation. Of these four actuated components, throttle control is the most straightforward. The Spider is wired at the factory for electronic throttle control through the engine control unit, so no additional work is necessary. For steering, where accurate tracking, no slip, and human drivability are all important design requirements, The Spider's steering column is directly actuated with a 1.1 hp MOOG 6000 series AC brushless motor and a pair of 1 : 1 spur gears that can be disengaged for human operation. A Celesco CLP-200 linear potentiometer is also installed directly on the tie rod for absolute steering angle measurements independent of the MOOG's encoders.

Like the steering system, the Spider's hydraulic disc brakes are actuated at the brake pedal to preserve human operability. The brake pedal is pulled toward the Spider's floor by a Maxon RE 40 brushed DC motor with a 43 : 1 GP 42C planetary gear head and sheet metal spool. This gear and spool system is capable of delivering 72 lbs. force to the brake pedal, but more importantly, it can only pull the brake pedal downward. It therefore does not resist human operation, and it cannot be driven in reverse. The system is also equipped with a HEDS 5540 500 count-per-turn digital optical encoder on the planetary gear head for feedback.

The final actuated component enables gear selection on the Spider's transmission. Like the other three actuation schemes, the primary consideration in design of the transmission actuator is to preserve human operability. In the Spider, the transmission shifter has been modified to a single-axis shift pattern so it can be directly driven by a Maxon EPOS 70-4 rotary motor. This Maxon is equipped with the same model optical encoder as the brake actuator for position feedback.

Commands are issued to each of these actuators through a Motorola HCS12 16-bit microcontroller over the CAN bus. The HCS12 acts as the interface between the Spider's hardware and its computer payload, running 150 Hz proportional, integral, derivative (PID) loops to convert computer-supplied steering angle and speed commands into brake, throttle, and steering wheel commands (Franklin, Powell, & Emami-Naeini, 2002). This configuration allows the hardware interface to be completely transparent to the computers, so path planning and hardware development can take place simultaneously.

The HCS12 also monitors vehicle health to keep the Spider running in several contingency situations. It has a direct interface to the Spider's engine control unit, so it can detect engine stalls and restart the Spider if necessary. It also monitors the generator's output to restart it if it stalls. Finally, the HCS12 timestamps the commands received from the path planning computer. If these commands are ever interrupted, the HCS12 centers the Spider's wheel and brings it to a stop while the path planner restarts.

12.4 Attitude and Position Estimation

Because the Grand Challenge route is specified as a set of absolute latitudes and longitudes, it is necessary for each Grand Challenge vehicle to be able to localize and orient itself with respect to the Earth. At first glance, both these tasks can be accomplished using one or more off-the-shelf Global Positioning System (GPS) receiver to measure vehicle position, velocity, and heading. Slow update rates, errors in the navigation solution, and signal interruptions, however, make it difficult for even the best differential GPS receivers to consistently localize a vehicle to sub-meter accuracy required to avoid obstacles. Early obstacle detection is also difficult without accurate position and orientation, as an obstacle detected at 20 m with $\pm 5^o$ heading uncertainty has at least ± 2 m uncertainty in location just due to heading uncertainty alone.

The above considerations make a more accurate system necessary for long-horizon planning and robust path tracking. The system must be more than accurate, however; it must also be smooth and consistent. A proven solution to such requirements is the combination of GPS with a 'strapdown' inertial navigation system (INS), where absolute GPS measurements are fused with vehicle accelerations and rotation rates measured at high frequency by an inertial measurement unit (IMU) (Ohlmeyer et al., 1997). The strapdown INS is self-contained and platform-independent, so it may be 'strapped down' at any location on any rigid body to estimate its position and orientation. This flexibility is an important benefit in the Spider's design, where space is limited. It also facilitates easy transfer between multiple test beds without the need for vehicle-specific dynamics models.

Although several off-the-shelf positioning systems are available, team Cornell opted to design its own for several reasons. First, off-the-shelf models are black boxes, so system faults arising from these systems are difficult to pinpoint. Second, high accuracy off-the-shelf models are expensive, whereas the team could obtain each individual sensor from sponsors cheaply. Finally, building the positioning system in house allowed it to be specialized for autonomous ground vehicle operation, thereby improving fault detection and recovery.

12.4.1 Sensors for Inertial Navigation

The Spider's fused position and orientation estimates are provided by three sensors: a Litton LN-200 inertial measurement unit (IMU), a Trimble Ag252 GPS receiver, and a vehicle speed sensor. Each of the three sensors affects the capabilities of the positioning system differently, and fusing them together brings the best qualities of each sensor to bear on the problem. The sensors are discussed below in turn.

The first, the LN-200 IMU, is a set of three-axis rate sensors and three-axis accelerometers. It provides measurements of vehicle accelerations and rates of rotation relative to an inertial frame. These measurements permit dead reckoning: numerical integration of accelerations and rates to obtain changes in position, velocity, and orientation. The IMU's 400 Hz-configured output makes it more than twice as fast as any other sensor on the Spider, and its tactical grade fiber optic rate sensors and silicon accelerometers have biases stable to within 10°/hr and 3 mg, respectively (Northrop Grumman Navigation Systems Division, 2000). These capabilities help the positioning system generate smooth outputs at high rates.

The second sensor, a Trimble AgGPS 252 GPS receiver, provides absolute position and velocity measurements so the vehicle may be localized on the surface of the Earth. The Trimble boasts a number of important features, including dual frequency carrier phase filtering and ionospheric corrections, OmniSTAR high precision (HP) differential corrections, and multipath mitigation. These combine for position accuracy within 10 cm and velocity accuracy within 5 cm/s when tracking the OmniSTAR signal, and sub-meter accuracy without it (Trimble, 2004). Although the Trimble only provides measurements at 10 Hz, its absolute measurements keep numerical integration errors bounded.

The final sensor, a speed sensor (SBS), is mounted directly to the Spider's transmission. The SBS primarily serves to augment the inertial sensors in the absence of GPS signals, where it slows the growth rate of numerical integration

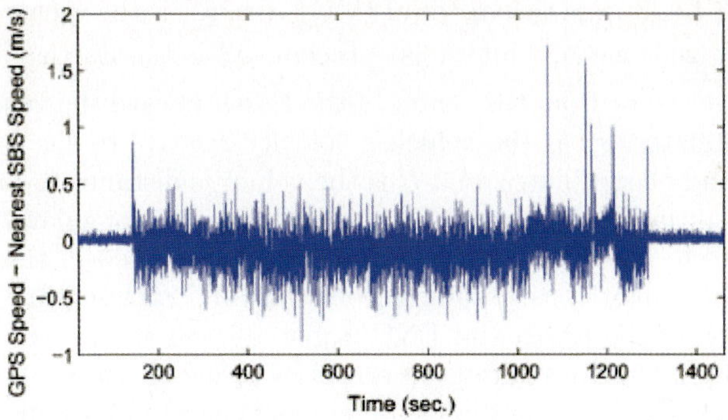

Fig. 12.3. Errors between the Spider's speed sensor (SBS) and the magnitude of GPS velocity in ideal GPS signal conditions

errors. This sensor was pre-installed on the Spider, so its characteristics were determined by experimentation. Figure 12.3 compares the most current SBS speed to the magnitude of GPS velocity during a field test in the presence of the OmniSTAR HP signal. Fortunately, the errors between the two signals are statistically unbiased, with an experimental mean of −0.0465 m/s in the data set of Figure 12.3. The standard deviation in the error signal, 0.1531 m/s, also gives an indication of the quality of the SBS measurements for tuning the positioning system.

12.4.2 Positioning System Equations of Motion

12.4.2.1 Attitude Dynamics

The attitude of a positioning system can be represented by one of several different sets of variables: Euler angles, quaternions, or a direction cosine matrix (DCM). The Cornell implementation uses a triad of ZY'X" Euler angles: yaw ψ, pitch ϕ, and roll θ, to represent the orientation of the Spider with respect to an East North Up (ENU) reference frame at the Spider's current latitude and longitude. The Euler angles are chosen over quaternions or a DCM for two reasons: they have an intuitive physical interpretation for quick debugging, and they are a minimal set of parameters. The latter point is important for Cornell's estimators, which cannot provide state estimates if the covariance matrix P is never invertible. The dynamic relations for the Euler angles defined for this problem are given below (Savage, 1998), (Triantafyllou & Hover, 2004):

$$\begin{pmatrix} \dot{\psi} \\ \dot{\phi} \\ \dot{\theta} \end{pmatrix} = \begin{pmatrix} 0 & s\theta/c\phi & c\theta/c\phi \\ 0 & c\theta & -s\theta \\ 1 & s\theta t\phi & c\theta t\phi \end{pmatrix} \cdot \left[\underline{\omega}_{IMU}^{B} - \underline{b}_{rg}^{B} - \underline{\eta}_{rg}^{B} - \underline{\omega}_{E}^{B} - \frac{\underline{e}_{R}^{B} \times \underline{v}^{B}}{R} \right] \quad (12.1)$$

using the abbreviations $s\cdot = sin(\cdot)$, $c\cdot = cos(\cdot)$, and $t\cdot = tan(\cdot)$, and letting superscripts denote the coordinate frame in which a particular vector is expressed.

In equation 12.1, $\underline{\omega}_{IMU}^{B}$ is the vector of IMU rates, $\underline{b}_{rg}^{B}$ is the vector of rate sensor biases, $\underline{\eta}_{rg}^{B}$ is a zero mean, white noise process, $\underline{\omega}_{E}^{B}$ is the Earth's angular velocity, $\underline{e}_{R}^{B}$ is a unit vector from the center of the Earth toward the vehicle expressed in body coordinates, $\underline{v}^{B}$ is the vehicle's velocity relative to the spinning Earth expressed in body coordinates, and R is the vehicle's distance from the center of the Earth. Intuitively, the bracketed vector term is the angular velocity of the vehicle relative to the ENU reference frame and expressed in the vehicle body (B) frame. This quantity is obtained from the IMU rate sensors by correcting for the rotation of the Earth and ENU frame, as well as bias and sensor noise. Note equation 12.1 defines rate gyro errors in terms of additive biases $\underline{b}_{rg}$ and white noise $\underline{\eta}_{rg}$. The additive biases are modeled as random walk processes:

$$\dot{\underline{b}}_{rg} = \underline{n}_{rg} \tag{12.2}$$

where $\underline{n}_{rg}$ is a zero mean, white noise process. Other gyro errors, such as g-sensitivity, scale factor, and misalignments, are not included in this model to improve computational speed. Despite these exclusions, it is shown below that the filter is still statistically significant, due in part to the quality of the fiber optic gyros.

12.4.2.2 Position and Velocity Dynamics

Like attitude, the position and velocity dynamics of a positioning system have several different representations, including latitude, longitude and altitude (LLA), Earth-Centered Inertial (ECI) coordinates, Earth-Centered Earth-Fixed (ECF) coordinates, and East North Up (ENU) coordinates. This implementation tracks vehicle position and velocity in ECF coordinates relative to the rotating Earth, primarily due to the simplicity of the dynamics equations. The position and velocity dynamics for this coordinate frame can be derived from the vector equations of velocity and acceleration relative to a moving coordinate frame (Moon, 1998), (Savage, 1998):

$$\dot{\underline{p}}^{ECF} = \underline{v}^{ECF} \tag{12.3}$$

$$\dot{\underline{v}}^{ECF} = \underline{a}_{IMU}^{ECF} - \underline{b}_{a}^{ECF} - \underline{\eta}_{a}^{ECF} + \underline{g}^{ECF} - \left(\underline{\omega}_{E}^{ECF} \times \underline{\omega}_{E}^{ECF} \times R\underline{e}_{R}^{ECF}\right) - 2\left(\underline{\omega}_{E}^{ECF} \times \underline{v}^{ECF}\right) \tag{12.4}$$

where $\underline{b}_{a}^{ECF}$ is a vector of accelerometer biases, $\underline{\eta}_{a}^{ECF}$ is a vector of zero mean white noise, $\underline{g}^{ECF}$ is the gravity vector, and the final two terms correct for centripetal and coriolis accelerations, respectively. Note equation 12.4 specifies an accelerometer error model consisting only of biases $\underline{b}_{a}^{ECF}$ and white noise $\underline{\eta}_{a}^{ECF}$. Like the rate gyro biases, the accelerometer biases are modeled as random walk processes:

$$\dot{\underline{b}}_{a} = \underline{n}_{a} \tag{12.5}$$

Other error terms, such as scale factors and misalignments, are not modeled. As with the rate gyros, it is shown below that statistical significance is still maintained within the filter. Gravity is also defined using a simple ellipsoidal model:

$$\underline{g}^{ECF} = -9.8 \cdot \frac{\begin{pmatrix} p_x & p_y & \frac{a_e^2}{b_e^2} p_z \end{pmatrix}^T}{\sqrt{p_x^2 + p_y^2 + (a_e^4/b_e^4) p_z^2}} \tag{12.6}$$

where a_e and b_e are the lengths of the semimajor and semiminor axes of the WGS-84 ellipsoid model of the Earth, and p_x, p_y, and p_z are the three components of the ECF position of the vehicle. It is shown below that the filter can be made statistically significant despite the gravity model. A more complicated spherical harmonic expansion was therefore not tried.

12.4.3 Fusing the Inertial Navigation Sensors

The three sensors discussed in section 12.4.1 are fused into an estimate of the Spider's position, velocity, and attitude using an extended Kalman Filter (EKF). The filter operates in two steps: a prediction and an update (Bar-Shalom, Rong Li, & Kirubarajan, 2001). In the prediction step, the estimates of attitude, position, velocity, and bias are integrated across one IMU sample interval ($dt = 0.0025s$) according to equations 12.1, 12.2, 12.3, 12.4, and 12.5 to make an *a priori* prediction of the estimates at the next time step. Because equations 12.1 and 12.4 have no general closed-form solution, a fourth order Runge-Kutta numerical integration is performed in place of an exact integration (Battin, 1999). During each numerical integration, the IMU accelerations and rates of rotation are assumed constant.

The filter also maintains an estimate of its estimation error covariance, $P(k)$ at each time index k. During the prediction step, the estimated error covariance is advanced in time to form an *a priori* estimate $\bar{P}(k+1)$ of the state error covariance. Because the attitude and velocity dynamics are nonlinear, a linearization and discretization of the continuous dynamics is made at each time step (Bar-Shalom et al., 2001):

$$\begin{aligned} \bar{P}(k+1) &\approx F(k)P(k)F^T(k) + \Gamma(k)Q(k)\Gamma^T(k) \\ F(k) &= I + dt \cdot J_x(k) + \frac{1}{2}dt^2 \cdot J_x^2(k) + \frac{1}{6}dt^3 \cdot J_x^3(k) \\ \Gamma(k) &= \left(dt \cdot I + \frac{1}{2}dt^2 \cdot J_x(k) + \frac{1}{6}dt^3 \cdot J_x^2(k) + \frac{1}{24}dt^4 \cdot J_x^3(k) \right) J_v(k) \end{aligned} \tag{12.7}$$

where $J_x(k)$ and $J_v(k)$ are the Jacobians of equations 12.1, 12.2, 12.3, 12.4, and 12.5 with respect to the state and noise inputs, respectively, evaluated at their current estimates, and $Q(k)$ is a matrix of white noise intensities assigned to the noise inputs during filter tuning.

In the update step, the predicted state $\bar{\underline{x}}(k+1)$ and its covariance $\bar{P}(k+1)$ are combined with sensor measurements $\underline{z}(k+1)$ of certain elements of the state in a linear minimum mean square estimator to form *a posteriori* state estimates $\hat{\underline{x}}(k+1)$ and their updated error covariance $P(k+1)$ (Bar-Shalom et al., 2001):

$$\begin{aligned}
\underline{z}(k+1) &= H(k+1)\underline{x}(k+1) + \underline{w}(k+1) \\
S(k+1) &= H(k+1)\bar{P}(k+1)H^T(k+1) + R(k+1) \\
W(k+1) &= \bar{P}(k+1)H^T(k+1)S^{-1}(k+1) \\
\hat{\underline{x}}(k+1) &= \bar{\underline{x}}(k+1) + W(k+1)\underline{\nu}(k+1) \\
P(k+1) &= \bar{P}(k+1) - W(k+1)S(k+1)W^T(k+1)
\end{aligned} \tag{12.8}$$

where $\underline{w}(k+1)$ is a vector of zero mean white measurement noise with covariance intensity $R(k)$, set during filter tuning, and $\underline{\nu}(k+1) = \underline{z}(k+1) - H(k+1)\bar{\underline{x}}(k+1)$ is called the 'innovation.'

The measurements $\underline{z}$ used in the update step are extracted from the three sensors that make up the positioning system. Position, velocity, and yaw (heading) are reported directly by the GPS receiver. Speed is reported by the SBS sensor, and a full ECF vector velocity measurement is created from it using vehicle attitude as an exogenous input. Pitch and roll measurements can also be extracted if the vehicle is at rest by using the accelerometers to measure the direction of the gravity vector in the body frame $\underline{e}_g^B$ and comparing it to a theoretical downward-pointing $\underline{e}_g^B$:

$$\underline{e}_g^B = \begin{pmatrix} s\phi \\ -s\theta c\phi \\ -c\theta c\phi \end{pmatrix} \approx \underline{a}_{IMU}^B \tag{12.9}$$

from which pitch ϕ and roll θ can be extracted, assuming $\frac{-\pi}{2} < \phi < \frac{\pi}{2}$. If the vehicle is moving, a similar calculation can be done using equation 12.4 to solve for $\underline{g}^{ECF}$, with $\dot{\underline{v}}^{ECF}$ approximated by a finite difference of GPS velocity. The moving calculation is noisier, however, as it depends on existing attitude estimates to transform $\underline{g}^{ECF}$ into $\underline{g}^B$.

While many EKF implementations use a single filter to estimate attitude, position, and velocity together, Cornell uses a pair of filters to estimate attitude and position / velocity separately. That is, one filter fuses equations 12.1 and 12.2 with measurements of yaw, pitch, and roll to estimate attitude states, and a second filter fuses equations 12.3, 12.4, and 12.5 with position and velocity measurements to estimate position and velocity states. To do so, the attitude estimator treats position and velocity as exogenous noisy inputs, and the position / velocity filter treats attitude as an exogenous noisy input. The motivation for this structure is computational: decoupling the states cuts the sizes of the matrices in equations 12.7 and 12.8 almost by a factor of four, enabling 400 Hz predictions on the 3.4 GHz Pentium IV attitude computer equipped with Windows Server 2003 that match the IMU rates.

The EKFs in Cornell's implementation are both Square Root Information Filters (SRIFs), a numerically robust implementation of the Kalman Filter algorithm (Bierman, 1977). The SRIF has two main advantages over standard KF implementations. First, matrix computations in the SRIF are performed using square root versions of the KF matrices, so numerical precision is effectively double that of the standard Kalman Filter. Second, computations are performed using the inverse of the covariance matrix, which allows a numerically-correct representation of infinite covariance during filter initialization. Bierman (1977) describes the SRIF in detail.

12.4.4 Filter Tuning and Verification

Although the main goal of the filters is to estimate the Spider's attitude, position, and velocity, a second goal of equal importance is to evaluate the accuracy of those estimates. That accuracy, stored in the state error covariance matrix $P(k)$, is essential for calculating correct filter updates (Bar-Shalom et al., 2001). In addition, accurate estimates of error statistics are necessary for identifying and localizing terrain in a probabilistically rigorous manner, which serves as the foundation of the Spider's terrain estimation algorithm.

One traditional method for evaluating the estimators' validity and statistical significance is to perform consistency tests (Bar-Shalom et al., 2001). These are statistical hypothesis tests of the assumptions made by the filters: whether the linearizations are valid, the dynamics and sensor error models are correct, and the reported covariance accurately describes the error statistics of the estimators. If these filter assumptions are correct, one consequence is that the innovations $\underline{\nu}(k)$ will be zero mean and white, with covariance $S(k)$ given in equation 12.8. A consistency test statistic for these hypotheses is based on the observed sequence of innovations:

$$\bar{\epsilon}_{\nu} = \frac{1}{N} \sum_{k=k_o}^{k_o+N-1} \underline{\nu}(k)^T S^{-1}(k) \underline{\nu}(k) \tag{12.10}$$

If the filter assumptions are valid, the test statistic follows a Chi-Square distribution:

$$\bar{\epsilon}_{\nu} \sim \chi^2_{N \cdot n_z} \tag{12.11}$$

where n_z is the length of the measurement vector $\underline{z}(k)$ (Bar-Shalom et al., 2001).

The statistical significance, correctness, and consistency of the Cornell filters are tested using observed innovation vectors taken during a test run in a parking lot. The values of the test statistics for the attitude filter are plotted in Figure 12.4 (left) and (right) for $N = 1$ and $N = 100$. Experimentally, approximately 84.7% of the test statistic values lie in the 95% interval for $N = 1$, suggesting that the attitude filter makes accurate assumptions and reports statistically significant estimates and covariances for the majority of the test run. The filter shows limitations for the more stringent $N = 100$ test, however, where only 21.4% of the values lie in the 95% interval. The attitude filter is therefore overconfident on average, but consistent for each individual measurement. Time constraints in the Grand Challenge prevented further experimentation to reduce the filter's overconfidence, though the time-correlations of the test statistic in Figure 12.4 (right) suggest the white noise assumption in GPS-derived attitude measurements might be to blame.

The values of the test statistic for the position / velocity filter are plotted in Figure 12.5 (left) and (right) for $N = 1$ and $N = 100$. The results are similar to

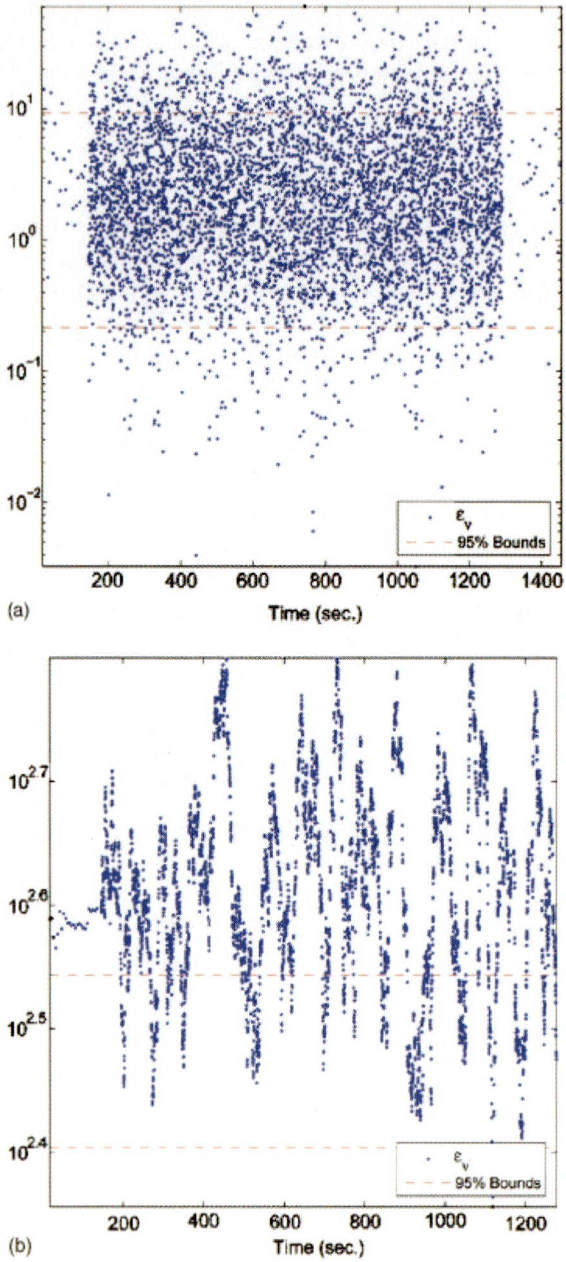

Fig. 12.4. (Left) Consistency test statistic $\bar{\epsilon}_\nu$ and 95% confidence bounds for a calibration run of the attitude estimator, with $N = 1$. (Right) $\bar{\epsilon}_\nu$ and 95% confidence bounds during the same trial with $N = 100$.

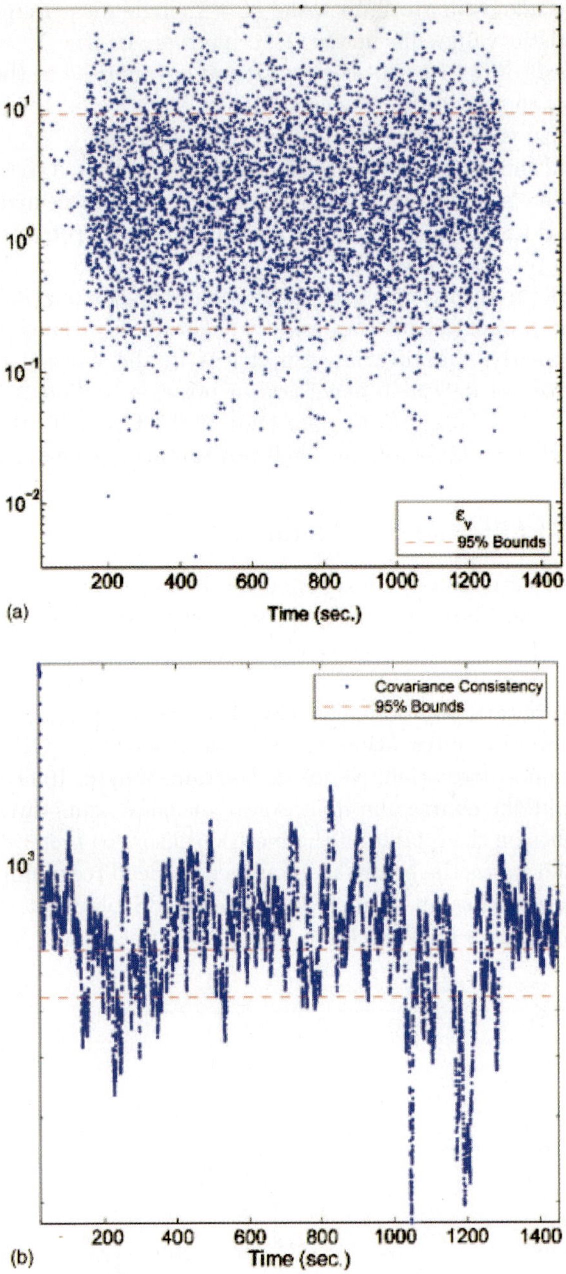

Fig. 12.5. (Left) Consistency test statistic $\bar{\epsilon}_\nu$ and 95% confidence bounds for a calibration run of the position / velocity estimator, with $N = 1$. (Right) $\bar{\epsilon}_\nu$ and 95% confidence bounds during the same trial with $N = 100$.

those of the attitude estimator. With the $N = 1$ single measurement test, 78.4% of the test statistic values lie in the 95% interval. In the $N = 100$ averaged measurement test, only 21.6% of the test statistic values lie in the 95% interval. The results show the position / velocity filter to report valid covariance matrices in most iterations, though it is overconfident on average.

The statistical consistency test described above is also used to tune the Cornell filters. That is, the process and measurement noise covariance matrices $Q(k)$ and $R(k)$ from equations 12.7 and 12.8 are tuned using test runs until the filter is as consistent as possible. The fact that the Cornell filters can be tuned to be consistent means simultaneously that the modeling approximations are valid, the filters are tuned properly, and they report accurate covariance matrices $P(k)$. Once tuned correctly, the covariance matrices in the Cornell filters typically report RMS errors of $0.02m$ in each axis in position, $0.05m/s$ in each axis in velocity, and 0.2^o in heading in the presence of the OmniSTAR signal, making them competitive with many off-the-shelf positioning systems available.

12.4.5 Special Filtering Considerations

Despite filter correctness and consistency, Cornell encountered two field events requiring special consideration: loss and reacquisition of the GPS signal. In principle, filtering theory has no difficulty handling these situations, as the IMU and speed sensor are used for repeated prediction steps until the GPS signal is reacquired. From a practical point of view, however, these events generate difficulties. In particular, numerical integration errors in the absence of a GPS signal grow in time without bound (Sukkarieh, Nebot, & Durrant-Whyte, 1999). With no prior information about the course and no absolute measurements during GPS blackouts, Cornell relies on the quality of the inertial sensors to keep estimation errors small. Figure 12.6 shows the errors accumulated by dead reckoning with the IMU and speed sensor during an artificially-created GPS blackout. Position errors reach only 4.8 m after a 60 s blackout, and the rate of growth is approximately

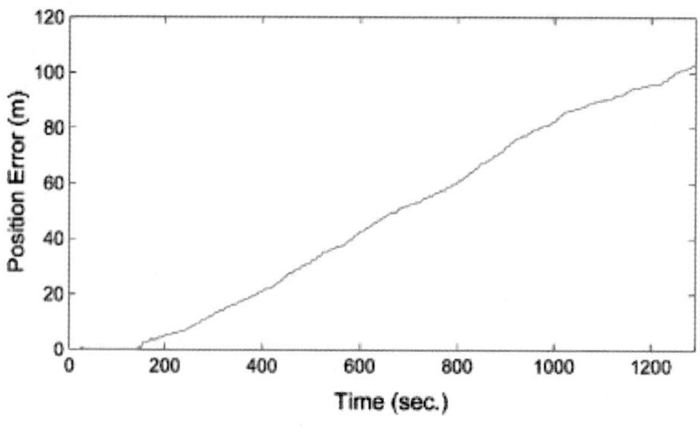

Fig. 12.6. Position errors from dead reckoning with Cornell's IMU and wheel speed sensor during an artificially-created GPS blackout starting at $t = 140$ s

constant. With the open desert environment, an average corridor width of more than 6.5 m, and average speeds nearing 9 m/s, this level of blackout tolerance is sufficient for blackout situations likely to be encountered in the Grand Challenge.

While GPS signal blackouts cause errors that grow over time, a more acute event occurs when the GPS signal is reacquired. The problem lies in the fact that the filter update step of equation 12.8 produces discontinuous jumps in state estimates, particularly position. Although these discontinuities are consistent with Kalman Filtering theory, they cause the vehicle to turn sharply if fed back in a path tracking controller. To limit the size of these discontinuities, a 'rate limiting' modification is performed in the position filter update step. The strategy works as follows:

1. Calculate the position and velocity discontinuities during a position filter update:
$$\begin{aligned} \Delta \underline{p}^{ECF}(k) &= \hat{\underline{p}}^{ECF}(k) - \bar{\underline{p}}^{ECF}(k) \\ \Delta \underline{v}^{ECF}(k) &= \hat{\underline{v}}^{ECF}(k) - \bar{\underline{v}}^{ECF}(k) \end{aligned} \tag{12.12}$$
Also calculate the magnitude d_p of the position discontinuity.
2. If d_p is larger than some tolerable position jump d_{max}, calculate the discount factor $c_o = \frac{d_{max}}{d_p}$.
3. Calculate the rate limited measurement update to limit the position discontinuity:
$$\begin{pmatrix} \hat{\underline{p}}^{ECF}(k) \\ \hat{\underline{v}}^{ECF}(k) \end{pmatrix} = \begin{pmatrix} \bar{\underline{p}}^{ECF}(k) + c_o \cdot \Delta \underline{p}^{ECF}(k) \\ \bar{\underline{v}}^{ECF}(k) + c_o \cdot \Delta \underline{v}^{ECF}(k) \end{pmatrix} \tag{12.13}$$
4. Calculate the rate limited filter covariance:
$$P(k) = [I - c_o \cdot W(k)H(k)]\, \bar{P}(k)\, [I - c_o \cdot W(k)H(k)]^T + c_o^2 \cdot W(k)R(k)W(k)^T \tag{12.14}$$
This rate limited covariance update must be used instead of equation 12.8 when position and velocity are rate limited if the filter is to be kept accurate and consistent (Bar-Shalom et al., 2001).

This rate limiting strategy modifies the filter gain $W(k)$ to ensure smooth reacquisition of GPS signals. It also allows the filter to be tuned to the vehicle controller, as the rate limiting threshold d_{max} may be set to ensure the maximum distance discontinuity per second does not exceed typical vehicle speeds. Figure 12.7 shows the effects of rate limiting on reacquisition after an artificially-created GPS blackout. Without the rate limiting, the position estimate undergoes an instantaneous jump of more than $5m$ for this example. In the field, a discontinuous jump of that magnitude would cause the vehicle to swerve and possibly roll. The rate limited position estimate, in contrast, smoothly converges toward the accurate GPS signal. The uncertainty in the estimate also decreases slowly.

Although effective at solving the GPS reacquisition problem, the rate limiting strategy has several drawbacks. First, the rate limited updates are not true

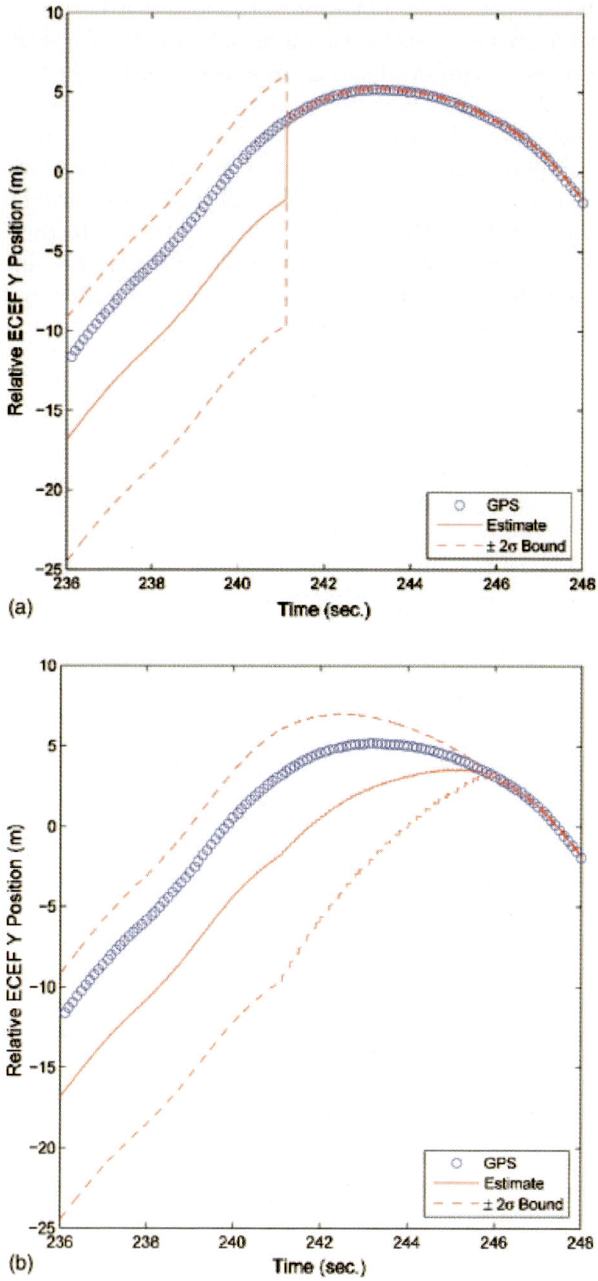

Fig. 12.7. (Left): Without rate limiting, the position estimate makes a discontinuous jump of more than 5 m during one GPS signal reacquisition. (Right): Rate limiting causes the position estimate and its uncertainty to converge smoothly.

minimum mean squared error (MMSE) estimates, even for estimation on a linear system. Second, the more computationally expensive equation 12.14 must be used to update filter covariance during rate limiting, as the update of equation 12.8 is not valid for arbitrary filter gains (Bar-Shalom et al., 2001). Finally, rate limiting has the potential to smear any absolutely-mapped objects. However, because equation 12.14 maintains accurate filter covariance matrices, the uncertainty in these objects decreases slowly as they converge to more accurate estimates. Without rate limiting, any absolutely-mapped objects make large discontinuous jumps on reacquisition of a GPS signal. Such discontinuities violate the slowly-changing world assumption required by Cornell's path planner, resulting in indecision and crashes during testing.

12.5 Terrain Sensing and Mapping

The 2005 DARPA Grand Challenge places great emphasis on the vehicles' ability to detect unsafe regions and unplanned obstacles inside the route boundaries. To distinguish between safe and unsafe regions of the route, each vehicle must be equipped with terrain sensors or obstacle detectors and an algorithm to localize the vehicle with respect to the sensed objects. Historically there have been a wide variety of approaches to this mapping problem, including maps of discrete obstacles or landmarks, grid-based approaches, statistical and stochastic models, and even polygonal object representations (Thrun, 2002).

Cornell's terrain representation is grid-based, similar to the representation in Olin and Tseng (1991). That is, terrain is divided into $40cm \times 40cm$ grid cells in a local ENU plane, with the data in each cell describing that cell's elevation. However, unlike Olin and Tseng (1991), Cornell uses a novel statistical terrain estimation algorithm to generate minimum mean squared error (MMSE) terrain estimates within each grid cell. This approach generates not only terrain estimates in real-time, but also their uncertainties, enabling paths to be selected based on statistical statements of traversability.

12.5.1 Terrain Sensors

Sensed terrain data for the terrain estimator is provided by three SICK LMS 291 laser rangefinders (LIDARs). Each of these LIDAR units scans along a single line and is configured to return 181 ranges spaced at half-degree increments over a 90° field of view, all at 75 Hz (SICK, 2003). Each LIDAR returns data via 500 kbps serial connection to the terrain computer, an AMD 2 GHz quad Opteron 846 server running Windows Server 2003, where the data is fused into terrain estimates. Figure 12.8 shows the three LIDAR units, each mounted to the hood of the Spider. The outer two LIDARs are mounted rigidly using a turnbuckle and rod ends to allow their pitch to be set without inducing yaw or roll. Both LIDARs face in the direction of vehicle travel; one is pitched at approximately 4° to scan 15 m in front of the vehicle, and the other is pitched at approximately 6° to scan the ground at 20 m. The central LIDAR sits on a two-axis gimbaled

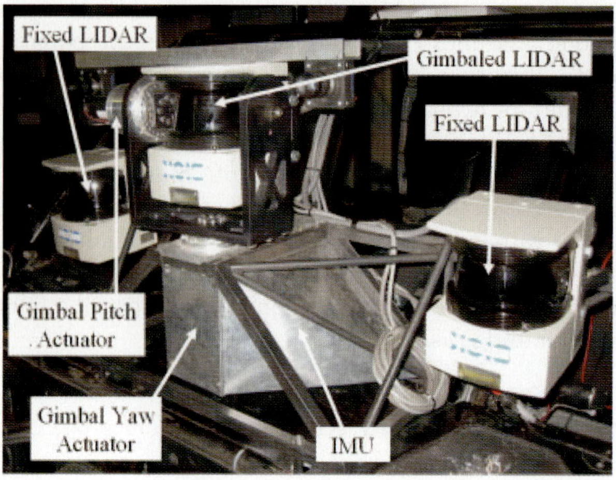

Fig. 12.8. The Spider's sensor platform and two-axis gimbal

platform. The platform is actuated in the yaw and pitch axes with two MAXON EPOS 70-4 rotary motors and separate Harmonic Drive gear trains, allowing the gimbaled LIDAR to be aimed by the path planner.

Cornell's sensor orientation is driven by two factors: budget constraints and terrain representation. Budget constraints limited the team to three LIDAR units, so each unit is positioned to sense terrain without risking damage upon collision. For this reason, the LIDARs are all mounted on the hood of the vehicle rather than near the ground. The LIDARs are aimed downward to provide the most useful information for the terrain estimation algorithm, which fuses data from all three LIDARs into a single persistent map. Because the LIDARs reinforce each other in their measurements of the terrain in front of the vehicle, the system is less sensitive to false obstacles than one in which each LIDAR is positioned to sense only one type of obstacle, such as an obstacle in the path of one of the vehicle's wheels.

12.5.2 Terrain Estimation Algorithm

The goal of the terrain estimation algorithm is to produce an elevation estimate within each grid cell, optimal in the sense of minimum mean-squared error (MMSE). Three steps are performed to generate these terrain estimates. First, a statistical representation of each sensor measurement is formed to account for multiple sources of error in a probabilistically rigorous manner. Second, each measurement is probabilistically assigned or 'associated' to one or more terrain cells in which it is likely to belong. Finally, the measurements assigned to each cell are fused in real-time into a single MMSE estimate of the elevation within each cell. These three steps are discussed in turn below.

12.5.2.1 Statistical Treatment of Sensor Measurements

The first step in the terrain estimation algorithm is to form a statistical representation of each LIDAR measurement to account for different errors in the sensing path. These errors include those due to the LIDAR itself, errors due to sensor orientation, errors due to the attitude estimator, and errors due to the position / velocity estimator (Huising & Pereira, 1998). The first of these error types are due to the LIDAR itself, and include range errors and errors due to expansion of the LIDAR beam. Range errors add directly to the ranges reported by the LIDAR, and have typical accuracy of ± 5 cm out to a range of 80 m when the LIDAR is configured in centimeter mode (SICK, 2003). Beam expansion errors, in contrast, are errors in detection angle due to the fact that LIDAR ranges are generated from reflections of a beam of light that expands as it travels from the LIDAR. Beam expansion accounts for approximately $\pm 0.24^o$ detection angle error for each LIDAR measurement (SICK, 2003).

The second source of sensing error is uncertainty in the sensors' orientations. For fixed sensors, this type of uncertainty can be eliminated by offline calibration routines against objects of known location. However, for the gimbaled LIDAR, orientation errors arise due to inaccuracies in the angle encoders on the MAXON motors actuating the gimbal. These errors affect the localization of measurements in the vehicle body frame.

The third and fourth sources of sensing error are uncertainty due to the attitude and position estimators. While these errors do not affect the sensor readings themselves, they introduce error in locating measurements on an absolute map. The statistical properties of these errors are captured by the attitude and position estimates $\underline{\hat{x}}(k)$ and their covariances $P(k)$ from equation 12.8.

In order to understand in a statistical sense how these four types of error affect each LIDAR measurement, it is first necessary to define the transformation in which they play a part. For this task, each valid sensor measurement is combined with the sensor's orientation, the vehicle's orientation, and the vehicle's location to express the measurement in the fixed ENU coordinates of the terrain map:

$$\underline{r}^{ENU} = \left(E^{ENU} \quad N^{ENU} \quad U^{ENU} \right)^T = f\left(\underline{p}, \underline{r}\right) \tag{12.15}$$

with:

$$\begin{aligned} \underline{p} &= \left(\underline{L}_i^T, \underline{x}_a^T, \underline{x}_p^T \right)^T \\ \underline{r} &= \left(\rho, \theta_D \right)^T \end{aligned} \tag{12.16}$$

where $\underline{L}_i$ is a vector of parameters describing the location and orientation of the ith LIDAR with respect to the vehicle body frame, $\underline{x}_a$ is the vehicle's attitude at the time of measurement, $\underline{x}_p$ is the vehicle's position at the time of measurement, ρ is a range returned by the LIDAR, and θ_D is the scalar angle of the measurement within the LIDAR's scanning plane. Intuitively, these parameters are divided into the vector $\underline{p}$ of LIDAR orientation parameters, and a measurement $\underline{r}$. Note in general the transformation function $f(\cdot)$ of equation 12.15 is a nonlinear function of these parameters. For the Cornell implementation, it is constructed from a set of 4×4 matrix transformations, with each matrix a function of one parameter as in Murray, Li, and Sastry (1994) and Moon (1998). A sample sequence of these transformations is as follows:

1. Rotate the coordinate frame about its Z axis by $-\theta_D$ to express the measurement $(\rho \quad 0 \quad 0 \quad 1)^T$ in LIDAR-centered coordinates, where ρ is a range reported by the LIDAR and θ_D is its detection angle.
2. Rotate the coordinate frame by $-S_\theta$ about its X axis, where S_θ is the LIDAR's roll angle with respect to the Spider.
3. Rotate the coordinate frame by $-S_\phi$ about its Y axis, where S_ϕ is the LIDAR's pitch angle with respect to the Spider.
4. Rotate the coordinate frame by $-S_\psi$ about its Z axis, where S_ψ is the LIDAR's yaw angle with respect to the Spider.
5. Translate the origin of the coordinate frame by $(-S_x \quad -S_y \quad -S_z)^T$, where S_x, S_y, and S_z represent the location of the LIDAR with respect to the vehicle body origin, centered at the IMU.
6. Rotate the coordinate frame by $-\theta$ about its X axis, where θ is the vehicle's roll angle.
7. Rotate the coordinate frame by $-\phi$ about its Y axis, where ϕ is the vehicle's pitch angle.
8. Rotate the coordinate frame by $-\psi$ about its Z axis, where ψ is the vehicle's yaw angle.
9. Translate the origin of the coordinate frame by $(-O_x \quad -O_y \quad -O_z)^T$, where O_x, O_y, and O_z represent the location of the vehicle with respect to the fixed ENU origin.

To augment each LIDAR measurement with an expression of its uncertainty, each of the parameters is modeled as corrupted by an additive measurement noise:

$$\begin{aligned} \underline{p} &= \underline{\hat{p}} + \delta \underline{p} \\ \underline{r} &= \underline{\hat{r}} + \delta \underline{r} \end{aligned} \tag{12.17}$$

where $\underline{\hat{p}}$ and $\underline{\hat{r}}$ are the measured values of $\underline{p}$ and $\underline{r}$, and $\delta\underline{p}$ and $\delta\underline{r}$ are measurement noise, assumed to be zero mean, Gaussian, and mutually uncorrelated. Using this representation, the terrain measurement $\underline{r}^{ENU}$ can be written:

$$\underline{r}^{ENU} = f\left(\underline{\hat{p}} + \delta\underline{p}, \underline{\hat{r}} + \delta\underline{r}\right) \tag{12.18}$$

A typical linearization can then be made to characterize the true measurement and the statistics of its uncertainty (Bar-Shalom et al., 2001):

$$\underline{r}^{ENU} \approx f\left(\underline{\hat{p}}, \underline{\hat{r}}\right) + J_{\underline{p}}\left(\underline{\hat{p}}, \underline{\hat{r}}\right)\delta\underline{p} + J_{\underline{r}}\left(\underline{\hat{p}}, \underline{\hat{r}}\right)\delta\underline{r} \tag{12.19}$$

where $J_{\underline{p}}(\cdot)$ and $J_{\underline{r}}(\cdot)$ are the Jacobians of the transformation $f(\cdot)$ with respect to the orientation parameters and measurement, respectively:

$$\begin{aligned} J_{\underline{p}}\left(\underline{\hat{p}}, \underline{\hat{r}}\right) &= \left.\frac{\partial f}{\partial \underline{p}}\right|_{\underline{p}=\underline{\hat{p}}, \underline{r}=\underline{\hat{r}}} \\ J_{\underline{r}}\left(\underline{\hat{p}}, \underline{\hat{r}}\right) &= \left.\frac{\partial f}{\partial \underline{r}}\right|_{\underline{p}=\underline{\hat{p}}, \underline{r}=\underline{\hat{r}}} \end{aligned} \tag{12.20}$$

Taking the expectation of equation 12.19 gives an approximate MMSE estimate of the terrain detection, accounting for all modeled sources of error in the sensing path (Bar-Shalom et al., 2001):

$$\hat{\underline{r}}^{ENU} = (\,\hat{E}^{ENU} \quad \hat{N}^{ENU} \quad \hat{U}^{ENU}\,)^T = f\left(\hat{\underline{p}}, \hat{\underline{r}}\right) \tag{12.21}$$

and the estimate has the associated mean squared error matrix:

$$P^{ENU} = J_{\underline{p}} Q_{\underline{p}} J_{\underline{p}}^T + J_{\underline{r}} Q_{\underline{r}} J_{\underline{r}}^T \tag{12.22}$$

where $Q_{\underline{p}}$ and $Q_{\underline{r}}$ are the covariance matrices of the orientation parameters' noise $\delta\underline{p}$ and the measurement's noise $\delta\underline{r}$, respectively. Note P^{ENU} is effectively an ENU covariance matrix for the LIDAR measurement estimate, taking into account the modeled sources of error in the sensing path.

12.5.2.2 Measurement Association

The second step in the terrain estimation algorithm is to assign each estimated terrain measurement $\hat{\underline{r}}^{ENU}$ to one or more grid cells from which it is likely to have originated. To accomplish this, the probability that each measurement belongs to each cell must be computed. Because each measurement's mean $\hat{\underline{r}}^{ENU}$ and covariance matrix P^{ENU} are available from equations 12.21 and 12.22, the joint ENU probability density $\mathcal{P}_{\hat{\underline{r}}}(E, N, U)$ of the measurement can be approximated as a multivariate Gaussian with vector mean $\hat{\underline{r}}^{ENU}$ and covariance P^{ENU}. The probability that the measurement belongs in a particular cell can then be evaluated by marginalizing the density with respect to U and integrating the resulting EN density $\mathcal{P}\hat{\underline{r}}(E, N)$ over the area of the cell. This integral can be approximated as the area of a single Riemann square:

$$p\left(\hat{\underline{r}}^{ENU} \in CELL\right) \approx (E_f - E_o) \cdot (N_f - N_o) \cdot \mathcal{P}_{\hat{\underline{r}}}\left(\frac{1}{2}(E_f + E_o), \frac{1}{2}(N_f + N_o)\right) \tag{12.23}$$

for the cell defined by $E_o \le E \le E_f$ and $N_o \le N \le N_f$. This gives the desired association probability p_i that the ith LIDAR measurement estimate $\hat{\underline{r}}_i^{ENU}$ belongs in a particular cell.

12.5.2.3 In-Cell Terrain Measurement Fusion

The final step in the terrain estimation algorithm is to fuse all measurements assigned to a particular cell into an estimate of the elevation distribution within that cell. First, each ENU measurement $\hat{\underline{r}}_i^{ENU} = (\,\hat{E}_i^{ENU} \quad \hat{N}_i^{ENU} \quad \hat{U}_i^{ENU}\,)^T$ assigned to a cell is transformed into a posterior estimate of the cell's elevation $\hat{U}_i$ by conditioning the measurement on the EN location of the cell (Bar-Shalom et al., 2001):

$$\hat{U}_i = \hat{U}_i^{ENU} + \left(P_i^{EN,U}\right)^T \left(P_i^{EN}\right)^{-1} \left[\frac{1}{2}\begin{pmatrix} E_f + E_o \\ N_f + N_o \end{pmatrix} - \begin{pmatrix} \hat{E}_i^{ENU} \\ \hat{N}_i^{ENU} \end{pmatrix}\right] \tag{12.24}$$

with associated conditional variance:

$$\sigma^2_{\hat{U},i} = P_i^U - \left(P_i^{EN,U}\right)^T \left(P_i^{EN}\right)^{-1} P_i^{EN,U} \tag{12.25}$$

where the ith measurement's covariance matrix P_i^{ENU} has been divided up into a 2×2 block P_i^{EN}, a 1×1 block P_i^U, a 2×1 block $\left(P_i^{EN,U}\right)^T$, and a 1×2 block $P_i^{EN,U}$:

$$P_i^{ENU} = \begin{pmatrix} P_i^{EN} & \left(P_i^{EN,U}\right)^T \\ P_i^{EN,U} & P_i^U \end{pmatrix} \tag{12.26}$$

With this, the task of fusing terrain measurements has been reduced to determining the univariate distribution of elevations within each cell. In this problem, each measurement $\hat{U}_i$ is effectively a terrain detection: an estimate of a piece of the terrain within a cell. From here it is convenient to assume that each of these measurement estimates corresponds to a different patch of terrain within the cell, so that no two measurements occur at precisely the same location. This assumption is justified by the fact that the terrain is continuous, the location of the terrain sensed by each measurement is uncertain, and the sensing platform is moving. Additionally, each terrain measurement is assumed equally likely; that is, there is no *a priori* terrain information. Finally, each cell is assumed to have one correct or 'dominant' elevation to be estimated, and that elevation is represented within the set of terrain measurements assigned to the cell. Under these assumptions, the posterior elevation distribution within the cell is a Gaussian sum or 'Gaussian mixture' constructed from the elevation estimates (Bar-Shalom et al., 2001), (Bishop, 1995):

$$\mathcal{P}(U|\hat{\underline{R}}^{ENU}) = \frac{\sum_{i=1}^{M} p_i \mathcal{N}\left(\hat{U}_i, \sigma^2_{\hat{U},i}\right)}{\sum_{i=1}^{M} p_i} \tag{12.27}$$

where $\hat{\underline{R}}^{ENU}$ is the set of all LIDAR measurements made so far, $\hat{U}_i$ from equation 12.24 is the ith elevation measurement estimate assigned to the cell, $\sigma^2_{\hat{U},i}$ from equation 12.25 is its associated conditional variance, p_i is the probability that the measurement belongs in the cell, and M is the number of measurement estimates assigned to the cell.

The distribution in equation 12.27 is the desired data-driven elevation distribution in the cell that takes into account all sources of uncertainty within the sensing path. However, the model is in general not computationally feasible, because the Gaussian mixture stored in each cell grows with the number of sensor measurements assigned to that cell. For real-time terrain estimates, an approximate model using a small set of information about each cell is desired: a set of data descriptive enough to be useful but small enough to be computationally feasible. It is proposed that the mean and variance of the elevation distribution are used. The expected value of the Gaussian mixture elevation distribution yields an approximate MMSE estimate of the characteristic or 'dominant' elevation of the jth cell:

$$\hat{U}_{GM,j} = \frac{\sum_{i=1}^{M} p_i \hat{U}_i}{\sum_{i=1}^{M} p_i} \approx E\left[U_j | \underline{\hat{R}}^{ENU}\right] \tag{12.28}$$

by the linearity of the expectation operator. Note this expectation is conditioned upon all the information available, as the elevation distribution itself is conditioned upon those measurements. Similarly, the second central moment of the Gaussian mixture gives the conditional mean square error of the estimate within the jth cell:

$$\sigma_{GM,j}^2 = \frac{\sum_{i=1}^{M} p_i (\hat{U}_i^2 + \sigma_{\hat{U},i}^2)}{\sum_{i=1}^{M} p_i} - \hat{U}_{GM,j}^2 \tag{12.29}$$

Equations 12.28 and 12.29 give the first two moments of the Gaussian mixture elevation distribution of the jth cell. Physically, equation 12.28 gives an estimate of the average or characteristic elevation of the cell with respect to the ENU reference origin derived from the available measurements. Mathematically, equation 12.28 is an approximate MMSE estimate of the elevation of the jth cell when taken with the assumptions discussed above. Equation 12.29, in contrast, may be interpreted as a measure of the roughness or spread of elevations within the cell, though it also stores information about the confidence of the mean elevation estimate. Equation 12.29 gives the second central moment of the measured elevations, effectively condensing the cell's elevation distribution into a Gaussian distribution with first and second moments matching those of the mixture model. The estimates of cell mean and variance can be used to make statistical statements about the elevations in the cell, taking into account all the noise present in the sensing path.

12.5.3 Terrain Estimator Performance

Cornell's terrain estimation algorithm has several unique advantages over traditional mapping strategies. First, the variance estimate within each cell gives information about the spread of elevations in that cell, allowing confidence intervals to be created over elevations in that cell. These confidence intervals hold more information than the standard binary obstacle representation of Martin and Moravec (1996), for example, without requiring the additional post-processing and interpolation of elevation-based terrain representations presented, for example, in Lohmann, Koch, and Schaeffer (2000) and Arakawa and Krotkov (1992).

A second advantage of this terrain model is that it can be generated and maintained in real-time. Recall from Section 12.5.2.2 that each measurement estimate is assigned to each cell according to the probability that the measurement lies in that cell. For finite numerical precision, however, only cells near the nominal measurement location derived from $\underline{\hat{r}}_i^{ENU}$ will be affected by that measurement estimate. As a result, each measurement estimate need only be applied to cells in a small neighborhood of the nominal measurement. This places limits on the number of cells to which each measurement can be applied, so the computational complexity is reduced to $O(k \cdot N)$ for N LIDAR measurements, each applied to a maximum of k terrain cells. Furthermore, if only the first two moments of the elevation distribution are retained, then each measurement can easily be fused with previous measurements. In fact, only the following four quantities are required for each cell:

$$\sum_{i=1}^{M} p_i, \qquad \sum_{i=1}^{M} p_i \hat{U}_i, \qquad \sum_{i=1}^{M} p_i \hat{U}_i^2, \qquad \sum_{i=1}^{M} p_i \sigma_{\hat{U},i}^2$$

where each of these quantities is itself a scalar. Also, because fusing a new measurement with the measurement history only requires knowledge of these four variables, the computational complexity and memory requirements of maintaining each cell are $O(1)$. That is, once sensor measurements have been used to update the terrain model, the original measurements can be discarded. The entire terrain map can therefore be maintained without storing a measurement history.

Fig. 12.9. Sample $\hat{U}_{GM} + 2\sigma_{GM}$ elevation map (in meters) resulting from the Spider's terrain estimation scheme. Warmer colors represent higher elevations.

Figure 12.9 shows a sample terrain map generated from LIDAR measurements as the Spider approaches several objects. In this example, the $\hat{U}_{GM} + 2\sigma_{GM}$ elevations from equations 12.28 and 12.29 (in meters) are plotted relative to an arbitrary ENU origin selected near the test site. These elevations are plotted across a color spectrum, and the axes in Figure 12.9 are to scale. The $\hat{U}_{GM}$ elevations for the cells near the 0.8 m tall trash cans are approximately 0.45 m higher than surrounding cells, and the $\hat{U}_{GM}$ elevations for the cells near the 0.46 m traffic cones are approximately 0.25 meters. These low elevation estimates reflect the fact that the cells containing these objects also contain some exposed portions of the ground plane. The uncertainties, however, are appropriately large

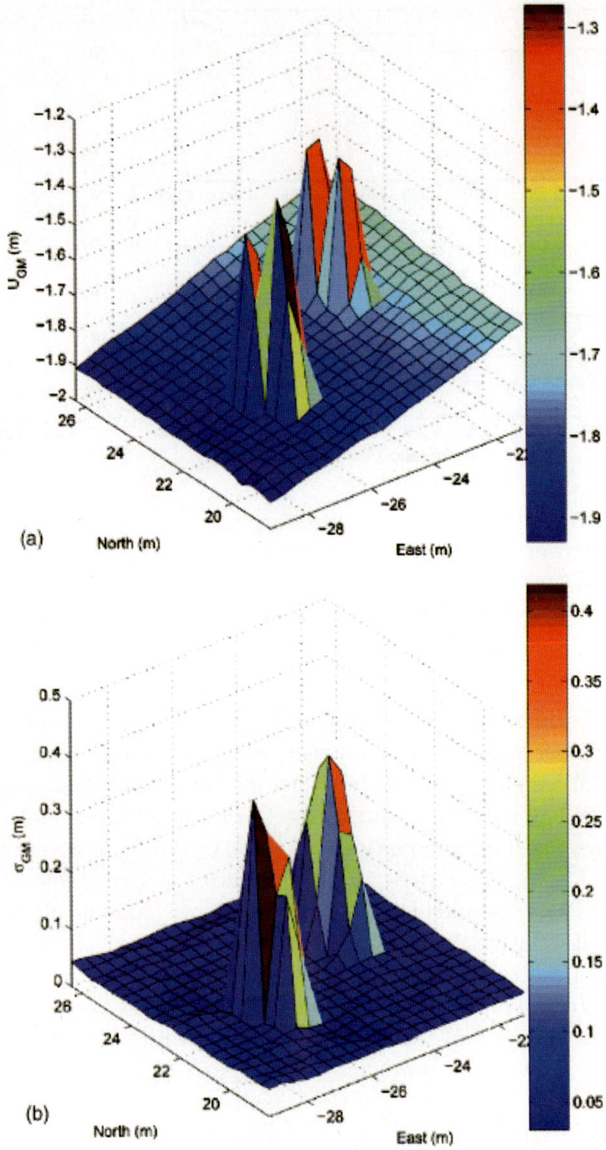

Fig. 12.10. (Left) Final $\hat{U}_{GM}$ map near the two 0.8 m tall trash cans. (Right) Final σ_{GM} map near the two 0.8 m tall trash cans.

for these cells: $\sigma_{GM} \approx 0.35$ m for the cells near the trash cans and $\sigma_{GM} \approx 0.17$ m for the cells near the traffic cones. Figure 12.10 shows the estimated elevation $\hat{U}_{GM}$ and estimation uncertainty σ_{GM} near the trash cans in greater detail.

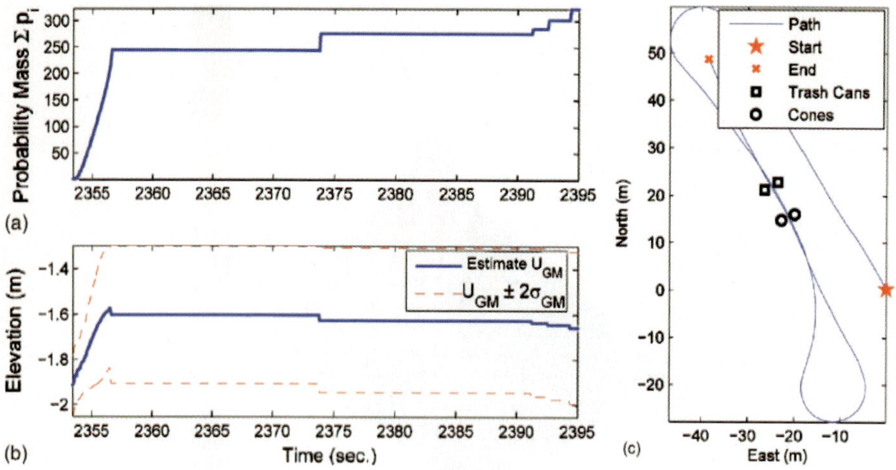

Fig. 12.11. (Left) Total association probability and $\hat{U}_{GM} \pm 2\sigma_{GM}$ bounds over time for a particular terrain cell containing a portion of a trash can. (Right) Vehicle ground track during real-time terrain experiment. The vehicle intersects the line connecting the trash cans at $t \approx 2358$, 2374, and 2395 seconds.

Figure 12.11 (left), in contrast, shows the evolution of the terrain estimate for one cell near one of the trash cans. Figure 12.11 (right) shows that during this test, the Spider passes near the trash cans three times: once with the trash cans in the periphery of the sensors' footprints at $t \approx 2358$ seconds, and twice directly between the two trash cans at $t \approx 2374$ and 2395 seconds. Figure 12.11 (left, top) shows the total association probability sum $\sum p_i$ accumulated for the cell over time. Figure 12.11 (left, bottom) shows the terrain elevation estimate $\hat{U}_{GM}$ and the $\pm 2\sigma_{GM}$ bounds for the same cell. Note that the elevation estimate for the terrain is relative to an arbitrary ENU origin, not the ground plane, as the location of the ground plane is unknown at the outset of the experiment. Figure 12.12 shows vehicle speed and heading during the test.

The elevation estimate in the particular cell shown in Figure 12.11 (left, bottom) fluctuates over time as more measurements are applied to it: some measurements, from the near-vertical face of the trash can, tend to increase the elevation estimate upward. Others, from the surrounding flat ground, tend to decrease the elevation estimate. These different measurements also affect the cell's variance: it is much higher than in surrounding cells, indicating rough terrain or obstacles, and it increases throughout the experiment. Finally, notice that the elevation estimate changes most on the Spider's first pass by the trash can, and that subsequent passes do not cause a substantial increase in total association probability of measurements assigned to that cell. The algorithm is therefore capable of producing accurate estimates in real-time, at reasonable speeds, and without revisiting old terrain.

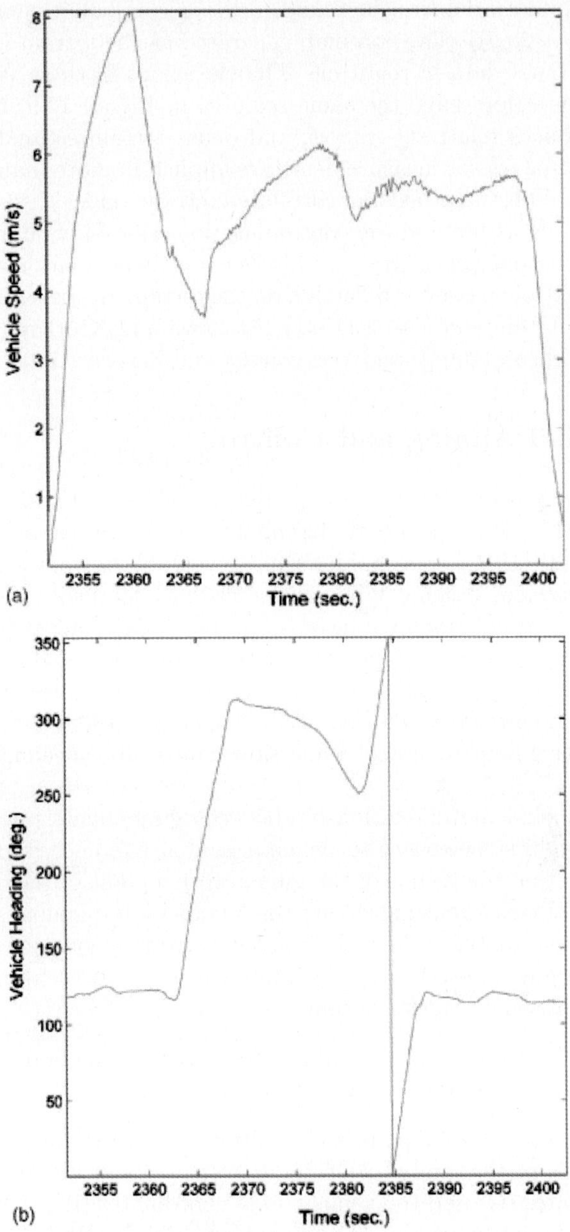

Fig. 12.12. (Left) Vehicle speed during real-time terrain experiment. (Right) Vehicle heading during real-time terrain experiment.

Additional information can be found in the accompanying video files, which show the $\hat{U}_{GM} + 2\sigma_{GM}$ elevation map generated as the terrain estimation algorithm fuses LIDAR data in real-time. The elevations in these video files are in meters, and the colors have the same scale as in Figure 12.9. Notice that the algorithm produces relatively smooth and dense terrain estimates; this occurs because individual sensor measurements are applied in more than one cell as per equation 12.27. This smoothed and correlated terrain model arises naturally from the estimation algorithm and the continuous posterior distribution of the errors in the sensors. It contrasts sharply with other terrain estimation algorithms that make use of multiple passes or recursion on the terrain estimate to smooth it out, such as those of Olin and Tseng (1991), Arakawa and Krotkov (1992), Hähnel, Burgard, and Thrun (2004), and Weingarten and Siegwart (2005).

12.6 Gimbal Aiming and Control

As discussed in section 12.5.1 and depicted in Figure 12.8, one of the Spider's terrain sensors is a two-axis actuated gimbal platform supporting a LIDAR unit. This gimbaled LIDAR is the Spider's most valuable terrain sensor, as it can scan in any direction. When combined in a feedback loop with the Spider's path planner, the gimbaled LIDAR is used to gather data along potential paths as they are generated. In order to point the LIDAR in a desired direction, however, a kinematic relationship must be derived between the actuated gimbal axes and the LIDAR's Earth-referenced orientation. This kinematic relationship can then be used for active gimbal control while simultaneously canceling motion of the vehicle.

To derive this kinematic relationship between gimbal yaw, pitch, and LIDAR orientation, it is first necessary to define a gimbal reference frame. This frame is defined by letting the X-axis of the gimbal frame point in the direction of the LIDAR's central measurement, letting the Y-axis be the gimbal pitch axis, and letting the Z-axis complete the right-handed coordinate system. In general, the vehicle body frame axes (B) may be rotated to the gimbal frame axes (G) by applying the following transformations:

1. Rotate by ψ_g, the gimbal yaw, about the Z-axis of the body frame.
2. Rotate by ϕ_g, the gimbal pitch, about the new Y-axis.

Next, define the LIDAR pointing direction $\underline{e}_l^G$ as the unit vector pointing along the X-axis of the gimbal coordinate system. Using the definitions of the gimbal coordinates (G) and the vehicle body coordinates (B), this unit vector $\underline{e}_l$ has the following representation in vehicle body coordinates:

$$\underline{e}_l^B = \begin{pmatrix} cos(\psi_g)cos(\phi_g) \\ sin(\psi_g)cos(\phi_g) \\ -sin(\phi_g) \end{pmatrix} \tag{12.30}$$

This unit vector $\underline{e}_l^B$ can then be compared to a unit vector $\underline{e}_d$ that represents the desired LIDAR pointing direction. The vector $\underline{e}_d$ may in general be constructed in ENU coordinates as a unit vector that points from the vehicle to some target of interest. When this vector is then expressed in vehicle body coordinates (B), define it to have the components $\underline{e}_d^B = (\, e_x^B \quad e_y^B \quad e_z^B \,)^T$. In order to point the gimbaled LIDAR in the desired direction $\underline{e}_d$, choose ψ_g and ϕ_g such that $\underline{e}_l^B = \underline{e}_d^B$:

$$\begin{pmatrix} cos(\psi_g)cos(\phi_g) \\ sin(\psi_g)cos(\phi_g) \\ -sin(\phi_g) \end{pmatrix} = \begin{pmatrix} e_x^B \\ e_y^B \\ e_z^B \end{pmatrix} \tag{12.31}$$

With the constraint $-\frac{\pi}{2} < \phi_g < \frac{\pi}{2}$, equation 12.31 has a unique solution:

$$\begin{aligned} \psi_g &= tan^{-1}\left(\frac{e_y^B}{e_x^B}\right) \\ \phi_g &= sin^{-1}(-e_z^B) \end{aligned} \tag{12.32}$$

Setting ψ_g and ϕ_g according to equation 12.32 aims the center of the gimbaled LIDAR in the desired direction $\underline{e}_d$. Because this solution accounts for the Spider's position and orientation, it also cancels the effects of vehicle motion.

In addition to canceling the motion of the Spider, this approach is also used by the Spider's path planner to gather terrain data about the path the Spider plans to traverse. The algorithm first walks forward from the Spider along the current planned path to some desired range R_d to select a point $(\, E_C \quad N_C \quad U_C \,)^T$ of interest. Because the algorithm may not know the elevation U_C at $(\, E_C \quad N_C \,)^T$, the Spider's current elevation is used. The desired pointing direction $\underline{e}_d$ is then computed by normalizing the vector from the LIDAR to this point of interest, and the gimbal control angles are then calculated using equation 12.32. A feedback loop is applied to the point of interest in subsequent iterations to ensure that the central measurements returned by the gimbaled LIDAR are near R_d:

$$U_C(k+1) = U_C(k) + K_1(R_d - R) - K_2 N \tag{12.33}$$

where $U_C(k+1)$ is the elevation of the point of interest at the next iteration, $U_C(k)$ is the elevation at the current iteration, R is the range measured at the center of the gimbaled LIDAR, N is the number of 'infinite' ranges reported by the LIDAR, and K_1 and K_2 are control gains. The K_1 term ensures data is gathered at the desired range R_d by adjusting the commanded elevation until it matches the terrain elevation at the desired range. The K_2 term is added to prevent the gimbaled LIDAR from aiming above the horizon, where it returns no terrain measurements. The form of the control law in equation 12.33 allows the gimbaled LIDAR to track the top of the terrain at the desired range along the currently planned path.

An example of the feedback loop's tracking capabilities is given in Figure 12.13, which shows the average of the central 20 ranges R measured by the gimbaled LIDAR plotted against a sinusoidal commanded range R_d as the Spider drives on a paved surface. The system tracks the commanded ranges accurately up to 25 meters, the approximate range limit of the LIDAR on pavement. Beyond this threshold the LIDAR typically returns ranges of infinity when measuring pavement, so the K_2 term of equation 12.33 causes the gimbal to pitch down. The LIDAR also occasionally receives one or two measurements from an object at a great distance; these objects cause the LIDAR to pitch down rapidly due to the K_1 term. The total effect of the K_1 and K_2 terms cause the gimbaled LIDAR to spend most of its time measuring the dense, informative terrain in front of the vehicle. In contrast, little time is spent aiming above the horizon or gathering sparse data far from the vehicle.

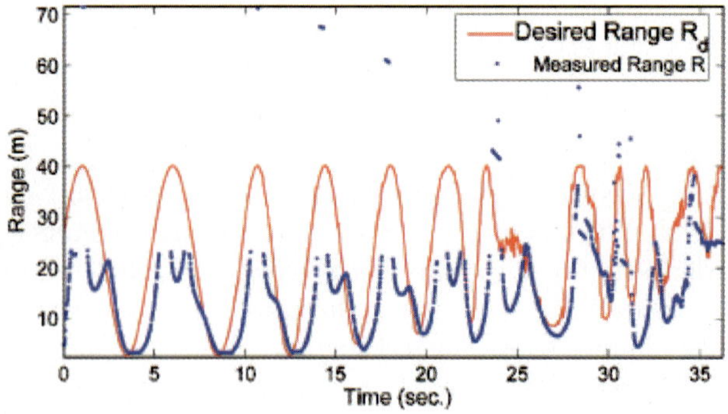

Fig. 12.13. Measured and commanded range of the center 20 measurements of the gimbaled LIDAR

Figure 12.13 also shows that the tracking loop has a significant time lag between the desired range and the measured range. This lag is not due to actuator dynamics: a first order lag is placed on the controller outputs to create smooth gimbal motion. Although the gimbal actuators are capable of tracking ranges at a higher bandwidth, it was found that higher frequency tracking induced vibrations in the sensor platform that produced invalid measurements from the other LIDAR units.

12.7 Robust Grid-Based Path Planning

One of the most difficult tasks for any autonomous ground vehicle is integrating vehicle capabilities, actuator states, position and orientation information, and sensed terrain maps to plan a path to traverse. The time-limited format of the Grand Challenge makes the task especially difficult, because critical path decisions must be made on-the-fly in unknown obstacle fields at average speeds of 7 m/s or more. To function under these constraints, a good path planner must have an efficient data representation, and it must be able to strike a balance between path quality and reaction time.

These two design attributes, data representation and search preference, generally categorize the behavior of path planners. The first attribute, data representation, describes how a path planner ranks potential paths. Existing path planners generally fall between two extremes: cost-based representations, and constraint-based representations. Cost-based representations, such as the one used in Stentz (1994), integrate all environment information into a scalar cost value for each potential path. Constraint-based representations, such as the one used in Spenko, Iagnemma, and Dubowsky (2004), store environment information as a set of constraints that must be satisfied for a path to be selectable. The other attribute, search preference, describes the types of paths a planner investigates. Global planners, such as the approach presented in Stentz (1994), choose paths over the entire course to be traversed. Local or reactive planners, in contrast, may select circular arcs or cubic splines that are only a few meters in length (Rosenblatt, 1997), (Urmson et al., 2004).

Cornell's Grand Challenge entry lies in between the extremes for each of these two attributes. Terrain information is represented as a scalar cost, while vehicle capabilities, position, and orientation are combined in a physics-based skidding constraint. Paths considered are cubic splines approximately 15 to 80 meters in length; they direct the Spider from its current location to a point that smoothly connects with the nominal route, beyond the Spider's sensor horizon. In practice these design choices allow the Spider's path planner to react quickly to obstacles, yet the paths selected are smooth and human-like.

12.7.1 Cost Representation for Path Planning

The terrain estimation algorithm from section 12.5 provides two pieces of environment information for path planning: elevation estimates μ_{GM}, and uncertainties σ_{GM} for each 40 cm $\times$40 cm grid cell of terrain. In the Cornell planner, this terrain data is used to assign a measure of traversability, in the form of a scalar cost, for each cell. The grid-based storage scheme and dense elevation and uncertainty estimates of the terrain estimation algorithm are ideal for evaluating the traversability of pieces of terrain according to the Spider's physical capabilities. In particular, the elevation estimates can be used to calculate finite difference gradients $\nabla\mu_{GM,i}$ for the ith cell:

$$|\nabla\mu_{GM,i}| \approx \frac{max_{j\in N(i)}\left(|\mu_{GM,i} - \mu_{GM,j}|\right)}{dx} \tag{12.34}$$

where $N(i)$ is the set of cells neighboring cell i, and dx is the width of the cell. These cell gradients are then converted to a gradient cost C_∇ for the ith cell based on the Spider's physical limits:

$$C_{\nabla,i} = 32000 \cdot \left(\frac{|\nabla\mu_{GM,i}|}{\nabla U_{max}}\right)^2 \tag{12.35}$$

where $\nabla U_{max} = 1.095$ is the Spider's maximum safe traversable gradient, calculated as an elevation change of 0.43 m across a single 0.4 m cell. The gradient cost in each cell is also multiplied by a scaling factor 32000 and converted to an integer to avoid additional floating point operations during path evaluations.

The uncertainty in each cell's elevation is also used to calculate a component of cost based on the Spider's physical limits:

$$C_{\sigma,i} = 2000 \cdot \left(\frac{\sigma_{GM,i}^2}{\sigma_{max}^2} \right)^2 \tag{12.36}$$

where $\sigma_{max} = 0.2$ m is chosen so the $2\sigma_{GM}$ elevation uncertainty in each cell is penalized if it exceeds the Spider's maximum traversable obstacle height of 40 cm. As with the gradient cost, the uncertainty cost is also multiplied by a scaling factor 2000 and converted to an integer.

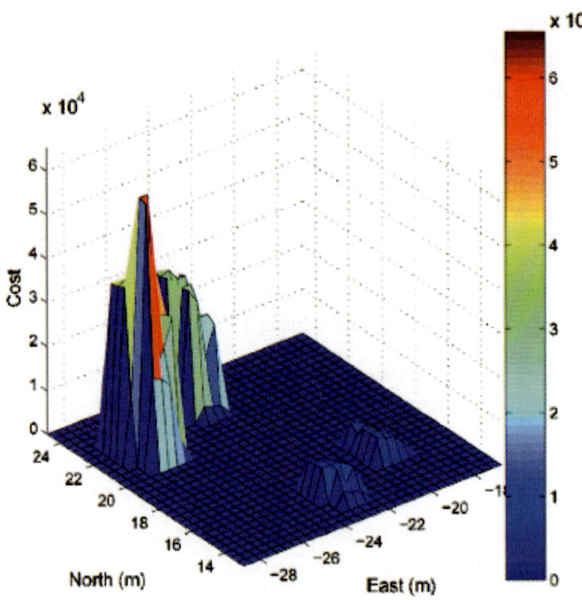

Fig. 12.14. Sample cost map generated from the parking lot elevation estimates from Figure 12.9 near the large trash cans and smaller traffic cones. Warmer colors indicate higher cost.

Figure 12.14 shows a cost map calculated from the sum of the gradient and uncertainty costs of the elevation estimates generated in Figure 12.9. This sum is used as a measure of the traversability of each cell, with less traversable or more uncertain cells having amplified costs. In addition, the quadratic form of equations 12.35 and 12.36 help attune gradients and uncertainty due to sensor noise. In the Cornell planner, these costs are used both to eliminate intraversable paths and to establish preferences if multiple paths are traversable.

12.7.2 Searching Preference in Path Planning: the Cubic Planner

While grid-based terrain estimates fit naturally into a cost framework, the remainder of the Spider's environment, including hardware constraints and vehicle position, velocity, and orientation, are not well-represented as costs. It is possible, for example, to penalize paths that violate vehicle constraints, turn sharply, or decelerate rapidly. These cost penalties do not guarantee compliance, however, so paths chosen to minimize this cost may not be physically realizable. Furthermore, with the cost function changing as fast as the vehicle's position, velocity, and orientation, the lowest-cost path may change drastically from one planning iteration to the next. In high-speed, high-dynamics tasks like the Grand Challenge, however, it is more desirable to find near-optimal local solutions that are consistent with vehicle dynamics. These points motivate Cornell's planner, which generates path segments from a small set of cubic splines.

Cornell's cubic planner uses several simplifying assumptions to generate a rich variety of paths using only three free parameters. First, it is noted that the Grand Challenge course is often narrow with respect to vehicle size, allowing little lateral motion. Second, the course must be drivable by commercial vehicles, so there is guaranteed to be a path from the start of the course to its end (DARPA, 2004). These two assumptions impose basic constraints on the radius of curvature and space required by candidate paths. These constraints can be checked in real-time with approximations to vehicle physics, allowing many candidate paths to be eliminated without lengthy numerical simulation. As a result, this verification-based technique can search among complicated paths that are more likely to be traversable over long distances. Figure 12.15 shows some of these candidate paths considered by the Cornell cubic planner.

Fig. 12.15. Candidate cubic paths generated for various turns. The paths are constructed from pairs of control points at the beginning and end of the path. Families of cubics are generated by varying the spacing of the two center control points.

12.7.2.1 Spline Construction

The Cornell path planner uses two-dimensional cubic Bézier curves to represent its candidate paths. Two-dimensional cubic Bézier curves are defined by four 'control points' $\{A, B, C, D\}$; see, for example, Weisstein (1999). Useful properties of the cubic Bézier for path planning are as follows:

1. The Bézier always passes through its first control point A. This point is set to the center of the Spider's back axle to represent the current planning origin.

2. The Bézier always ends at its last control point D. Point D therefore represents the end of the cubic spline, and it should be inside the course boundary at all times.
3. The Bézier is always tangent to the line segment $\overline{AB}$ at point A. Line segment $\overline{AB}$ is therefore set in the current direction of motion, according to the Spider's yaw angle ψ. Control point B is therefore fully-defined by a scalar distance: the forward distance from A in the direction of travel.
4. The Bézier is also always tangent to the line segment $\overline{CD}$ at point D. The Cornell planner chooses the line segment $\overline{CD}$ such that it is parallel to the DARPA-specified path at the end of the cubic Bézier. As a result, control point C, like B, is entirely specified by a single parameter: its distance from D toward the vehicle.

With these constraints, the cubic search space is fully defined by five pieces of information: the vehicle's current location, its current yaw angle ψ, the location of the terminal point D, the offset of control point B from A in the direction of vehicle travel, and the offset of the control point C from point D backward toward the vehicle. Note that of these five parameters, only the two scalar offset lengths and the path termination point are free parameters.

In Cornell's implementation, the location of the terminal point D is determined by selecting a look-ahead distance along the nominal DARPA route based on desired vehicle velocity. The look-ahead distance is set at a minimum distance of 15 m so the vehicle does not consider a turn it cannot achieve. The terminal point D is then generated by projecting the vehicle's location onto the nominal DARPA route and walking forward by the look-ahead distance. In this implementation, four additional terminal points are also created from equidistant lateral projections off the original terminal point. An example of paths generated to each of these five lateral offsets is shown in Figure 12.16.

In addition, more complicated paths can be created by varying the locations of points B and C by changing the 'offset lengths': the lengths of segments $\overline{AB}$ and $\overline{CD}$. Intuitively, changing the offset length of the control point B determines when and how severely the vehicle reacts as it approaches a turn, as shown in

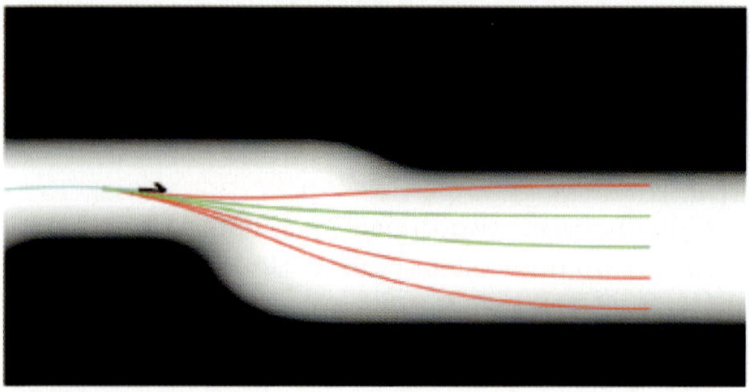

Fig. 12.16. The five lateral offsets considered for a single terminal point

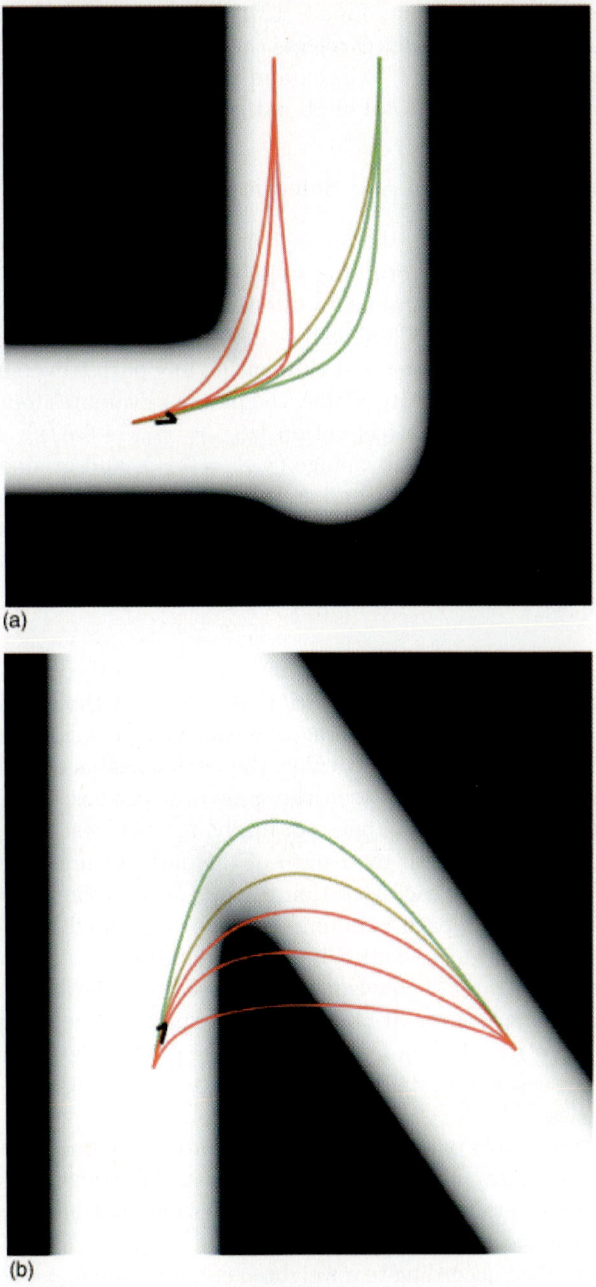

Fig. 12.17. (Left): Variations in the B offset affect the lag in the start of a turn. (Right): Variations in the C offset affect the duration of a turn.

Figure 12.17 (left). In contrast, Figure 12.17 (right) shows that changing the offset length of the control point C affects the duration of a turn. In the Cornell planner, three B offsets, four C offsets, and five terminal points are considered at each planning cycle, for a total of 60 paths.

12.7.2.2 Path Evaluation and Selection

In order to select a final path, the planner determines the best candidate spline in a thresholding step and an ordering step. The first step, thresholding, applies cost and physics-based constraints to eliminate infeasible paths. The planner applies cost constraints by walking each candidate spline in the cost grid, computing a line integral C_{LI} of cost and eliminating any path traversing impassable obstacles. For the remaining candidates, the planner evaluates four potential vehicle speeds v_{des}: emergency deceleration ($v_{des} = v_{cur} - 6m/s$), gradual deceleration ($v_{des} = v_{cur} - 3m/s$), no change ($v_{des} = v_{cur}$), and gradual acceleration ($v_{des} = v_{cur} + 3m/s$). The paths and speeds are then evaluated using a lateral skidding test, which imposes a minimum curvature constraint based on speed (Spenko et al., 2004):

$$min(R_c) > \frac{v_{des}^2}{\mu g} \tag{12.37}$$

where $min(R_c)$ is the smallest radius of curvature along the path, μ is the coefficient of friction, and g is the acceleration of gravity. Although equation 12.37 approximates lateral skidding assuming constant vehicle speed and turning angle, it is inexpensive to test. In particular, the path's radius of curvature can be evaluated in each grid cell it crosses at the same time the cost line integral is calculated, so the computational expense of applying both cost and physics-based constraints grows linearly with the length of the path. In practice, the 60 paths and 4 speeds evaluated in each iteration of the Cornell planner are completed in less than 40 ms on the path planning computer equipped with dual 1.8 GHz AMD Opteron 244 processors and Windows Server 2003.

The skidding constraint of equation 12.37 is used to remove paths that are intraversable due to sharp turns. It also affects the Spider's macroscopic preference toward smoother, straighter, and more human-like paths. The Cornell implementation uses a conservative $\mu = 0.15$, for example, to discourage last-minute turns to avoid obstacles.

After all intraversable paths have been eliminated, the final selection is made by optimizing a performance metric over the remaining traversable paths. First, the Cornell planner retains only the fastest traversable speed for each cubic path, as finishing time is most important when all paths are traversable. The remaining paths are then ranked according to a weighted combination $C_{LI} + 600\Phi_{LI}$ of total cost C_{LI} and total steering effort Φ_{LI}, calculated during the cost line integral as the 1-norm of changes in steering wheel angle from cell to cell along the path. The final path chosen minimizes this weighted combination, representing a balance between avoiding small changes in cost and preferring paths with less steering effort. If no paths are deemed traversable, the Cornell planner chooses

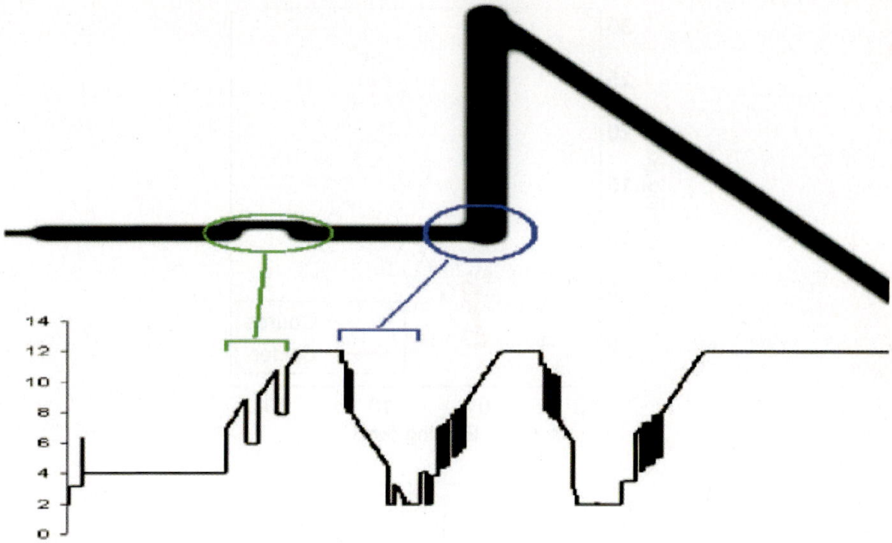

Fig. 12.18. Speed selection on a simulated course. The bottom plot shows desired vehicle speed selected by the path planner, in m/s. Note how the Spider decelerates slowly for the soft turn in the beginning and faster for the two sharp turns later on.

the path with the least worst cost while performing a sharp deceleration. This contingency behavior permits the Spider to gather more terrain data, so a better path can be found in the next planning cycle.

Examples of the path planner's final speed selections are shown in Figure 12.18 for a simulated course. Notice that the planner slows the vehicle down while taking turns, picking up speed as the vehicle exits the turn. Although speeds are tested in discrete 3 m/s increments, Figure 12.18 shows that they form a feedback loop with the Spider. Sudden speed changes are effectively smoothed out by the 10 Hz replan rate of the path planner and the lag in the Spider's engine.

12.8 Grand Challenge Performance and Failure Analysis

The Spider began the 131.8-mile Grand Challenge course along with the other competitors in Primm, Nevada, in the morning of October 8, 2005. Team Cornell faced several mechanical difficulties just in bringing the Spider to the starting line, including a failing torque converter and a generator that broke its bearings the day before. The Spider made a successful start despite the mechanical difficulties, traveling 9.0 miles before hitting a concrete wall on an overpass. The vehicle was subsequently stopped by DARPA officials. Figure 12.19 (left) shows the total Grand Challenge course, with the Spider's path marked.

Figure 12.19 (right) shows a magnified view of the Grand Challenge course and the Spider's path as it drifted from the center of DARPA's route into the

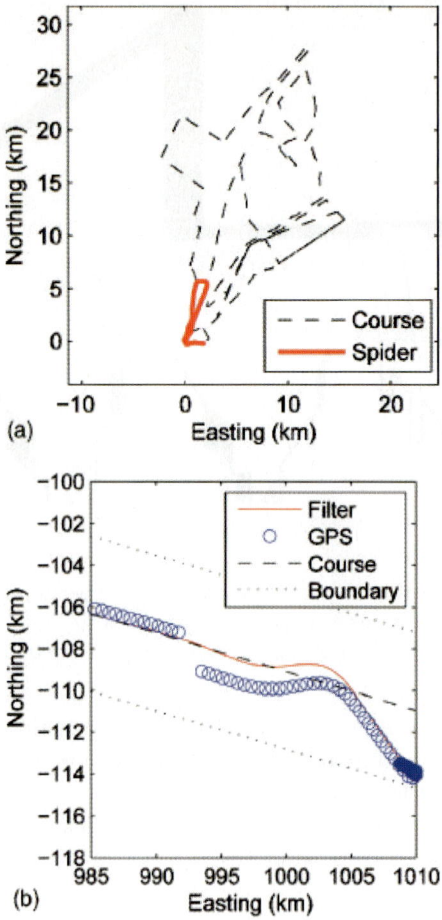

Fig. 12.19. (Left) Grand Challenge course, with the Spider's final progress marked. (Right) Magnified view of the Spider's point of failure.

concrete wall. This reveals two odd features of the Spider's final moments in the race. First, the solution reported by the GPS receiver experienced a jump of approximately 2m, and thereafter it reported an apparent error of more than 1m until the Spider hit the wall. This offset is evident from the fact that the Spider's filtered position closely followed the center of DARPA's route, which was confirmed by the GPS solution prior to the anomaly. A further inspection of race data reveals that the GPS jump was due to reacquisition of the OmniSTAR HP signal, which the Spider had lost approximately 175 seconds earlier. Rate limiting was in effect, however, so the Spider did not swerve to follow the incorrect HP signal. In fact, the Spider turned gently as the rate-limited filter converged to the HP signal.

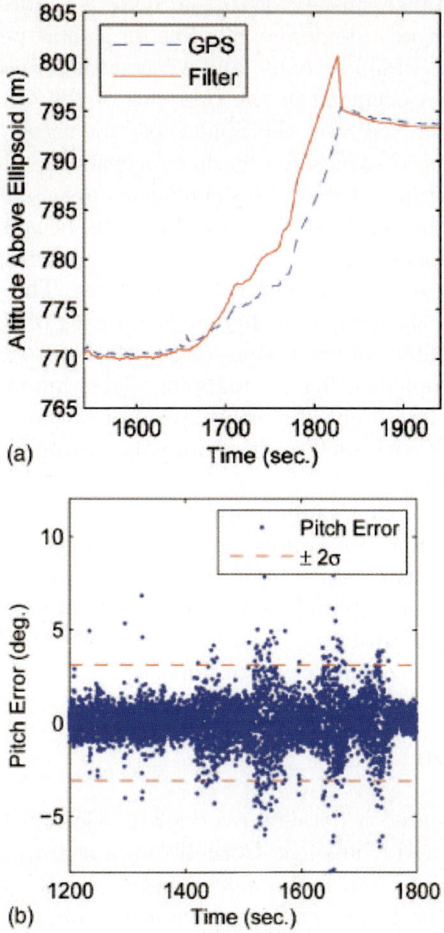

Fig. 12.20. (Left) Several minutes before it failed in the Grand Challenge, the Spider developed significant errors in its altitude estimate. (Right) The altitude errors are believed to have been caused by four instances of significant pitch error at $t = 1440$, 1540, 1655, and 1735 seconds into the race, just prior to the Spider's failure.

The second anomaly at the Spider's point of failure is the fact that it chose to turn into the wall. The only explanation for such behavior is that the Spider detected an obstacle in front of it, though post-race investigation revealed no obstacles on the overpass. This behavior suggests obstacle detection errors, and Figure 12.20 reinforces that conclusion. Figure 12.20 (left) shows that the Spider developed nearly 8m of error in its altitude estimate over the last few minutes leading up to its decision to hit the wall. Figure 12.20 (right) shows the likely cause with a comparison of the attitude estimator's pitch estimate to pitch derived from GPS velocity. The difference between these two signals shows four neighboring instances just prior to the Spider's failure during which pitch error consistently strayed outside the $\pm 2\sigma$ line for a short period of time: at 1440 seconds, 1540 seconds, 1655 seconds, and 1735 seconds into the race. The first two of these instances occurred in the presence of the OmniSTAR HP signal, while the last two occurred after the Spider lost the signal. The fact that these four instances are spaced so closely together suggests the pitch estimate became incorrect during this time interval. This conclusion also explains the Spider's decision to stray from the center of the DARPA route, because pitch errors would cause the Spider to localize LIDAR measurements incorrectly: even flat terrain in the center of the route would appear as obstacles. This problem could have been mitigated either through online hypothesis testing of GPS measurements as in Sukkarieh et al. (1999), online estimator health monitoring as in Bar-Shalom et al. (2001), or by implementing the mapping algorithm of section 12.5 relative to the vehicle and independent of absolute position. Unfortunately, the problem did not manifest itself with such severity until the Grand Challenge itself, where it could not be repaired.

The location of the Spider's failure was unfortunate. Reverse driving was not implemented due to time constraints, so the Spider could not recover from its collision. Instead, it was disabled as it struggled to turn free from the wall. If it had chosen to turn in a place where there were no concrete walls, it would have recovered without problem. Even if it strayed from the course boundary, the DARPA officials may have let it continue.

12.9 Conclusion

This paper has presented a detailed overview of Cornell University's entry in the 2005 DARPA Grand Challenge. Cornell's approach to the Grand Challenge divided the problem into four main tasks: designing and building a robust platform, localizing the platform, sensing its environment, and navigating through that environment. This approach had the benefit that each subsystem could be developed independently once the information flow between them was defined.

Cornell's entry in the Grand Challenge was based upon a Spider Light Strike Vehicle, a robust military platform able to tolerate mistakes made by its autonomous components. The platform was localized with dual square root information filters for attitude and position / velocity that fused separate inertial and navigation sensors. This filtering scheme was shown to be as accurate as many of the more expensive off-the-shelf positioning systems. A rate limiting scheme was also presented to handle GPS signal loss and reacquisition, two difficult situations for path tracking and obstacle detection in autonomous ground vehicles. A novel terrain estimation algorithm was also presented to combine the Spider's three laser rangefinders into accurate terrain estimates and their uncertainties. This terrain estimate provided a real-time statistical interpretation of the Spider's environment, permitting robust path planning without heuristic obstacle identification. A method for creating a feedback loop between these terrain estimates and the path planner was presented, and the feedback loop was shown to be able to gather terrain data at the desired range along the planned path. A cubic spline-based path planner was also presented to utilize these dense terrain estimates to generate paths consistent with the physical capabilities of the Spider. It was shown that simple path constraints based on skidding, radius of curvature, and terrain gradient macroscopically affected the final path chosen, making the resulting path smoother and more human-like.

Cornell's Spider was one of the 195 original entrants into the Grand Challenge, and one of only 23 to make it to the starting line of the final Grand Challenge event. Despite experimental validation of each subsystem and the fully-integrated vehicle, the Spider's performance in the Grand Challenge was disappointing. Last-minute generator and torque converter failures tested the reliability of the system, while faulty navigational estimates in a highly-constrained portion of the course ultimately brought about the Spider's failure in the Grand Challenge. These failures stress the importance of additional online health monitoring and contingency behaviors that would have been added to the Spider if there were more time.

Acknowledgments

The authors acknowledge the hard work and dedication of all members of Team Cornell from the summer of 2004 through the spring of 2006. The authors also thank the sponsors of Team Cornell, especially Singapore Technologies Kinetics, for their support of this research.

References

Arakawa, K., & Krotkov, E. (1992). *Fractal surface reconstruction with uncertainty estimation: Modeling natural terrain* (Tech. Rep. No. CMU-CS-92-194). Pittsburgh: School of Computer Science, Carnegie Mellon University.

Bar-Shalom, Y., Rong Li, X., & Kirubarajan, T. (2001). *Estimation with applications to tracking and navigation: Theory, algorithms and software.* New York: John Wiley & Sons, Inc.

Battin, R. (1999). *An introduction to the mathematics and methods of astrodynamics, revised edition.* Reston: American Institute of Aeronautics and Astronautics.

Bierman, G. (1977). *Factorization methods for discrete sequential estimation.* New York: Academic Press.

Bishop, C. (1995). *Neural networks for pattern recognition.* Oxford: Oxford University Press.

DARPA. (2004). *Darpa grand challenge 2005 rules.* Defense Advanced Research Projects Agency. (http://www.darpa.mil/grandchallenge05/Rules_8oct04.pdf)

Franklin, G., Powell, D., & Emami-Naeini, A. (2002). *Feedback control of dynamic systems* (4th ed.). Upper Saddle River: Prentice-Hall.

Hähnel, D., Burgard, W., & Thrun, S. (2004). Learning compact 3d models of indoor and outdoor environments with a mobile robot. *Elsevier Science Special Issue Eurobot '01*, 1 - 16. (http://www.informatik.uni-freiburg.de/ burgard/postscripts/ haehnel.ras-03.pdf)

Huising, E., & Pereira, L. (1998). Errors and accuracy estimates of laser data acquired by various laser scanning systems for topographic applications. *ISPRS Journal of Photogrammetry and Remote Sensing, 53*, 245–261.

Lohmann, P., Koch, A., & Schaeffer, M. (2000). Approaches to the filtering of laser scanner data. *International Archives of Photogrammetry, Remote Sensing and Spatial Information Sciences, 33.* (Working Group III-2)

Martin, M., & Moravec, H. (1996, March). *Robot evidence grids* (Tech. Rep. No. CMU-RI-TR-96-06). Pittsburgh: The Robotics Institute, Carnegie Mellon University.

Moon, F. (1998). *Applied dynamics with appilcation to multibody and mechatronic systems.* New York: John Wiley and Sons.

Murray, R., Li, Z., & Sastry, S. (1994). *A mathematical introduction to robotic manipulation.* Boca Raton: CRC Press.

Northrop Grumman Navigation Systems Division. (2000). *Ln-200 fiber optic inertial measurement unit* (Brochure No. 22939/02-06/2000/Crawford). (http://www.nsd.es.northropgrumman.com/Html/LN-200/brochures/ln200.pdf)

Ohlmeyer, E., Pepitone, T., Miller, B., Malyevac, D., Bibel, J., & Evans, A. (1997, August). Gps-aided navigation system requirements for smart munitions and guided missiles. In *Aiaa guidance, navigation, and control conference collection of technical papers* (p. 954–968). Reston: American Institute of Aeronautics and Astronautics.

Olin, K., & Tseng, D. (1991, August). Autonomous cross-country navigation: An integrated perception and planning system. *IEEE Expert: Intelligent Systems and Their Applications, 6*(4), 16–30.

Rosenblatt, J. (1997). *Damn: A distributed architecture for mobile navigation.* Phd thesis, The Robotics Institute, Carnegie Mellon University.

Savage, P. (1998, January - February). Strapdown inertial navigation integration algorithm design part 1: Attitude algorithms. *Journal of Guidance, Control, and Dynamics, 21*(1), 19–28.

SICK. (2003, June). *Lms 200, lms 211, lms 220, lms 221, lms 291 laser measurement systems technical description* (Technical Description No. 8 008 970/06-2003).

Singapore Technologies Kinetics. (2006). *Spider lsv 1.04* [Brochure]. (http://www.stengg.com/upload/306MJIVTWhk0eRlE2kH.pdf)

Spenko, M., Iagnemma, K., & Dubowsky, S. (2004, September). High speed hazard avoidance for mobile robots in rough terrain. In G. Gerhart, C. Shoemaker, & D. Gage (Eds.), *Proceedings of the spie* (Vol. 5422, p. 439–450). Bellingham: The International Society for Optical Engineering.

Stentz, A. (1994, May). Optimal and efficient path planning for partially-known environments. In *Proceedings of the ieee international conference on robotics and automation* (p. 3310–3317). Los Alamitos: Institute of Electrical and Electronic Engineers.

Sukkarieh, S., Nebot, E., & Durrant-Whyte, H. (1999, June). A high integrity imu/gps navigation loop for autonomous land vehicle applications. *IEEE Transactions on Robotics and Automation, 15*(3), 572–578.

Team Cornell. (2005, December). *Technical review of team cornell's spider: Darpa grand challenge 2005* [Technical Review]. (http://www.darpa.mil/grandchallenge05/TechPapers/TeamCornell.pdf)

Thrun, S. (2002, February). *Robotic mapping: A survey* (Tech. Rep. No. CMU-CS-02-111). Pittsburgh: School of Computer Science, Carnegie Mellon University.

Triantafyllou, M., & Hover, F. (2004, Fall). *Maneuvering and control of marine vehicles*[Maneuvering and Control of Surface and Underwater Vehicles OpenCourseWare Website]. (http://ocw.mit.edu/NR/rdonlyres/Ocean-Engineering/13-49Fall-2004/C8072E70-A4F4-466A-98DF-A3EC2865A835/0/lec1.pdf)

Trimble. (2004, February). *Aggps 252 receiver user guide version 1.00, revision a* [Technical Manual].

Urmson, C., Anhalt, J., Clark, M., Galatali, T., Gonzales, J., Gowdy, J., et al. (2004, June). *High speed navigation of unrehearsed terrain: Red team technology for grand challenge 2004* (Tech. Rep. No. CMU-RI-TR-04-37). Pittsburgh: The Robotics Institute, Carnegie Mellon University.

Weingarten, J., & Siegwart, R. (2005). Ekf-based 3d slam for structured environment reconstruction. In *Proceedings of the ieee / rsj international workshop on intelligent robots and systems* (p. 3834–3839). Los Alamitos: Institute of Electrical and Electronic Engineers.

Weisstein, E. (1999). *Bézier curve.* [MathWorld - A Wolfram Web Resource]. (http://www.mathworld.wolfram.com/BezierCurve.html)

13

A Mixture-Model Based Algorithm for Real-Time Terrain Estimation

Isaac Miller* and Mark Campbell**

Sibley School of Mechanical and Aerospace Engineering
Cornell University, Ithaca NY, 14853

Summary. A real-time terrain mapping and estimation algorithm using Gaussian sum elevation densities to model terrain variations in a planar gridded elevation model is presented. A formal probabilistic analysis of each individual sensor measurement allows the modeling of multiple sources of error in a rigorous manner. Measurements are associated to multiple locations in the elevation model using a Gaussian sum conditional density to account for uncertainty in measured elevation as well as uncertainty in the in-plane location of the measurement. The approach is constructed such that terrain estimates and estimation error statistics can be constructed in real-time without maintaining a history of sensor measurements. The algorithm is validated experimentally on the 2005 Cornell University DARPA Grand Challenge ground vehicle, demonstrating accurate and computationally feasible elevation estimates on dense terrain models, as well as estimates of the errors in the terrain model.

13.1 Introduction

A problem of critical importance to modern robotics and remote sensing alike is mapping: the task of constructing a computational representation of an environment from available sensor data. These data-driven environment maps have a number of classical applications, including robot localization, exploration, target tracking, and planning (Thrun, 2002). Continued improvements in sensor accuracy and computational power have also spurred the use of automated mapping techniques for non-traditional applications, including aerial surveying, reconnaissance, and model generation (Ackermann, 1999), (NIMA, 2000), (Arakawa & Krotkov, 1992), (El-Hakim, Whiting, Gonzo, & Girardi, 2005).

The mapping approaches adopted for these applications are as diverse as the applications themselves. In the robotics field, common approaches include tracking features or beacons, building belief maps or evidence grids of obstacles, building $2\frac{1}{2}$D elevation grids, and building Monte-Carlo sampled environments (Thrun, 2002), (Martin & Moravec, 1996), (Pagac, Nebot, & Durrant-Whyte,

* Graduate Research Fellow, Sibley School of Mechanical and Aerospace Engineering, Cornell University

** Senior Member of AIAA, Associate Professor, Sibley School of Mechanical and Aerospace Engineering, Cornell University

M. Buehler, K. Iagnemma, and S. Singh (Eds.): DARPA'05, STAR 36, pp. 407–436, 2007.
springerlink.com

1998), (Olin & Tseng, 1991), (Lacroix et al., 2002), (Leal, 2003). These approaches are generally constructed in real-time for robotic navigation, and they make tradeoffs between the richness of their models and computational time. Beacon-based representations provide statistically rigorous map estimates, for example, but the maps are not dense. Occupancy grid approaches are dense, but they are fundamentally limited to binary obstacle identification. Monte-Carlo and elevation grid approaches generate dense maps, but they must either resample or perform heuristic interpolations to avoid computational bottlenecks. Initial research in representing robotic sensor data using true three dimensional terrain maps has recently been investigated (Thrun, Burgard, & Fox, 2000), (Hähnel, Burgard, & Thrun, 2004), (Weingarten & Siegwart, 2005). These approaches either maintain large histories of sensor measurements or attempt to identify planar structures within their sensor data. They are currently limited to structured environments or constrained sensor geometries.

Mapping techniques associated with the remote sensing field, in contrast, are used to build digital elevation models for accurate surveying and geographical studies. These techniques often consist of a data collection phase with significant post processing to generate maps offline (Axelsson, 1999), (Ackermann, 1999), (Lohmann, Koch, & Schaeffer, 2000), (Satale & Kulkarni, 2003). Uncertainty in these digital elevation models is typically characterized in terms of errors sampled from a known set of reference survey points; therefore, this type of model fundamentally cannot be generated or maintained in real-time. Instead, these techniques are most commonly used for generating extremely precise digital elevation models for land surveying purposes.

While a unified mapping approach across these diverse applications is almost never adopted, an increasing interest in autonomous vehicle navigation and real-time reconnaissance suggests the merits of such an approach to a real-time terrain model. This paper presents one such possibility, using Gaussian sum representations of terrain uncertainty to generate and maintain an accurate and statistically rigorous real-time elevation model. The approach is unique in several regards. First, error transformation techniques are used to treat sensor measurements with a statistical model rather than as point clouds or inputs to an interpolation scheme. This allows the terrain estimation algorithm to handle multiple sources of uncertainty during data collection rigorously. It also allows estimates of the errors in the terrain map to be generated on the fly rather than in post processing steps. In addition, the Gaussian sum representation allows the full terrain model to be built, stored, and maintained cheaply in real-time using a standard desktop computer. The specific case of generating an elevation model from a ground vehicle and laser rangefinders is presented, though the approach is general to all navigation problems such as airborne platforms and other sensors.

The work that follows derives the Gaussian sum algorithm for real-time terrain estimation along with experimental results obtained in a practical setup. Section 13.2 describes the terrain estimation problem and derives the Gaussian sum algorithm for terrain estimation, including steps for rigorous statistical analysis of terrain sensor measurements and assigning measurement locations within the terrain model. Section 13.3 gives a one-dimensional simulated example of the terrain estimation algorithm to describe its behavior. Section 13.4 presents experimental results of applying the algorithm on a full-sized ground vehicle operating at realistic speeds.

13.2 Terrain Estimation Algorithm

In representing terrain or digital elevation models, one common approach is to store terrain data as a 'Cartesian height map' or 'raster grid', where a region of the Earth's surface is parameterized by a location in the XY plane and an associated elevation relative to an origin of interest (Olin & Tseng, 1991), (Stoker, 2004). This same parameterization is used to store elevation maps in this study. That is, it is assumed there exists an origin of interest, described according to its latitude, longitude, and altitude (LLA) with respect to the WGS-84 ellipsoid model of the Earth (Kaplan, 1996). The reference plane for this study is then calculated as an East-North-Up (ENU) plane tangent to this ellipsoid model at the origin of interest, where the X-axis points East along a line of latitude, the Y-axis points North along a line of longitude, and the Z-axis completes the coordinate frame. This reference plane is divided into N_c grid cells, with the jth cell extending in the East direction from E_{j-} to E_{j+} and in the North direction from N_{j-} to N_{j+}. The goal is to develop an algorithm to estimate the elevations of each of these grid cells in real-time in the presence of multiple sources of noise. More specifically, the goal is to develop elevation estimates in real-time that are optimal in the sense of the minimum mean square error (MMSE).

The proposed terrain estimation algorithm has three separate steps to accomplish this goal. First, a statistical representation of each sensor measurement is formed, in order to account for multiple sources of sensing error in a probabilistically rigorous manner. Second, each sensor measurement is assigned or 'associated' to one or more grid cells to which it is likely to correspond. Finally, the measurements assigned to each grid cell are fused in real-time into an optimal elevation estimate for that grid cell. These three steps are discussed in turn below.

13.2.1 Statistical Treatment of Sensor Measurements

The first step of the terrain estimation algorithm is to form a statistical representation of each sensor measurement in order to account for all sources of sensor error. These sensor errors are commonly due to four general sources: 1) errors due to the sensor itself, 2) errors due to uncertainty in the sensor's orientation and location on the sensing platform, 3) errors due to uncertainty in the orientation of the sensing platform itself, and 4) errors due to uncertainty in the location of the sensing platform (Huising & Pereira, 1998). The first type of error, due to the sensor itself, describes the sensor's accuracy. This type of error is a function of the method by which the sensor makes measurements, and it is generally independent of the sensor's orientation. The second type of error, sensor orientation and location error, arises because the terrain map is often not built in a sensor-centric coordinate frame. As a result, the sensor measurements must be transformed to other coordinate frames, and these transformations may introduce errors. For rigidly-mounted sensors, these errors can be approximately eliminated or reduced to inconsequential values with offline calibration against objects of known location. For actuated sensors or sensors subject to platform vibration, however, these errors must be considered as statistical quantities. The third and fourth types of errors are due to imperfect knowledge of the orientation and position of the sensing platform. For moving platforms, these errors may be reported by an inertial navigation system or another position estimation scheme. The statistical contributions of platform orientation and location errors affect all sensors on the platform in an identical manner.

In order to understand in a statistical sense how these four sources of error affect each sensor measurement, it is first necessary to transform each sensor measurement and its uncertainties into a common coordinate frame for terrain estimation. To begin, each raw sensor measurement $\underline{r}$ is expressed in the fixed ENU coordinate frame of the terrain map:

$$\underline{r}^{ENU} = \begin{pmatrix} E \\ N \\ U \\ 1 \end{pmatrix} = f\left(\underline{p}, \underline{r}\right) \tag{13.1}$$

where $\underline{r}$ is the raw sensor measurement, $\underline{p}$ is the set of parameters that describe the sensor's orientation with respect to the ENU map frame, and $f(\cdot)$ is the function that transforms the raw measurement into the ENU frame. Note that $\underline{r}^{ENU}$ is expressed as a four-element vector. The fourth element of each measurement, always 1, permits the use of 4×4 rotation and translation matrices to express the transformation function $f(\cdot)$ as a series of matrix multiplications (Moon, 1998), (Murray, Li, & Sastry, 1994). Although other representations of the transformation may be used, the sequential matrix representation will be shown to be particularly useful for real-time implementations.

Each transformed measurement $\underline{r}^{ENU}$ is a terrain detection generated from a single raw measurement. Each terrain detection gives a measurement of the elevation and location of a small patch of terrain near the sensing platform. These elevation measurements are built up from both the original raw sensor measurement $\underline{r}$ and the sensor's orientation parameters $\underline{p}$ at the time the measurement was produced. As discussed previously, the orientation parameters $\underline{p}$ and raw measurement $\underline{r}$ are uncertain; they are more accurately modeled using estimates of their true values with associated errors:

$$\begin{aligned} \underline{p} &= \hat{\underline{p}} + \delta\underline{p} \\ \underline{r} &= \hat{\underline{r}} + \delta\underline{r} \end{aligned} \tag{13.2}$$

where $\underline{p}$ and $\underline{r}$ are the true orientation parameters and noise free measurement, $\hat{\underline{p}}$ and $\hat{\underline{r}}$ are the values of the parameters reported by the sensors or state estimators, and $\delta\underline{p}$ and $\delta\underline{r}$ are the errors in those reported values. Under this formulation, the values $\delta\underline{p}$ and $\delta\underline{r}$ can be due to any of the four error sources discussed in the beginning of section 13.2. It is important to note that the values $\underline{p}$ and $\underline{r}$ are not known perfectly unless there are no sources of error in any aspect of the sensing system. In general, only $\hat{\underline{p}}$ and $\hat{\underline{r}}$ are known.

To form a statistical representation taking into account all errors $\delta\underline{p}$ and $\delta\underline{r}$, equation 13.2 is substituted into equation 13.1. This expresses the true measurement in terms of the available sensor and estimator outputs:

$$\underline{r}^{ENU} = f\left(\hat{\underline{p}} + \delta\underline{p}, \hat{\underline{r}} + \delta\underline{r}\right) \tag{13.3}$$

Notice that equation 13.3 accounts for any potential sources of error arising either from the sensor or from the transformation to the ENU frame. If the elements of $\underline{p}$ and $\underline{r}$ fully describe the transformation from sensor measurements to terrain detections, equation 13.3 takes into account all sources of error.

To continue, a Taylor expansion of equation 13.3 is made about the observed measurement and orientation parameters. Then, assuming the estimation errors are small, the expansion is truncated at first order to make the expression more tractable:

$$\begin{aligned} \underline{r}^{ENU} &\approx f\left(\hat{\underline{p}}, \hat{\underline{r}}\right) + \left.\frac{\partial f}{\partial \underline{p}}\right|_{\underline{p}=\hat{\underline{p}}, \underline{r}=\hat{\underline{r}}} \delta\underline{p} + \left.\frac{\partial f}{\partial \underline{r}}\right|_{\underline{p}=\hat{\underline{p}}, \underline{r}=\hat{\underline{r}}} \delta\underline{r} \\ &= f\left(\hat{\underline{p}}, \hat{\underline{r}}\right) + J_{\underline{p}}\left(\hat{\underline{p}}, \hat{\underline{r}}\right)\delta\underline{p} + J_{\underline{r}}\left(\hat{\underline{p}}, \hat{\underline{r}}\right)\delta\underline{r} \end{aligned} \tag{13.4}$$

where $J_{\underline{p}}(\cdot)$ and $J_{\underline{r}}(\cdot)$ are the Jacobians of the transformation function $f(\cdot)$ with respect to the sensor orientation parameters and raw measurement, respectively. Additionally, assume the parameter estimates $\hat{\underline{p}}$ and raw measurement $\hat{\underline{r}}$ are both unbiased and conditioned upon all available orientation and sensor information $\mathcal{I}$. Under these assumptions, a posterior estimate of the terrain detection may be formed by taking the expectation of equation 13.4 conditioned on all available information (Bar-Shalom, Rong Li, & Kirubarajan, 2001):

$$\hat{\underline{r}}^{ENU} \equiv \begin{pmatrix} \hat{e} \\ \hat{n} \\ \hat{u} \\ 1 \end{pmatrix} = E\left[\underline{r}^{ENU} \middle| \mathcal{I}\right] \approx f\left(\hat{\underline{p}}, \hat{\underline{r}}\right) \tag{13.5}$$

Note that if some elements in the parameter estimates $\hat{\underline{p}}$ and $\hat{\underline{r}}$ are biased, the biases enter into the value of $\hat{\underline{r}}^{ENU}$ when taking the expectation of equation 13.4. In general, however, such biases can be removed through more refined calibration procedures. After such procedures the residual measurement errors will have a bias that is statistically indistinguishable from zero.

Continuing with the error analysis, the mean square error (MSE) matrix of the posterior measurement estimate yields the desired statistical measure of the measurement's uncertainty. Note this uncertainty takes into account all sources of error contained in the estimates $\hat{\underline{p}}$ and $\hat{\underline{r}}$:

$$\begin{aligned} P_{\hat{\underline{r}}} &= E\left[\left(\underline{r}^{ENU} - \hat{\underline{r}}^{ENU}\right)\left(\underline{r}^{ENU} - \hat{\underline{r}}^{ENU}\right)^T \middle| \mathcal{I}\right] \\ &\approx J_{\underline{p}}\left(\hat{\underline{p}}, \hat{\underline{r}}\right) E\left[\delta\underline{p} \cdot \delta\underline{p}^T \middle| \mathcal{I}\right] J_{\underline{p}}^T\left(\hat{\underline{p}}, \hat{\underline{r}}\right) + J_{\underline{r}}\left(\hat{\underline{p}}, \hat{\underline{r}}\right) E\left[\delta\underline{r} \cdot \delta\underline{r}^T \middle| \mathcal{I}\right] J_{\underline{r}}^T\left(\hat{\underline{p}}, \hat{\underline{r}}\right) \\ &= J_{\underline{p}}\left(\hat{\underline{p}}, \hat{\underline{r}}\right) Q_{\underline{p}} J_{\underline{p}}^T\left(\hat{\underline{p}}, \hat{\underline{r}}\right) + J_{\underline{r}}\left(\hat{\underline{p}}, \hat{\underline{r}}\right) Q_{\underline{r}} J_{\underline{r}}^T\left(\hat{\underline{p}}, \hat{\underline{r}}\right) \end{aligned} \tag{13.6}$$

where $Q_{\underline{p}}$ and $Q_{\underline{r}}$ are the mean square error matrices for the estimators used in determining $\hat{\underline{p}}$ and $\hat{\underline{r}}$. This representation includes the statistical effects of the orientation and sensor errors up through their second moments, mapped through a linear approximation of the actual transformation. These approximate linearized statistical techniques are common and work well in nonlinear estimation problems (Bar-Shalom et al., 2001). Techniques that preserve higher order statistical effects through nonlinear transformations, such as the Unscented transform or Monte Carlo methods, could also be used (Julier & Uhlmann, 1996), (Arulampalam, Maskell, Gordon, & Clapp, 2002). These are ignored in the present study, however, due to higher computational costs.

Notice the structure of equation 13.6 assumes the sensor orientation errors and raw measurement errors are uncorrelated. If that is not the case, it is straightforward to account for cross correlation by creating a stacked vector $\underline{s}$ of all uncertain parameters and their errors:

$$\underline{s} = \begin{pmatrix} \underline{p} \\ \underline{r} \end{pmatrix} = \begin{pmatrix} \hat{\underline{p}} + \delta \underline{p} \\ \hat{\underline{r}} + \delta \underline{r} \end{pmatrix} = \hat{\underline{s}} + \delta \underline{s} \tag{13.7}$$

The MSE matrix $P_{\hat{\underline{r}}}$ is now defined as:

$$P_{\hat{\underline{r}}} = J_{\underline{s}} \left(\hat{\underline{s}} \right) Q_{\underline{s}} J_{\underline{s}}^T \left(\hat{\underline{s}} \right) \tag{13.8}$$

where $Q_{\underline{s}}$ is a covariance matrix with $Q_{\underline{p}}$ and $Q_{\underline{r}}$ as its diagonal blocks and any cross correlations between $\delta \underline{p}$ and $\delta \underline{r}$ as its off-diagonal blocks.

Two additional comments about this statistical representation are in order. First, the analysis is only affected by elements of $\underline{p}$ and $\underline{r}$ that are uncertain. Any known elements in these vectors have no statistical variance and no correlation with any other elements of $\underline{p}$ or $\underline{r}$, so the rows of the Jacobian matrices corresponding to any known parameters have no effect on $P_{\hat{\underline{r}}}$. Second, the analysis is also independent of the lengths of $\underline{p}$ and $\underline{r}$; the MSE matrix $P_{\hat{\underline{r}}}$ is always 4×4. It is noted, however, that the MSE matrix $P_{\hat{\underline{r}}}$ will not be full rank unless the Jacobian matrix and covariance matrix that comprise it both have a rank of at least 4.

With the posterior measurement estimate $\hat{\underline{r}}^{ENU}$ defined in equation 13.5, equation 13.6 gives its 4×4 mean square error matrix $P_{\hat{\underline{r}}}$. This matrix describes the relative size and correlation of the terrain measurement errors in the elements of $\hat{\underline{r}}^{ENU}$ and behaves as a covariance matrix. The useful partition of the matrix $P_{\hat{\underline{r}}}$ is the upper-left 3×3 block P^{ENU}, which contains the ENU mean square errors and cross correlations for the measurement $\hat{\underline{r}}^{ENU}$. The remaining elements of the matrix, which involve correlations with the always-unity fourth term of $\hat{\underline{r}}^{ENU}$, are not used. That is,

$$P^{ENU} = \begin{pmatrix} I_{3\times 3} & 0_{3\times 1} \end{pmatrix} \cdot P_{\hat{\underline{r}}} \cdot \begin{pmatrix} I_{3\times 3} \\ 0_{1\times 3} \end{pmatrix} \tag{13.9}$$

Equations 13.5 and 13.9 model each measurement as a nominal terrain measurement $\hat{\underline{r}}^{ENU}$ with uncertainty characterized by ellipsoidal equiprobability surfaces. This strategy is itself an atypical representation of sensor measurements in remote sensing applications, where they are typically modeled only as discrete points with no associated error (Stoker, 2004). Horizontal and vertical errors are instead computed in post-processing validation against truth models (Huising & Pereira, 1998). In addition, errors in sensor orientation or sensor platform position and orientation are commonly treated as constants using offline calibration techniques (Morin, 2002).

More recently, Ref. (Stoker, 2004) suggested representing each sensor measurement as a voxel, a three-dimensional pixel with volume characterized by its error along three directional axes. However, this representation cannot support cross-correlation between coordinate axes, so it must necessarily make conservative error estimates to capture uncertainty. In contrast, the benefit of the statistical representation of equations 13.5 and 13.9 over typical representations is that it is dynamical: the error in each measurement is estimated in real-time based on the current accuracy of all the systems used to generate $\underline{\hat{p}}$ and $\underline{\hat{r}}$.

13.2.2 Measurement Association

The second step of the terrain estimation algorithm is to use the statistical analysis of section 13.2.1 to assign each terrain measurement to one or more grid cells from which it is likely to have originated. This problem of determining which measurements belong in each cell arises only because the full uncertainty in the measurements is considered during processing. That is, measurement correspondence on a cell-by-cell basis is uncertain due to the uncertainty in each measurement's in-plane location as well as its elevation. This problem is similar to the problem of association in target tracking literature such as (Bar-Shalom, Kirubarajan, & Lin, 2003) and (Kirubarajan & Bar-Shalom, 2004). The technique used here to assign measurements to cells is similar to those discussed in (Bar-Shalom et al., 2003) for assigning measurements to a single target. The present application is significantly different from traditional target tracking, however, because all measurements correspond to some portion of the fixed terrain map.

For this step of the terrain estimation algorithm, it is assumed that sensor measurement data has been processed according to equations 13.5, 13.6, and 13.9 to generate a set of N measurement estimates $\underline{\hat{r}}_i^{ENU} = (\, \hat{e}_i \quad \hat{n}_i \quad \hat{u}_i \,)^T$ and their associated MSE matrices P_i^{ENU} for $i \in [1, \ldots, N]$. To begin, this information is used to calculate the probability that the ith measurement $\underline{r}_i^{ENU}$ belongs in a cell of interest. Assuming the ENU probability distribution $\mathcal{P}_i\,(E, N, U)$ of the measurement is known, the probability can be evaluated explicitly by integrating over the area covered by the cell. First, the vertical coordinate is integrated out to yield the in-plane marginal distribution:

$$\mathcal{P}_i\,(E, N) = \int_{-\infty}^{\infty} \mathcal{P}_i\,(E, N, U)\, dU \tag{13.10}$$

This in-plane distribution of the measurement can then be integrated over the area of the jth cell to yield the probability that the measurement corresponds to that cell:

$$P\left(\underline{\hat{r}}_i^{ENU} \in j\right) = \int_{E_{j-}}^{E_{j+}} \int_{N_{j-}}^{N_{j+}} \mathcal{P}_i\,(E, N)\, dN dE \tag{13.11}$$

where the jth cell is defined by the rectangle $E_{j-} \leq E \leq E_{j+}$ and $N_{j-} \leq N \leq N_{j+}$. To complete this integral, the posterior measurement $\hat{\underline{r}}_i^{ENU}$ and its associated MSE matrix P_i^{ENU} are used to approximate the joint ENU distribution as a multivariate Gaussian:

$$\mathcal{P}_i\left(E, N, U\right) \approx \mathcal{N}\left(\hat{\underline{r}}_i^{ENU}, P_i^{ENU}\right) \tag{13.12}$$

This distribution is exact for the case of linear transformations, Gaussian noise sources, and unbiased estimates of the sensor orientation parameters and raw measurements. The Gaussian approximation is also commonly made for nonlinear, non-Gaussian cases using either the linearization presented above or the first and second moments implied by the Unscented Transform (Julier & Uhlmann, 1996). Applying the Gaussian transform, the EN joint probability distribution can be written in closed form:

$$\mathcal{P}_i\left(E, N\right) = \mathcal{N}\left(\hat{\underline{r}}_i^{EN}, P_i^{EN}\right) \tag{13.13}$$

where $\hat{\underline{r}}_i^{EN}$ is the first two components of the measurement estimate $\hat{\underline{r}}_i^{ENU}$ and P_i^{EN} is the upper left 2×2 EN block of the MSE matrix:

$$P_i^{ENU} = \begin{pmatrix} P_i^{EN} & P_i^{EN,U} \\ P_i^{U,EN} & P_i^{U} \end{pmatrix} \tag{13.14}$$

Using the given measurement estimate and its MSE matrix, the integral of equation 13.11 can now be computed. For real-time computational purposes, this integral is approximated as a single Riemann square (Marsden & Weinstein, 1985):

$$p_i \equiv P\left(\hat{\underline{r}}_i^{EN} \in j\right) \approx (E_{j+} - E_{j-})(N_{j+} - N_{j-}) \cdot \mathcal{P}_i\left(\frac{1}{2}(E_{j+} + E_{j-}), \frac{1}{2}(N_{j+} + N_{j-})\right) \tag{13.15}$$

which gives the approximate probability p_i that the measurement belongs to the jth cell. Equation 13.15 can then be used with equation 13.13 to approximate the probability p_i that a measurement estimate $\hat{\underline{r}}_i^{ENU}$ corresponds to a particular cell.

13.2.3 In-Cell Terrain Measurement Fusion

The final step of the terrain estimation algorithm is to fuse all the terrain measurements into an optimal terrain estimate. Because the measurements are assigned to grid cells according to the probability that they belong to those cells, this task is equivalent to determining the distribution of elevations in each cell given all measurements assigned to it. There are, however, two competing objectives in this task. First, a more accurate representation of the elevation distribution creates a better estimate within each cell. With this objective, ideally all individual measurements are retained separately to preserve the full set of data reported by the sensors. The tradeoff with this type of accurate representation is a second competing objective: computational feasibility. Memory and computational requirements scale worst case as $O(N_c \cdot N)$ with N_c the number of cells and

N the number of individual measurements retained, so it very quickly becomes infeasible to retain all measurements separately for any reasonably-sized terrain grid. As a result, a more compact representation is desired, one that yields useful terrain information without sacrificing computational feasibility.

To develop this computationally tractable, real-time representation, recall that the original measurement estimates $\hat{\underline{r}}_i^{ENU}$ are uncertain in three axes: East, North, and Up. What is desired in the adopted grid representation, however, is the univariate distribution of elevations in a particular cell at a particular East and North location. This knowledge can be used to form a posterior univariate elevation estimate $\hat{U}_i = E\left[\hat{u}_i \middle| E, N\right]$ for the ith measurement by conditioning on the East and North location of the jth cell (Bar-Shalom et al., 2001):

$$\hat{U}_i = \hat{u}_i + P_i^{U,EN} \left(P_i^{EN}\right)^{-1} \cdot \left[\frac{1}{2}\begin{pmatrix} E_{j+} + E_{j-} \\ N_{j+} + N_{j-} \end{pmatrix} - \begin{pmatrix} \hat{e}_i \\ \hat{n}_i \end{pmatrix}\right] \tag{13.16}$$

where $(\,E \quad N\,) = \frac{1}{2}\,(\,E_{j+} + E_{j-} \quad N_{j+} + N_{j-}\,)$ is the center of the jth cell. This estimate $\hat{U}_i$ has conditional variance:

$$\sigma^2_{\hat{U}_i} = P_i^U - P_i^{U,EN} \left(P_i^{EN}\right)^{-1} P_i^{EN,U} \tag{13.17}$$

In forming this elevation estimate $\hat{U}_i$ and its variance $\sigma^2_{\hat{U}_i}$ for association to the jth cell, each measurement is assumed to originate from the center of that cell, $\frac{1}{2}\begin{pmatrix} E_{j+} + E_{j-} \\ N_{j+} + N_{j-} \end{pmatrix}$, as per equation 13.16. This approximation approaches the exact continuous-terrain solution as cell size decreases, presenting a tradeoff between terrain map resolution and computational feasibility. Experimentally, it is found that cell size is commensurate to the smallest terrain features that can be detected.

Each of the measurements $\hat{U}_i$ is effectively a terrain detection: a measurement of a piece of the terrain within a particular cell. It is convenient to assume that each of these measurements corresponds to a different patch of terrain within the cell, so that no two measurements occur at precisely the same location. This assumption is justified by the fact that the terrain is continuous, the location of the terrain sensed by each measurement is uncertain, and often the sensing platform is moving. Two other assumptions are also made. First, each terrain measurement is assumed equally likely; that is, there is no *a priori* terrain information. This uniform, uninformative prior is adopted for convenience, as the uniform prior acts as a constant removed by normalization in the final distribution. Second, it is assumed that each cell has one correct or 'dominant' elevation to be estimated, and that elevation is represented within the set of terrain measurements obtained.

Several conclusions follow from these assumptions. First, the likelihood of each elevation measurement $\hat{U}_i$ is dependent only on the set of sensor orientation parameters $\hat{\underline{p}}_i$ and raw sensor measurement $\hat{\underline{r}}_i$ used to generate it. In particular, the likelihood of each elevation estimate conditioned on all measurements made so far is just equal to the likelihood of that elevation estimate conditioned on only the parameters used to generate it. Second, the likelihood of each measurement being a measure of the correct elevation for a particular cell is equal to the probability that the measurement came from that cell, i.e. its association probability p_i. Third, because the *a posteriori* distributions of the elevation measurements are all Gaussian, the elevation distribution within the jth cell $\mathcal{P}\left(U_j|\underline{\hat{p}}_{1...M}, \underline{\hat{r}}_{1...M}\right)$ conditioned on the full set of measurements $\underline{\hat{p}}_{1...M} = \left\{\underline{\hat{p}}_1, \underline{\hat{p}}_2, \ldots, \underline{\hat{p}}_M\right\}$ and $\underline{\hat{r}}_{1...M} = \{\underline{\hat{r}}_1, \underline{\hat{r}}_2, \ldots, \underline{\hat{r}}_M\}$ is a Gaussian sum or 'Gaussian mixture' constructed from the elevation estimates (Bar-Shalom et al., 2001):

$$\mathcal{P}(U_j|\underline{\hat{p}}_{1...M}, \underline{\hat{r}}_{1...M}) = \frac{\sum_{i=1}^{M} p_i \mathcal{N}\left(\hat{U}_i, \sigma^2_{\hat{U}_i}\right)}{\sum_{i=1}^{M} p_i} \tag{13.18}$$

where $\hat{U}_i$ from equation 13.16 is the ith elevation measurement estimate, $\sigma^2_{\hat{U}_i}$ from equation 13.17 is its associated conditional variance, p_i from equation 13.15 is the probability that the elevation measurement belongs in the cell, and M is the number of measurements assigned to the cell.

Equation 13.18 represents a desired data driven elevation distribution in the cell which takes into account all sources of uncertainty present in the system. However, the model is not computationally feasible, because the Gaussian mixture stored in each cell grows with the number of sensor measurements assigned to that cell. For real-time terrain estimation, a smaller set of information about each cell is desired. This set of data must be descriptive enough to exceed raw measurements in usefulness but small enough to be computationally feasible. The mean and variance of the elevation distribution satisfy both these requirements. Taking the expected value of the elevation distribution yields an approximate MMSE estimate of the characteristic or 'dominant' elevation of the jth cell:

$$\hat{U}_{GM,j} = \frac{\sum_{i=1}^{M} p_i \hat{U}_i}{\sum_{i=1}^{M} p_i} \approx E\left[U_j|\underline{\hat{p}}_{1...M}, \underline{\hat{r}}_{1...M}\right] \tag{13.19}$$

by the linearity of the expectation operator. Similarly, the second central moment of the elevation distribution gives the mean square error of the elevation estimate within the jth cell (Bar-Shalom et al., 2001):

$$\begin{aligned} \sigma^2_{GM,j} &= E\left[(U_j - \hat{U}_{GM,j})^2|\underline{\hat{p}}_{1...M}, \underline{\hat{r}}_{1...M}\right] \\ &= E\left[U_j^2|\underline{\hat{p}}_{1...M}, \underline{\hat{r}}_{1...M}\right] - \hat{U}^2_{GM,j} \\ &= \frac{\sum_{i=1}^{M} p_i(\hat{U}_i^2 + \sigma^2_{\hat{U}_i})}{\sum_{i=1}^{M} p_i} - \hat{U}^2_{GM,j} \end{aligned} \tag{13.20}$$

Equations 13.19 and 13.20 give the first two moments of the elevation distribution of the jth cell. Physically, equation 13.19 is an estimate of the characteristic elevation of the cell. Mathematically, it is an approximate MMSE estimate of the elevation of the jth given the assumptions discussed above. Equation 13.20

may be interpreted as a measure of the roughness or spread of elevations within the cell, though it also stores information about the confidence of the mean elevation estimate. The estimates of cell mean and variance can be used to make statistical statements about the elevations in the cell, taking into account the noise present in the entire system.

13.2.4 Algorithm Benefits and Real-Time Implementation

The proposed statistical representation of the terrain has several unique advantages over more traditional mapping strategies. First, the variance estimate within each cell gives information about the spread of elevations likely to be encountered within that cell, enabling real-time statements of elevation confidence intervals. These confidence intervals arise naturally as part of the estimation process, because the terrain estimation algorithm includes error estimates. The error estimates thus represent a richer set of information and physical meaning than standard binary obstacle detection algorithms without requiring the additional post processing of typical surveying representations.

A second advantage of this terrain model is that it can be generated and maintained in real-time. Recall from sections 13.2.2 and 13.2.3 that each measurement estimate is assigned to each cell according to the probability that the measurement lies in that cell. For practical implementations with finite numerical precision, however, only cells near the nominal in-plane measurement location derived from $\hat{\underline{r}}_i^{ENU}$ are affected by that measurement estimate. As a result, each measurement estimate need only be applied to cells in a small neighborhood of the nominal measurement. To counteract this issue, measurements can also be thresholded based on the probability that they belong to a particular cell to provide further reduction in computational complexity. These steps place limits on the number of cells to which each measurement can be applied, so the computational complexity is reduced from $O(C \cdot N)$ spent applying N raw measurements to all C cells in the entire terrain map, to $O(k \cdot N)$ spent applying each measurement to a maximum of k terrain cells. Furthermore, if only the first two moments of the elevation distribution are desired, then each measurement can easily be fused with previous measurements. In fact, only the following four quantities are required for each cell:

$$\sum_{i=1}^{M} p_i, \qquad \sum_{i=1}^{M} p_i \hat{U}_i, \qquad \sum_{i=1}^{M} p_i \hat{U}_i^2, \qquad \sum_{i=1}^{M} p_i \sigma_{\hat{U}_i}^2$$

where each of these quantities is itself a scalar. Also, because fusing a new measurement with the measurement history only requires knowledge of these four variables, the computational complexity and memory requirements of maintaining each cell are $O(1)$. This makes it possible to fuse measurements from many sensors at once in a distributed multithreaded architecture built from standard desktop computers, all while maintaining the terrain map in real-time. Finally, it is important to note that once sensor measurements have been used to update the appropriate cells in the terrain model, the original measurements can be discarded. In other words, the entire terrain map can be maintained without storing any measurement history. This makes it possible to maintain a full terrain map over many square miles of terrain in memory on a standard desktop computer.

Further computational savings can also be made in the measurement and MSE transformations of equations 13.5 and 13.6. In general, both transformations require a large number of multiplications and trigonometric evaluations. However, the transformation function $f(\cdot)$ from equation 13.1 can be written as a series of matrix multiplications, as suggested above:

$$f\left(\hat{\underline{p}}, \hat{\underline{r}}\right) = T_n\left(\hat{\underline{p}}[n]\right) \cdot \ldots \cdot T_2\left(\hat{\underline{p}}[2]\right) \cdot T_1\left(\hat{\underline{p}}[1]\right) \cdot R\left(\hat{\underline{r}}\right) \tag{13.21}$$

where $\hat{\underline{p}}[l]$ is the lth element of $\hat{\underline{p}}$, $T(\cdot)$ is a 4×4 transformation matrix, and $R\left(\hat{\underline{r}}\right)$ is a matrix representation of the raw sensor measurement $\hat{\underline{r}}$. This matrix-based representation is useful because each transformation matrix $T_i(\cdot)$ is a function of only one orientation parameter. Many of these orientation parameters are common to more than one measurement or constant in time. As a result, many of the matrix multiplications can be precomputed once and cached for future measurements.

The Jacobian calculations of equation 13.6 can also be computed efficiently by utilizing the structure of equation 13.21. In general calculating the Jacobians $J_{\underline{p}}(\cdot)$ and $J_{\underline{r}}(\cdot)$ of the transformation function require repeated application of the chain rule. However, because each independent orientation parameter appears in only one matrix in equation 13.21, the chain rule is not necessary. With this representation, each element of the Jacobian matrices is reduced to matrix multiplications and the differentiation of a single matrix. The matrix multiplications can be cached in the manner described above to reduce the expense of calculating the Jacobians. In essence, the algorithm benefits from the fact that many measurements share all but one or two orientation parameters.

13.3 A Simple Example

The behavior and mechanics of the terrain estimation algorithm presented in section 13.2 are best illustrated with a simple example. Consider a cart moving forward at constant speed of 5 m/sec along one dimension. The cart is equipped with a typical rangefinder that returns noisy ranges between it and the terrain below. The rangefinder is mounted on a platform elevated 1.5 m above the bottom of the vehicle wheels. Furthermore, the rangefinder is pitched forward by an approximate angle of 5° with respect to the horizontal. That is, the rangefinder is angled down to scan along the ground as the vehicle moves forward. Figure 13.1 shows the hypothetical setup for this one-dimensional simulated vehicle.

The rangefinder is modeled after a SICK LMS 291 laser rangefinder or 'LIDAR' (*LMS 200, LMS 211, LMS 220, LMS 221, LMS 291 Laser Measurement Systems Technical Description*, 2003). In this simple one-dimensional example, the sensor returns one range per scan and scans at 75 Hz. The LIDAR is modeled with three sources of noise: raw ranging error due to sensor noise, uncertainty in sensor pitch due to encoder noise, and uncertainty in vehicle location due to GPS noise. These three sources of noise are specific instances of uncertainty of

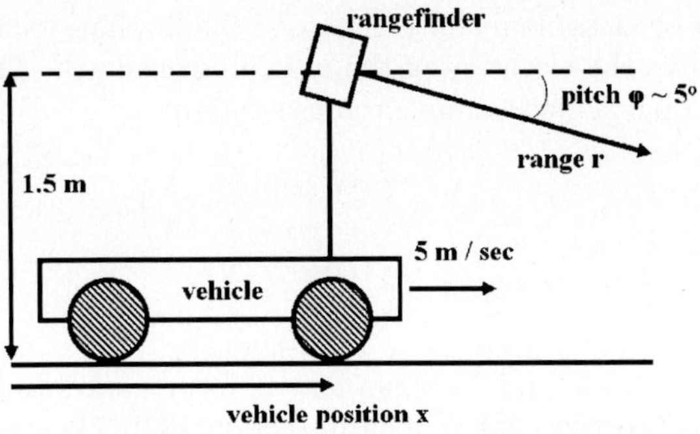

Fig. 13.1. Problem geometry of the simulated one-dimensional cart and rangefinder

types 1, 2, and 4 from section 13.2.1. The first noise source, raw ranging error due to sensor noise, is assumed corrupted by additive Gaussian noise so that the measured ranges are:

$$\hat{r} = r + \delta_r \tag{13.22}$$

where r is the true range and $\delta_r \sim \mathcal{N}\left(0, \sigma_r^2\right)$ is the additive Gaussian noise. The other uncertainties, pitch and vehicle position, are incorporated into the sensor orientation parameters $\underline{p}$,

$$\underline{p} = \begin{pmatrix} \phi \\ x \end{pmatrix} \tag{13.23}$$

where ϕ is the sensor's pitch angle downward from horizontal, and x is the vehicle's position relative to the arbitrarily-set origin. These parameters are assumed corrupted by additive Gaussian noise:

$$\underline{\hat{p}} = \underline{p} + \underline{\delta}_p \tag{13.24}$$

with $\underline{\delta}_p \sim \mathcal{N}\left(\underline{0}, diag\left[\sigma_\phi^2, \sigma_x^2\right]\right)$, where σ_ϕ^2 is the variance in the sensor's pitch measurements and σ_x^2 is the variance in the vehicle's position measurements. The noise sources are set according to typical accuracy of a LIDAR, angle encoder, and differential GPS receiver:

$$\begin{aligned} 3\sigma_r &= 0.1m \\ 3\sigma_\phi &= 0.5^\circ \\ 3\sigma_x &= 0.6m \end{aligned} \tag{13.25}$$

The simulation is constructed from a true terrain model entered at the program's start. Ranges are calculated according to truth values, and then the 'measured' values of the range, sensor pitch, and vehicle location are corrupted with appropriate noise for use in the terrain estimation algorithm. To proceed, note the transformation $f(\cdot)$ to produce the terrain measurement is:

$$\underline{\hat{r}}^{ENU} = \hat{r} \cdot \begin{pmatrix} \cos(\hat{\phi}) \\ -\sin(\hat{\phi}) \end{pmatrix} + \begin{pmatrix} 0 \\ 1.5 \end{pmatrix} + \begin{pmatrix} \hat{x} \\ 0 \end{pmatrix} \tag{13.26}$$

where $\underline{\hat{r}}^{ENU}$ is a two-element vector estimate of the in-plane location of the measurement as well as the elevation of the measurement itself. The Jacobian matrices also have a particularly simple representation:

$$J_{\underline{p}}\left(\hat{\underline{p}}, \hat{r}\right) = \begin{pmatrix} -\hat{r} \cdot \sin(\hat{\phi}) & 1 \\ \hat{r} \cdot \cos(\hat{\phi}) & 0 \end{pmatrix}$$
$$J_{\underline{r}}\left(\hat{\underline{p}}, \hat{r}\right) = \begin{pmatrix} \cos(\hat{\phi}) \\ -\sin(\hat{\phi}) \end{pmatrix} \tag{13.27}$$

Using the transformation and Jacobian matrix from equations 13.26 and 13.27, the computations proceed as described in section 13.2. The only caveat is that the equations are modified from a full three-dimensional East-North-Up representation to a two-dimensional East-Up representation. In addition, nominal measurement locations $\hat{E}$ are rounded to the nearest cell center for simplicity. The terrain is divided into 1 m grid cells, and measurements are applied up to 3 cells away from their nominal location.

The simulation features a hypothetical ground with flat terrain ($0 \leq x < 100$), a small bump ($100 \leq x < 150$), a small ditch ($150 \leq x \leq 155$), a large vertical face ($155 < x \leq 170$), and a plateau ($170 < x \leq 200$). Figure 13.2 shows the true ground surface along with the estimated elevation $\hat{U}_{GM}$ and its associated $\pm 2\sigma_{GM}$ bounds, in 1 m increments. The terrain estimation algorithm reveals several points of interest. First, grid cells 1 to 13 have no terrain estimate because the vehicle starts at $x = 0$ and the rangefinder is angled such that no data is ever gathered near these cells. Figure 13.3 (left) gives the total association probability and number of measurements assigned to each cell to confirm sensor measurement density in more detail. Figure 13.3 (left) also shows that more

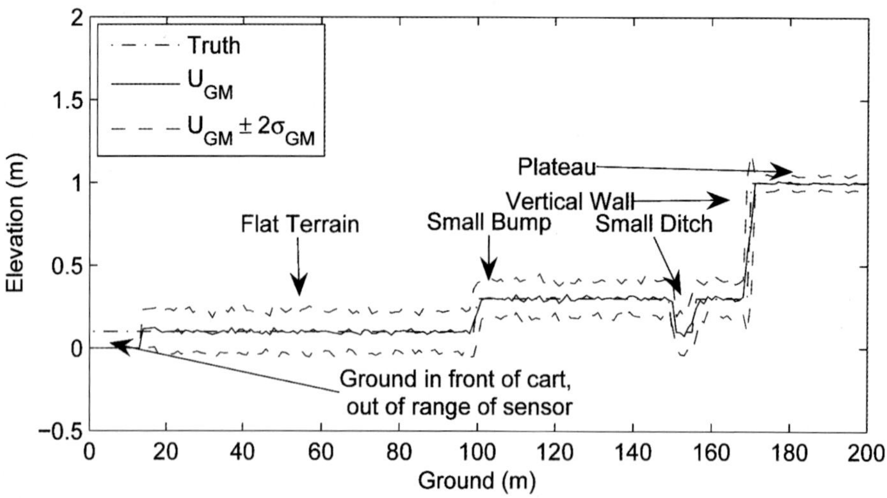

Fig. 13.2. True terrain, terrain estimate $\hat{U}_{GM}$, and $\pm 2\sigma_{GM}$ bounds for a one-dimensional terrain example with cart moving at 5 m/s

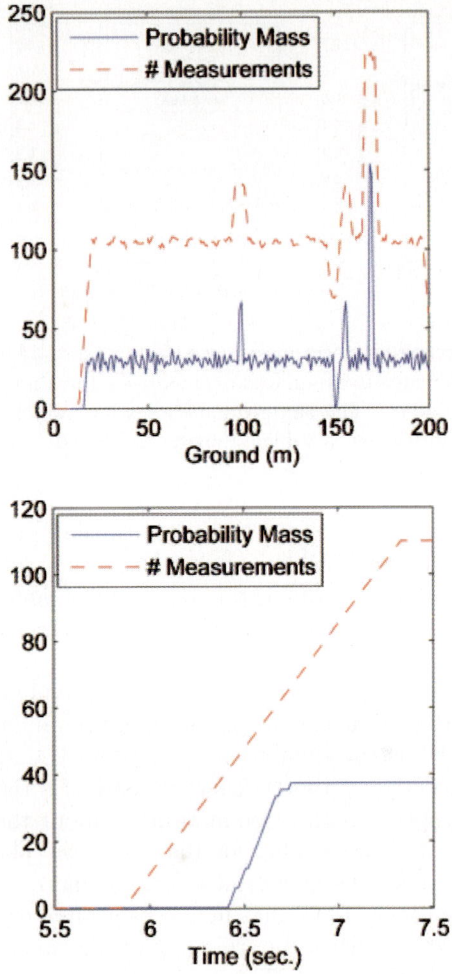

Fig. 13.3. (Left) Terrain estimate measurement probability mass levels and number of measurements assigned for one-dimensional terrain example. High probability mass in this example indicates the presence of a vertical or near-vertical face. (Right) Terrain estimate measurement probability mass level and total number of measurements accumulated over time for terrain cell $j = 50$.

measurements are assigned to cells near $j = 100$, 155, and 170, which contain near-vertical faces. Cells near $j = 150$ receive less measurements, in contrast, because a portion of the ditch is occluded. A second feature of the terrain estimation algorithm is that it estimates the sensed portions of the terrain accurately. Figure 13.4 (left) shows the terrain estimate for grid cell $j = 50$ (on the flat ground) as it evolves during the simulation. The estimate is observed to have no information until $t \approx 5.85$ seconds, when the first measurements are applied to

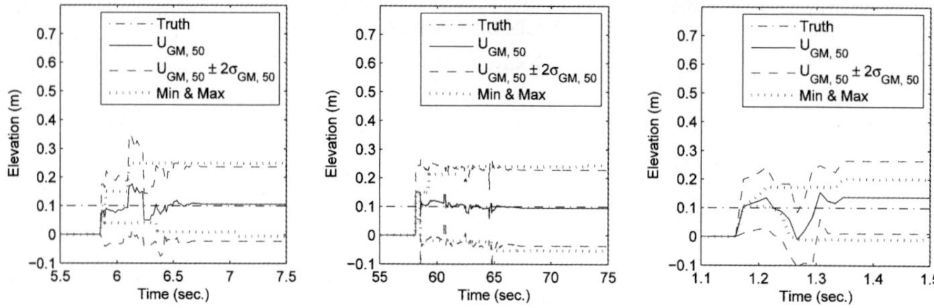

Fig. 13.4. (Left): True terrain, terrain estimate $\hat{U}_{GM,j}$, $\pm 2\sigma_{GM,j}$ bounds, and minimum and maximum measurements accumulated over time for terrain cell $j = 50$ with vehicle moving at 5 m/s. (Middle): The same quantities with vehicle moving at 0.5 m/s. (Right): The same quantities with vehicle moving at 25 m/s.

this particular cell. From $t \approx 5.85 \rightarrow 7.33$ seconds, the rangefinder continues to sweep over the cell, delivering new measurements to the terrain estimation algorithm. During this time, the terrain estimate is refined, and the variance within the cell fluctuates based on the quality of the measurements received. Because the terrain is flat and the measurements are not particularly noisy, the variance for this cell is dominated by the variances of the individual measurements assigned to it. Figure 13.3 (right) shows the level of probability mass $\sum p_i$ of the estimate of cell $j = 50$ as the simulation progresses. The approximately linear increase in probability mass / time is characteristic of a rigid sensor mounting and constant vehicle speed. A third point to note about the simulation is that the estimation errors σ_{GM} are smaller for the top of the plateau than they are for the bottom. This is a consequence of sensor geometry: the elements of the Jacobian matrix in equation 13.27 that map sensor pitch error into vertical terrain uncertainty are smaller when the true ranges r are smaller, which occurs for taller terrain relative to the sensor. This particular point emphasizes the effects of the statistical representation over *ad hoc* approaches, as it utilizes geometrical relationships to improve estimation accuracy where appropriate.

The vertical face present in the terrain in cell $j = 170$ is also estimated appropriately. Figure 13.2 shows that the estimated elevation of the cell is approximately half the elevation of the plateau, though the variance of the estimate near these cells is quite large. In these cells, the variance is dominated by the spread of the terrain measurements, which are taken all along the vertical face of the plateau. In addition, the probability mass along the vertical face of the plateau is significantly higher than at any other point in the simulation, as shown in Figure 13.3 (left). This property is particularly useful for gathering information about objects protruding from the natural ground plane.

As a point of reference, the terrain estimation algorithm is compared against a naïve algorithm that simply reports the maximum and minimum elevation measurement assigned to each cell. Note this naïve algorithm is effectively a

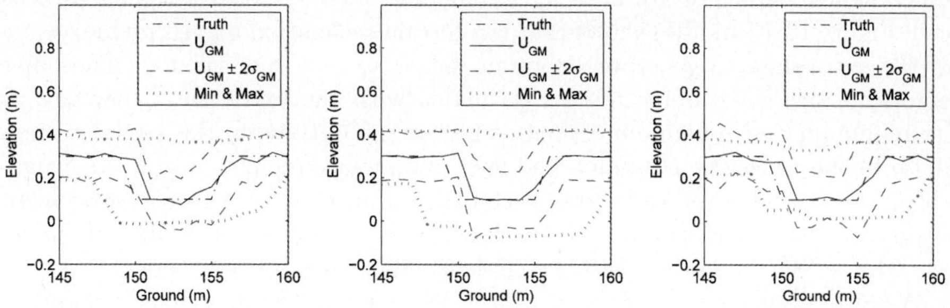

Fig. 13.5. (Left): True terrain, terrain estimate $\hat{U}_{GM,j}$, $\pm 2\sigma_{GM,j}$ bounds, and minimum and maximum measurements near a simulated ditch with vehicle moving at 5 m/s. (Middle): The same quantities with vehicle moving at 0.5 m/s. (Right): The same quantities with vehicle moving at 25 m/s.

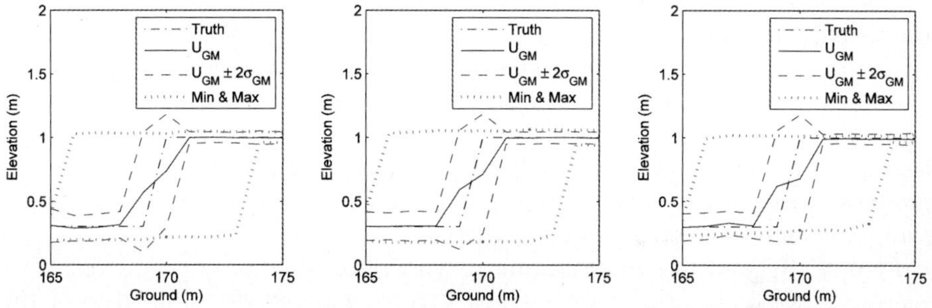

Fig. 13.6. (Left): True terrain, terrain estimate $\hat{U}_{GM,j}$, $\pm 2\sigma_{GM,j}$ bounds, and minimum and maximum measurements near a simulated wall with vehicle moving at 5 m/s. (Middle): The same quantities with vehicle moving at 0.5 m/s. (Right): The same quantities with vehicle moving at 25 m/s.

convolution using both a maximum and minimum operator over the same ranges of cells considered by the statistical algorithm. The maximum and minimum measurement lines for the naïve algorithm are plotted in dotted green in Figure 13.4(left) for terrain cell $j = 50$. For 5 m/s cart speeds and standard flat terrain like cell $j = 50$, the naïve approach behaves similarly to the $\pm 2\sigma_{GM,j}$ elevation bounds. This supports the use of the terrain estimation algorithm for statements of statistical confidence about the terrain. However, Figure 13.4 (left) shows that the statistical algorithm requires fewer measurements to converge to its final uncertainty $\sigma_{GM,j}$, because it considers each measurement as a statistical quantity. The naïve approach, in general, requires larger sample sizes to establish reasonable minimum and maximum elevation bounds.

A second simulation, performed on the same terrain but with a cart speed of 0.5 m/s, is shown in Figure 13.4 (middle). In this simulation the cart moves much slower, so there are a larger number of measurements assigned to each cell. Figure 13.4 (middle) shows that the terrain estimation algorithm makes use of the extra data to generate smoother estimates and uncertainties. The naïve algorithm produces worse terrain estimates with the extra data, however, as its minimum and maximum estimates reflect the outliers in the sensed terrain data. In this sense the minimum and maximum measurements in the naïve algorithm diverge monotonically from each other, regardless of sensor measurement statistics. The statistical algorithm, in contrast, provides a well-defined framework for statements of statistical confidence about the terrain model.

A third simulation, performed on the same terrain with a vehicle speed of 25 m/s, is shown in Figure 13.4 (right). In this simulation the cart moves very fast, so very few measurements are gathered in each cell. In this case the estimates from the statistical terrain algorithm are not as smooth as they were in the cases with more data. They do, however, still report uncertainties similar to those reported in the previous examples. The output of the naïve algorithm is not as stable as the amount of data changes. In Figure 13.4 (right), for example, the naïve algorithm consistently reports bounds that are smaller than the bounds reported in the previous two simulations. The statistical algorithm generates accurate terrain uncertainties after only one measurement, while the naïve algorithm consistently underestimates uncertainty. This simulation shows the robustness of the statistical terrain estimation algorithm even with much less data than in the other simulations. The naïve algorithm is much worse: it produces optimistic estimates from a small number of measurements, and pessimistic estimates as more data is gathered.

Figures 13.5 and 13.6 show magnified views of the final terrain estimates near the ditch and wall, respectively. These Figures support the robustness of the terrain estimation algorithm in comparison to the naïve algorithm. In Figure 13.5, for example, the naïve algorithm blurs over the small ditch at each simulated cart speed. These estimates could be improved in the naïve algorithm at slow speeds by only applying a measurement in its nominal cell. At faster speeds this will reduce the naïve algorithm's performance, however, due to the decreased number of measurements in each cell. The statistical algorithm, which uses association weights to form optimal estimates, suffers from neither of these problems. The statistical terrain estimates remain accurate at all speeds, and improve as the amount of data increases.

A similar comparison can be seen in Figure 13.6, which shows a magnified view of the final terrain estimates near the wall. As before, the naïve algorithm grows the boundaries of the wall due to its *ad hoc* method of applying measurements to cells. Figure 13.6 (middle) shows the naïve algorithm counterintuitively produces worse terrain estimates as the number of measurements increases, as it is sensitive to outliers. Figure 13.6 (right) also confirms that the algorithm produces overly optimistic estimates as the number of measurements decreases. The statistical algorithm, in contrast, produces consistent estimates across all the simulated cart speeds.

13.4 Real World Application: A Moving Ground Vehicle Equipped with Multiple Sensors

The terrain estimation algorithm described in section 13.2 is capable of generating and maintaining a terrain model with confidences in real-time. In this section, the capabilities of the terrain estimation algorithm are applied to a more challenging and realistic problem: constructing and maintaining a real-time terrain map using multiple sensors in a dynamical environment. In particular, this section tests the algorithm on Cornell University's 2005 DARPA Grand Challenge autonomous ground vehicle (Team Cornell, 2005).

In the DARPA Grand Challenge and the National Qualifying Event, the terrain estimation algorithm provided estimates to a path planner, which used the terrain estimates to determine traversable areas of the course. In both the Grand Challenge and the National Qualifying Event, the algorithm helped the vehicle successfully distinguish traversable flat ground, speed bumps, and shallow inclines from intraversable features such as rocks, large vegetation, and tank traps. Miller et al. (2006) discusses the method by which the terrain estimates were used for autonomous path planning, which will not be addressed further in this paper. In the present study, the Cornell vehicle is used as a means to validate the terrain estimation algorithm against a known, surveyed landscape. To begin, the Cornell vehicle is built upon a 'Spider Light Strike Vehicle' from Singapore Technologies Kinetics, shown in Figure 13.7.

Fig. 13.7. The Spider Light Strike Vehicle used to test the terrain estimation algorithm

The Spider is equipped with a Trimble Ag252 GPS receiver and an inertial navigation system for attitude and position determination, as well as three SICK LMS 291 LIDARs for terrain estimation. All three LIDAR units are fixed to the

Fig. 13.8. The Spider's three laser rangefinders (LIDARs) and two-axis gimbal platform

front hood of the Spider, as shown in Figure 13.7. The left and right LIDARs are mounted rigidly and pitched to intersect the ground plane approximately 15 m and 20 m in front of the vehicle. The center LIDAR is mounted on a two-axis gimbal, with yaw and pitch actuated by a pair of MAXON EPOS 70-4 rotary motors and Harmonic Drive gear boxes. Figure 13.8 shows the Spider's actuated gimbal and LIDAR units.

All types of error discussed in section 13.2.1 are modeled in this experiment. Individual LIDAR errors are modeled as ranging error with a typical accuracy of ±5 cm, and an angular error with coning angle of ±0.24° due to the expansion of the detecting light pulse (*LMS 200, LMS 211, LMS 220, LMS 221, LMS 291 Laser Measurement Systems Technical Description*, 2003). These two error sources affect terrain measurements differently: the range error yields uncertainty in the sensing direction regardless of detection range, while the angular error term yields higher uncertainty at longer ranges for this problem geometry. The second type of error, sensor orientation error, is modeled as error up to ±0.7° on the yaw and pitch angles reported by the gimbal encoders. This term only affects the gimbaled LIDAR; the two fixed LIDARs' orientation parameters are determined via offline calibration against the gimbaled LIDAR. The third and fourth sources of sensing error are due to the attitude and position estimator. Orientation errors of less than 0.2° and position errors of less than $0.2m$ are common in this system, though the exact MSE matrices used are those reported by the estimators in real-time.

The statistical analysis and Jacobian transformations discussed in section 13.2.1 are easily applied to the sensors on the Spider to model these four sources

Fig. 13.9. Example sensor axes and ENU reference axes that determine the transformation from sensor to terrain measurements

of error. To begin, each raw LIDAR range measurement $\underline{r}$ is defined in its own measurement axes:

$$\underline{r} = \begin{pmatrix} \rho \cdot \cos\left(\theta_{\mathrm{D}}\right) \\ \rho \cdot \sin\left(\theta_{\mathrm{D}}\right) \\ 0 \\ 1 \end{pmatrix} \tag{13.28}$$

where ρ is a scalar range reported by the LIDAR and θ_{D} is the reported detection angle. In this measurement frame, the x-axis is chosen to point out the front of the LIDAR, the z-axis is chosen perpendicular to the LIDAR detection plane, and the y-axis completes the right-handed coordinate system. The origin of the coordinate axis is the origin of the measurement: the point at which $\rho = 0$. From here, the measurement is transformed into a terrain detection in the ENU reference frame. These two coordinate axes are represented graphically in Figure 13.9. Mathematically, the transformations are performed using 12 separate orientation parameters for each LIDAR:

1. LIDAR Euler angles S_ψ, S_ϕ and S_θ with respect to the vehicle body.
2. LIDAR position elements S_x, S_y, and S_z with respect to the inertial navigation system.
3. Vehicle Euler angles ψ, ϕ, and θ with respect to the ENU reference frame.
4. Vehicle position O_x, O_y, and O_z with respect to the ENU reference origin.

The vector $\underline{p}$ of sensor orientation parameters for this particular application is therefore:

$$\underline{p} = \begin{pmatrix} S_\psi & S_\phi & S_\theta & S_x & S_y & S_z & \psi & \phi & \theta & O_x & O_y & O_z \end{pmatrix}^T \tag{13.29}$$

where all elements of this vector except for S_x, S_y, and S_z are modeled as uncertain.

These parameters are used along with the raw range measurement from equation 13.28 to generate measurements and an associated mean square error (MSE) matrix as described in section 13.2.1. All uncertain orientation parameters are treated as corrupted with additive zero mean, Gaussian white noise. The maximum magnitude of each error source, discussed in the beginning of section 13.4, is taken as its 3σ value. The exceptions are the attitude and position covariance matrices, which are taken from the attitude and position estimator as the algorithm runs. The covariance matrices $Q_{\underline{p}}$ and $Q_{\underline{r}}$ for the experimental setup are block diagonal:

$$\begin{aligned} Q_{\underline{p}} &= diag\left[\sigma^2_{S_\psi}, \sigma^2_{S_\phi}, \sigma^2_{S_\theta}, \sigma^2_{S_x}, \sigma^2_{S_y}, \sigma^2_{S_z}, P_{\psi\phi\theta}, P_{xyz}\right] \\ Q_{\underline{r}} &= diag\left[\sigma^2_\rho, \sigma^2_{\theta_D}\right] \end{aligned} \tag{13.30}$$

where $P_{\psi\phi\theta}$ and P_{xyz} are the attitude and position MSE matrices reported by the attitude and position estimators at the time each measurement is taken.

The remainder of the sensor fusion algorithm was implemented according to sections 13.2.2 and 13.2.3. A list of algorithm steps is given below:

1. Obtain current encoder, attitude, and position measurements to construct $\hat{\underline{p}}$.
2. Obtain a scan line of raw measurements $\hat{\underline{r}}_i$ from a LIDAR.
3. Construct and cache the transformation and Jacobian matrices for the measurements in the obtained scan line.
4. For each raw measurement $\hat{\underline{r}}_i$:
   - Calculate a terrain measurement estimate $\hat{\underline{r}}_i^{ENU}$ and MSE matrix P_i^{ENU} using equations 13.5 and 13.6 and the cached transformation and Jacobian matrices.
   - For each cell (up to k) near the nominal measurement $\hat{\underline{r}}_i^{ENU}$:
     - Calculate the association probability p_i for this measurement in this applied to the cell using equation 13.15.
     - Calculate a posterior elevation measurement $\hat{U}_i$ and variance $\sigma^2_{\hat{U}_i}$ from $\hat{\underline{r}}_i^{ENU}$ for the cell, using equations 13.16 and 13.17.
     - Fuse $\hat{U}_i$, $\sigma^2_{\hat{U}_i}$, and p_i with existing measurements in the cell by updating the 4 numbers discussed in section 13.2.4 stored for the cell.
5. When elevation and uncertainty estimates are desired, calculate them using equations 13.19 and 13.20.

For this particular implementation, grid cells were set to be 40 cm by 40 cm squares. This cell size was motivated by the size of the Spider's wheels and the desired resolution of the landscape's features (Miller et al., 2006). An arbitrary latitude and longitude near the experiment were chosen as the reference origin. For simplicity, the nominal East - North location of each measurement was rounded to the nearest cell center.

To ensure computational feasibility, measurements were applied to cells at most 2 m from the nominal measurement location. This threshold was used because the measurement MSE matrix P^{ENU}, when accounting for all sources of error, yielded no significant effect in cells farther than 2 m from the nominal measurement. In addition, a probability threshold of $p_i \geq 10^{-4}$ was also enforced, so measurements were applied only to cells in which they were likely to belong. While this did not reduce the computational search space of this particular implementation, it did tend to eliminate the effects of rare faulty measurements.

The entire terrain estimation algorithm was implemented on a commercial four-processor server running Microsoft Windows Server 2003. Each processor in the server was a 2.0 GHz AMD Opteron 846. Each of the three rangefinders was

Fig. 13.10. Sample $\hat{U}_{GM} + 2\sigma_{GM}$ elevation map resulting from the Spider's sensor fusion scheme

run on its own event-driven reader thread at the LIDARs' scanning frequency of 75 Hz. Data from each LIDARs was read across a 500 kbps serial connection. LIDAR scans were queued for sequential fusion into the terrain map, so that a full scan from each LIDAR was processed before moving on to the next. Each LIDAR measurement was transformed to a terrain detection using all 12 LIDAR orientation parameters. Transformation matrices common to a particular LIDAR scan were computed once per scan and applied to all measurements in that scan. Common components of the Jacobian matrices were also computed once per scan. These optimizations enabled the Spider to process all 40725 measurements per second delivered by the 3 LIDARs using less than 10% of a single processor's computing resources. These results suggest the algorithm could be run on a much smaller processor, or scaled across a parallel architecture for a large number of distributed sensors. The algorithm required storage space linear in the number of cells: only 4 floating point numbers per cell. It could thus maintain many miles of mapping regions in active memory. In contrast, a system retaining the same measurements without a fusion scheme would require storage space of nearly 0.95 Mb per second.

Figure 13.10 shows a sample terrain model generated from the terrain estimation algorithm as the Spider approaches several objects in an experimental run. In this example, the $\hat{U}_{GM} + 2\sigma_{GM}$ elevations from equations 13.19 and 13.20 (in meters) are plotted relative to an arbitrary ENU origin selected near the test site. These elevations are plotted across a color spectrum, and the axes in Figure 13.10 are to scale. Like the example in section 13.3, the algorithm generates valid estimates and accurate confidence bounds. For example, the $\hat{U}_{GM}$ elevations for the cells near the 0.8 m tall trash cans are approximately 0.45 m higher than

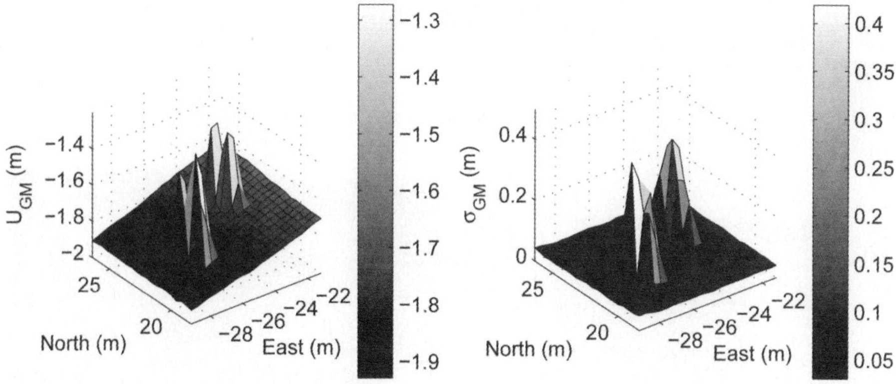

Fig. 13.11. (Left) Final $\hat{U}_{GM}$ map near the two 0.8 m tall trash cans. (Right) Final σ_{GM} map near the two 0.8 m tall trash cans.

surrounding cells, and the $\hat{U}_{GM}$ elevations for the cells near the 0.46 m traffic cones are approximately 0.25 meters. These lower elevation estimates reflect the fact that the cells containing these objects also contain some exposed portions of the ground plane. The uncertainties, however, are appropriately large for these cells: $\sigma_{GM} \approx 0.35$ m for the cells near the trash cans and $\sigma_{GM} \approx 0.17$ m for the cells near the traffic cones. Figure 13.11 shows the estimated elevation $\hat{U}_{GM}$ and estimation uncertainty σ_{GM} near the trash cans in greater detail.

Figure 13.12 (left), in contrast, shows the evolution of the terrain estimate for one cell near one of the trash cans. Figure 13.12 (right) shows that during this test, the Spider passes near the trash cans three times: once with the trash cans in the periphery of the sensors' footprints at $t \approx 2358$ sec., and twice directly between the two trash cans at $t \approx 2374$ sec. and $t \approx 2395$ sec. Figure 13.13 shows vehicle speed and heading during the test. Notice that the average speed of 5.4 m/s is comparable to speeds at which a human drives, confirming the real-time capabilities of the algorithm. Figure 13.12 (left, top) shows the total association probability sum $\sum p_i$ accumulated for the cell over time. Only points at which the probability increases are included. Notice that most of the association probability is assigned on the Spider's first pass by the trash can, suggesting that repeated passes by the trash can are not necessary for the algorithm to generate accurate results in real-time. The bottom left subplot of Figure 13.12 shows the terrain elevation estimate $\hat{U}_{GM}$ and the $\pm 2\sigma_{GM}$ bounds for the same cell. Note that the elevation estimate for the terrain is relative to an arbitrarily-set ENU origin, not the ground plane, so the absolute elevation scale on the vertical axis is non-intuitive. The variance in this particular cell is also much higher than in surrounding cells, indicating the presence of variegated terrain or obstacles in comparison with other nearby cells. The elevation estimate in this cell also fluctuates over time as more measurements are applied to it: some measurements, from the near-vertical face of the trash can, tend to increase the

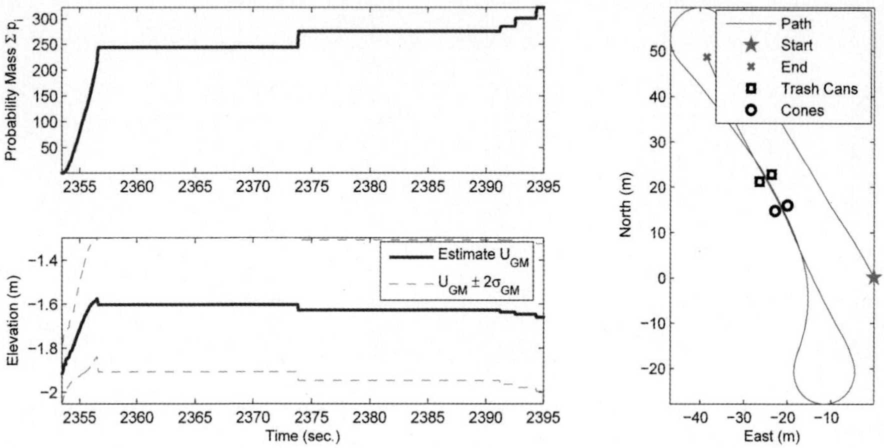

Fig. 13.12. (Left) Total association probability and $\hat{U}_{GM} \pm 2\sigma_{GM}$ bounds over time for a particular terrain cell containing a portion of a trash can. (Right) Vehicle ground track during real-time terrain experiment. The vehicle intersects the line connecting the trash cans at $t \approx 2358$ sec., $t \approx 2374$ sec. and $t \approx 2395$ sec.

elevation estimate upward. Others, from the surrounding flat ground, tend to decrease the elevation estimate. The effect of these different measurements is also seen in the cell's variance, which increases throughout the experiment. Finally, notice that the elevation estimate changes most on the Spider's first pass by the trash can, and that subsequent passes do not cause a substantial increase in total association probability of measurements assigned to that cell. The algorithm is therefore capable of producing accurate estimates in real-time, at reasonable speeds, and without revisiting old terrain. Notice that the algorithm produces relatively smooth and dense terrain estimates; this occurs because individual sensor measurements are applied in more than one cell as per equation 13.18. This smoothed and correlated terrain model arises naturally from the estimation algorithm and the continuous posterior distribution of the errors in the sensors. It contrasts sharply with other terrain estimation algorithms that make use of multiple passes or recursion on the terrain estimate to smooth it out (Olin & Tseng, 1991), (Arakawa & Krotkov, 1992), (Hähnel et al., 2004), (Weingarten & Siegwart, 2005).

While this algorithm produces relatively consistent estimates, it does have several drawbacks. First, it is not integrated with the attitude and position estimator, so it is directly subject to incorrect attitude and position estimates. For example, large discontinuous jumps resulting from update steps in an Extended Kalman Filter produce 'phantom' walls and obstacles due to incorrectly-placed LIDAR scans. These types of phantom walls can also arise from unmodeled timing delays and inaccurate noise models, depending on the severity of the modeling error. Figure 13.14 shows a sample phantom obstacle resulting from a

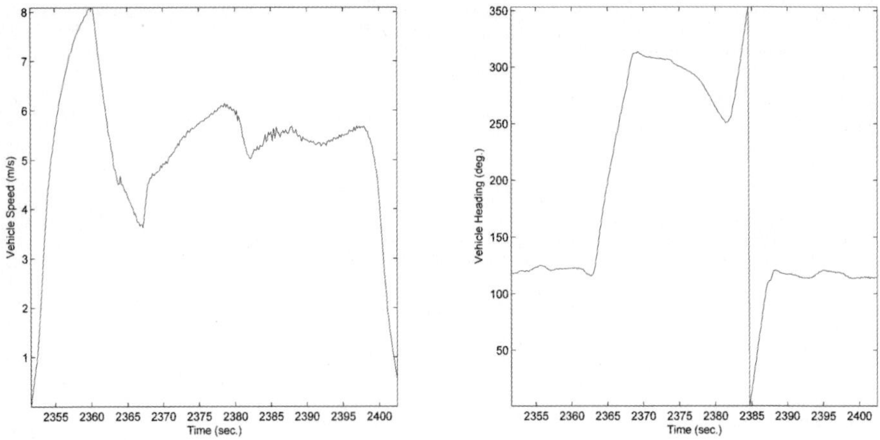

Fig. 13.13. (Left) Vehicle speed during real-time terrain experiment. (Right) Vehicle heading during real-time terrain experiment.

mismodeled noise source. This result was generated in a second driving experiment in which the gimbaled LIDAR was set to a sinusoidal pitching pattern, suggesting that mismodeled noise in the gimbal encoders may be the cause. In general these anomalies can be fixed or minimized with the addition of more accurate models of the offending sources of error. Error checks can also be performed to determine whether entire LIDAR scans are likely to be faulty.

A second drawback of this and other terrain estimation schemes is that all the sensors must be calibrated correctly. Such a task is particularly difficult with LIDAR units mounted on a vehicle, as precise positions and orientation angles are very difficult to measure. For this implementation, Cornell's Grand Challenge team used the Spider's attitude and position estimator along with a section of flat airplane runway to calibrate the gimbaled LIDAR with respect to the ground. Then, the fixed LIDAR orientations were found in software through a greedy search over their orientation angles by comparing their terrain estimates to those produced by the gimbaled LIDAR (Russell & Norvig, 2003). This automatic calibration technique produced terrain estimates consistent to within one or two centimeters between the LIDAR units.

A final drawback of this terrain estimation scheme is the fact that only the first two moments of each terrain cell are retained. As a result, the algorithm can give misleading terrain estimates inside tunnels, or in cells in which there are distinct clusters of elevations such as tree canopies. These arise because the algorithm makes no attempt to characterize or classify macroscopic features of the terrain, such as vegetation or terrain solidity, across multiple cells. One alternative explored in the literature is the idea of expanding the terrain estimate to a 3D set of binary obstacle / no-obstacle evidence grid cells (Martin & Moravec, 1996). However, computational resources tend to limit the practicality of such 3D grids. Furthermore, path searches through a 3D environment

Fig. 13.14. 'Phantom' walls in the $\hat{U}_{GM} + 2\sigma_{GM}$ map arising from high variance due to incorrectly-modeled gimbal encoder noise

take large amounts of time, especially considering the vehicle is effectively constrained to the ground. The given approach instead attempts to concentrate on representations that preserve the computational efficiency and searchability of the 2D framework without sacrificing the richness of the representation.

13.5 Conclusions

This paper presents a real-time terrain mapping and estimation algorithm using Gaussian sum height densities to model terrain variations in a planar gridded elevation model problem. Techniques for statistical analysis and estimation of each individual sensor measurement are used to account for multiple sources of error in the transformation from sensor measurements to terrain detections in a rigorous manner. Linearization techniques are used to approximate the uncertainty of each sensor measurement with a nominal measurement and a joint Gaussian distribution in the physical East, North, and Up directions. Measurements are associated to multiple locations in the elevation model using a Gaussian sum conditional density to account for uncertainty in measured elevation as well as uncertainty in the in-plane location of the measurement.

The accuracy and interpretation of the algorithm is evaluated using a simple one-dimensional example with a hypothetical range sensor. Results show accurate statistics as the vehicle traveled over a bump, ditch, large vertical face, and plateau. Consistent statistical estimates were also verified across a range of vehicle velocities. Straightforward use of minimum / maximum elevations measured directly from the sensors produced comparatively inconsistent results, with degrading performance as the number of sensor measurements increased.

The algorithm was demonstrated in a practical application of real-time mapping with multiple sensors on a platform moving at realistic speeds. The vehicle, Cornell's DARPA Grand Challenge entry, included multiple LIDAR sensors, GPS, and attitude sensors. The algorithm was shown capable of producing statistically accurate and computationally feasible elevation estimates on dense terrain models, as well as estimates of the errors in the terrain model. This type of dense terrain map holds more information than traditional real-time mapping approaches, and it has the potential to be useful in a number of autonomous or remote mapping applications.

Acknowledgments

The authors would like to thank Sergei Lupashin, Brian Schimpf, Noah Zych, Singapore Technologies Kinetics, and the members of Team Cornell for helpful discussion, implementation assistance, and the use of Team Cornell's Spider Light Strike Vehicle for data collection. This material is based upon work supported under a National Science Foundation Graduate Research Fellowship.

References

Ackermann, F. (1999). Airborne laser scanning - present status and future expectations. *ISPRS Journal of Photogrammetry and Remote Sensing*, *54*, 64-67.

Arakawa, K., & Krotkov, E. (1992). *Fractal surface reconstruction with uncertainty estimation: Modeling natural terrain* (Tech. Rep. No. CMU-CS-92-194). Pittsburgh: School of Computer Science, Carnegie Mellon University.

Arulampalam, M., Maskell, S., Gordon, N., & Clapp, T. (2002, February). A tutorial on particle filters for online nonlinear/non-gaussian bayesian tracking. *IEEE Transactions on Signal Processing*, *50*(2), 174 - 188.

Axelsson, P. (1999). Processing of laser scanner data - algorithms and applications. *ISPRS Journal of Photogrammetry and Remote Sensing*, *54*, 138-147.

Bar-Shalom, Y., Kirubarajan, T., & Lin, X. (2003). Probabilistic data association techniques for target tracking with applications to sonar, radar, and eo sensors. *IEEE Systems Magazine*.

Bar-Shalom, Y., Rong Li, X., & Kirubarajan, T. (2001). *Estimation with applications to tracking and navigation: Theory, algorithms and software*. New York: John Wiley & Sons, Inc.

El-Hakim, S., Whiting, E., Gonzo, L., & Girardi, S. (2005, August). 3-d reconstruction of complex architectures from multiple data. In *Proceedings of the isprs working group v/4 workshop*. Venice-Mestre, Italy: International Society for Photogrammetry and Remote Sensing.

Hähnel, D., Burgard, W., & Thrun, S. (2004). Learning compact 3d models of indoor and outdoor environments with a mobile robot. *Elsevier Science Special Issue Eurobot '01*, 1-16.

Huising, E., & Pereira, L. (1998). Errors and accuracy estimates of laser data acquired by various laser scanning systems for topographic applications. *ISPRS Journal of Photogrammetry and Remote Sensing*, *53*, 245-261.

Julier, S., & Uhlmann, J. (1996, November). *A general method for approximating nonlinear transformations of probability distributions* (Tech. Rep.). Oxford: Department of Engineering Science, University of Oxford.

Kaplan, E. (Ed.). (1996). *Understanding gps: Principles and applications.* Boston: Artech House.

Kirubarajan, T., & Bar-Shalom, Y. (2004, March). Probabilistic data association techniques for target tracking in clutter. *Proceedings of the IEEE, 92*(3).

Lacroix, S., Mallet, D., Bonnafous, G., Bauzil, S., Fleury, S., Herrb, M., et al. (2002). Autonomous rover navigation on unknown terrains: Functions and integration. *Interjational Journal of Robotics Research, 21*(10 - 11), 917 - 942.

Leal, J. (2003). *Stochastic environment representation.* PhD Thesis, University of Sydney. (Australian Centre for Field Robotics)

Lms 200, lms 211, lms 220, lms 221, lms 291 laser measurement systems technical description (Technical Description No. 8 008 970/06-2003). (2003, June).

Lohmann, P., Koch, A., & Schaeffer, M. (2000). Approaches to the filtering of laser scanner data. *Interanational Archives of Photogrammetry and Remote Sensing, 33.*

Marsden, J., & Weinstein, A. (1985). *Calculus 1* (2nd ed.). New York: Springer-Verlag.

Martin, M., & Moravec, H. (1996, March). *Robot evidence grids* (Tech. Rep. No. CMU-RI-TR-96-06). Pittsburgh: The Robotics Institute, Carnegie Mellon University.

Miller, I., Lupashin, S., Zych, N., Moran, P., Schimpf, B., Nathan, A., et al. (2006). Cornell university's 2005 darpa grand challenge entry. *Journal of Field Robotics.* (To appear)

Moon, F. (1998). *Applied dynamics with application to multibody and mechatronic systems.* New York: John Wiley and Sons.

Morin, K. (2002). *Calibration of airborne laser scanners.* Unpublished doctoral dissertation, University of Calgary.

Murray, R., Li, Z., & Sastry, S. (1994). *A mathematical introduction to robotic manipulation.* Boca Raton: CRC Press.

NIMA. (2000). *Digital terrain elevation data (DTED).* (http://www.fas.org/irp/program/core/dted.htm)

Olin, K., & Tseng, D. (1991, August). Autonomous cross-country navigation: An integrated perception and planning system. *IEEE Expert: Intelligent Systems and Their Applications, 6*(4), 16-30.

Pagac, D., Nebot, E., & Durrant-Whyte, H. (1998, August). An evidential approach to map-building for autonomous vehicles. *IEEE Transactions on Robotics and Automation, 14*(4).

Russell, S., & Norvig, P. (2003). *Artificial intelligence: A modern approach* (2nd ed.). Upper Saddle River: Pearson Education, Inc.

Satale, D., & Kulkarni, M. (2003). Lidar in mapping. In *Proceedings of the 2003 map india poster session.* New Delhi, India: GISdevelopment.

Stoker, J. (2004, May). Voxels as a representation of multiple-return lidar data. In *Asprs annual conference proceedings.* Bethesda, MD: American Society for Photogrammetry and Remote Sensing.

Team Cornell. (2005, December). *Technical review of team cornell's spider: Darpa grand challenge 2005* [Technical Review]. (http://www.darpa.mil/grandchallenge05/TechPapers/TeamCornell.pdf)

Thrun, S. (2002, February). *Robotic mapping: A survey* (Tech. Rep. No. CMU-CS-02-111). Pittsburgh: School of Computer Science, Carnegie Mellon University.

Thrun, S., Burgard, W., & Fox, D. (2000). A real-time algorithm for mobile robot mapping with applications to multi-robot and 3D mapping. In *Proceedings of the ieee international conference on robotics and automation (icra).* San Francisco, CA: IEEE.

Weingarten, J., & Siegwart, R. (2005). Ekf-based 3d slam for structured environment reconstruction. In *Proceedings of the ieee / rsj international workshop on intelligent robots and systems* (p. 3834 - 3839). Los Alamitos: Institute of Electrical and Electronic Engineers.

14

Alice: An Information-Rich Autonomous Vehicle for High-Speed Desert Navigation

Lars B. Cremean, Tully B. Foote, Jeremy H. Gillula, George H. Hines, Dmitriy Kogan, Kristopher L. Kriechbaum, Jeffrey C. Lamb, Jeremy Leibs, Laura Lindzey, Christopher E. Rasmussen, Alexander D. Stewart, Joel W. Burdick, and Richard M. Murray

California Institute of Technology, Control and Dynamical Systems, 1200 East California Boulevard, Pasadena, California 91125

Summary. This paper describes the implementation and testing of *Alice*, the California Institute of Technology's entry in the 2005 DARPA Grand Challenge. Alice utilizes a highly networked control system architecture to provide high performance, autonomous driving in unknown environments. Innovations include a vehicle architecture designed for efficient testing in harsh environments, a highly sensory-driven approach to fuse sensor data into speed maps used by real-time trajectory optimization algorithms, health and contingency management algorithms to manage failures at the component and system level, and a software logging and display environment that enables rapid assessment of performance during testing. The system successfully completed several runs in the National Qualifying Event, but encountered a combination of sensing and control issues in the Grand Challenge Event that led to a critical failure after traversing approximately 8 miles.

14.1 Introduction

Team Caltech was formed in February of 2003 with the goal of designing a vehicle that could compete in the 2004 DARPA Grand Challenge. Our 2004 vehicle, Bob, completed the qualification course and traveled approximately 1.3 miles of the 142-mile 2004 course. In 2004-05, Team Caltech developed a new vehicle to participate in the 2005 DARPA Grand Challenge. Through a Caltech course in multi-disciplinary project design, over 50 undergraduates participated in conceiving, designing, implementing and testing our new vehicle, named "Alice" (Figure 14.1). The team consisted of a broad range of students from different disciplines and at different skill levels, working together to create a sophisticated engineering system. The final race team completed the implementation and optimization of the system over the summer as part of the Caltech Summer Undergraduate Research Fellowship (SURF) program.

Alice's design built on many standard techniques in robotics and control, including state estimation using Kalman filters, sensor-based terrain estimation and fusion, optimization-based planning through a "map" of the terrain, and feedback control at multiple levels of abstraction. A novel aspect of the design

M. Buehler, K. Iagnemma, and S. Singh (Eds.): DARPA'05, STAR 36, pp. 437–482, 2007.
springerlink.com

Fig. 14.1. Caltech's 2005 DARPA Grand Challenge Entry, Alice

compared with many robots built prior to the grand challenge was the high-speed nature of the system: Alice was designed to travel through unstructured environments at speeds of up to 15 m/s (35 mph) using multiple cameras and LADARs across a network of high performance computers. The raw data rates for Alice totaled approximately 350 Mb/s in its race configuration and plans were computed at up to 10 Hz. This required careful attention to data flow paths and processing distribution, as well as the use of a highly networked control architecture. In addition, Alice was designed to operate in the presence of failures of the sensing and planning systems, allowing a high level of fault tolerance. Finally, Alice was designed to allow efficient testing, including the use of a street legal platform, rapid transition between manual and autonomous driving and detailed logging, visualization and playback capabilities.

This paper describes the overall system design for Alice and provides an analysis of the system's performance in desert testing, the national qualification event, and the 2005 Grand Challenge race. We focus attention on three aspects of the system that were particularly important to the system's performance: high-speed sensor fusion, real-time trajectory generation and tracking, and supervisory control. Data and measurements are provided for a variety of subsystems to demonstrate the capabilities of the component functions and the overall system.

Alice's design built on many advances in robotics and control over the past several decades. The use of optimization-based techniques for real-time trajectory generation built on our previous experience in receding horizon control for motion control systems (Milam, 2003; Murray *et al.*, 2003) and extended that work to include efficient methods for cost evaluation along a terrain with sometimes sparse data (Kogan, 2005). Our sensor fusion architecture and the use of speed maps built on work at JPL (Goldberg et al., 2002) and we benefited greatly from the work of Dickmanns (Dickmanns, 2004). The supervisory control architecture that we implemented also relied heavily on concepts developed at JPL (Rasmussen, 2001).

The design approach for Alice was shaped by the lessons learned from fielding a team for the 2004 Grand Challenge race, and by the shared experiences of other teams in that event, notably the technical report published by the Red Team (Urmson, 2005; Urmson et al., 2004) and the relative overall success of path-centric versus behavior-based approaches.

The deliberative approaches to solving the Grand Challenge centered on building a grid-based or obstacle-based map of the environment, and performed a search through that map for an optimal path. The field of finalists for the 2005 race partially reflected a convergence of system-level architectures to this approach; 17 of the 23 team technical papers (including those from the five vehicles that completed the course) describe various deliberative implementations (Defense Advanced Research Projects Agency, 2005).

Based on the technical papers, three teams (Axion, Virginia Tech's Cliff, and IVST) implemented a primarily behavior-based navigation architecture, and Princeton University implemented a purely reactive architecture. These alternative approaches are a source of valuable experience and experimental data, and might provide some insight into the relative merits of different approaches.

The description of Caltech's approach proceeds as follows: Section 14.2 describes our system architecture, from the vehicle and associated hardware to the software design. Section 14.3 details the specifics of the vehicle actuation and trajectory-following controller. Our mapping and planning algorithms are explained in Sections 14.4 and 14.5, and our higher-level control and contingency management is described in Section 14.6. Experimental results that illustrate the performance of our system are presented in Section 14.7.

14.2 System Architecture

Caltech's 2004 entry in the DARPA Grand Challenge utilized an arbiter based planning architecture (similar to that in Rosenblatt (1998)) in which each terrain sensor "voted" on a set of steering angles and speeds, based on the goodness determined by those sensors. These votes were then combined by an arbiter, resulting in a command-level fusion. This approach had advantages in terms of development (each voter could be developed and tested independently), but it was decided that this approach would not be able to handle complex situations involving difficult combinations of obstacles and terrain Cremean (2006). Hence a more sophisticated architecture was developed for the 2005 race, based on optimization-based planning. In addition, a high level system specification was used to guide the operation of each component in the system.

14.2.1 System Specification

Team Caltech's strategy for the race was embedded in its overall system specification, which described the performance characteristics for Alice and the team's operations. This system specification was used to drive the specifications for individual components. The final system specification contained the following requirements:

- (S1) 175 mile (282 km) range, 10 hours driving, 36 hours elapsed (with overnight shutdown and restart).
- (S2) Traverse 15 cm high (or deep) obstacles at 7 m/s, 30 cm obstacles at 1 m/s, 50 cm deep water (slowly), 30 cm deep sand and up to 15 deg slopes. Detect and avoid situations that are worse than this.
- (S3) Operate in dusty conditions, dawn to dusk with up to 2 sensor failures.
- (S4) Average speed versus terrain type:

Terrain type	Distance (%)		Speed (mph)			Expected		Time (hr)
	Min	Max	Min	Max	Exp	mi	%	
Paved road	1	10	20	40	30	18	10%	0.6
Dirt road	40	60	10	40	25	132	75%	5.3
Trails	20	30	5	15	10	18	10%	1.8
Off-road	5	20	1	5	5	5	3%	1
Special	n/a	n/a	2	2	2	2	1%	1
Total			1	40	25	175	100%	9.7

- (S5) Safe operation that avoids irreparable damage, with variable safety factor.
- (S6) Safety driver w/ ability to immediately regain control of vehicle; 20 mph crash w/out injury.
- (S7) Commercially available hardware and software; no government supported labor.
- (S8) $120K total equipment budget (excluding donations); labor based on student enrollment in CS/EE/ME 75abc (multi-disciplinary systems engineering course) and 24 full-time SURF students.
- (S9) Rapid development and testing: street capable, 15 minute/2 person setup.

The speed versus terrain type specification (S4) was updated during the course of development. Expected mileages by terrain type were updated from analyses of the 2004 Grand Challenge course, and expected speeds were selected to find a "sweet spot" that balances the trade-off between chance of completing the course (whose trend generally decreases with increased speed, and favors lower speeds) and minimizing completion time (which favors higher speeds).

One of the most important specifications was the ability to do rapid development and testing (S9). This was chosen to to take advantage of being within a few hours drive of desert testing areas: our goal was to test as much as possible in race-like conditions. Our vehicle from the 2004 Grand Challenge, Bob, had to be towed by trailer to and from test sites. We had also removed Bob's steering wheel, which meant that he had to be driven using a joystick. This added unnecessary complication and effort to the testing process. With Alice, the transition

(a) (b)

Fig. 14.2. Vehicle features: (a) front differential, suspension, and front bumper/LADAR mount and (b) the rear workstation area with server array visible (as viewed through the rear doors)

from a human driving to autonomous testing only required the operator to insert a single pin to engage the brake mechanism and then flip a few switches to turn on the other actuators. This meant that we could transition from human driving to an autonomous test in only 5 minutes. Alice also supports a complement of four connected but independent interior workstations for development and testing, a vast improvement over Bob's design.

14.2.2 Vehicle Selection

To help achieve ease of development, a Ford E350 van chassis was obtained and modified for off-road use by Sportsmobile West of Fresno, CA. A diesel engine, augmented by a 46-gallon fuel tank, was selected since it was well suited to the conventional operating envelope of generally slow speeds and long idle times during periods of debugging.

Sportsmobile's conversion is primarily intended for a type of driving known as rock-crawling: low-speed operation over very rough terrain. A four-wheel-drive system distributes power through a geared transfer case (manufactured by Advanced Adapters, Inc.) and the suspension rests in deep compression, allowing the unsprung components to drop very quickly over negative terrain discontinuities (see Figure 14.2(a)). Another key suspension feature is the use of a reverse shackle design in the implementation of the front leaf spring suspension. As opposed to traditional designs, the reverse shackle method places the extensible end of the leaf spring toward the rear of the vehicle. This provides better tracking at speed. The high-knuckle front axle design decreases minimum turning circle by 6% to 45 feet in diameter.

The vehicle was purchased as a "stripped chassis," which means that the vehicle was bare behind the front seats. This allowed a custom interior design that included a central enclosure for the server array and four racing seats, two of which replaced the stock Ford captain's chairs in the front of the cabin. A removable table was placed in front of the two rear seats, and five terminals were positioned throughout the vehicle: one for each seat and one at the side door, accessible from outside the vehicle. Figure 14.2(b) shows a view of the interior of Alice from the rear. Thus equipped, the vehicle is capable of supporting three developers and safety driver while mobile, or four developers while stationary.

The electrical system consists of a 3 kilowatt generator mounted to the rear of the vehicle, producing 120 VAC. This power is directed to two 1500 watt inverter/chargers. Power is passed through these inverters directly to the loads without loss when the generator is running, with the remaining power being diverted to charge the auxiliary battery bank. The battery bank consists of four 12 volt marine gel cell batteries rated at 210 amp-hours each, positioned beneath the floor at the rear of the vehicle to keep the center of gravity low. Power from the charged gel cell batteries is diverted back through the inverters to power AC loads when the generator is not functioning, or to the 12 volt systems powering the LADARs, IMU, GPS and actuators. The power system was tested and verified to operate in high-shock, high-vibration environments for up to 22 hours between generator refuelings.

14.2.3 Computing and Networking

One of the initial specifications for Team Caltech's second vehicle was to be able to survive the failure of one or more computers. The design of the vehicle computing and networking systems was developed based on this specification, even though this functionality was not implemented during the race (due to shortness of time and the strong reliability of the hardware in testing).

The computing platform consisted of 6 Dell PowerEdge 750 servers with 3 GHz, Pentium 4 processors and a single IBM eServer 326 with dual 2.2 GHz dual-core AMD 64 processors. All machines were configured to run Linux; the Gentoo distribution (Vermeulen et al., 2005) was selected based on performance testing early in the system design. Each machine had two 1 Gb/s Ethernet interfaces, although only one interface is used in the race configuration. Originally a serial device server was used to allow any machine to talk to any actuator (all of which had serial interfaces), but the serial device server had reliability problems and was eventually removed from the system. A wireless bridge and wireless access point were also used in the system during testing to allow remote access to the vehicle network.

To allow software components to communicate in a machine-independent fashion, a custom messaging and module management system called "Skynet" was developed. Skynet was specified to provide inter-computer communication between programs on different computers or on the same computer completely transparently. The code run on Alice is therefore broken down into discrete functional modules, each derived from a Skynet class to allow Skynet to start and stop modules. This is required because Skynet was also designed be able to start, stop, and redistribute modules between computers based on computational resources available, assuming hardware failures of any type were possible. Skynet's communication capabilities are built on top of the Spread group communication toolkit (Amir et al., 2004).

In the race configuration, the ability to redistribute and run modules on different computers was not implemented and module restart was accomplished using runlevel functionality provided by Linux (Vermeulen et al., 2005). A custom runlevel was created to put each computer into race-ready mode upon entry, by running a series of scripts to start the modules appropriate for each machine. Should a given module exit or die for any reason, it is immediately restarted using the respawn setting in /etc/inittab. Paired with other scripts for recompiling and distributing our modules and configuration files, this proved to be an extremely efficient way of upgrading software and switching to autonomous operations during development and testing.

14.2.4 Software Architecture

A diagram of our general system architecture is shown in Figure 14.3. Environmental sensors (stereo vision and LADAR) are used to create an elevation map of the terrain around the vehicle. Our range sensor suite consisted of multiple LADAR units and stereo camera pairs. The data from these sensors, in combination with our state estimate, creates an elevation map in the global reference frame. The elevation map consists of a grid of cells, centered on the vehicle, where the value of each cell corresponded to the elevation of that cell. The map moves along with the vehicle, so as cells move some distance behind us, new cells are created in front of the vehicle.

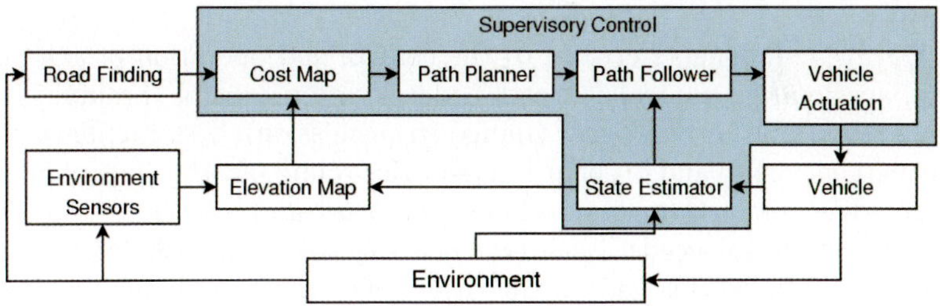

Fig. 14.3. Overall System Architecture for Alice

This elevation map is then converted into a cost map by considering aspects such as elevation gradient, quality of data in that cell, etc. The cost map establishes a speed limit for each location in the terrain around the vehicle. In addition to terrain data, the speed limits set in the Route Definition Data File (RDDF) and information from road finding algorithms (Rasmussen and Korah, 2005) are integrated at this point. The speed limit-based cost map allows the vehicle to traverse rough sections of the terrain (slowly) and insures that the vehicle attempts to make forward progress unless there was an insurmountable obstacle (speed = 0) in its path. The elevation and cost mapping algorithms are discussed in more detail in Section 14.4.

Once the cost map is generated, it was passed onto the planner where a time-optimal path is created that satisfies vehicle and map speed constraints (Kogan, 2005). This path is sent onto the path follower, which in turn computes and sends appropriate actuation commands (brake, throttle, and steering) to the vehicle actuation system. The vehicle control systems are discussed in Section 14.3 and the path planning algorithm is discussed in Section 14.5.

The supervisory control module serves to detect and manage higher-level system faults that other individual modules cannot. This includes scenarios such as losing and reacquiring GPS, and being stuck on an undetected obstacle. The supervisory control module is also responsible for maintaining forward progress in unusual conditions, such as the case of running up against barbed wire (the failure mode for Team Caltech's 2004 entry). This system is described in more detail in Section 14.6.

14.3 Control System Design

The lowest level of the vehicle control system consists of the vehicle's actuation system, its state sensing hardware and software, and the trajectory tracking algorithms. Like many teams in the Grand Challenge, these systems were custom designed based on commercially available hardware. This section describes the design decisions and implementation of the control system for Alice.

14.3.1 Vehicle Actuation Systems

There are five components critical to the control and operation of any vehicle: steering, acceleration control (throttle), deceleration control (braking), engine start and stop (ignition) and gear change (transmission). Two auxiliary components, emergency stop and on-board diagnostics round out the control interfaces in use on Alice. Team Caltech developed actuators and interfaces to meet each of the aforementioned needs. Each actuator was designed with safety and performance in mind. To this end, in the case of electrical, mechanical or software failure, critical actuation subsystems automatically bring the vehicle to a quick stop without needing external commands. To achieve the desired overall performance, each actuator is designed by specification to be able to react at least as fast as a human driver.

Each actuator uses its own RS-232 serial interface to communicate with the computing system. The original specification required the ability to survive one or more computer failures which dictated that any actuator had to be available to at least two computers at any time. This led to the use of an Ethernet enabled serial device server. However, through testing it was determined the mean time between failures for the serial device server was shorter than other components in the actuator communication chain. As a result the specification was dropped and direct serial connections were made to the computers.

Each actuation subsystem is controlled through individual switches accessible to a safety driver sitting in the driver's seat (Figure 14.4(a)). This configuration

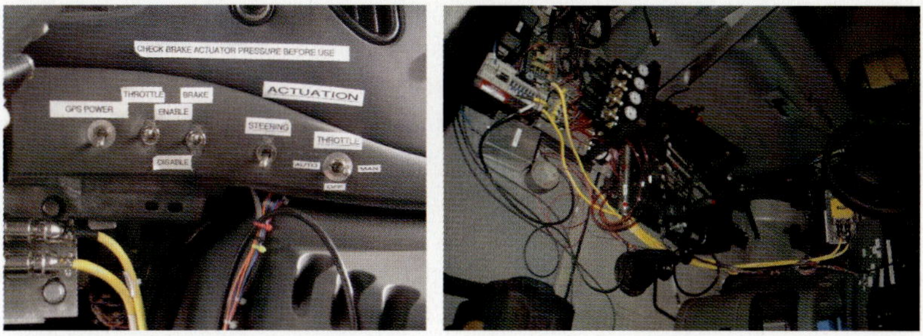

(a) (b)

Fig. 14.4. Vehicle actuation systems: (a) dashboard controls and (b) pneumatic braking system

allows a great deal of flexibility in testing, including allowing the safety driver to control some functionality of the system while the software controlled other functions. The dashboard controls are also used to allow the safety driver to take over control of the vehicle during testing.

Steering Actuator. The steering system used on Alice was originally developed for Bob and used during the 2004 Grand Challenge. It was adapted and improved for use in the 2005 competition and proved to be a highly reliable component. The basis of this system is a Parker-Hannefin 340 volt servo motor and a GV-6UE controller interfaced to the computing cluster via RS-232. This servo-motor is capable of providing 8 Nm of torque while moving and up to 14 Nm stall torque. Attached to this motor is a 3:1 single stage planetary gearbox which is attached to the steering column by means of steel chain providing a further 1.8:1 reduction. The result is a system capable of providing a continuous torque of more than 40 Nm to the steering column, which is sufficient to turn the wheels should the power steering system of the vehicle fail. The variable speed motor is tuned to allow the wheels to traverse from one end of travel limit to the other in approximately 1.5 seconds, which is slightly faster than is possible for a human. Higher speeds were tested, but despite the reduced delay, they were found to contribute to an overly aggressive behavior in the vehicle control system, as well as causing heavier wear on all components.

Brake Actuator. Following Team Caltech's less than satisfactory experience with electrical linear actuators for braking in the 2004 Grand Challenge, a pneumatic actuation system was chosen for use this year. This choice provided a faster response while still providing enough force to bring the vehicle to a complete stop. The five piston mechanical portion of the actuator was designed and implemented by Chris Pederson of competing team A. I. Motorvators. As shown in Figure 14.4(b), the actuator consists of five pistons of incrementally increasing bore arrayed in parallel with their piston rods acting in tension attached to a single pivot point. A one-to-one ratio pivot positioned beneath the driver's seat changes the retracting motion of the pistons to a compressive motion applied to the brake pedal by means of a removable pushrod. Four primary pistons are used for articulate control of braking pressure during autonomous operation. Each piston may be applied and released independently by means of 5 volt active solenoid control valves, each of which is in turn controlled by solid-state relays linked to an Atmel AVR microprocessor, interfaced to the control computer. In case of electrical or air compressor failure, the fifth cylinder is automatically deployed using a separate air reserve. The entire system runs on 70 psi compressed air provided by two 2.4 cfm, 12 volt air compressors. Air is compressed, piped to a reservoir, and then distributed to the brake and sensor cleaning systems. A pressure switch located at the main manifold will close and deploy the reserve piston if the main air system pressure falls below a safe level.

Throttle Actuation. Caltech's 2005 E-350 van shipped with a 6.0L Powerstroke diesel engine that is entirely electronically controlled, including control of engine acceleration, referred to here as "throttle". The accelerator pedal in this vehicle is simply an electrical interface to the Powertrain Control Module (PCM), which is the Ford vehicle computer. Using specifications provided by engineers from Ford, an interface was designed that closely approximates the response of the accelerator pedal. Using a microcontroller and digital to analog converters, the accelerator pedal was replaced with an effective interface that was controlled via RS-232. The stock pedal was left in the vehicle, and can be used by flipping a switch to disable the aftermarket actuator. Unfortunately, the throttle actuation solution was not perfect. The PCM requires that three sampled input lines agree to within 2.1% otherwise an error condition is declared. In this case the PCM puts the vehicle into a mode called "Limited Operating System" which restricts the engine to idle. The only method to clear the LOS condition is to restart the engine, which requires ignition actuation. Despite these minor problems, the throttle actuation performed admirably, exhibiting no measurable delay between command and actuation, as well as a high degree of operational robustness.

Ignition and Transmission Actuation. Ignition and transmission actuation, while two completely separate sub-systems on the vehicle, were bundled together in one actuator controller out of convenience, as neither required high communication bandwidth nor was processing intensive for a microcontroller. Ignition control was achieved through the use of three 50 amp solid state relays to control three circuits: powered in run, powered in start, or powered in run and start. The ignition actuator as developed is tri-state: Off, Run or Start. The vehicle protects the ignition system so that the starter motor cannot operate if the engine is already running, and so that the engine cannot be started unless the vehicle is in park or neutral.

Transmission actuation was achieved through the use of an electronic linear actuator connected to the transmission by means of a push-pull cable. A four position controller was used to provide automated shifting into Park, Reverse, Neutral or Drive. The ignition and transmission can also be controlled by a human in one of two ways. Included in the actuator design is a control box that allows the human driver to take control and operate the vehicle by means of a push-button ignition system and a multi-position switch knob for transmission control. Alternatively each can be controlled manually after disabling the actuators.

On Board Diagnostics. To gather additional real-time data about the vehicle's performance, Alice's CAN bus was accessed using a commercial OBD II reader. Data values such as wheel speed, throttle position, transmission position, and engine torque are gathered to provide more information for contingency management and state estimation. However the data had a large amount of latency due to poor interfaces, with overhead on the order of 10 times the data throughput. To alleviate the bandwidth limitations, asymmetric polling methods are implemented. By prioritizing how often data fields are polled, performance was increased from 0.5 Hz updates to 2 Hz for the time critical value of vehicle speed, which is still significantly less than the desired 10 Hz. Data rates for other fields have been decreased to as slow as once every 8 seconds for the lowest priority information.

Emergency Stop. Critical to the safety of operators and observers is a method to stop the vehicle when it is autonomous. DARPA provided the basis for this remote safety system in the form of an "E-Stop" system consisting of a transmitter-receiver pair manufactured by Omnitech Robotics. Using a 900 MHz transmitter, an operator outside the vehicle and up to 11 miles (line of sight) away can send one of three commands: RUN, PAUSE, or DISABLE. RUN allows normal operation of the vehicle, PAUSE brings the vehicle to a full and complete stop with all components still functioning, and DISABLE powers down the throttle, turns off the engine, and deploys full braking pressure. The receiver inside the vehicle is connected to a microcontroller interfaced to the computing cluster. This device also takes input from several PAUSE and DISABLE switches located throughout the vehicle, and then passes the most restrictive state to the vehicle interface, called Adrive, which is described in the next section. This system is built with many failsafes to ensure safe operation over a variety of failure modes, including loss of power and communication. The emergency stop system is designed so that even in the event of a full computing system or failure, the vehicle can be remotely brought to a safe stop.

14.3.2 Vehicle Interface

A software module, called Adrive, was specified to provide an abstracted network interface between all the vehicle actuators and computer control. The primary role for the module was to listen for commands on the network then execute them within 50 ms. The second role was to report regularly each actuator's current state (status and position). And a third role of the abstraction was to protect the vehicle from being damaged by control logic or system failures.

All the current state data as well as configuration settings were contained in a hierarchical data structure. Adrive used a multi-threaded design to allow multiple interrupt driven tasks, such as serial communication, to operate concurrently. For optimal performance the command threads would only execute on new commands when the actuator is ready, instead of queuing commands. This allows subsequent commands to have less lag. The maximum delay of the caching system is only

$$\text{Max Delay} = 1/f_{\text{actuator}} + 1/f_{\text{command}},$$

versus the minimum possible delay

$$\text{Optimal Delay} = 1/f_{\text{actuator}},$$

where f_{actuator} is the frequency of the actuator update and f_{command} is the frequency of the incoming command. So as long as f_{command} is much larger than f_{actuator} this approach approximates optimal.

Within Adrive, two levels of fault tolerance were implemented. At the command level every incoming command was run through a set of rules to determine if the command was safe to execute. For example, when the vehicle is in park and command is received to accelerate, the command will not be executed. In addition to physical protection these low level of rules also encompassed the procedures for PAUSE and DISABLE conditions.

At the actuator level there was a supervisory thread which periodically checks each of the actuators for reported errors or failure to meet performance specifications. If the error occurred on a critical system the supervisory thread will automatically PAUSE the vehicle and attempt to recover the failed subsystem. This can be seen in the case where the engine stalls. The OBD-II will report the engine RPM is below a threshold. The supervisory thread will detect the low RPM as below acceptable and put the vehicle into PAUSE mode, which immediately stops the vehicle. The command will then be sent to restart the engine. When the OBD-II reports the RPM above the threshold the PAUSE condition will be removed and operation will return to normal.

The use of Adrive to abstract simultaneous communication between multiple software modules and multiple actuators worked well. The computer resources required are usually less than 1% CPU usage and required less 4MB of memory. Response times range from 10 ms to 100 ms. However the maximum bandwidth of approximately 10 Hz is limited by the serial communications. The final structure proved to be a relatively adaptable and scalable architecture to allow many asynchronous control interfaces accessible to multiple processes across a network.

14.3.3 State Sensing

The estimate of vehicle state (X) is made up of the vehicle's global position (northing, easting, altitude), orientation (roll, pitch, yaw), both the first and second derivatives of these values, as well as the precise time-stamp for when this measurement was valid. The requirements and specifications of tolerances for planning, trajectory tracking and sensor fusion dictate how timely and accurate estimation of Alice's state needs to be. The main specification we believed necessary for "safe driving" was to be able to detect a 20 cm obstacle at 80 m. Doing so requires our system to estimate the terrain with no more than 10 cm of relative error over 80 m of travel, and orientation estimates accurate to within 0.001 radians (based on the approximation $\arctan(0.1/80) \approx 0.001$).

We chose to approach the state estimation problem by combining the outputs of an inertial measurement unit (IMU), global positioning system (GPS), and on-board velocity measurement (OBD II). The output of the IMU (Z^{imu}) is integrated forward from the current position estimate according to the standard inertial navigation equations:

$$X_{t+1}^{imu} = Nav(X_t, Z_{t+1}^{imu}).$$

Unfortunately, while the IMU provides smooth relative position that is accurate over a short time, errors in the initial estimate, IMU biases, and scale factors lead to a solution that drifts from the true position quadratically.

A measurement of the true position or velocity (Z^{true}) can be read directly from the GPS unit or OBD II system. In the case of OBD II velocity, time delays are such that only the stationary condition is used, still allowing us to eliminate drift when stationary, even in the absence of a GPS signal. By taking the difference between our IMU and GPS or OBD II based state estimates, we arrive at a measurement of our error (ΔX):

$$Z_t^{\Delta X} = X_t^{imu} - Z_t^{true}.$$

We then use an extended Kalman filter to estimate ΔX, using the self-reported covariance of the GPS measurement as the covariance of $Z^{\Delta X}$. ΔX is propagated forward in time according to a linearization of the equations of motion being integrated in the inertial solution, and an estimate of the errors in the IMU readings. Corrections are then applied to the true state estimate by subtracting these estimated errors from the inertial solution, effectively zeroing out the error values:

$$\begin{aligned} X_t &= X_t^{imu} - \Delta X_t \\ \Delta X_t &= 0. \end{aligned}$$

An additional constraint largely influencing the construction of the state estimator was the fact that our maps are built in a global reference frame. The problem is that previous state errors are cemented into the regions of the map where we are not getting new terrain measurements. This has large ramifications on correcting position errors, since updating our state to the correct location creates a jump discontinuity, which in turn creates a sharp, spurious ridge in the map. As such, there was a trade-off to be made between reporting the most accurate state at a given time, and the most "useful" estimate of state from the point of view of the other modules.

We took two approaches to solving this problem. One solution is to smooth the applied corrections. Rather than applying a correction all at once, a fraction of the correction is applied at each time step, leading to a smoothed out exponential decay in error rather than a sharp discontinuity:

$$\begin{aligned} X_t &= X_t^{imu} - c\Delta X_t \\ \Delta X_t &= (1-c)\Delta X_t. \end{aligned}$$

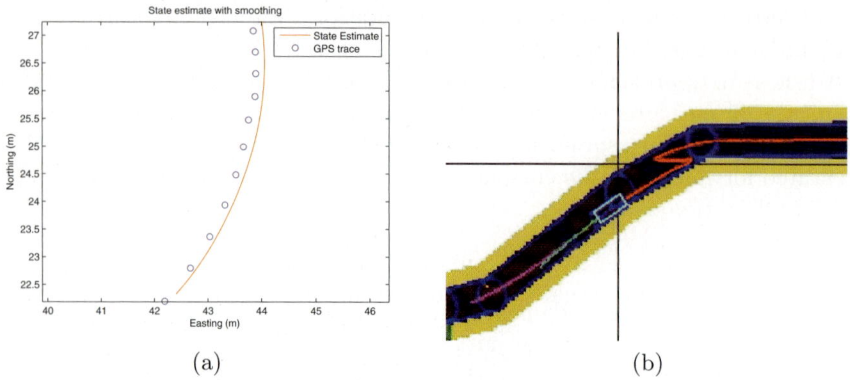

Fig. 14.5. Vehicle state estimate: (a) smoothing of small GPS signal and (b) sample GPS "jump": the trace shows the estimated vehicle state, with a correction while the vehicle is stopped (from second NQE run, just after the tunnel)

For small jump-discontinuities this is sufficient to avoid problems with discontinuities, but increases the amount of tracking error, as shown in Figure 14.5(a).

However, occasionally, when new GPS satellites are picked up or lost (such as when going through a tunnel, or under power-lines), the GPS reading itself can jump by many meters. To deal with this problem we pre-process the GPS data to determine when it is "safe" to incorporate new measurements. As long as the effective jump is below a threshold, GPS data can be continuously applied, making small corrections and keeping the state estimate from ever drifting. If the GPS jumps by a large enough amount (2 meters in the race configuration), we temporarily ignore it, assuming it is more likely a glitch in GPS. During this time, however, the covariance in positional error begins to grow. If the covariance crosses an empirically determined threshold, we deem it necessary to start incorporating GPS measurements again, and bring the vehicle to a pause while we apply the new corrections. The vehicle is allowed to return to forward motion when the state estimator indicates that the estimate has converged, as determined by looking at the magnitude of the differences between subsequent state estimates. Because of the use of a global representation for the cost map (and hence the implicit location of obstacles), the elevation and cost maps for the system are cleared at this point. This "jump" functionality is implemented through the use of a supervisory control strategy as described in Section 14.6. A sample state jump is shown in Figure 14.5(b).

14.3.4 Trajectory Tracking

The design specification for the trajectory tracking algorithm is to receive a trajectory from a planning module and output appropriate actuator commands to keep Alice on this trajectory. The inputs to the algorithm are the current state of the vehicle (position and orientation, along with first and second derivatives) and the desired trajectory (specified in northing and easting coordinates, with their first and second derivatives). From these inputs, the algorithm outputs steering and brake/throttle commands to Adrive. Goals for accuracy were +0/-10% for velocity tracking, and ±20 cm perpendicular y-error at 5 m/s, with larger errors allowable at higher speeds. These performance criteria needed to be met on any terrain type found in the system specifications, at speeds up to 15 m/s.

System Characterization. Before designing the controller, it was necessary to characterize the open-loop dynamics of the system. With this characterization, a mapping from actuator positions to accelerations was obtained. They showed that Alice understeers, and allowed the determination of safe steering commands at various speeds, such that the vehicle would remain in the linear response region. In this region, the feedforward term will be reasonable, and possibly dangerous roll angles/sliding are avoided. Additionally, system delays were determined by examination of the time between commands leaving this module and the resulting vehicular accelerations.

Control Law Design. Although not entirely independent, the lateral and longitudinal controllers are treated separately in the system design. Longitudinal (throttle and brake) control is executed by a feedback PID loop around error in speed plus a feedforward term based on a time-averaged vehicle pitch, to reduce steady-state error when traveling up or down hills.

The trajectories received as input to the trajectory follower encoded first and second derivative data as well as geometry of the path, so that desired velocity and acceleration are encoded. For the longitudinal controller, we decided not to use a feedforward term associated with acceleration based on the input trajectory. This was determined by experience, as there were occasions where the feedforward term would overpower the feedback, and simultaneous tracking of speed and acceleration was not achievable. For example, the vehicle might not correct error associated with going slower than the trajectory speed if the trajectory was slowing down.

The lateral control loop includes a feedforward term calculated from curvature of the path along with a PID loop around a combined error term.

The error for the lateral PID is a combination of heading and lateral errors:

$$C_{\mathrm{err}} = \alpha \tilde{Y}_{\mathrm{err}} + (1 - \alpha)\theta_{\mathrm{err}},$$

where $\tilde{Y}_{\mathrm{err}}$ is the lateral position error (saturated at some maximum value $Y_{\max}$), θ_{err} is the heading error and β is a scale factor. This form was motivated by a desire to obtain stability at any distance from the path. Using this error term, the (continuous) vector field in the neighborhood of a desired path will be tangent to the path as $Y_{\mathrm{err}} \to 0$ and will head directly toward the path at distances greater than $Y_{\max}$ away from the path.

Note that the use of this error term requires an accurate estimate of the vehicle heading. Systematic biases in this estimate will result in steady-state error from the desired path.

The lateral feedforward term is generated by determining the curvature required by the input trajectory, and applying the mapping for steering position to curvature, which yields

$$\phi_{\mathrm{FF}} = \arctan\left(L\frac{\dot{N}\ddot{E} - \dot{E}\ddot{N}}{(\dot{N} + \dot{E})^{\frac{3}{2}}}\right), \tag{14.1}$$

where N is the northing position, E is the easting position and L is the distance between front and rear wheels. For this calculation, it is assumed that Alice behaves as described by the bicycle model (McRuer, 1975).

To avoid oscillations, an integral reset was incorporated in both the lateral and longitudinal controllers, when the relevant error was below some acceptable threshold. In testing, the lateral integral term rarely built up to any significant amount, since performance of the system maintained modest errors comparable to the threshold. For the longitudinal controller, resetting helped to alleviate the overshoot associated with transferring from hilly to flat ground.

To compensate for system delays, a lookahead term was added. This chose the point on the trajectory that lateral feedforward and longitudinal feedback would be computed from.

System Interfacing. Many of the difficulties in implementing the tracking controller were due not to the actual control algorithm but to interface issues with the trajectory planner and the state estimation software. The controller is relatively simple, but has to make certain assumptions about the data and commands it is receiving from the other modules. Working out these interfaces made the most difference in performance. The trajectory tracking module (trajFollower) interfaced with three other modules: the trajectory planner (plannerModule), the state estimator (Astate) and the software that directly controlled the actuators (Adrive).

Since the planner has no concept of the actual topography of the area, a feasibility evaluator was implemented to check final commands against vehicle state. This is broken down into two components: a "bumpiness" sensor (DBS) and a dynamic feasibility estimator (DFE). The DBS algorithm analyzes the frequency components of the state signal to determine how rough the road is, and adjusts the reference velocity accordingly. In testing on dirt trails, this component choose speed limits comparable to a human's judgment (within 1 m/s). The DFE algorithm places hard limits on safe steering angles, dependent on our speed. These limits were chosen based on where Alice began to understeer.

In the final implementation, the quality of the state estimate provides the most variability in trajectory following performance. Small jumps in state are detrimental to stability, as they lead to an immediate jump in proportional error, and thus steering command. More problematic are the intermittent difficulties with poor heading estimates, which lead to constant y errors in tracking the trajectory.

The interface between the planner and trajectory following has to make assumptions about how aggressive trajectories may be and what type of continuity they possess. Obviously, the sequential plans need to be continuous for the controller to be stable. However, this leaves open the question of what they are continuous with respect to. Initially the plans originated from the current vehicle state, but this led to oscillations due to the similarity of the rates between the two loops (and the consequent phase delays in the loop transfer function). Instead, the final race configuration used the planner to update plans based on new terrain, and information continuity was then established with the initial portion of the previous trajectory, both in space and in time. One side effect of this approach is that it is possible for situations to occur where the vehicle physically cannot do what is commanded. To handle these cases, it is necessary for the planner to reset and replan from vehicle state whenever thresholds for position or velocity errors are exceeded.

Tracking Performance. Figure 14.6 shows representative results of the performance of the trajectory tracking algorithms for longitudinal and lateral motion. The algorithm was tested extensively prior to the qualification event and demonstrated excellent performance in testing and during the multiple qualifying event runs.

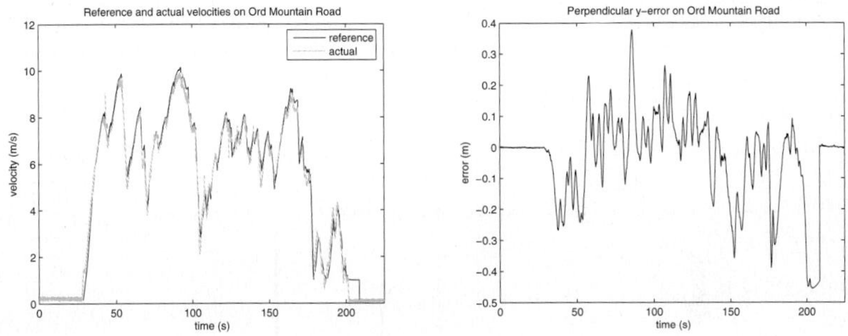

(a) (b)

Fig. 14.6. Trajectory tracking results: (a) longitudinal control and (b) lateral control. The resetting of the error at the end of the trajectory is due to the planner replanning from current vehicle state when the vehicle was externally paused.

14.4 Sensing and Fusion

Alice uses a sensor suite of an IMU, GPS, range sensors and monocular vision to perceive its own state and that of the environment around it (see Table 14.1 for the full list). Obstacle detection is performed using a combination of several SICK and Riegl LADAR units, as well as two pairs of stereovision cameras. Figure 14.7 shows Alice's sensor coverage. This combination of disparate sensors was chosen to allow Alice to be robust to different environments as well as to multiple sensor failures—a critical requirement in our specification for autonomous desert driving.

To accurately navigate through its environment, it is necessary to provide Alice with a method of fusing the sensor data from its range sensors into a single representation of its environment that can capture information concerning both

Table 14.1. Sensors used on Alice

Sensor Type	Mounting Location	Specifications
LADAR (SICK LMS 221-30206)	Roof	180° FOV, 1° resolution, 75 Hz, 80 m max range, pointed 20 m away
LADAR (SICK LMS 291-S14)	Roof	90° FOV, 0.5° resolution, 75 Hz, 80 m max range, pointed 35 m away
LADAR (Riegl LMS Q120i)	Roof	80° FOV, 0.4° resolution, 50 Hz, 120 m max range, pointed 50 m away
LADAR (SICK LMS 291-S05)	Bumper	180° FOV, 1° resolution, 80 m max range, pointed 3m away
LADAR (SICK LMS 221-30206)	Bumper	180° FOV, 1° resolution, 80 m max range, pointed horizontally
Stereovision Pair (Point Grey Dragonfly)	Roof	1 m baseline, 640x480 resolution, 2.8 mm focal length, 128 disparities
Stereovision Pair (Point Grey Dragonfly)	Roof	1.5 m baseline, 640x480 resolution, 8 mm focal length, 128 disparities
Road-Finding Camera (Point Grey Dragonfly)	Roof	640x480 resolution, 2.8mm focal length
IMU (Northrop Grumman LN-200)	Roof	1–10° gyro bias, 0.3–3 mg acceleration bias, 400 Hz update rate
GPS (Navcom SF-2050)	Roof	0.5 m CEP, 2 Hz update rate
GPS (NovAtel DL-4plus)	Roof	0.4 m CEP, 10 Hz update rate

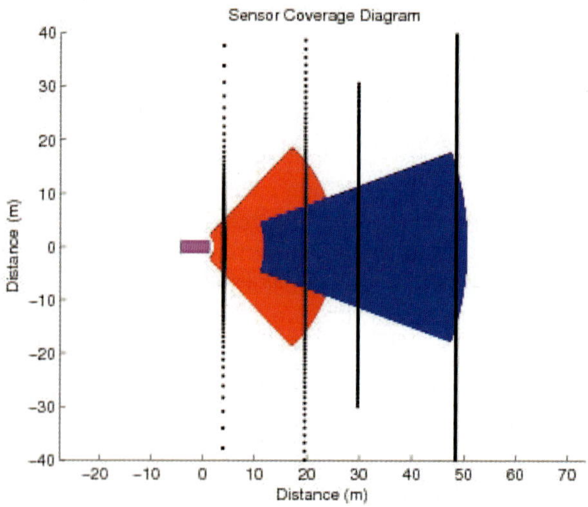

Fig. 14.7. Alice's sensor coverage. Black dotted lines indicate the intersection of LADAR scan planes with the ground, the shorter, wider cone indicates ground covered by the short-range stereovision pair, and the longer, narrower cone indicates coverage by the long-range stereovision pair. The box to the left is Alice, for scale.

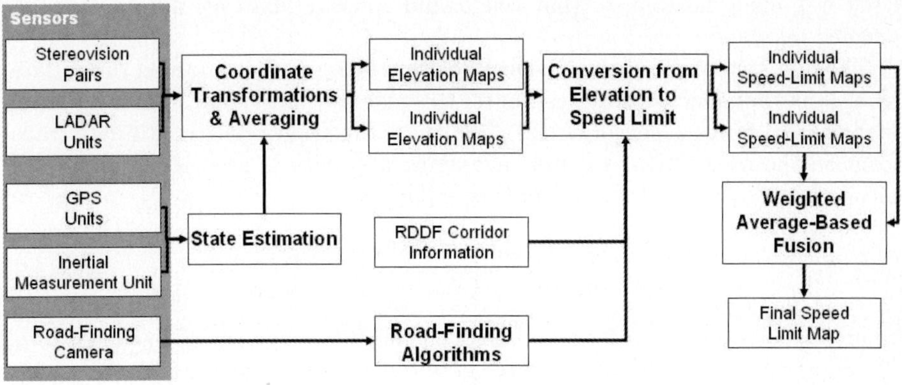

Fig. 14.8. The sensor fusion framework

where it would be safest and fastest for it to drive. The decision of what sort of representation to use was driven primarily by the requirements of the path-planning software: it expected as its input a speed limit map, where each cell in the map contained a maximum speed intended to limit the speed at which the vehicle could drive through that cell. The problem of sensor fusion is thus naturally broken down into two sub-problems: how the incoming sensor data should be fused, and how the speed limit data should be generated from the raw sensor data. The resulting algorithm involves three basic steps (Figure 14.8):

1. For every range sensor, incoming data is transformed into global coordinates (UTM for x and y, and altitude in meters for z) using the vehicle's state estimate, and then averaged into the map along with any existing data from that sensor. This creates several individual elevation maps (one per sensor).
2. For every elevation map, a conversion is performed to transform the elevation data into speed limit data, based on a set of heuristic measures of goodness.
3. Matching cells from all of the speed limit maps are averaged together to produce a single speed limit map, which is passed to the path-planning software.

A more detailed description of each stage of the process, as well as an analysis of some of the strengths and weaknesses of our approach, is given below.

14.4.1 Map Data Structure

The map data structure we use is a simple 2.5D grid with fixed cell length and width (40 cm per side for short-range sensors, 80cm for long-range sensors), as well as fixed overall map length and width (200 m per side for all sensors). The map is referenced using global coordinates (UTM coordinates for x and y, altitude for z), and is scrolled along with the vehicle's movement such that the vehicle always remained at the center of the map. The data structure of the map itself was made flexible, so that cells could contain whatever information was deemed necessary.

The first step of the sensor fusion algorithm is to use the incoming range data, as well as the vehicle's state estimate, to create an individual digital elevation map (DEM) for each sensor. After performing the appropriate coordinate transform on the range data, a simple averaging approach is used to integrate new measurements with old ones, using the equation:

$$\begin{aligned} z_{ij} &\leftarrow (z_{ij} + z_m)/(n_{ij} + 1) \\ n_{ij} &\leftarrow n_{ij} + 1, \end{aligned} \tag{14.2}$$

where z_{ij} corresponds to the estimate of the height of a given cell in the grid, n_{ij} is a count of the number of measurements that fall into the cell i, j, and z_m is a new height measurement for that cell. Due to time constraints, we were not able to take into account error models of either the range sensors or the state estimates when creating our elevation maps. This is one of the reasons we chose to perform sensor fusion in the speed limit domain: it dramatically reduced the sensitivity of the system to cross-calibration errors in the range sensors. Measurements are fused by the system as they come in, and after each scan (or frame) the resulting changes are sent on to the next stage of the algorithm.

14.4.1.1 Elevation to Speed Limit Conversion

In the next stage of the algorithm, the system needs to convert elevation data into speed limit data. We use two independent measures for this task. The first measure is the variance in elevation of the cell: the larger the variance, the larger the spread in elevation measurements that fall into that cell, the more likely that cell contains some sort of vertical obstacle, and thus the more dangerous the cell is to drive through. If the variance is greater than a certain threshold, the cell is identified as an obstacle, and the speed limit s_1 of that cell is set to zero.[1] Otherwise, the speed limit is set to some maximum value. Thus the variance acts as a very rough binary obstacle detector, able to detect obstacles that are completely contained within a single cell (e.g., a fence post, which is thinner than a cell width or length, but certainly tall enough to pose a danger to the vehicle).

The second measure we use for determining the speed limit of a cell is a discrete low-pass filter over a window surrounding that cell. The precise equation for the low-pass filter for a cell at row r and column c is

$$l(r, c) = \frac{1}{n} \sum_{\substack{i=-K \\ i \neq 0}}^{K} \sum_{\substack{j=-K \\ j \neq 0}}^{K} |(z_{r+i,c+j} - z_{r,c}) * G_{i,j}(\sigma)| \tag{14.3}$$

where K is the half-width of the window (e.g., the window has $2K + 1$ cells per side), $z_{r,c}$ is the elevation of the cell at row r and column c, n is the number of cells

[1] In fact, a small non-zero value is used to avoid numerical difficulties in the path planner.

used within the window (at most $(2K+1)^2$), and $G_{i,j}(\sigma)$ corresponds to the value at cell i, j of a discrete Gaussian centered at the origin with standard deviation σ. In essence then, the filter acts as a Gaussian weighted average of the difference in height between the central cell and its neighbors within the window described by K. Of course, not all of the neighboring cells necessarily have an elevation estimate. For example, a large obstacle may cast a range shadow, blocking the sensor from making measurements behind the obstacle. When such a cell is encountered in the sum, it is discarded and n is reduced by one. Thus we have $n = (2K+1)^2 - 1 - m$, where m is the number of cells without elevation data inside the window. Additionally, if n is smaller than some threshold, then the output of the filter is not computed, based on the principle that the system did not have enough information about the neighborhood surrounding the center cell to make a reasonable estimate of the area's roughness.

Assuming the output is computed, however, we still need to transform it into a speed limit. To accomplish this transformation, we pass the output through a sigmoid of the form

$$s_2 = h * \tanh(w * (l(r,c) - x)) + y, \tag{14.4}$$

where $l(r,c)$ is the response of the low-pass filter given above, and the parameters h, w, x, and y are tuned heuristically by comparing sample filter responses to the real-life situations to which they corresponded. Then, to compute the final speed s_f, we simply take the minimum of s_1, the speed generated using the variance, and s_2, the speed generated using this roughness filter.

Of course, the computations described here are not computationally trivial; to do them every time a single new elevation data point is generated would be infeasible given that some of the sensors are capable of generating hundreds of thousands of measurements per second. Additionally, over a small time interval (e.g., less than one second) measurements from a sensor will frequently fall into a small portion of the entire map. To take advantage of this redundancy, and to increase the computational speed of the algorithm, speed limits are generated at a fixed frequency instead of on a per-measurement basis. At some predefined interval the algorithm sweeps through the map, regenerating the speed limit of any cell whose elevation value has changed since the last sweep. Additionally, any cells whose neighbors have changed elevation values are also re-evaluated, as the response of the roughness filter could be different. To speed this process up, the algorithm uses a dynamic list to keep track of which specific cells have changed. Any time a cell receives a new measurement, the algorithm adds that cell to the list of cells to update during the next sweep (if the cell has not been added already). The result is an algorithm that takes advantage of the redundancy of subsequent measurements from a single sensor to transform a digital elevation map into an accurate speed limit map quickly and efficiently.

14.4.2 Speed Limit-Based Fusion

After computing a separate speed limit map for each sensor, all that remains is to fuse these different maps together into a single final speed limit map. We

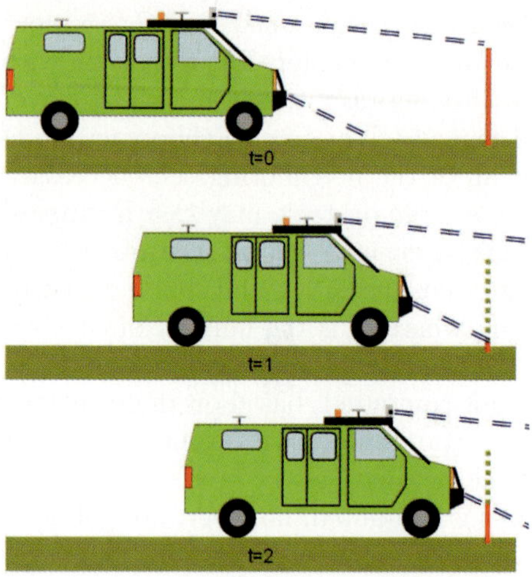

Fig. 14.9. The disappearing obstacle problem

chose to use a simple weighted average, using heuristics to determine the weights for each sensor. Thus the equation for determining the final fused speed limit $s(r, c)$ of a cell at row r and column c is

$$s(r, c) = \frac{\sum_i s_i(r, c) * w_i}{\sum_i w_i}, \tag{14.5}$$

where $s_i(r, c)$ is the speed limit of the corresponding cell in the speed limit map of sensor i, and w_i is the weight associated with the sensor i. However, this algorithm has a serious flaw: obstacles occasionally disappear and then reappear as they entered the range of each sensor (see Figure 14.9). This phenomenon is due to the weights used for fusing the sensors: sensors that are pointed closer to the vehicle have higher weights than sensors that are pointed further away, on the principle that the closer a sensor is pointed, the less susceptible its output is to synchronization errors between range measurements and the vehicle's state estimate. As a result, a chain of events can occur in which:

1. Sensor A, a long-range sensor, picks up measurements from an obstacle and assigns the cells that contain it a low speed.
2. The very bottom of the obstacle comes into the view of sensor B, a short-range sensor.
3. The cell containing the leading edge of the obstacle is set to a higher speed than was there before because: sensor B sees only the part of the obstacle near to the ground; seeing only the part near the ground, sensor B perceives only a very small obstacle (if it detects one at all); and finally, sensor B has a higher weight than sensor A.

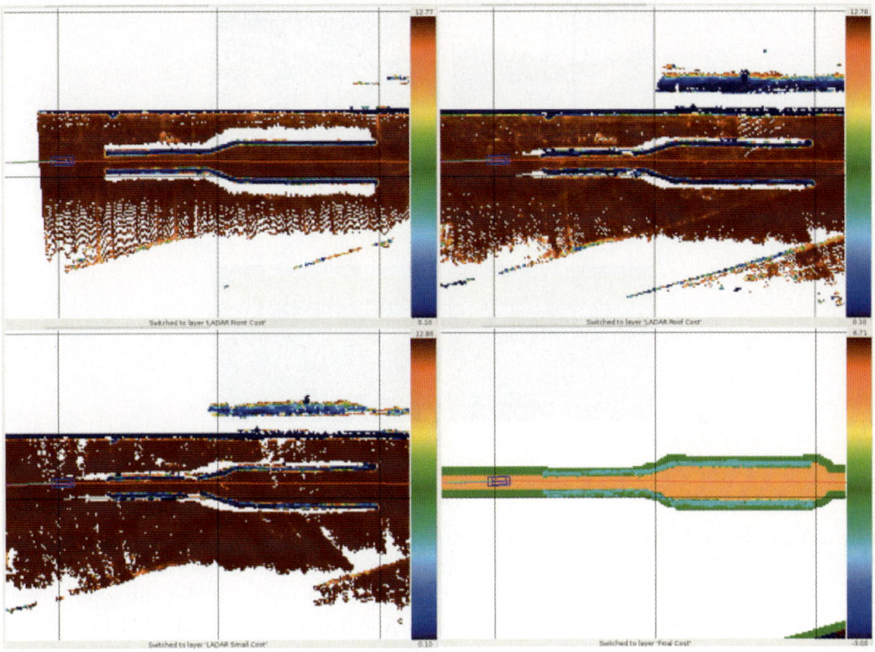

Fig. 14.10. Example input speed limit maps and a final output map. In each case, the small rectangle to the left is the vehicle (which is traveling from right to left) and the line extending behind it is its previous path. (For scale, the grid lines are 40 m apart.) From bottom-left, moving clockwise: speed limit map generated by a midrange LADAR; speed limit map generated by a short-range LADAR; speed limit map generated by a second midrange LADAR; the combined speed limit map. The combined speed limit map has also taken into account information about the race corridor.

4. Sensor B sees the entirety of the obstacle, and the cell containing the obstacle is once again set to a low speed. At this point, it is usually too late for the vehicle to react, and a collision occurs.

To solve this problem, we introduce an additional term into the weighted average that served to favor cells that have more measurements (and thus were more complete). The equation becomes

$$s(r,c) = \frac{\sum_i s_i(r,c) * w_i * n_i(r,c)}{\sum_i w_i * n_i(r,c)}, \tag{14.6}$$

where $n_i(r,c)$ is the number of measurements by sensor i that were contained in the cell at (r,c). As with the cost-generation algorithm, cost-fusion is done at a fixed frequency instead of on a per-cell basis, and only cells whose speed limits have changed since the last sweep are updated.

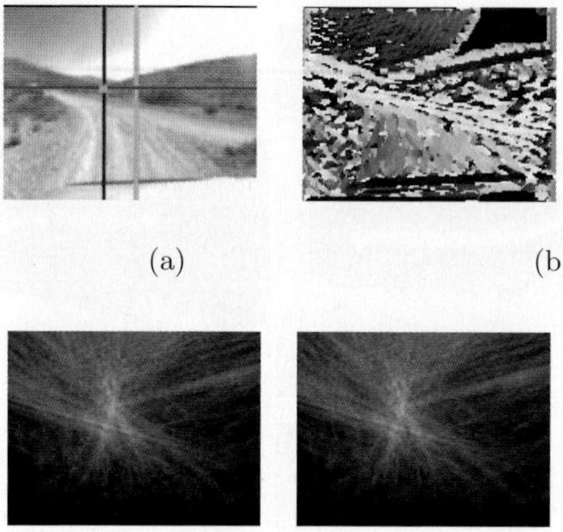

(a) (b)

(c) (d)

Fig. 14.11. Sample road follower output on test run: (a) computed road direction and Alice direction (image projections of diagonal, vertical lines, respectively, in (d)); (b) dominant orientations ($[0, \pi]$ angle proportional to intensity); (c) vanishing point votes $\mathbf{I}_{VP}$; (d) overhead view of estimated road region (green/lightly shaded area): Alice is the rectangle, bumper LADAR returns are the scattered points, LADAR-derived obstacle function $h(x)$ is shown in at bottom.

14.4.3 Additional Information: Roads, RDDF Corridors, and No Data

Although the sensor fusion problem makes up the bulk of the work, additional information is available for integration into the final speed limit map.

Road Following. Alice's road following algorithm (Rasmussen, 2006) uses a single calibrated grayscale camera and a bumper-mounted SICK LADAR. The best-fit vanishing point of the road ahead is repeatedly extracted from the pattern of parallel ruts, tracks, lane lines, or road edges present in the camera image, yielding a target direction in the ground plane. This visual information is augmented with estimates of the road width and Alice's lateral displacement from its centerline, derived from the LADAR returns, to populate a "road layer" in the combined cost map used by the planner. Additionally, a failure detection process classifies images as road or non-road to appropriately turn off the road layer and monitors for sun glare or excessive shadow conditions that can confuse the vanishing point finder. Figure 14.11 illustrates the operation of the algorithm.

To obtain the road vanishing point, the dominant texture orientation at each pixel of a downsampled 80×60 image is estimated by convolution with a bank of 12×12 complex Gabor wavelet filters over 36 orientations in the range $[0, \pi]$. The filter orientation $\theta(\mathbf{p})$ at each pixel $\mathbf{p}$ which elicits the maximum response implies that the vanishing point lies along the ray defined by $\mathbf{r}_\mathbf{p} = (\mathbf{p}, \theta(\mathbf{p}))$. The instantaneous road vanishing point for all pixels is the maximum vote-getter in a Hough-like voting procedure in which each $\mathbf{r_p}$ is rasterized in a roughly image-sized accumulation buffer $\mathbf{I}_{VP}$. Using $\mathbf{I}_{VP}$ as a likelihood function, we track the road vanishing point over time with a particle filter.

The road region is computed by projecting LADAR hit points along the vision-derived road direction onto a line defined by Alice's front axle, deweighting more distant points, to make a 1-D "obstacle function" $h(x)$. The middle of the gap in $h(x)$ closest to the vehicle's current position marks Alice's lateral offset from the road midpoint; it is tracked with a separate particle filter. If sufficient cues are available, this road-finding system is able to identify a rectangular section of the area in front of Alice as a road. Using the assumption that roads are generally safer to drive on (and thus the vehicle should drive faster on them), this data is integrated into the existing speed limit map using a heuristic algorithm as a speed bonus. As a result, roads appear as faster sections in the map, enabling the path-planning algorithm to keep Alice on roads whenever possible.

RDDF Corridor. The final (and perhaps most important) piece of information that is incorporated is the race corridor. Incorporating this information into the map is trivial: if a given cell falls inside the race corridor, then the value of that cell is set to be the minimum of the corridor speed limit and the terrain and road-based speed limit. If a cell falls outside the corridor, its speed is set to zero. In effect, the race corridor acts simply as a mask on the underlying terrain data: no inherent assumptions are made in the mapping software that rely on the corridor information. In fact, testing was frequently performed in which the corridor speed limit was essentially set to infinity, and the corridor's width was set to be dozens of meters wide. Even with these loose bounds, the mapping software was easily able to identify drivable areas and generate accurate speed limit maps.

No-Data Cells. One final topic that we have alluded to, but not covered in detail, is what we call the no-data problem: in many cases, not all of the cells in a given area of the map will contain elevation measurements. For example, no range measurements will be returned when a sensor is pointed off the side of a cliff, or from behind an obstacle that is obstructing the sensor's field of view. This problem can be partially alleviated by making some assumptions about the continuity of the terrain in the environment, and interpolating between cells accordingly. Our software performs some interpolation (by filling in small numbers of empty cells based on the average elevation of surrounding cells that did have data). Despite this, there are still situations where large patches of the map simply have no data, and interpolating across them could lead to incorrect estimates of the nature of the surrounding terrain. The question, then, is how the mapping software should treat those areas: should it be cautious and set them to have a low speed limit? Or should it make assumptions about the continuity of the surrounding terrain and set them to have a high speed limit? We found that for robust performance, a compromise between these two extremes was necessary.

Clearly, setting no-data cells to a high speed limit would be inviting trouble. For example, the vehicle may decide it prefers to drive off a cliff because the open area beyond the cliff is a high-speed no-data region. However, setting no-data cells

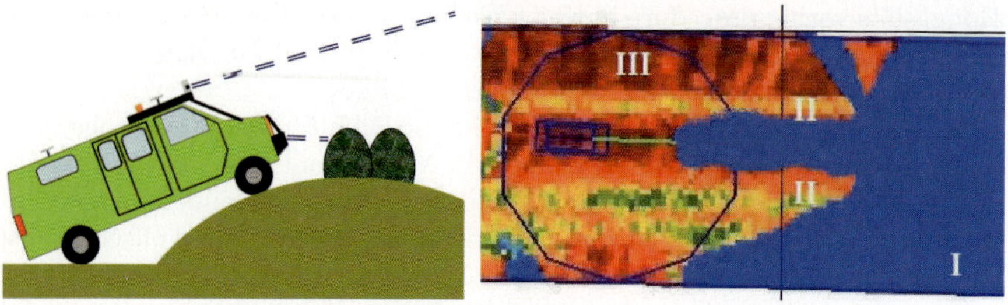

(a) (b)

Fig. 14.12. Example no-data problem. (a) As the vehicle crests the hill, its sensors will detect the bushes lining the road before they detect the road itself. (b) A corresponding sample no-data map (with the vehicle traveling from left to right). If area I is no-data, and is a lower speed than area II (corresponding to obstacles) and area III (corresponding to flat terrain), then the vehicle may prefer to drive in no-data areas (I) instead, even if they correspond to dangerous obstacles.

to a speed limit of zero can also result in undesirable behavior. Consider a set of steep rolling hills with a dirt road down the center, and bushes or boulders lining the sides. As the vehicle crests a hill, the first thing its sensors will detect are the bushes lining the side of the road (Figure 14.12(a)). In particular, these bushes may be detected and placed into the map prior to the vehicle detecting the dirt road below, resulting in a speed limit map that looks something like Figure 14.12(b). In this situation, if the no-data speed is set too low (that is, lower than the speed assigned to the cells containing the bushes) the vehicle may prefer to drive through the bushes instead of stay on the road. Clearly this is undesirable—as a result, a compromise must be made where the no-data speed is low (so as to avoid the vehicle barreling across uncharted terrain at high speed) but not lower than the speed associated with most obstacles. The interpretation of no-data cells is performed by the path planning module, described in the next section.

14.5 Navigation and Planning

The navigation system for Alice was designed as two discrete components: a planner that generates trajectories for the vehicle to follow, and a controller that follows those trajectories. The controller was described in Section 14.3.4; this section will describe the design and implementation of the trajectory planner.

14.5.1 Planner Approach

The top level design goal of the planning system is to be able to traverse the DARPA-specified course while satisfying DARPA's speed limits and avoiding obstacles. An ideal planner will produce trajectories that are "best," in some sense. Thus a numerical optimization method lies at the core of Alice's planning system. Since the DGC race is over 100 miles long, it is both computationally

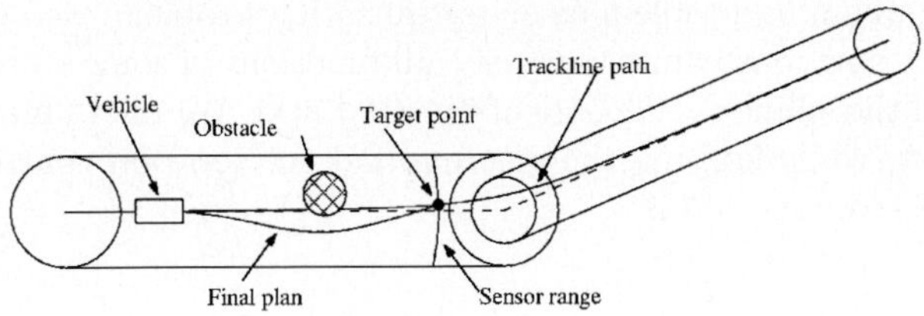

Fig. 14.13. An illustration of the receding horizon framework. A very inaccurate (but easily computable) path is used as the global plan. Here this is a spline based on the trackline. The solver that produces a higher-quality plan performs its computation to reach a point on the lower quality path, some distance ahead. In this figure, that distance (the horizon) is set to the range of the sensors of the vehicle.

prohibitive and unnecessary to plan the vehicle's trajectory to the finish line. An appropriate solution to deal with this problem is to run the planner in a receding horizon framework. A receding horizon scenario is one where plans are computed not to the final goal, but to a point a set horizon (spatial or temporal) ahead on some rough estimate of the path towards the goal (Figure 14.13).

The vehicle model used by the planning system is a rear-centered, kinematic model:

$$\begin{aligned} \dot{N} &= v\cos\theta \\ \dot{E} &= v\sin\theta \\ \dot{v} &= a \\ \dot{\theta} &= \frac{v}{L}\tan\phi, \end{aligned} \tag{14.7}$$

where N, E are Cartesian spatial coordinates, θ is yaw (measured from north to east), v is the scalar speed, a is the vehicle acceleration, L is the vehicle wheelbase and ϕ is the steering angle.

To apply a numerical optimization scheme to the path planning problem, we have to represent the space of all potential trajectories as a vector space. To cut down on the dimensionality of this vector space, we represent the spatial part of the trajectory as a quadratic spline of $\theta(s)$ where s is length along the trajectory normalized to the total length of the trajectory, S_f, and the temporal part as a linear spline of $v(s)$. With this representation, we can represent all variables in the system model (14.7) as functions of derivatives of $\theta(s)$, $v(s)$, S_f:

$$\begin{aligned} N(s) &= N_0 + S_f\int_0^s \cos(\theta(s))ds \\ E(s) &= E_0 + S_f\int_0^s \sin(\theta(s))ds \\ \dot{N}(s) &= S_f\cos(\theta(s))v(s) \\ \dot{E}(s) &= S_f\sin(\theta(s))v(s) \end{aligned} \qquad \begin{aligned} a &= \frac{v}{S_f}\frac{dv}{ds} \\ \dot{\theta} &= \frac{v}{S_f}\frac{d\theta}{ds} \end{aligned} \qquad \begin{aligned} \tan\phi &= \frac{L}{S_f}\frac{d\theta}{ds} \\ \dot{\phi} &= \frac{LS_f\frac{d^2\theta}{ds^2}}{S_f^2 + \left(L\frac{d\theta}{ds}\right)^2}. \end{aligned} \tag{14.8}$$

To pose the planning problem as an optimization problem, we need an objective to optimize and constraints to satisfy, all functions of some state vector. Our state is S_f and the spline coefficients of $\theta(s)$ and $v(s)$. We try to find trajectories that are fastest, while keeping the steering and acceleration control effort low. Thus our objective function is

$$J = S_f \int_0^1 \frac{1}{v(s)} ds + k_1 \left\| \dot{\phi}(s) \right\|_2^2 + k_2 \left\| a(s) \right\|_2^2 . \tag{14.9}$$

Besides the initial and end condition constraints, we also satisfy dynamic feasibility constraints:

$$\begin{array}{lcccc} \text{Speed limit :} & & & v & < v_{\text{limit}} \\ \text{Acceleration limit :} & a_{\min} & < & a & < a_{\max} \\ \text{Steering limit :} & -\phi_{\max} & < & \phi & < \phi_{\max} \\ \text{Steering speed limit :} & -\dot{\phi}_{\max} & < & \dot{\phi} & < \dot{\phi}_{\max} \\ \text{Rollover constraint :} & -\frac{gW}{2h_{\text{cg}}} & < & v^2 \frac{\tan\phi}{L} & < \frac{gW}{2h_{\text{cg}}} . \end{array} \tag{14.10}$$

In the rollover constraint expression, W is the track of the vehicle (distance between left and right wheels), h_{cg} is the height of the center of gravity of the vehicle above ground, and g is the acceleration due to gravity. This expression is derived from assuming flat ground and rollover due purely to a centripetal force. In reality, on many surfaces sideslip will occur much before rollover, so this constraint has an adjustment factor.

Obstacle avoidance enters into the problem as the speed limit constraint, with the RDDF and terrain data processed to produce a combined, discrete map that represents a spatially dependent speed limit. What this means is that the areas outside of the RDDF and those that lie inside obstacles have a very low speed limit, and thus any trajectory that goes through those areas is not explicitly infeasible, but *is* suboptimal. This representation allows the obstacle avoidance and quick traversal conditions to be cleanly combined.

The processing of v_{limit} from the discrete map generated from the sensors as described in Section 14.4, to a C_1 surface was a nontrivial matter. A lot of thought went into the design of this component, but its details are beyond the scope of this paper. More information is available in (Kogan, 2006).

Since the numerical optimizer is only locally optimal, and the problem is only locally convex, choosing a good seed for the solver is an important prerequisite to a successful solution cycle. The seeding algorithm used for Alice's planner consists of a coarse spatial path selector, planning several times beyond Alice's stopping distance. This approach evaluates quickly, introduces a needed element of long-term planning, and works well to select a good local basin of attraction.

14.5.2 System Interfacing

During the development of the planning system the challenges were not limited to the planning system itself: a substantial amount of effort went towards interfacing the planner with the other system components.

One issue that needed to be addressed was the interaction of the controller and the planner, and the location of the feedback loop. If a feedback loop is contained in both the controller (through tracking error feedback) and the planner (by planning from the vehicle state), these two feedback loops interact unfavorably and produce spurious oscillations in the closed loop system performance. Since these two feedback loops work on the same time scale, these interactions are natural, and one of the feedback loops has to be eliminated to address the problem. The controller feedback can be eliminated by using only feedforward control input. This creates a very clean system, but requires relatively quick replanning, and a high fidelity model for the trajectory generation. Since there are no theoretical guarantees for the numerical optimizer convergence, this option was rejected in favor of moving all of the feedback into the controller: the controller feeds back on its tracking errors, and the planner plans from a point on its previously computed plan. In this way, the vehicle motion does not directly affect the planner's operation, and the oscillations are eliminated. This can result in a potential issue of runaway poor tracking. To address this concern, the planner *does* plan from the vehicle's state if the tracking errors grow beyond some predefined threshold.

An issue observed during the testing of the planner was an indecisiveness about which way to go to avoid a small obstacle. The local optimality of the planner could repeatedly change the avoidance direction between planning cycles, resulting in an ungraceful, late swerve to avoid an obstacle that has been detected long before. Since the seeding algorithm chooses the basin of attraction, the fix was placed there. By adding a slight element of hysteresis to the seeding algorithm to lightly favor solutions close to the previous solution, the balance between the two avoidance directions was broken, and a consistent direction was favored.

A third interaction challenge that was discovered during the planner development was the treatment of no-data cells in the map. This issue was briefly mentioned in Section 14.4, but the planning-side issues are described here. While all map cells that the planner gets from the mapping components represent a specific speed limit that the planner should satisfy, this is not as clear for no-data cells. To avoid being overly cautious, or overly brazen around no-data, a dual interpretation of no-data was adopted. First, a threshold distance is computed as the larger of twice the stopping distance or 10 m. Then, any no-data cell closer than the threshold distance to the vehicle is treated as a very slow cell, while no-data beyond the threshold distance is treated as twice-the-current-speed. This representation attempts to avoid or slow down through no-data cells where they could be dangerous (at the vehicle), but does not let no-data negatively influence long-term navigation decisions.

14.5.3 Planner Results

With these issues taken care of, the planning problem was solved by an off-the-shelf numerical optimizer, SNOPT. This optimization-based approach provides a very clean planning solution, and avoids most heuristic aspects of the planners

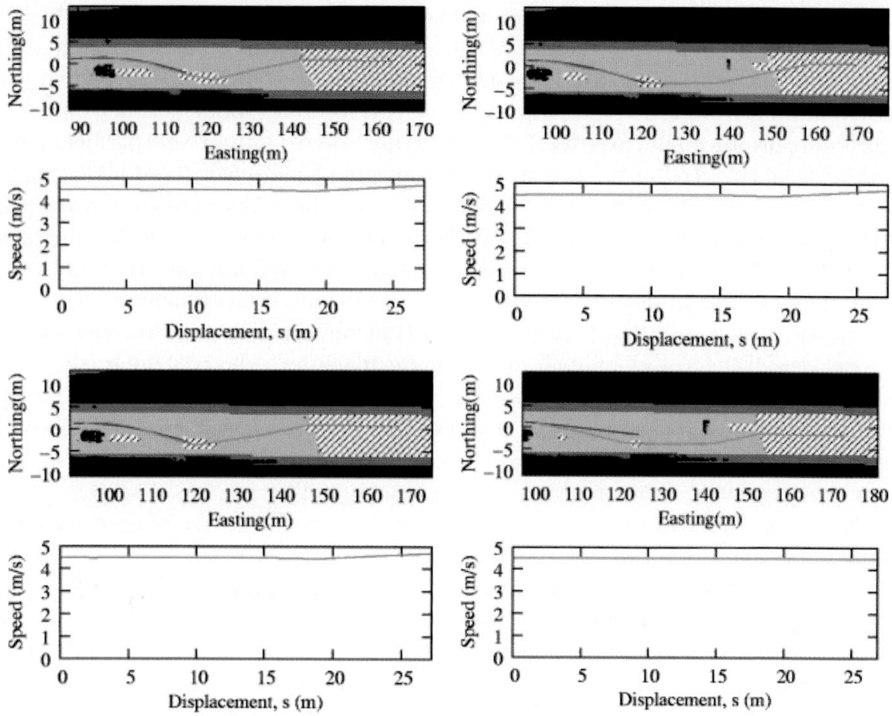

Fig. 14.14. Non-consecutive planner iterations showing Alice avoid a car at the NQE. The vehicle is traveling east. The planner seed is the lighter, long path and the final solution the darker (and shorter) path. Grayscale colors are the speed limits in the map, with hashed regions representing no-data. Cars south and straight ahead of the vehicle are visible, along with the sensor shadow of no-data for the vehicle straight ahead. We can see distant no-data being treated favorably. We can also see a change of plans to avoid the newly detected second car.

that were so ubiquitous in the planning systems of most other teams. Optimizer convergence speed is a potential issue, but on Alice's 2.2 GHz Opteron, an average rate of 4.28 plans/second was achieved during the race. Further, this system produces very drivable and precise trajectories, which allows for very tight obstacle avoidance in potentially difficult situations (more than sufficient for the DGC). Some planner performance from the National Qualifying Event is illustrated in Figure 14.14.

14.6 Contingency Management

Experience has shown that a major challenge in the successful deployment of autonomous systems is the development of mechanisms to allow these systems to continue to operate in unanticipated (non-nominal) conditions. An important subset of these non-nominal conditions is system-level faults. These are situations in which all the individual components of the system are operating correctly, that is to say as detailed in their specifications, yet the system as a whole is unable to achieve its stated goals.

In small-scale, comparatively simple systems, a state machine is often used as the system level controller. The principal advantages of this approach are the ability of the designer to specify through a state-table exactly how the system can evolve so that it responds effectively to the specific faults they envisage. As the state-table defines exactly how the system can evolve, only the stability of the system in the scenarios described by the state table need be considered, reducing the scale of the robustness problem and typically leading to a comparatively short development time. However the state machine approach does not scale well with system complexity or size, and as it cannot execute multiple states simultaneously is not suited to situations in which multiple connected faults may have occurred.

A recent approach combining state-feedback with multiple rule-based filtering stages is demonstrated in JPL's Mission Data System (MDS) (Rasmussen, 2001). MDS seeks to provide a very flexible and expandable architecture for large and complex systems, such as unmanned space missions, that is capable of managing multiple faults simultaneously. However, the disadvantages of this approach are the absence of a known method to verify system stability and robustness, hence significant time is required to develop or enhance the system while maintaining previous empirically proven system robustness.

14.6.1 Problem Specification

The primary objective of the supervisory controller, SuperCon, is to ensure that where physically possible Alice continues to make forward progress along the course as defined by the RDDF. In situations where forward progress is not immediately possible (e.g., intraversible obstacle in front of Alice's current position), SuperCon should take all necessary action to make it possible. This objective encompasses the detection, and unified response to, all system-level faults from which it is possible for the system to recover given its designed functionality, performance, constraints and assumptions. SuperCon is not required to consider specific responses to hardware failure since these are handled at a lower level. It should also be noted that Alice does not carry any backups for its actuators. It is however a requirement of the mapping software that it be resistant to up to two sensor failures. In addition, Alice's computer cluster is configured such that in the event that any machine powers down it will restart (if physically possible), and if any module crashes it will be immediately restarted.

The secondary objective of SuperCon is to modify the overall system configuration/status if required by a module under certain conditions and manage the effects of these changes upon the different components of the system. SuperCon should also verify its own actions, and make decisions based on the accuracy of the information available to it. SuperCon must also be stable, robust, easy to expand to respond to new issues as they are discovered through testing, fast to test and verify and quick to initially assemble.

In essence the presence of SuperCon should only increase the performance, functionality, stability and robustness of the complete system and so increase Alice's chances of winning, it should never lead to a decrease in system performance by impeding correct operation.

14.6.2 Hybrid Architecture

The hybrid architecture developed has a state table and a series of rule-based filtering layers that are re-evaluated each cycle and stored in a diagnostic table. The resulting framework shares the general characteristics with MDS, described above. The state machine is composed of ten strategies, each of which consists of a sequence of stages to configure the system to achieve a specified goal (e.g., reversing requires changing gear in addition to the reconfiguration of several affected software modules). Each stage is composed of a series of conditional tests that must be passed in order for the (single) SuperCon action (e.g., change gear) to be performed, and hence the stage to be marked as complete. If a stage is not completed, and the current strategy is not updated, the next time the state machine steps forward it will re-enter the (same) stage. This allows the state machine to loop until conditions are met while always using the current version of the diagnostic table. If the test was a transition condition (which determines whether SuperCon should move to another strategy) and evaluates to true, the corresponding strategy change is made, for execution in the next cycle of the state machine.

Half of the strategies are stand-alone strategies that are responsible for managing Alice's response to special case scenarios (e.g., the safe reincorporation of GPS after a sustained signal outage). The other half form an interconnected ring that can be used to identify and then resolve any identified scenarios in which Alice would be unable to make forward progress along the course. The hybrid architecture effectively expands the capabilities of the state machine approach through rule-based filtering, and a strategy-driven approach making it easier to design for large, complex systems while retaining its desirable robustness and ease of expansion properties.

The SuperCon hybrid structure consists of two permanent threads, the state update thread maintains the live version of the state table that holds the current and selected history data for all SuperCon state elements and is purely event-driven (by the arrival of new messages). The deliberation thread by contrast is prevented from executing above a defined frequency (10Hz) to prevent unnecessary processor loading.

The SuperCon state machine only attempts to complete one stage each time it "steps forwards". Once it has finished processing a stage (regardless of whether it was successfully completed) the state machine suspends execution and requests a new copy of the state table. Once the new copy of the state table has been received, and the diagnostic rules table has been re-evaluated, the state machine takes another step forward.

A significant advantage of the hybrid approach detailed above is that as it is modular, it is comparatively easy to test as the components can be verified independently in sequence.

In order to make SuperCon robust, every time the SuperCon state machine takes an action (e.g., sends a request to change gear), before any additional action is taken SuperCon verifies that any action associated with the previous stage has been successfully completed. Each of these verification stages has an associated time-out. If the time-out is exceeded then SuperCon assumes a fault has occurred, and reverts to the Nominal strategy, from which SuperCon is reset and can re-evaluate which strategy should be active and then transition to it.

14.6.3 Strategies

The role of the strategies is to create a logical structure that can fulfill the primary and secondary objectives for SuperCon. Corresponding to the primary objective, to ensure that Alice always continues to make forward progress along the RDDF where possible, a list of possible scenarios (from experience and analysis) that could prevent such progress was produced. By definition, this list was coined as No Forward Progress (NFP) scenarios. Analysis of the list produced five distinct families of NFP scenarios, prototypical examples for each of which are shown in Figure 14.15.

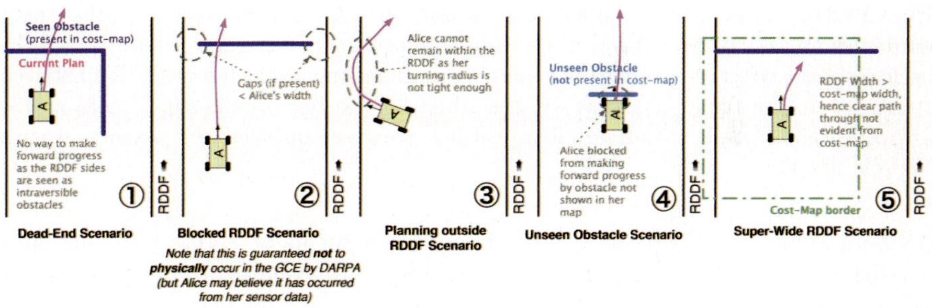

Fig. 14.15. Diagram showing prototypical examples of the five general NFP (No Forward Progress) scenarios identified

Due to the limited perceptive and cognitive abilities of our system (as compared to a human) it is often not initially possible to distinguish between the five groups shown in Figure 14.15. Hence in order for SuperCon to be robust an integrated array of strategies that can identify and resolve any arbitrary scenario from any of the groups is required.

After extensive analysis of the responses that a human would take to attempt to resolve the prototypical examples (assumed to be at least on average the best response) and the information they used to do so, the NFP Response Cycle shown in Figure 14.16 was produced. The cycle consists of five strategies including the

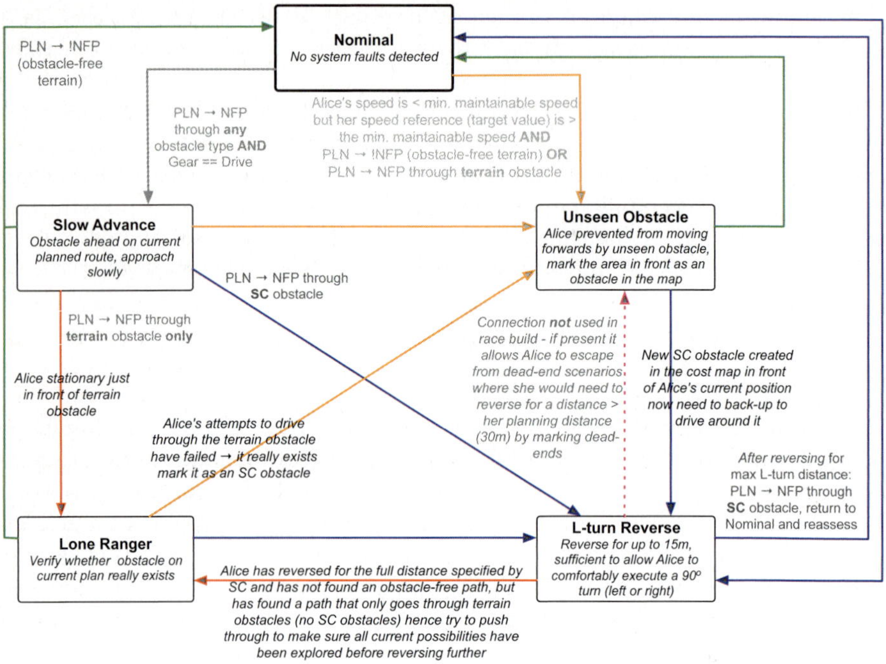

Fig. 14.16. No-Forward-Progress (NFP) Response Cycle, showing the constituent strategies and purposes of each of the interconnections between them to produce the cycle response effect. Note that the general conditions under which a transition between strategies occurs (colored arrows) is described by the text of the same color as the arrow. Specific details of individual transitions are given in black text associated with the relevant arrow.

nominal strategy, which the system should remain in or revert to if no fault conditions are detected. The following strategies are used in the NFP Response Cycle:

Nominal. No system faults/requests detected and the current plan does not pass through any obstacles. Verify that the system is correctly configured for nominal operation (gear = drive etc) and correct any exceptions found.

Slow Advance. The planner's current plan now passes through an obstacle (terrain or SuperCon, including outside the RDDF corridor) when it did not pass through any obstacles previously (nominal strategy). Hence limit the max speed to ensure the sensors can accurately scan the terrain ahead, but allow Alice to continue along the current plan, whose speed profile will bring Alice to a stop in front of the obstacle if it is not cleared from the map by new sensor data.

Lone Ranger. The planner's current plan passes through a terrain obstacle, which has not been cleared by subsequent sensor data, even after Alice has driven up to it. As Alice would always avoid any terrain obstacles if there was a way to do so (that did not pass through SuperCon obstacles), verify that the terrain obstacle really exists by attempting to push through it.

Unseen Obstacle. Alice is stationary when it does not intend to be (TrajFollower's reference speed less than minimum maintainable speed), its engine is on and it is actively attempting to move forwards (positive accelerator command and zero brake command) yet an external (assumed) object in front of it is preventing it from doing so. Hence mark the terrain directly in front of its current position as a SuperCon obstacle, as it has been investigated and found to be intraversible.

L-turn Reverse. The planner's current plan passes through a SuperCon obstacle, these are avoided in preference of all other terrain where possible and indicate areas that have been verified as intraversible (or are outside the RDDF). Hence reverse along the previous path (trail of state data points) until the current plan passes through no obstacles, up to a maximum of 15 m (the distance required for Alice to perform a 90-degree left/right turn plus an error margin) per call of the L-turn reverse strategy. If the current plan passes through terrain obstacles after reversing for 15 m then go forwards and investigate them to verify their existence. If it still passes through SuperCon obstacles then recall L-turn Reverse and hence initiate another reversing action.

A significant benefit of the NFP response cycle design is that it only requires a very small amount of additional information to be computed on top of what is already available in the system (prior to SuperCon). The main piece of information used is the value of the speed limit in the lowest speed cell (evaluated over Alice's footprint) through which the current plan evaluated through the most recent version of the cost map speed-layer passes, referred to as the minimum speed cell (MSC). The current plan is defined as the plan currently being followed by the TrajFollower. The planner that Alice uses (as detailed in Section 14.5) is a non-linear optimizer. As a result, it is not necessary for SuperCon to know the location at which the minimum speed cell occurs, as the plan is optimal for the cost function it represents and the best option available in terms of its obstacle avoidance. There are three ranges (of values) in the speed layer of the cost map: no-obstacle, terrain obstacle and SuperCon obstacle and outside the RDDF. The range that the current minimum speed cell value belongs to is used by SuperCon when determining what action (if any) to take. There is a fundamental difference between what is known about terrain and SuperCon obstacles. Terrain obstacles are identified by the sensor suite as having an elevation, and/or gradient outside of Alice's safe limits. SuperCon obstacles represent areas where Alice attempted to drive through (irrespective of whether they were identified as terrain obstacles) and was unable to do so. As such SuperCon obstacles are verified terrain obstacles. Note that all "obstacles" are theoretically intraversible by definition.

The process is analogous to a typical feedback controller, where the planner is the process whose output is the minimum speed cell value, SuperCon is the

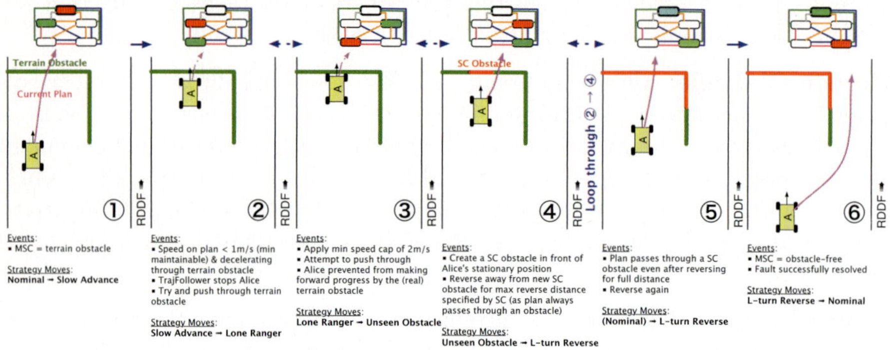

Fig. 14.17. Example of Alice resolving a dead-end scenario fault, showing the events and strategy transitions at each point

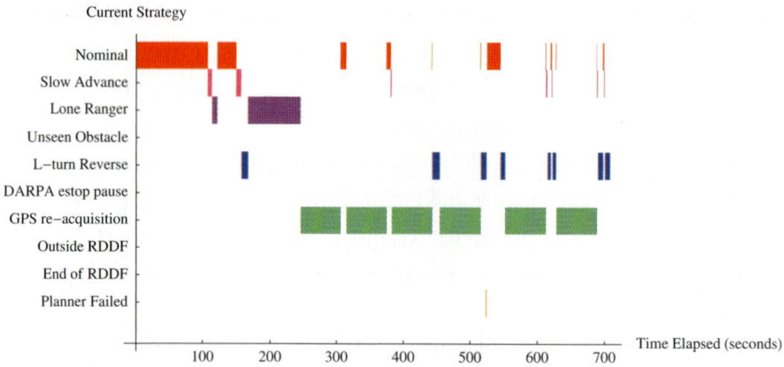

Fig. 14.18. Strategies used during NQE Run #1. The GPS re-acquisition intervals are execution of the design requirement to safely stop the vehicle in the face of large corrections of state estimate errors.

controller and the reference signal is the range of obstacle-free terrain values. The error signal then becomes non-zero when the current plan passes through an obstacle, which prompts SuperCon to take corrective action through the use of the NFP response cycle.

Figure 14.17 shows an example of the NFP response cycle enabling Alice to successfully navigate its way out of a dead end scenario, listing the actions taken by SuperCon at each stage, and indicating when each strategy is active. Stages 2–4 in the response are repeated until SuperCon obstacles have been placed as shown in stage 5 (assuming the terrain dead-end obstacle really exists). Note that the RDDF border on the right hand-side of Alice does not need to be verified, as outside the RDDF the terrain is equivalent to an SuperCon obstacle by default. A miniaturized version of the NFP response cycle is also shown for each stage, with the current strategy highlighted in red (darker shading), and the next in green (light shading). In case 5, blue (upper center state) is used to denote the most probable current strategy at the start of the stage).

Figure 14.18 depicts the evolution of SuperCon strategies that enabled the forward progress made during Alice's first NQE run.

14.7 Experimental Results

In the previous six sections we have described in detail Team Caltech's approach to designing its unmanned ground vehicle, Alice. Unfortunately, due to the nature of the Grand Challenge, very few controlled experiments were performed on the vehicle as a whole. Although many subsystems were individually tested to make sure they met certain specifications, the measure of success for the end-to-end system was a more qualitative one, focused on making sure that the vehicle could drive autonomously in a wide variety of situations including open desert, parking lots, rainy and wet conditions, dirt roads of varying degrees of roughness, rolling hills, winding roads and mountain passes. The following sections describe the nature and results of the testing of Alice leading up to the National Qualifying Event, for the National Qualifying Event itself, and in the Grand Challenge Event. Although the data presented is somewhat qualitative, we believe it makes clear Alice's capabilities (and weaknesses) as an autonomous vehicle.

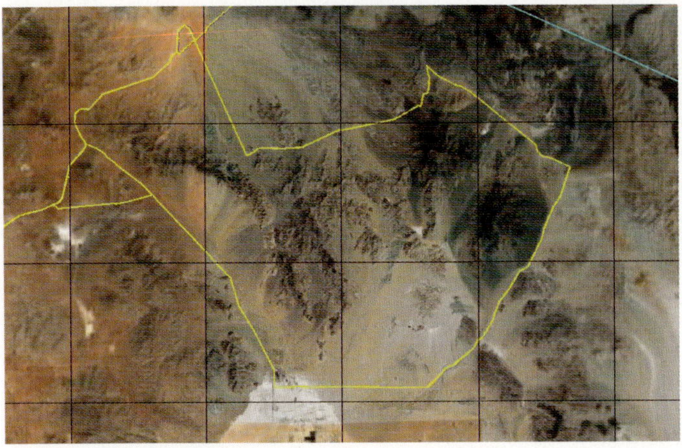

Fig. 14.19. A compilation of several RDDFs used during testing within the Stoddard Wells area of the Mojave Desert. The grid lines are 10 km apart. The RDDF leaving the top of the figure is the 2004 race RDDF, and the line cutting across the top right corner is the boundary of the no-go zone given by DARPA. These RDDFs covered a wide variety of desert terrains, including dirt roads, rocky hills, dry lake beds, bumpy trails, smooth pavement, and mountain passes. During the majority of testing the RDDF speed limit over the entire RDDF was set to be unlimited, and the vehicle's speed was chosen automatically as it traveled.

14.7.1 Desert Testing

Team Caltech documented over 300 miles of fully autonomous desert driving with Alice from June 2005 to the National Qualifying Event in Fontana in September, all in the Mojave Desert. Figure 14.19 shows some of the RDDFs used during testing.

Approximately 33 of these miles were driven on the 2004 Grand Challenge race route during the week of June 13th, 2005. Alice traversed these miles with a testing team of four people inside, scrutinizing its performance and making software improvements and corrections. Over the course of three days, Alice suffered three flat tires including a debilitating crash into a short rocky wall that blew out the inside of its front left tire and split its rim into two pieces. This crash was determined to be caused primarily by a lack of accurate altitude estimates when cresting large hills. Along with several related bug fixes, an improved capability to estimate elevation was added to the state estimator.

The majority (approximately 169 miles) of autonomous operation for Alice took place in the two weeks leading into the National Qualifying Event. This operation included a full traversal of Daggett Ridge at 4 m/s average speed, and significant operation in hilly and mountainous terrain (see Figure 14.20). The top speed attained over all autonomous operations was 35 mph. The longest uninterrupted autonomous run was approximately 25 miles.

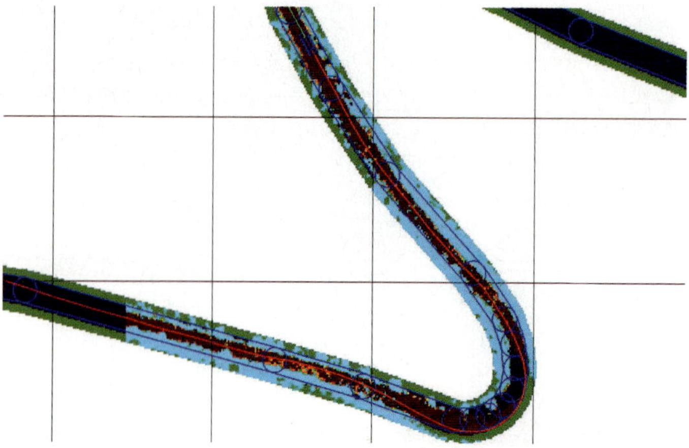

Fig. 14.20. A sample speed limit map taken in Daggett Ridge during testing on the 2004 Grand Challenge course. For scale, the grid lines are 40m apart. The light-colored areas along the sides of the corridor are obstacles (cliff-faces taller than the vehicle, or berms about 0.5m high), the actual road is the darker area in the center, and the confetti-like coloring of some parts of the road indicates bumpier sections. Despite the narrowness of the corridor (around 3-4m wide) and the difficulty of the terrain, Alice (the rectangle near the top of the figure) was able to pick out a safe course (the line extending back down the figure and to the left) at an average speed of 4m/s.

14.7.2 National Qualifying Event

As one of 43 teams at the California Speedway in Fontana, Alice successfully completed three of its four qualifying runs. Several of its runs were characterized by frequent stopping as a result of the performance of the state estimator under conditions of intermittent GPS. Specifically, under degraded GPS conditions its state estimate would drift considerably, partially due to miscalibration of IMU angular biases (especially yaw) and partially due to lack of odometry inputs to the state estimator. However, zero-speed corrections were applied to the Kalman filter when Alice was stopped, which served to correct errors in its state estimate due to drift quite well.

During Alice's first NQE run, zero-speed corrections were not applied in the state estimator. Accordingly, drift accumulating in the state estimator was not corrected adequately when Alice stopped. Travel through the man-made tunnel produced drift substantial enough for Alice's estimate to be outside the RDDF, which at the time resulted in a reverse action. Alice performed a series of reverse actions in this state, going outside the actual corridor, and resulting in DARPA pause and the end of its first run.

Zero-speed corrections were added to the state estimator after Alice's first run, enabling it to successfully complete all subsequent runs, clearing all obstacles and 137 out of a total 150 gates. Completion times were slow for runs 2, 3 and 4 partially due to frequent stopping as a result of state estimator corrections. Figure 14.21 provides a summary of Alice's NQE run performances. Figure 14.22 shows a snapshot of Alice's map and followed paths during the third NQE run.

Alice was one of only eleven teams to complete at least three of four NQE runs. As a result of Alice's performance at the NQE, it was preselected as one of ten teams to qualify for a position on the starting line in Primm, Nevada.

14.7.3 Grand Challenge Event

When Alice left the starting line on October 8th, 2005, all of its systems were functioning properly. However, a series of failures caused it to drive off course, topple a concrete barrier and disqualify itself from the race as a result. Although

Run	Gates	Obst	Time	Issues
1	21/50	0/4	DNF	State estimate problems after tunnel
2	44/50	4/4	12:45	
3	49/50	5/5	16:21	Multiple PAUSEs due to short in E-Stop wiring
4	44/50	5/5	16:59	Frequent stops due to state drift

Fig. 14.21. Alice on the National Qualifying Event course (left). Table of results from Alice's four runs at the NQE (right).

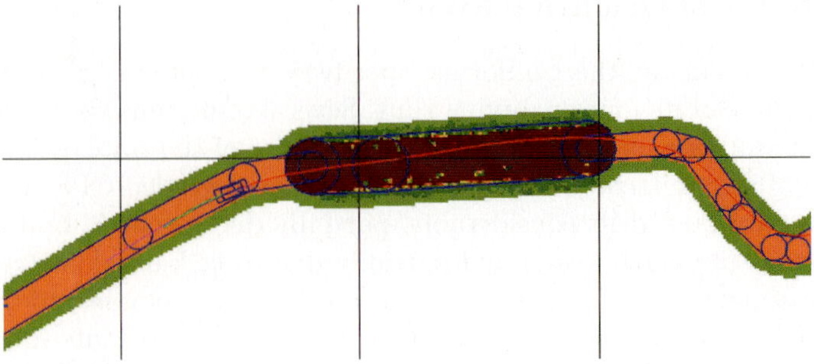

Fig. 14.22. A sample speed limit map taken during the third NQE run. For scale, the grid-lines are 40m apart. The solidly-colored areas in the center are the intended corridor, the lighter areas above and below the central section are lines of hay bales (around 0.5m tall), and the small spots near those hay bales are additional obstacles laid out by DARPA for Alice to avoid (such as wooden planks and tires). Alice (the rectangle to the left) was easily able to detect and avoid the obstacles (even though it likely could have traversed them easily)—the line extending from Alice to the right of the figure is the path that was followed. (Direction of travel is from right to left.)

as mentioned above, the system we have described performed well over the course of hundreds of miles of testing in the desert prior to the Grand Challenge, we believe the pathological nature of this particular failure scenario demonstrates a few of the more important weaknesses of the system and exemplifies the need for further ongoing research. We will begin by providing a brief chronological timeline of the events of the race leading up to Alice's failure, followed by an analysis of what weaknesses contributed to the failure.

Alice's timeline of events in the Grand Challenge is as follows:

- Zero minutes into the race, Alice leaves the starting line with all systems functioning normally.
- Approximately four minutes into the race, two of its midrange LADARs enter an error mode from which they cannot recover, despite repeated attempts by the software to reset. Alice continues driving using its long and short-range sensors.
- Approximately 30 minutes into the race, Alice passes under a set of high-voltage power lines. Signals from the power lines interfere with its ability to receive GPS signals, and its state estimate begins to rely heavily on data from its Inertial Measurement Unit (IMU).
- Approximately 31 minutes into the race, Alice approaches a section of the course lined by concrete barriers. Because new GPS measurements are far from its current state estimate, the state estimator requests and is granted a stop from supervisory control to correct approximately 3 meters of state drift. This is done and the map is cleared to prevent blurred obstacles from remaining.
- GPS measurements report large signal errors and the state estimator consequently converges very slowly, mistakenly determining that the state has

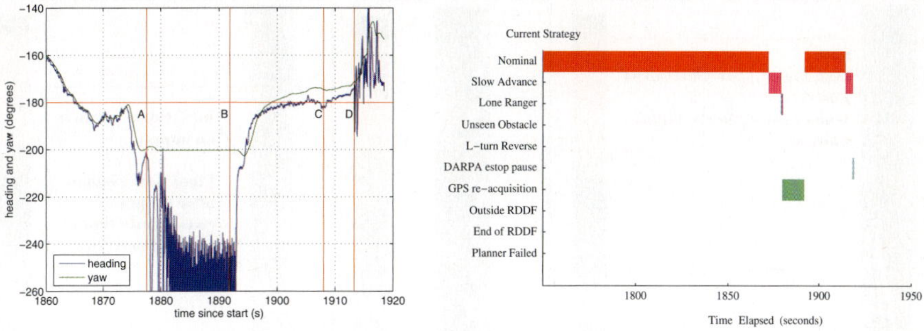

(a) (b)

Fig. 14.23. (a) Alice's estimated heading and yaw, both measured clockwise from north, in its final moments of the GCE. Yaw is from the state estimator and heading is computed directly as the arctangent of the easting and northing speeds of the rear axle. The times at which the state estimator requests a vehicle pause (A), the map is cleared and pause is released (B), Alice speeds up after clearing a region of no data (C) and impact (D) are shown. Between A and B the direction of motion is noisy as expected as Alice is stopped. Between B and D the state estimate is converged around 180 degree yaw, which we know to be about 8 degrees off leading into the crash. (b) The supervisory controller mode (strategy) during the same period.

converged after a few seconds. With the state estimate in an unconverged state, Alice proceeds forward.

- A considerable eastward drift of the state estimate results from a low confidence placed on the GPS measurements. This causes the velocity vector and yaw angle to converge to values that are a few degrees away from their true values. Based on the placement of the north-south aligned row of K-rails in the map by the short-range LADAR (see Figure 14.24(a)), Alice's actual average yaw for the 5 or so seconds leading into the crash—between times C and D on Figure 14.23—appears to be about 8 degrees west of south (−172 degrees). For the same period, our average estimated yaw was about −174 degrees and our average heading (from $\dot{N}$, $\dot{E}$) was about −178 degrees. Roughly, as Alice drives south-southwest, its state estimate says it is driving due south, straight down the race corridor (Figure 14.24(a)).
- Alice's long-range sensors detect the concrete barriers and place them improperly in the map due to the error in state estimate. Alice's mid-range sensors are still in an error mode. Alice picks up speed and is now driving at 10–15 mph.
- At 32 minutes, because it is not driving where it thinks it is, Alice crashes into a concrete barrier. Its short-range sensors detect the barrier, but not until it is virtually on top of it (Figure 14.24(b) and Figure 14.25).
- Almost simultaneously, DARPA gives an E-Stop pause, the front right wheel collides with the barrier, the power steering gearbox is damaged, and the driving software detects this and executes its own pause also, independent of

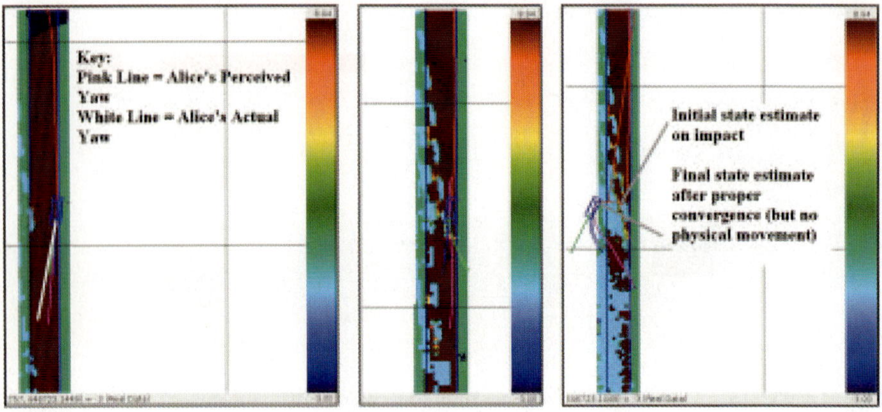

Fig. 14.24. Alice's speed limit maps over the last few seconds of the race, with notation added. (For scale, the grid lines are 40m apart.) For all three diagrams, the center dark area is the road, the light vertically-oriented rectangular areas are the concrete barriers, the hollow rectangle is Alice (traveling downward on the page), and the light-colored line is its perceived yaw (direction of travel). The color-bar on the right indicates the speed limit that was assigned to a given cell, where brown is highest and blue is lowest. The leftmost diagram indicates Alice's expected yaw and Alice's actual yaw during the last few seconds of the race. The center diagram is Alice's speed limit map less than one second before it crashed, indicating the differing placement of the obstacles by the short and long-range sensors—indicated as two parallel lines of concrete barriers (light-colored rectangles). In fact, there was only a single line of them, and that line went directly north-south, not angled as Alice perceived. The rightmost diagram is Alice's final estimate of its location: accurate, but thirty seconds too late.

that from DARPA. The strong front bumper prevents Alice from suffering any major damage as it drives over the barrier.

- Once Alice has come to a stop, the state-estimation software once again attempts to re-converge to get a more accurate state estimate—this time it corrects about 9.5 meters and converges close to the correct location, outside the race corridor. Alice is subsequently issued an E-Stop DISABLE (Figure 14.24(c)).

Figure 14.23(b) shows the SuperCon state during this final segment, including a Slow Advance through some spurious obstacles, a brief Lone Ranger push to clear them, the GPS reacquisition while stopped, and the final Slow Advance only after Alice has picked up speed and is near to crashing.

It is clear that while Alice's ultimate demise was rooted in its incorrect state estimates (due to poor GPS signals), other factors also contributed to its failure, or could conceivably have prevented it. These include the mid-range LADAR sensor failures, the lack of a system-level response to such failures, and the high speeds assigned to long range sensor data even in the face of state uncertainty. Additionally, in race configuration the forward facing bumper LADAR sensor

Fig. 14.25. Alice as it "clears" a concrete barrier during the GCE

was only used to assist in detection of the boundaries of the roads for the road following module. This could have helped in assigning appropriate speeds in the map for the row of K-rails.

14.8 Lessons Learned and Future Work

Even after many miles of desert testing and satisfactory performance at the NQE, Alice was not able to demonstrate its full capabilities in the race. Despite this failure, the full capabilities of the system provide an excellent testbed for future research and there are many open challenges that remain to be addressed. Accomplishments of the team include demonstration of a highly networked vehicle architecture with large amounts of sensor data, an optimization-based planner capable of rapid computation of feasible trajectories that maximize speed along the course, and a supervisory control layer capable of aggressively maintaining forward motion in the presence of complex driving conditions.

A critical short term need is to improve the state estimation capability in the presence of noisy signals. While other teams did not appear to have problems with state estimation, the combination of Alice's reliance on accurate state estimates (for placing obstacles) and supervisory logic for maintaining fault tolerance made it susceptible to poor GPS quality. These problems did not surface during desert testing since most testing was done in the open desert, away from the types of electromagnetic interference experienced at the NQE and the race.

The mapping system was also partly to blame for Alice's "grand finale." One of the major shortcomings that contributed to the failure was that the algorithm did not factor in any measure of confidence in the state estimate when determining the locations of obstacles. This is a major source of error because the state estimate must be used to place the sensor measurements, which are taken in a local coordinate frame, into the map, which is in a global coordinate frame. Thus, when the state estimate is very inaccurate (as was the case in the situation described above) the resulting map is also inaccurate. If the algorithm had made a probabilistic estimate of the location of obstacles instead of a hard estimate, then it is likely that the resulting map would have made the concrete barriers seem much larger than they were. (This is because the low confidence in the state estimate would have propagated down to become a low confidence in their placement in the map, which would in turn have caused them to occupy a greater portion of the map.) Had this been the case, the vehicle might have avoided a collision.

The second shortcoming of the mapping system was that it did not factor in any measure of the confidence in the sensor measurements when determining the speed of a given cell in the map. In other words, although one sensor was able to outweigh another when placing data, the map still did not reflect which sensor took a given measurement, and how accurate (or inaccurate) that sensor was when it made that measurement. As a result, the map did not reflect that two of the midrange sensors (which usually performed the bulk of the mapping) were inoperative. Instead, the map was filled with semi-inaccurate data from the long-range sensor, which was not differentiated from any other data that might have been present (such as more accurate data from the midrange sensors, had they been functioning correctly). As a result, the path-planning algorithm commanded a speed of between 10 and 15 mph, which prevented the vehicle from reacting in time when the obstacles were more accurately placed by the short-range sensor.

Team Caltech's experience indicates the need for future research in constructing efficient, timely, reliable and actionable models of the world from a rich sensor suite. Of particular interest are ideal sensor coverage algorithms and proper high-level response to loss of particular regions of sensor coverage, as well as formal analysis of the interplay between sensor data longevity (for more complete maps) and degradation of the state estimate (which tends to produce misregistered maps).

Ongoing work on Alice includes demonstration of more sophisticated sensor fusion techniques (Gillula, 2006), a networked architecture of combining sensors for state sensors without the need for offline calibration (Leibs, 2006), and model-based road following using LADAR data (Cremean, 2006; Cremean and Murray, 2006). Alice will also be used during the summer of 2006 for additional SURF projects in autonomous systems.

Acknowledgments

Team Caltech received generous support from a number of organizations and individuals. We would like to thank our primary sponsors—the California Institute of Technology, Northrop Grumman Electronic Systems, and Sportsmobile West, Inc—for their generous support throughout the project. In addition, we received financial support and donations from many sponsors, including AEK Technology, Aluminess, Barry Controls, Blanchard Associates, CommTech, Evolution Robotics, Ford, Horsepower Engineering, IBM, ITU Ventures, Johnson Controls, Lonseal, Mohr Davidow Ventures, Motion Picture Marine, Navcom, Northrop Grumman Space Technology, NovAtel, OmniSTAR, SICK, Santa Anita Park, Space X, Systems Technology Incorporated, Woodhead and Wright Tool.

The team also benefited from the wisdom and experience of many individuals, including Elliott Andrews, Ole Balling, Lynn Burgess, Mani Chandy, Mike Elder, Marionne Epalle, Alex Fax, Tony and Sandie Fender, Will Heltsley, Melany Hunt, John Kalmbach, Maria Koeper, Larry Matthies, Duane McRuer, Kenny Meyer, Bob Rasmussen, Doug Robinson, Jessica Stephens, Claire Tomlin and Dave van Gogh. We would also like to thank Steven Koonin, former Provost at Caltech, for his early support of the project.

Finally, this paper could not have been written without the hard work of the rest of the members Team Caltech, including Sterling Algermisson, Lyudmil Antnov (VPI), Henry Barnor, Ryan Cable, Eric Cady, Kevin Cossel, Adam Craig, Joe Donovan, Kevin Duncklee, Scott Elder, Ryan Farmer, Josh Feingold, Ken Fisher, Laura Fishman, Marie Göransson (Lund), Ike Gremmer, Rob Grogan, Lily Gruenke, Barrett Heyneman, Sue Ann Hong, Haomiao Huang, Cedric Jeanty, Erik Johannesson (Lund), Tony Kelman, Gustov Lindström (Lund), Kyle Miller, Lisa Nyström (Lund), Pietro Perona, Sam Pfister, Ben Pickett, Harmon Pollack, Tami Reyda, David Rosen, Ben Sexson, Alan Somers, Dan Soudek, Phillip Welch, Chris Wetzel, Lusann Yang (Princeton) and Jason Yosinski.

References

Amir, Y., Danilov, C., Miskin-Amir, M., Schultz, J., and Stanton, J. (2004). The spread toolkit: Architecture and performance. Technical Report CNDS-2004-1, Johns Hopkins University.

Cremean, L. B. (2006). *System architectures and environment modeling for high-speed autonomous navigation.* PhD thesis, California Institute of Technology, Mechanical Engineering.

Cremean, L. B. and Murray, R. M. (2006). Model-based estimation of off-highway road geometry using single-axis ladar and inertial sensing. In *Proc. IEEE International Conference on Robotics and Automation.*

Defense Advanced Research Projects Agency (2005). Grand Challenge 2005 team technical papers. `http://www.darpa.mil/grandchallenge05/techpapers.html`.

Dickmanns, E. D. (2004). Dynamic vision-based intelligence. *AI Magazine*, 25(2):10–30.

Gillula, J. H. (2006). A probabilistic framework for real-time mapping on an unmanned ground vehicle. Senior thesis, California Institute of Technology.

Goldberg, S. B., Maimone, M. W., and Matthies, L. (2002). Stereo vision and rover navigation software for planetary exploration. In *Proceedings of the 2002 IEEE Aerospace Conference*, Big Sky, MT.

Kogan, D. (2005). Realtime path planning through optimization methods. Master's thesis, California Institute of Technology.

Kogan, D. (2006). Optimization-based navigation for the DARPA Grand Challenge. *Conference on Decision and Control (CDC)*. Submitted.

Leibs, J. (2006). Learning positional and conformational state in multi- sensor networks. Senior thesis, California Institute of Technology.

McRuer, D. T. (1975). Measurement of driver-vehicle multiloop response properties with a single disturbance input. *IEEE Transactions on Systems, Man, and Cybernetics*, SMC-5(5).

Milam, M. B. (2003). *Real-Time Optimal Trajectory Generation for Constrained Dynamical Systems.* PhD thesis, California Institute of Technology.

Murray *et al.*, R. M. (2003). Online control customization via optimization-based control. In Samad, T. and Balas, G., editors, *Software-Enabled Control: Information Technology for Dynamical Systems.* IEEE Press.

Rasmussen, C. and Korah, T. (2005). On-vehicle and aerial texture analysis for vision-based desert road following. *cvpr*, 3:66.

Rasmussen, C. E. (2006). A hybrid vision + LADAR rural road follower. In *Proceedings of the 2006 International Conference on Robotics and Automation*, Orlando, FL.

Rasmussen, R. D. (2001). Goal-based fault tolerance for space systems using the Mission Data System. In *Proceedings of the 2002 IEEE Aerospace Conference*, Big Sky, MT.

Rosenblatt, J. (1998). Utility fusion: Map-based planning in a behavior-based system. In *Field and Service Robotics*, pages 411–418. Springer-Verlag.

Urmson, C. (2005). Navigation regimes for off-road autonomy. Technical Report CMU-RI-TR-05-23, Robotics Institute, Carnegie Mellon University.

Urmson, C., Anhalt, J., Clark, M., Galatali, T., Gonzalez, J. P., Gowdy, J., Gutierrez, A., Harbaugh, S., Johnson-Roberson, M., Kato, H., Koon, P. L., Peterson, K., Smith, B. K., Spiker, S., Tryzelaar, E., and Whittaker, W. (2004). High speed navigation of unrehearsed terrain: Red team technology for grand challenge. Technical Report TR-04-37, Robotics Institute, Carnegie Mellon University.

Vermeulen, S., Marples, R., Robbins, D., Houser, C., and Alexandratos, J. (2005). Gentoo linux x86 handbook. Available online at `http://www.gentoo.org/doc/en/handbook/handbook-x86.xml`

15

MITRE Meteor: An Off-Road Autonomous Vehicle for DARPA's Grand Challenge

Robert Grabowski, Richard Weatherly, Robert Bolling, David Seidel, Michael Shadid, and Ann Jones

The MITRE Corporation, 7525 Colshire Drive, McLean VA, 22102

Summary. The MITRE Meteor team fielded an autonomous vehicle that competed in DARPA's 2005 Grand Challenge race. This paper describes the team's approach to building its robotic vehicle, the vehicle and components that let the vehicle see and act, and the computer software that made the vehicle autonomous. It presents how the team prepared for the race and how their vehicle performed.

15.1 Introduction

In 2004, the Defense Advanced Research Projects Agency (DARPA) challenged developers of autonomous ground vehicles to build machines that could complete a 132-mile off-road course. 195 teams which applied – only 23 qualified to compete. Qualification included demonstrations to DARPA and a ten-day National Qualifying Event (NQE) in California. The race took place, on October 8 and 9, 2005 in the Mojave Desert, over a course containing gravel roads, dirt paths, switchbacks, open desert, dry lakebeds, mountain passes, and tunnels.

The MITRE Corporation decided to compete in the Grand Challenge in September 2004 by sponsoring the Meteor team. They believed that MITRE's work programs and military sponsors would benefit from an understanding of the technologies that contribute to the DARPA Grand Challenge.

This paper describes the MITRE Meteor team's approach to building its robotic vehicle, the resulting vehicle and associated sensors and systems, and the results of their efforts in the Grand Challenge race.

15.2 Approach

The Meteor team first decided on some underlying approaches to building the robot:

- Create a robot that could act as a test vehicle for missions, such as convoy leader/ following (Cheok, Smid, Kobayashi, Overholt & Lescoe, 1997), surveillance (Saptharishi, Bhat, Diehl, Dolan & Khosla, 2000), unmanned transport (Sotelo, Rodriguez & Magdalen, 2000), and cooperative robot missions (Sato et al., 2004).

M. Buehler, K. Iagnemma, and S. Singh (Eds.): DARPA'05, STAR 36, pp. 483–516, 2007.
springerlink.com

Fig. 15.1. The MITRE Meteor starting the finals of the 2005 DARPA Grand Challenge

- Create a robot that represents an affordable solution for sponsors, such as the Department of Defense and Department of Homeland Security.
- Remember that the robot must be built and tested in less than ten months.
- Focus on the most challenging aspect of the race – sensing and maneuvering.
- Employ COTS solutions wherever possible don't create where commercial solutions suffice.
- Use an incremental model-simulate-test approach. Build a model suitable to the current task. Verify and tune the model using simulation and replay. Test the model and system in real situations, and then use the results of the testing to adjust the model as necessary.
- Actively manage risk. Fundamentally, the challenge is straightforward and does not require a complex solution. Rather, the contest exposes the need to manage interdependencies among multiple systems.

15.3 Vehicle

The first step in producing a rugged autonomous vehicle is platform selection. Several vehicle types were considered: Racing buggy, all-terrain vehicle, four-wheel drive (4WD) sports-utility vehicle, and 4WD pickup. Investigation of the 2004 Grand Challenge results pointed to a platform capable of off-road travel, able to support multiple sensors and processors, and able to withstand desert temperatures.

The vehicle should also be easy to transport and operate. To reduce the cost of deployment, it should be drivable by a human operator. To improve safety during testing, a human operator should be able to override computer controls.

A commercially available vehicle that could be retrofitted with drive-by-wire capabilities was preferred – this would let the team focus more quickly on relevant issues. Similarly, a solution that could be adapted to any vehicle with drive-by-wire capability (such as, Humvees or tracked vehicles) was preferred.

The team chose a 2004 Ford Explorer Sport Trac pickup from a local dealer (Figure 15.1). It has reasonable off-road capability and a sufficiently cooled interior for computing equipment. Ford was selected because it is most understood by the selected drive-by-wire vendor.

15.3.1 Vehicle Modifications

The Grand Challenge required three major modifications to the vehicle: A drive-by-wire capability, an expanded electrical power system, and a chassis lift.

The Electronic Mobility Controls Corporation (EMC) designed and installed the drive-by-wire capability. They modified the transmission and steering column (Figure 15.2). A servo, mounted inline with the steering wheel, accomplishes turning. A second servo, attached to the firewall, controls the throttle and brakes. The installation also included an electrical override console with controls for the steering, throttle, and brake. Using this, a human safety operator can take control of the vehicle with the touch of a button. The capability also provides a safety override so that an operator in the driver's seat can always operate the brakes manually.

The second modification was installing a heavyduty 220-A alternator and power bus. This eliminated the need for additional batteries or an external generator, and easily handled the load of the sensors and processors. Power passes from the alternator to the rear processing rack via a high-current dc bus. A 3000 W dc-to-ac converter provides 120 V for computers and sensors.

The last major modification was the installation of a lift kit on the vehicle. An analysis of the 2004 route raised concerns about whether the vehicle could clear moderate-sized rocks and debris (Urmson et al., 2004; Ozguner, Redmill &

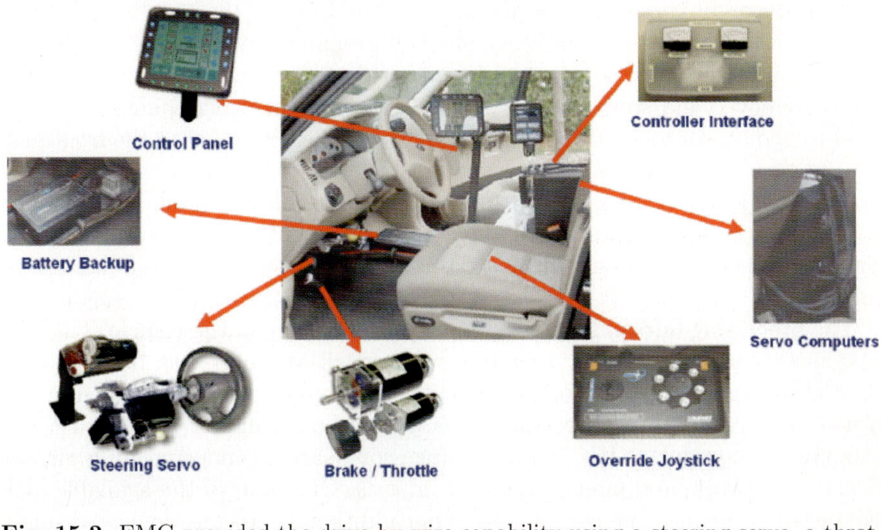

Fig. 15.2. EMC provided the drive-by-wire capability using a steering servo, a throttle/brake servo, a control console, two servo computers, a battery backup, and a manual override

Broggi, 2004). So, a 4 in. suspension lift was donated by SuperLift, an off-road retrofitter. Additionally, four 30 in., offroad Super Swamper tires were installed to increase traction in rugged terrain.

By employing these COTS components, the team was able to start testing on a reliable platform within two months.

15.3.2 Sensor and Processor Suite

Similarly, the team selected COTS sensors and computers. The sensing centers around eight laser sensors were oriented in different directions and mounted on the top and front of the vehicle (Figure 15.3). Laser range finders provide a two-dimensional range/distance map out to 40 m with a 100° field of view. These lasers are the staple of the robotics community and were used by almost every entrant in the Grand Challenge. Meteor utilized two models of the laser sensor from SICK; one operating at 5 scans per second, the other at 75 scans per second.

The laser range finders were divided into classes based on their orientation. Two vertically mounted lasers provide information about the ground plane. Three lasers, mounted horizontally at different angles, provide both short- and long-range detection of road obstacles. A sixth horizontal laser is mounted on gimbals, and points dynamically to compensate for vehicle pitch and terrain variations. Finally, two downward-looking lasers are mounted on the roof to detect road characteristics.

Two sensor racks were built to mount the sensors to the outside of the vehicle. The largest sensor rack is mounted on top of the vehicle. It provides the platform for mounting the global positioning system (GPS) receivers, inertial navigation system (INS), and magnetic compass. This positioning allows the greatest visibility to satellites and moves the sensors away from vehicle noise. Additionally, the two down-looking lasers are mounted at the front of the rack to give them the greatest clearance and visibility of small ground obstacles near the vehicle. The rack is held to the roof struts using four U-bolts, so that the entire assembly can be removed and stored in the vehicle cargo bay during shipping.

A second sensor rack utilizes a grill extension. Shelves added to the front-grill guard house the horizontal and vertical obstacle and road scanners. The rack is mounted rigidly a few inches in front of the grill to allow engine air flow. The headlight guards were removed to prevent misalignment in the event of collision. The headlights were expendable and the sensor system should prevent a head-on collision.

The processing infrastructure is mounted in the cabin of the vehicle. The rear passenger seat was partially removed, and a rack assembly was built to hold the processing and display components. The entire rack is shockmounted in five places. The computing infrastructure is a multiboard militaryhardened VME computer array of 1.8 GHz Pentium processor boards, connected by a gigabit Ethernet network, and mounted in a 19 in. rack. Only four of the available nine slots were needed for the competition. One additional card was dedicated solely to logging and display during testing. The rear passenger area also contains a

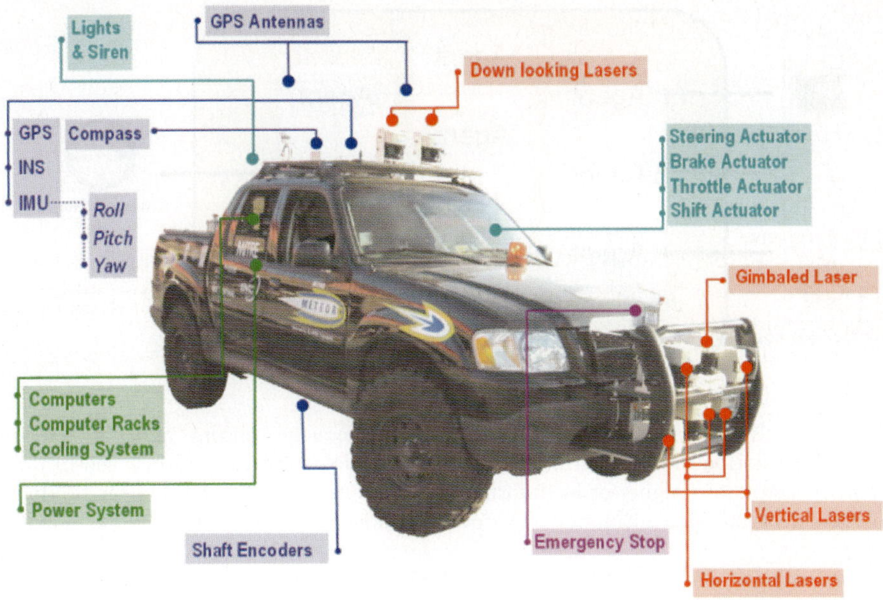

Fig. 15.3. MITRE Meteor layout. Meteor has an array of COTS components. Positioning is by three GPS units, an INS unit and a compass on the roof. Four horizontal lasers detect road obstacles. Two vertically-mounted lasers provide terrain information. Two down-looking lasers detect small objects and negative spaces. A computer assembly is in the back seat.

rack-mounted monitor, keyboard, and mouse, to permit an observer to oversee and interact with the system during testing.

15.4 Software Architecture

Experience shows that software evolves continually until it is retired. Characterizing axes of change reveals what can and cannot be specified by software architecture. For this effort, two axes of change were clear: (1) The controller would have a number of quasi-independent activities, whose algorithms could change during the development process, and (2) the population and relationship between these activities would evolve. Therefore, an agent-based architecture (Minsky, 1985) was chosen.

Experience also shows that building the software scaffolding (test harnesses, stubs for yet-to-be-built components, verifiers, etc.) surrounding a software product can be time consuming. To reduce this overhead cost, two themes were applied to the design: Location transparency and employment transparency. That is, use the same piece of code in as many different places as makes sense, and for as many different purposes as is reasonable.

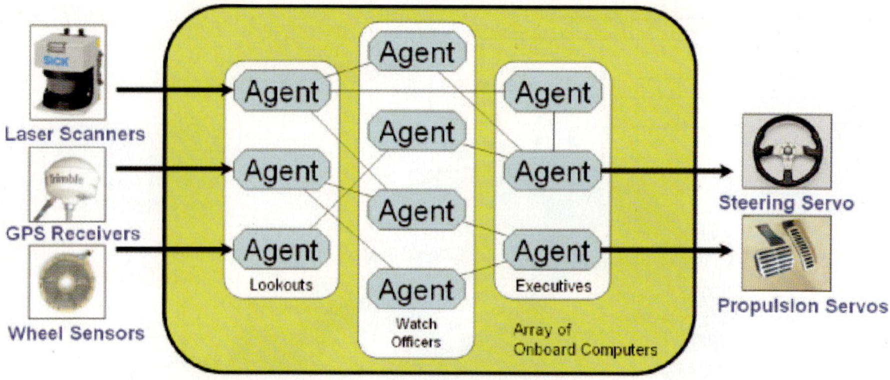

Fig. 15.4. Basic agent makeup. Lookout agents manage sensors, and convert raw sensor information into system messages. Watch Officers process and fuse sensor data from Lookouts into higher-order information. Executives use Watch Officer products to make decisions that ultimately control the actions of the Meteor.

15.4.1 Agent-Based Approach

The entire software architectures consists of agents that receive state information, add value to that input, and pass it to the remainder of the system (Figure 15.4). Agents are of three basic types: "Lookouts," "Watch Officers," and "Executives." Lookout agents manage sensors and convert raw sensor information into system messages. Watch Officers process and fuse sensor data from one or more Lookouts to provide higher-level information, such as vehicle pose, obstacle definitions, and ground plane estimates. Executives use Watch Officer products to make decisions that ultimately control the actions of Meteor.

Specialized "executives" in the program map the vehicle state to motion (Figure 15.5). A "Captain" takes the provided Route Definition Data File (RDDF) and generates a lane in which the vehicle can operate. A "Navigator" takes the lane and obstacles and generates a viable path. It then uses the path to determine a desired speed and direction based on the vehicles current state. Desired speed and direction are passed to a "Helmsman" which converts these parameters into commands that adjust the two voltages that drive the EMC system. An "Admiral" agent gives a final go/no-go signal based on the state of the emergency stop system. While components within this architecture evolved, the basic architectural design remained unchanged.

15.4.2 Location Transparency

Vehicle control is achieved by agents that communicate with each other using UDP messages. Raw sensor input and aggregate state are manipulated as Java classes, serialized for transmission as messages, and then shared between computers via a fast Ethernet switch. The agent state is mirrored on each computer and kept current through a constant assertion methodology. The vehicle state,

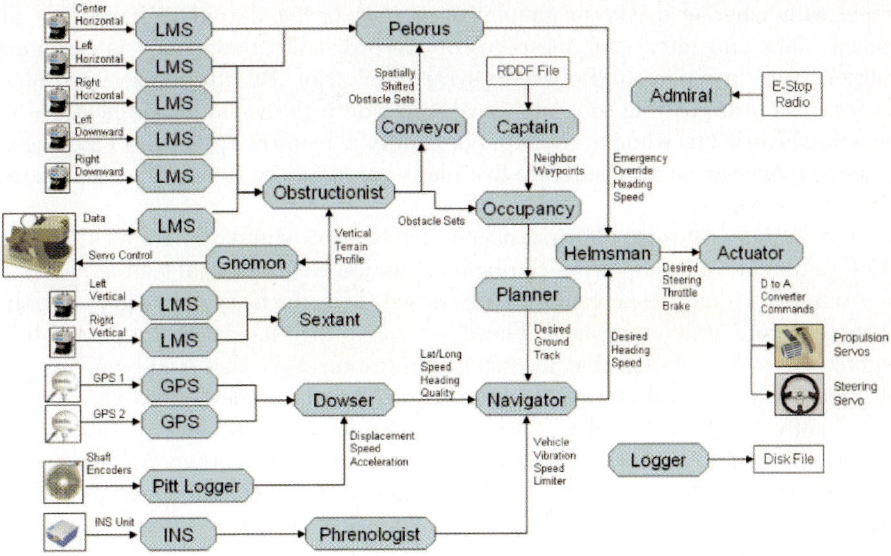

Fig. 15.5. Software architecture implementation. An agent-based architecture converts raw sensor information into voltages to drive the robot. "Lookouts" convert raw sensor data into system messages. "Watch Officers" turn raw information into information, such as, obstacles and position. The "Captain" takes the RDDF and defines the next achievable point and the lane in which the vehicle can operate. The "Planner" uses the lane and obstacles generated by the lasers to generate a plan. The "Navigator" determines desired speed and direction based on the plan and the vehicle's state. The "Helmsman" converts the speed and direction into commands that cause the EMC system to produce motion. The "Admiral" monitors the E-Stop radio and gives the master-go signal.

sensor values, and obstacle inventory are calculated and reported periodically. The advantage of employing a message transfer protocol is that agents can be located anywhere without modifying the agent code. This permits load balancing among multiple computers. Initially, all of the components required to drive the vehicle ran on a single laptop computer. As functionality was added, agents were distributed among several computer boards. High demand agents, such as the occupancy map Watch Officer, were allocated their own computers. The motion model and obstacle filters require less computation and share the same computer. One of the biggest processor loads is the graphical display software used by the operator to monitor progress during development — it was allocated its own computer.

15.4.3 Employment Transparency

A unique feature of this architecture is that it is insensitive to the source of input. Because the system state is continuously shared, the system behaves nearly identically whether driven by simulation, replay, or live data. During testing, all sensor data and interagent messages are recorded. Posttest analysis examines performance in replay, and observes system behavior. By filtering selected messages, it is also possible to exploit real-world data to evaluate modifications to existing algorithms while in the lab. For example, by filtering out path messages, a new planner agent can generate live plans based on the information generated from a previous test.

The system can also be driven completely from simulated sources (Figure 15.6). Simulated sensors produce raw information from external sources, such as a ground truth map. These signals are passed through the system producing the same control voltage messages. These control voltage messages are passed to a simulated vehicle model, that in turn generates messages that feed back into the system, such as vehicle location, steering angle, and encoder values.

Because the operational software functions in the same way in simulation, replay, or actual use, each of these modes can support the other. As new agents were developed, they were first run in the simulator. When behavior was acceptable, they were tested in the field. Data from live runs were used to tune the agents and simulation models.

Location and employment transparency reduced testing time by recording all aspects of a run, evaluating the results, and testing resultant changes in simulation. Even though only 10 days of desert testing was conducted, the design permitted significant improvements to the system.

15.4.4 Development Environment

The software was developed for the Fedora Core 3 Linux operating system using the Java 5 programming language. The development environment employed Sun

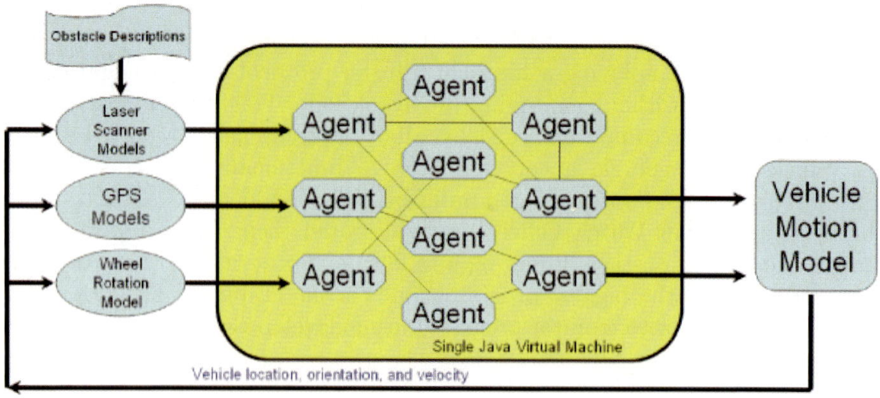

Fig. 15.6. Employment transparency. Employment transparency means that the system is unaware of whether it is being driven by real, simulated, or replay data. Here, a simulated vehicle model adds a feedback mechanism to allow simulation of events.

JDK 1.5, Subversion, Ant, and Eclipse. Each of these technologies, is proven, readily available, and carries a low maintenance and registration overhead.

15.5 Maneuvering and Navigation

The most rudimentary task facing a vehicle in the Grand Challenge is the ability to follow a series of waypoints. DARPA defined the route for the race as a series of about 3,000 connected latitude/longitude positions that wind through the desert. DARPA specified that the points would define a path with no intervening impassable features. They also defined the maximum lane width and maximum safe speed for each segment. Each vehicle then needed to compute its own position, the position of the next point, and a strategy for getting there.

15.5.1 GPS Positioning

Positioning for Meteor is accomplished by three GPS units. The primary GPS receivers are two Trimble AgGPS132 differential GPS units with differential corrections from the Omnistar subscription service. With proper sky visibility, the Omnistar service improves position accuracy to within 5-10 centimeters at a rate of ten times per second.

Each Trimble unit reports signal strength and signal-to-noise ratio. These measurements determine whether a GPS unit is reporting accurate readings. Testing showed that the units were not consistent with one another; so to accept positions as valid, both measurements were required to be above a chosen threshold. In most cases, the measurements dropped quickly enough to indicate a loss of valid position before the system has deviated too far.

Figure 15.7(a) shows the GPS signal-to-noise and signal strength measurements for a run during the NQE. Here, the GPS dropped below threshold many times along the track, with the tunnel being the longest. Circles correspond to the five selected points (pluses) in the data with the second being immediately before the tunnel. The chosen threshold for the signal-to-noise ratio was 9 and the signal strength was 110. During the NQE, both values were raised slightly.

The GPS units seemed reliable during testing up to the NQE, including testing in the deserts of Yuma and Primm. However, after anomalies during the first NQE run, their output was re-examined. During practice testing and replay of the first run, the vehicle position jumped a few meters and returned even when the vehicle was stopped. Between Runs 1 and 2, static GPS tests were conducted. Figure 15.7(b) (top) shows the results of taking 2,500 GPS samples with the vehicle stationary. Instead of positions randomly centered about the origin of the robot, several correlated clusters formed a pattern about the origin (dotted circles). Moreover, the plotted histories of the pattern (Figure 15.7(b) bottom) showed that, instead of being randomly distributed, they dwelled at one point before jumping to the other. When running, the vehicle had enough time between jumps to try to compensate for the new errors. The vehicle appeared to jog in and out of the lane as it was driving.

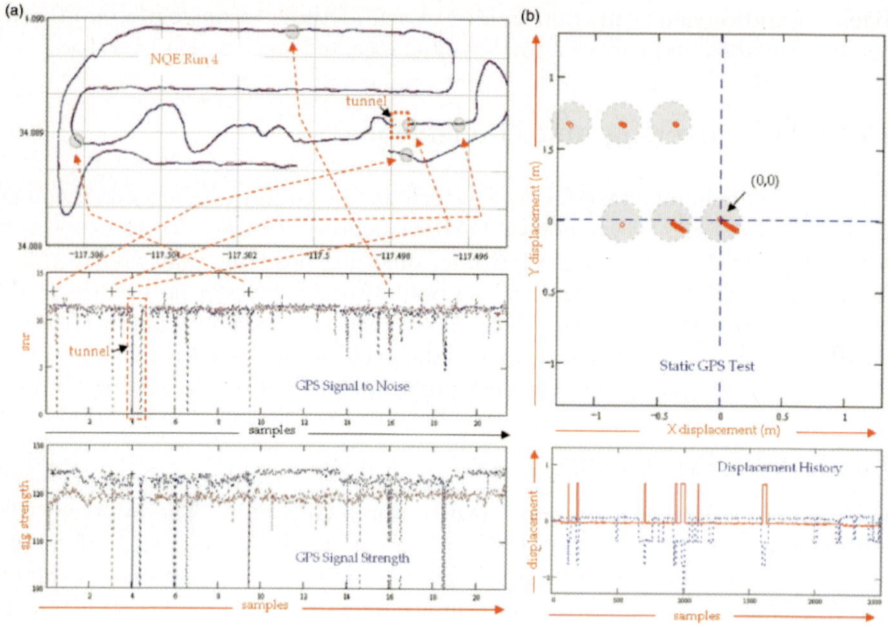

Fig. 15.7. GPS verification. GPS reliability is determined by assessing signal strength and signal-to-noise ratio reported by GPS units. If the ratios drop below a set threshold, their reported positions are not trusted, and Meteor is guided by a dead reckoning model. (a) Signal-to-noise ratio and signal strength plots during a run at the NQE. Circles correspond to selected points on the plots. Note the complete drop out inside the tunnel. (b) 2,500 GPS readings taken while the robot was stopped during testing at the NQE. Note the apparent spatial periodicity implying multipath. Below is the same series as a function of time and displacement.

The regular pattern of clusters suggested a strong influence from multipath, perhaps from metal structures, such as, fences and bleachers that surrounded the NQE. The other interesting phenomenon was the correlation between readings within the clusters. Even with multipath, the readings should have been random about a point. Instead, they seemed smeared along a line oriented in a south/south east direction. No explanation was discovered for this behavior.

A third GPS is a MIDG-2 INS. This GPS unit is augmented by an internal inertial measurement unit that maintains some location ability during GPS outages. If both Trimble GPS units are above their threshold, the MIDG GPS unit serves as a tiebreaker. Tiebreaking prevents arbitrarily jumping back and forth between units. Even though both are accurate, they differ slightly and jumping back and forth introduces high-frequency position noise.

Heading is determined by comparing successive positions. This method proved to be very accurate at speeds above 3 miles per hour, but ineffective at lower speeds. Initially, a Honeywell magnetic compass provided an alternative source of heading. However, it was abandoned because it required calibration in different locations (Nevada and California), had high latency (degrees per minute), and drifted as much as 10° with the vehicle stopped.

15.5.2 Motion Model

Even with differential GPS in wide-open spaces, GPS experienced periodic and sometimes sustained losses. To overcome these losses, a simple bicycle model was used to predict the vehicles' location.

Vehicle displacement is measured via a multielement quadrature shaft encoder, mounted directly to the vehicle drive shaft. We assume no differential slip or tire slip. A similar encoder is mounted to the steering column, and provides steering wheel angle. This is input to a lookup table that maps steering column angle to turn radius. The table was populated experimentally. The turn radius and displacement measurements are passed to the bicycle model, which produces a new position estimate. Since the bicycle model is described in many robotic papers, we omit it here.

If either Trimble GPS unit produces a valid reading (above the detection threshold), the deadreckoning position is not needed, and its current estimate is set to the measured GPS position. If no valid GPS reading is available, the dead-reckoning model becomes the position provider.

Testing showed that, for most of the surfaces, the dead-reckoning model could run for about 100 meters without a GPS lock and keep the vehicle within the boundaries of a road. This distance drops quickly if the vehicle must perform many turns. While not sufficient to track position for a long distance, the model was sufficient for the types of GPS losses expected in the course.

15.5.3 Steering Gain

Once the vehicle is able to determine its own position, it has a context to follow a route. A popular approach for route following is to continuously point the vehicle toward the next waypoint in the route definition. This approach has subtle but significant drawbacks. The error between the actual steering angle (the way the vehicle is pointing) and the derived steering angle (the angle to the waypoint) increases as the vehicle nears the waypoint. This effect, termed as steering gain by the team, can lead to oscillations when the vehicle is close to a waypoint. The phenomenon is exacerbated by small perturbations in reported vehicle position. Instability increases quickly with speed and can lead to unsettling performance above 20 mph. Steering instabilities can lead to other effects, such as excessive roll, which can lead to errors in obstacle detection.

To address this, a "carrot" mechanism was introduced to regulate the steering gain of the robot (Figure 15.8a). The carrot acts as a virtual waypoint that moves back and forth along the intended path of the robot. Instead of steering toward the next waypoint, the vehicle steers toward the carrot. Since steering gain is a function of vehicle speed, carrot distance is a function of vehicle speed. As vehicle speed increases, the carrot moves farther from the vehicle. As the vehicle slows, the carrot moves closer.

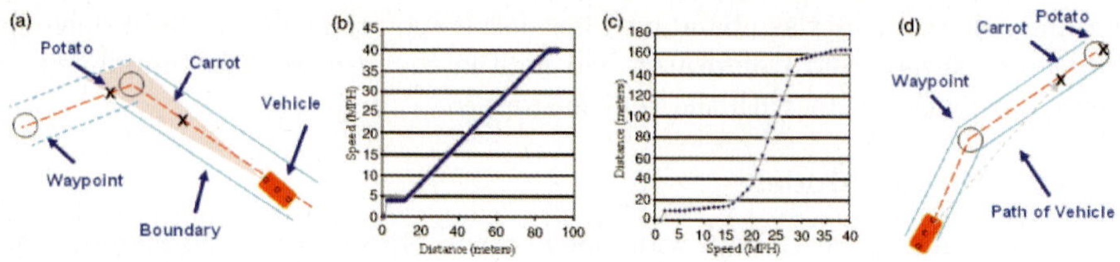

Fig. 15.8. Carrot maneuvering. Steering stability is achieved using a carrot and potato. (a) The carrot maintains its distance in front of the vehicle based on vehicle speed. Maximum vehicle speed is a function of path visibility and is determined by the potato. The potato defines visibility, the carrot follows the path, and the vehicle follows the carrot. (b) A curve mapping the distance to the potato to maximum vehicle speed. (c) A curve mapping actual vehicle speed to carrot distance. (d) An example where a large carrot distance can cause a slight shaving of the defined route.

The carrot mechanism provides stable steering across the range of vehicle speeds but does not determine target speed. This is accomplished by the "potato" which, like the carrot, is a point that moves along the intended path of the robot. The potato stays ahead of the vehicle on the path at a point that represents the limit of vehicle path visibility. The term path visibility captures the notion of how far the vehicle can see down a 2-dimensional pipe defined by the intended path. Notwithstanding the effect of obstacles, the vehicle sets its speed based on the distance to the potato (Figure 15.8b). If the path is long and straight, the potato is far away and the vehicle can travel swiftly. If the path is windy or contains a sharp curve, the potato is close and the vehicle slows.

Path visibility distance is determined by summing the distance along the path starting from the vehicle, while simultaneously summing the absolute angle between those segments. When the absolute sum of the angles exceeds a threshold, path distance summation stops. The summation method accounts for sharp curves, as well as the cumulative effect of shallow curves.

The functions that map vehicle speed to carrot distance (Figure 15.8c) and potato distance to target vehicle speed are constructed of linear segments. These mappings were determined first in the system simulator and then experimentally on a dry lakebed in Nevada. For a given speed, the carrot distance was manually adjusted out until the vehicle steering stability (degree of overshoot under a perturbation) was satisfactory. Next, the potato speed mapping was adjusted to assure that the vehicle slowed properly when approaching a turn. Aside from traction problems, entering a turn too fast forces the carrot around the corner too soon and causes the vehicle to shave the inside of the turn (Figure 15.8d).

The biggest advantage of using path visibility is that the vehicle naturally slows as it approaches large turns, and then speeds up as it passes through the turn. Note that this path is generated by the planner (discussed later), and not the path given by the route definition (RDDF file).

15.5.4 Additional Speed Limiting

In addition to path geometry, several additional mechanisms regulate vehicle speed. While speed is important for competing in the Grand Challenge, it increases risks inherent in a large moving vehicle (Gillula, 2005). Simply increasing the speed without addressing safety, stability, and sensor range fails to recognize the dangers inherent in large robots. Higher speed reduces the distance available to react to an obstacle, decreases sensor fidelity as samples are taken over a larger area, and consequently decreases confidence in a selected action. At higher speeds, vehicles are more likely to tip over or swerve off the road from an unexpected steering correction. In the event of a collision, higher speed increases the momentum of a vehicle; increasing the likelihood of damage. This is evident from the damage to the parked cars at the NQE from robots colliding with them at low speeds. DARPA defined the maximum safe speed for each segment of the route. The team chose to respect this value.

As discussed above, path visibility is the primary factor for setting the speed along a segment. It determines speed based on how far it can "see" along the geometry of the path and slows the vehicle before entering sharp curves or a complex calculated path.

Another speed consideration is congestion where the vehicle slows when it perceives that it is entering an area with high density of objects, even if it has determined a clear path through it. The robot should not drive through an obstacle field at high speeds, but should anticipate danger and react appropriately. Congestion is calculated by summing the number of occupied cells (discussed in Section 15.7.2) about a narrow rectangular space along the path in front of the vehicle. If the path is congested, the vehicle slows.

A similar safety mechanism slows the vehicle if it senses excessive vibration. Not only can excessive vibration damage sensitive components, such as computer disk drives and connectors, but it can also increase stopping distance and add steering instabilities. Moreover, vibration can be a precursor to other dangers. Testing in the desert showed that seemingly small desert shrubs could accumulate large amounts of dirt at their base. Meteor hit one at 25 mph and the resultant jolt broke off one of the shock mounts and forced the vehicle several meters off the road. Thus, vehicle speed is reduced based on frequency and magnitude measurements from the INS vertical accelerometers. Like many such mappings, a transfer function was defined and tuned based on driving over a range of different road conditions in the desert.

The last speed control mechanism is engaged by the reactive system. Normally, the vehicle follows a path generated by the planner, and speed is based on path visibility and congestion. However, a reactive system engages when the planner misses an obstacle and it endangers the hull. The reactive systems steers the vehicle away from the obstacle along the nearest open path. At the same time, it slows the vehicle to provide a better opportunity to maneuver. When the reactive system engages, it slows the vehicle to 4 mph until it has cleared the obstacle. A small dead band and hysteresis prevents premature triggering and release.

These methods of reducing speed in anticipation of danger cause the robot to err on the side of caution. Because of limited development time and access to desert terrain, speed control mechanisms were only roughly tuned. As a result, the vehicle was able to achieve its maximum speed only few times during the NQE and race. However, these speed mechanisms were the likely reason that the vehicle suffered no major incidents or collisions during testing and competition.

15.6 Obstacle Processing

The primary function of the laser rangefinders is to detect obstacles that lie directly in the path of the vehicle. Initially three fixed lasers were mounted on the front at slightly different angles to determine obstacles immediately in the path of the robot (Figure 15.9a). One laser pointed directly out while the other two were oriented about 3 degrees above and below the horizon. Different angles were selected to maintain acceptable sensing distances even as the vehicle was driving over and down elevated terrain. Early tests showed that the small pitches in terrain could cause the laser to miss obstacles as large as trashcans as it passed over the object or struck the ground just before it. Just before deployment, a steerable laser was developed that could scan up and down, reducing the need for fixed lasers. However, time constraints prevented comparison between the two approaches and both were used.

15.6.1 Obstacle Generation

Each laser produces a series of 400 range readings with an angle increment of 0.25 degrees. With four lasers reporting at 5 times a second and the remaining

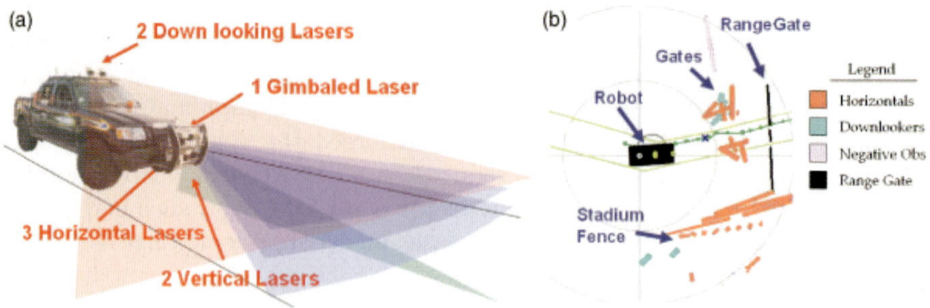

Fig. 15.9. Sensor types. (a) Meteor carries four classes of laser sensors. Three fixed horizontal lasers mounted at slightly different angles detect obstacles directly in front of the vehicle. A gimbaled horizontal laser tracks the horizon providing the greatest sensing distance. Two vertically-mounted lasers detect road terrain and provide dynamic range gating to the horizontal lasers. Two down-looking lasers detect small ground obstacles and negative terrain. (b) Line objects generated by each laser are displayed during testing. This is a view of Meteor as it is about to pass through the first gate at the NQE.

four reporting at 75 times a second, raw message traffic is large. To reduce message traffic, each laser scan is converted to a series of line segments before being passed through the system. Conversion from raw laser data to line segments provides a more compact representation - generally less than 10 segments per scan. This representation also reduces the processing complexity of downstream agents analyzing these segments. In fact, it increased the performance of occupancy map generation by an order of magnitude. Reducing the fidelity of laser scans also reduces the detail and look of the generated map, but it had little effect on the ability to represent traversable terrain.

Objects detected by lasers fit into three categories – point obstacles, line obstacles, and vistas. Objects are generated during each pass of a laser scan by clustering continuous readings together using a simple state machine. Contiguous readings that are less than the maximum range are clustered into obstacles and represent solid obstacles. Contiguous readings that are all at maximum range are clustered as vistas and represent open space.

A post-scan filter was also added to remove some objects based on their projected size. That is, small line and point obstacles close to the vehicle indicate objects that are only a few centimeters in width and are more likely to be noise or debris like falling leaves.

Figure 15.9b shows the line obstacles and vistas produced by each of the lasers as Meteor passes between two gates at the beginning of the NQE. Each is coded based on their source. This instantaneous view suggests how the vehicle is responding to the most current event and is an essential tool for understanding and development.

15.6.2 Vertical Lasers and Dynamic Range Gating

A complication of mounting horizontal sensors low in front of the vehicle is that the sensor scan might hit the ground. A horizontal laser pointing downward will probably return non-maximal readings across the entire scan making it difficult to distinguish between an obstacle and the ground. If uncorrected, the robot would always try to avoid the ground in front of it. A naive model would assume that the world was flat and horizontal then calculate where the laser would intersect the ground and reject all readings greater than that value. However, this assumption limits the terrain where the vehicle could operate. Small crests or depressions would produce unwanted results. Moreover, passing over rough terrain could introduce large perturbations in pitch and roll which have a similar effect to changing terrain for the laser scanners.

To compensate for the effects of non-flat terrain, two lasers were mounted vertically to the front grill (Figure 15.10d). Each vertical laser produces a scan that starts from directly beneath the vehicle and scans outwards along its major axis (Figure 15.10e left). In the absence of obstacles, the scan represents the contour of the ground directly in front of the vehicle (Figure 15.10c).

The vertical scan is used by the horizontal lasers to generate a range gate – a single value that represents the distance from that sensor to the ground plane. Horizontal range values beyond the range gate are likely to be caused by intersection

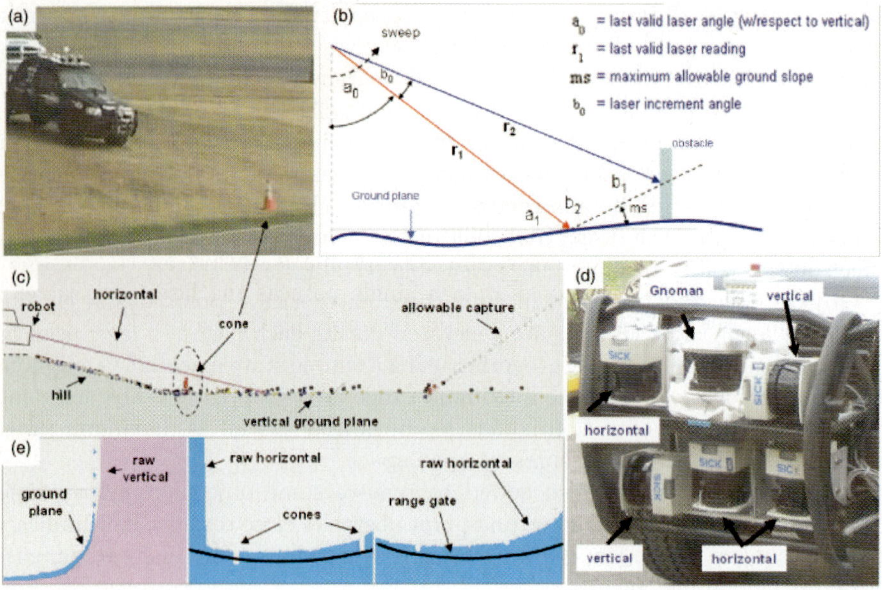

Fig. 15.10. Vertical lasers. (a) Meteor travels down a steep hill at the NQE. (b) A slope filter removes obstacle readings from the vertical scan based on the slope change between readings to generate a ground plane estimate for range gating. (c) A projection of the filtered ground plane just before Meteor begins to drive down the hill. Note that the cone at the bottom of the hill (dotted ellipse) has been removed from the scan. (d) The physical mounting of the lasers. A vertical laser collinear to the horizontal lasers it services. This alignment makes the range gate a simple geometric lookup in the vertical scan. (e) Raw vertical laser scan (left pane, shaded) and projected ground plane (bowed line). Raw range scans from two horizontal lasers (center and left pane shaded). The crossing lines are the dynamic range gate. Note that the cones on either side of the bottom of the hill are visible as gaps in the raw scans, and are starting to show up as obstacles in two of the horizontals.

with the ground and are rejected (Figure 15.10e right). Those shorter than the range gate are likely to be the result of a true obstacle. Since both the vertical and horizontal lasers are mounted rigidly and the vertical lasers are mounted beside the corresponding horizontal lasers, the dynamic range gate is insensitive to whether the changing profile is caused by terrain changes or vehicle transients. Moreover, by mounting them collinear to one another, range gate calculations are reduced to a simple index lookup and make fewer assumptions about the geometry of the vehicle.

To provide an accurate range gate, the vertical laser must filter out any obstacles that lie directly along the axis of the vehicle. Without this filter, the obstacles would be incorporated into the ground plane estimate, and could result in an erroneous range gate that blinded the horizontal scans of true obstacles.

Obstacles are rejected by looking at the rate at which the ground plane height (a function of range) changes over the scan (Figure 15.10b). If the height changes by a large amount, it is probably an obstacle. If the rate change is small, it is likely the road contour. Based on experiments in the desert, a rate change was selected that corresponds to a road slope of 15 degrees (the largest test measurement of road slope was 9 degrees).

To reduce computational complexity and latency, this rate change is expressed in terms of the previous range value and current sweep angle. If the scan is started from close to the vehicle, the road profile is always monotonic. This allows filtering to be accomplished in a single pass as the laser scan is read. We calculate the minimum range reading by applying the Law of Sines (Figure 15.10b, Equation 15.1).

$$r_2 = r_1 \cdot \frac{\sin(b_2)}{\sin(b_1)} = r_1 \cdot \frac{\sin(90 - ms + a_0)}{\sin(90 + ms - a_0 - b_0)} \tag{15.1}$$

If there are no obstacles on the road, each successive range reading should exceed this minimum threshold. If the threshold is not met, the ground plane is estimated (Equation 15.2). Since no information about the terrain behind an obstacle is available, the ground shadowed by any obstacles is assumed horizontal.

$$\begin{aligned} &\text{if } (\mathrm{r}_{\text{measured}} > r_2)\left\{\, a_0 = a_0 + b_0 \;:\; r_1 = r_2 \;:\; \mathrm{r}_{\text{ground}} = r_2 \,\right\} \text{ (measured)} \\ &\text{else}\left\{ \mathrm{r}_{\text{ground}} = \frac{h_{laser}}{\cos(a_0 + b_0)} \right\} \text{ (estimated)} \end{aligned} \tag{15.2}$$

Since an obstacle may shield many points, the estimate will continue until a valid ground measurement is obtained. However, since the ground could have continued to rise in this shielded region; the acceptable threshold is slowly increased to allow it to deal with rising terrain beyond the obstacle. However, this too was tempered. For example, at the NQE the first series of gates shadowed a large region beyond the vertical lasers as it passed over them. The next readings that were not shadowed were from the bottom of the bleachers directly behind the gates (another obstacle). These points were low enough that they could be mistaken for a slow rise starting at the gates and cause an erroneous indication of ground plane. To alleviate this situation, the ability to recapture the ground plane beyond large occlusions was limited.

Figure 15.10 shows the vertical lasers in action as the vehicle is about to travel down the steep hill just before entering the hay bales (Figure 15.10a). Figure 15.10c shows the projection of both vertical laser scans reaching below and outward from the vehicle. The dots projecting away from the vehicle show the points accepted as part of the ground plane except those in the dashed circle that were rejected as an obstacle. In this view, the reading was taken at the top of the hill looking down about 5 meters from the bottom at a slight angle to the road path. The vertical scan passed across one of the cones. Because the instantaneous ranges to these points do not increase fast enough, they were rejected. The solid line projecting out of the robot represents the horizontal lasers and shows where they intersect the ground plane. Figure 15.10e (left) shows the raw scan from one of the vertical lasers (light shade) and the corresponding ground plane estimation (darker line). Figure 15.10e (center and right) shows two of the raw horizontal scans (shaded) with the corresponding range gate drawn as a solid black curve across the scan. Note that the range gate is just above the main part of the scan effectively eliminating all the ground readings but is low enough to indicate the presence of the two cones located at the bottom of the hill on either side of the lane.

15.6.3 Scanning Laser

An articulated laser, called the Gnomon, was developed that could scan up and down using a simple 4-bar mechanism. The articulated laser could be adjusted to point up or down 10 degrees about the horizontal.

To maximize sensing distance, the Gnomon was pointed just above the horizon. The Gnomon agent determines the horizon by scanning the ground plane developed by the vertical lasers to find the point where the range readings reach their maximum. Since the ground plane reading is filtered, this maximal reading represents the farthest point in front of the vehicle. The Gnomon is then steered a few degrees above that number to clear any ground clutter.

Steering to the horizon is not always the best action for a laser. As the robot enters a large valley, the horizon may actually be on the opposite crest. If the laser is pointed toward the horizon, it could miss an obstacle on the road at the bottom. However, the vehicle is also equipped with multiple fixed lasers that protect the vehicle. Future work will explore the addition of a maximal height filter to limit the degree of scan based on the maximum distance of its laser plane with the ground plane. This work could reduce the number of lasers needed on the front of the vehicle.

A drawback to a mechanical steering device is rapid transient response. Most terrain that the robot must traverse has a slow rate changes with respect to the Gnomon speed (full range deflection of -10 degrees to +10 degrees in about 250ms). However, as the vehicle travels over rougher terrain, the pitches and rolls can require compensation that exceeds the Gnomon's ability to react. Since the range gate produced by the vertical sensors is much quicker than the mechanical reaction time, it corrects for range errors introduced by the transient.

15.6.4 Down-Looking Lasers

While horizontal lasers are good at detecting large objects in front of the vehicle, they are not oriented to detect objects that lie close to the ground. Several teams addressed this issue by mounting horizontal lasers low to the ground where they are more susceptible to fouling and damage from low-lying obstacles and debris. Even so, horizontally mounted lasers are inadequate for detecting large holes or drop-offs in the terrain. Reports from the previous year (Urmson et al., 2004) and testing in the desert proved the need to detect these low-lying and negative

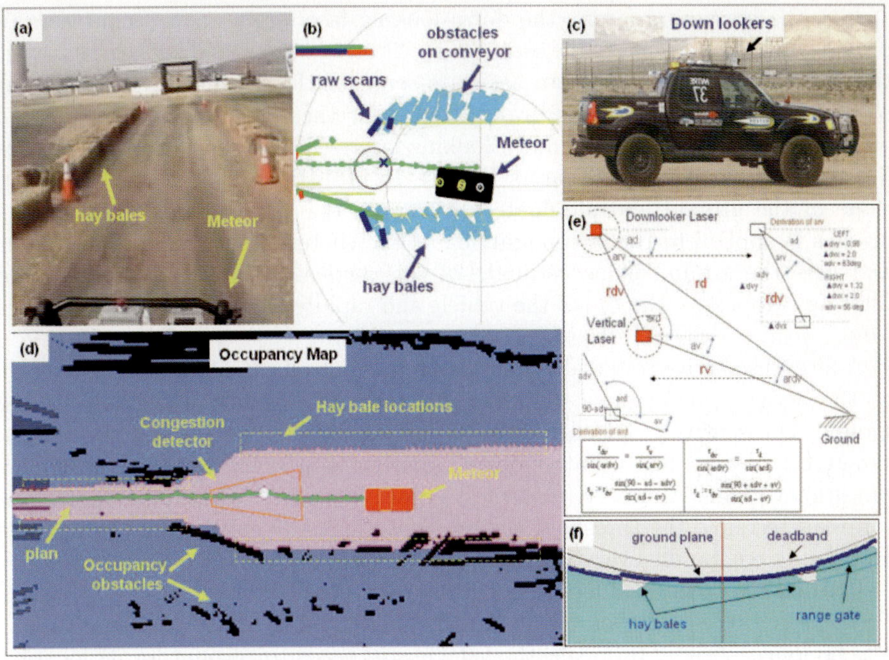

Fig. 15.11. Down looking. (a) Meteor travels down the center of low-lying hay bales. (b) Line segments generated by the down-looking lasers clearly show hay bales. Note Meteor correcting slightly right. (c) Down-looking lasers mounted to the top of the roof rack. (d) Occupancy map generated in the middle of the hay bales. The lane crosses the center image and narrows from right to left. Light areas indicate open space, dark segments are obstacles. Note that hay bales do not show up in the occupancy map. (e) Like the vertical lasers, the down-looker ground plane filter is based on the law of sines. (f) The raw output of a down-looking laser (shaded). The dark line riding on top of the raw readings is the horizontal ground plane. The hay bales are clearly seen in the scan. The ground plane shows that Meteor slowly rolled to the left.

obstacles. Roads in the mountains of the desert southwest are dominated by narrow paths with occasional 30-meter drop offs. They are also littered with large boulders (Figure 15.15d). Even in the relatively flat terrain of a lakebed, desert shrubbery can trap mounds of dirt.

To address these concerns, two additional lasers were mounted to the roof of the vehicle pitched down at 10 and 11 degree angles (Figure 15.11c). This down-looker scanned the road 10-15 meters in front of the vehicle. Using two scanners with slightly different angles increased the likelihood of seeing narrow obstacles in the road. The heightened roof position provided a better scan angle for sweeping across the ground. Lower positions would have required a shallower scan angle and would have made the scan distance highly dependent on vehicle pitch.

Like the horizontal lasers, the down-looking lasers rely on vertical lasers to provide an initial ground plane estimate for filtering out ground strikes. However, the vertical lasers only give an accurate sense of the terrain directly in front of the vehicle and do not accurately represent the terrain to either side. To overcome this limitation, the down-looking lasers employ their own rate change filter similar to the vertical lasers (Figure 15.11e). Both look at the change in slope of the ground and eliminate any ranges that change too quickly. This process is applied to the sweep starting at the center of the scan radiating out and uses the ground plane estimate of the verticals as a seed. This form of filtering is robust to the roll of the vehicle and rapidly changing terrain on either side.

A filter insensitive to terrain changes, pitch, and roll can be tuned very closely to the actual ground plane allowing detection of obstacles lying low to the ground. On relatively flat surfaces, such as parking lots and roads, the system can detect positive and negative obstacles as small as 6 inches above and below the ground plane. However, to be less sensitive to large vehicle perturbations (off-road terrain), the threshold was set to 10 inches. Vehicle ground clearance and large tires handle everything below that threshold.

Because the down-looking lasers point so sharply downward, they only see an obstacle as it passes in front of the vehicle once the scan moves past an obstacle, it is no longer visible. However, the vehicle still needs to account for its location to avoid steering into it. To maintain persistence, 3-dimensional line segments are generated to represent scanned obstacles and placed on a structure similar in concept to a conveyor belt.

Since the primary consumer of down-looker obstacles is the reactive system (discussed in Section 15.7.3), down-looker obstacles are represented in a relative coordinate system with respect to the front of the vehicle. As a consequence, this representation must be modified to account for movement. At each time step, obstacles placed on the conveyor belt are translated and rotated (about the nose of the vehicle) opposite to its motion to give their new positions relative to the vehicle. When obstacles clear the rear axle, they are removed from the conveyor belt.

Unlike objects generated by the horizontal lasers, once a down-looker passes over an obstacle, that obstacle is not sensed again. If the obstacle is sensed correctly, there is no issue. However, if the obstacle is incorrectly generated from noise or a ground strike, there is no mechanism for correcting it. The robot will try to avoid it and if no path is found, may stop completely.

To prevent oversensitivity to these kinds of failures, two mechanisms clear the conveyor belt. First, it is cleared whenever the robot is stopped. This assumes that obstacles at issue are small with respect to the path, and is consistent with the purpose of the reactive system. However, this means that if the robot were to approach a large drop-off across the road, it would first stop and then probably continue. This failure mode is outside the scope of the Grand Challenge.

The second mechanism is to react only to objects that are moving towards the vehicle head on. The down looker lasers scan directly in front of the vehicle but also scan several meters to either side. If the vehicle is headed into a turn, this means that one side of the laser scan is running directly along the path instead of across it. As the vehicle then turns along that curve, it is looking at the previous obstacle from a poor perspective that may not clearly represent a true ground obstacle. To resolve this phenomenon, obstacles on the conveyor belt that must be rotated about the nose of the vehicle more than 45 degrees are removed.

The down-looking lasers proved themselves in the hay bale maze at the NQE. Just before the tunnel, two rows of hay bales about 50m long were placed on the either side of a narrowing lane. As the lane got closer to the tunnel, the lane boundaries shrank from 30m to 10m. Unless the GPS units were performing extremely well, it was difficult to travel down the center of the hay bale corridor without sensing them. As several teams discovered, it is extremely difficult to steer if a hay bale is lodged under a wheel.

Figure 15.11a shows the view from a camera mounted on top of Meteor as it enters the narrowing section of the hay bale maze just before reaching the tunnel. Figure 15.11f shows the raw output of the down-looker polar plot. The three parallel curved lines represent the uncorrected ground plane from the vertical laser and the dead band used to determine positive and negative obstacles. The dark line riding on top of the raw measurements is the lateral ground plane generated by the down-looker filter. The shape of the ground plane shows that the vehicle was rolling slightly to the left. Objects that exceeded 10 inches above this ground plane were treated as an obstacle. Depressions sensed below were treated as negative obstacles. In this plot, the hay bales on either side are evident whereas the occupancy map generated by the horizontal lasers does not show any hay bales (Figure 15.11d).

Figure 15.11b shows the down-looker obstacles that have been placed on the conveyor belt as the vehicle moves through the maze. The dotted lines show the boundaries defined by the route definition. The darker segments are obstacles from the current down-looker scan and the lighter segments are obstacles that are being retained and managed by the conveyor belt. Figure 15.11d shows the corresponding occupancy map generated by the horizontal lasers. Most hay bales do not show up in the occupancy map. Without the down-looking lasers, it is likely that the vehicle would not have stayed centered within the hay bale rows. As it turned out, the vehicle was able to navigate the hay bales without collision or incident.

15.7 Planning and Reactive Layers

Meteor employs a three-layer approach to maneuvering and navigation: waypoint following, path planning and hull protection. Waypoint following consists of mechanisms for determining speed and steering commands based on the defined route and was discussed in section 15.5. Path planning modifies the path of the vehicle based on a perceived obstacle field. Hull protection protects the hull of the vehicle whenever the path planning fails. These layers balance making progress down the road with avoiding collision and damage.

15.7.1 Occupancy Map

To effectively plan through an obstacle field, a robot must have a representation that shows obstacle locations and a safe passage between them. It must be able to accept and fuse information from sensors operating at different rates and restrictions. Moreover, since the vehicle is moving down a road, most of the sensors point forward and have a limited sensor footprint. Without a world representation, the vehicle may attempt to turn into objects that are present but no longer seen by the sensors.

Meteor uses an occupancy map a tessellation of space into equally sized grid cells. Each cell represents the likelihood that the corresponding space is either occupied or open on a continuous scale between 0 and 1. 0.0 represents open space with full confidence, 1.0 represents an obstacle with full confidence. 0.5 indicates no bias either way. Given no a priori information, cells of the map are initialized to 0.5. Figure 15.9d shows a typical occupancy map. The dark areas represent detected obstacles. The lighter areas represent open space. Uncertain space is the shade between.

Through a corresponding sensor model, each participating sensor can update cell probabilities based on how it perceives occupancy. For lasers, the sensor model is a triangular wedge that couples the defined line segment to the source of the laser that generated it. Two types of laser objects are fused into the map: line obstacles that represent solid objects and vistas that represent open space. Cells in the occupancy map that lie directly along the line segment of a line obstacle are updated to indicate a greater likelihood of being occupied. All cells that lie within the interior of the wedge are less likely to be occupied and are updated accordingly. A vista is similar to a line object except that only its interior is updated.

Occupancy maps have the ability to correct and arbitrate between inconsistent readings based on confidence. Sensors may erroneously report the presence of obstacles due to noise or an erroneous range gate. Those errors are projected onto the map to indicate impassible regions. However, new evidence from other readings is constantly fused into the map. As more correct readings appear, errors are reduced. If there is more confidence in a sensor reading, the correction occurs more quickly. Because the sensors report more false positives (obstacles that are not present) than false negatives (missing a real object), the sensor model is biased toward clearing the map more quickly than populating it.

To minimize processor load and complexity, a 2-dimensional occupancy map represents the terrain that the vehicle must pass through. Though lasers are 2-dimensional scans through a 3- dimensional space, their results can be projected onto a 2-dimensional map. However, each scans at a different angle and not all obstacles are the same height. As a result, a laser pointing slightly up might miss an object where another might see it. That puts the fusion mechanism in a dilemma since the confidence of both is high. This problem increases when one laser is updating at 75 times per second and the other at 5 times a second. The higher rate laser would dominate.

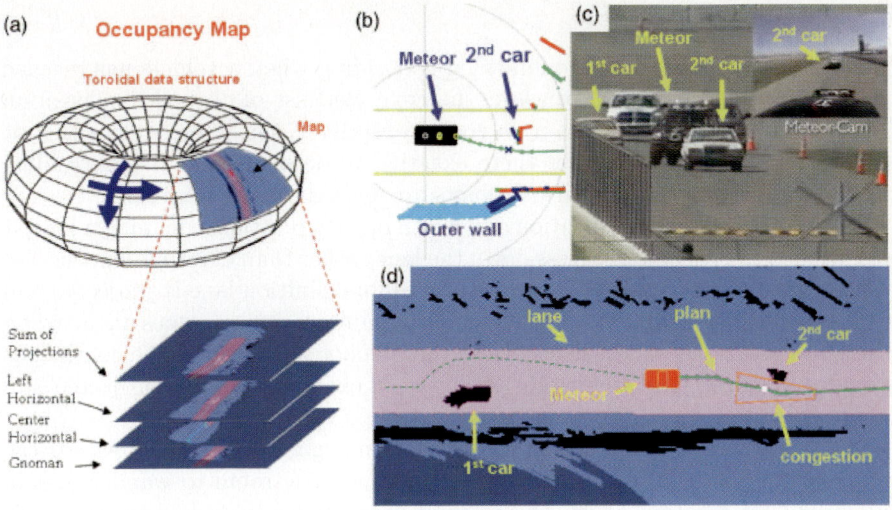

Fig. 15.12. Planning. (a) An occupancy map composed of multiple layers and summed. The structure is set up as a torus in that as Meteor moves, new information is added on one end and dropped from the other. (b) Laser information is converted to line segments and coded based on their source. (c) Meteor passes between two parked cars at the NQE. (d) The occupancy map produced while driving between the cars. The light areas are open space, the darker segments are obstacles, unknown space is the shade between. The route crosses through the center of the map. The line emanating from Meteor is its derived path around the car. The wedge is the congestion region. Note the two cars are shown in the lanes.

To solve this issue, projections are summed rather than projecting the sums. That is, the world is represented by multiple occupancy maps, one for each participating laser (Figure 15.12a bottom). Each laser builds its own map, fusing laser readings over time at its own rate. All maps are fused together 10 times per second to build a composite representation of the environment.

Another limitation of the occupancy map is the number of cells needed to represent an environment. The first year of the Grand Challenge spread over a large area and would have required a very large grid and memory footprint (Urmson et al., 2004). This storage problem was solved by building an occupancy map that acts like a moving window in space. As the vehicle moves, information behind the vehicle is shifted off the map and lost. Since the vehicle can shift in both x and y, the map can be envisioned as a large toroidal data structure (Figure 15.12a top). This representation has a fixed size and computational complexity. We chose 360 x 360 cells with each cell spanning 0.25m. That occupancy map provided a radial sensing distance of 45m. The dimensions of the map were a balance between processor load, memory footprint and the smallest size of obstacle it could represent. With these parameters, the occupancy map could be updated 10 times per second with a 30% processor load.

15.7.2 Route Projection

One of the difficulties of route following is deciding when a vehicle has reached one waypoint and should go towards the next. Because of obstacles or position error, a vehicle rarely reaches a waypoint exactly. When it recognizes this, it might circle back to try to get there exactly. Accommodations like expanding the radius of a point fail when waypoints are spaced closely together.

Since DARPA specified location and width of path positions, defining a lane, it was decided to optimize progress down the lane rather than determining whether Meteor reached a waypoint. Therefore, the route definition lane is projected and a plan is sought within it. This freedom eliminates many issues in deciding whether a waypoint was reached. Here the planner need only make continuous progress along the lane. Figure 15.12c shows an example of a lane projected into occupancy space.

A drawback to this method is that it does not guide the robot towards the center of the lane. On narrow segments, there is little room to wander, but in the event of a location offset, the method prevents the vehicle from trying to regain what it thinks is the center of a lane but is actually the edge. On wider segments, such as an open lakebed, the robot could wander several meters to either side of the route. Normally, this is not a concern because those routes tend to be wide open. To reduce wandering, projected lane widths were limited to 20 meters.

15.7.3 Planning and Verifying

Planning finds a viable path through occupancy space. The planner must select a route that not only avoids colliding with an obstacle but also respects the width of the vehicle and is resilient to changes.

Meteor employs an iterative greedy planner. The planner builds a plan in steps of 1.5 meters. At each planning step, it evaluates and selects the best next step from the end of the existing plan from choices fanned out in fixed angle increments. Each segment is verified by testing pixels adjacent to either side of the segment equivalent to the width of the vehicle along its length. If no obstacles are present, the segment is viable. For each planning iteration, the segment that makes the farthest progress down the lane is chosen. With no obstacles present, the selected segment is always the one parallel to the path. If the plan remains valid, it grows to the maximum range of the sensors: 40m. As the vehicle moves forward, the path shrinks and the planner builds it back up.

In addition to generating new segments, the planner also verifies the entire length of the existing plan 10 times per second. Though laser range finders can see beyond 30 meters, noise and geometry degrade the reliability of open and closed space calculations. As the vehicle gets closer to an object, the existing path might pass through an obstacle. Rather than invalidate the entire plan, the plan is truncated to a point just before the anomaly and is regrown from that point. The effect of this process looks like a snake constantly moving in front of the vehicle and around obstacles.

Figure 15.12d shows an example of planning through occupancy space. The wide line running down the center from left to right represents the lane defined by the route definition and restricts the planning space. The vehicle (center) is passing between the two parked cars in the later part of an NQE run (Figure 15.12c). The line extending from the vehicle toward the right is the current plan showing the individual planning segments. The two dark objects in the center of the lane (both front and back) are parked vehicles. Clearly, the planner has detected the second vehicle and has planned around it. The white dot along the path in front of the robot is the carrot. The thinner line extending behind the robot is the path the robot followed to this point and was added for illustration.

Finally, the planner provides a mechanism to feed congestion back to the speed regulator to slow the vehicle around obstacles. Testing showed that the vehicle moved up to a field of obstacles at full speed and attempted to quickly zigzag between them. Congestion represents the density of the obstacle field in the region in front of and to the sides of the current plan. It is measured by summing the number of obstacle pixels in a wedge-shaped region in front of the vehicle centered about the carrot. If the count exceeds a threshold, the robot is commanded to slow. The relationship was empirically determined through testing.

15.7.4 Reactive Approach

Maneuvering and navigation are primarily accomplished by planning a path through occupancy space. However, this has some drawbacks. First, fusion and integration of multiple sensor readings introduces latency in identification of obstacles. Several scans may be required for an object to reach the planner's obstacle threshold. In other cases, the object may lie on the inside of a turn and might not be seen until the vehicle is very close. This is especially bad if the space was reported as open or closed. Testing showed that an obstacle might not be detected until the vehicle is less than 10 meters away. In these cases, the planner would attempt to re-plan but the vehicle could not execute the new plan fast enough to avoid collision. One test site had a rapid drop immediately after one of the turns. Obstacles strategically placed just below this turn did not show up in the sensor sweeps until the vehicle had made the turn and was dropping into the depression. The vehicle often collided with these suddenly discovered obstacles. Though the elevation issue was solved by adding sensors, several cases remain where a similar phenomenon could occur.

A second complication results from heading and location error on the accurate integration of multiple sensor readings over space and time. When GPS is unreliable, a dead reckoning model is adequate to make vehicle progress but its fidelity is insufficient to maintain accuracy in the occupancy map. Laser measurements cannot be reliably correlated and so a blurred map and a less reliable plan result. This is especially true when the GPS signal does not completely drop out but degrades slowly over time.

Testing in desert terrain called for a more reactive mechanism to protect the nose of the vehicle. A purely reactive system does not rely on derived position or

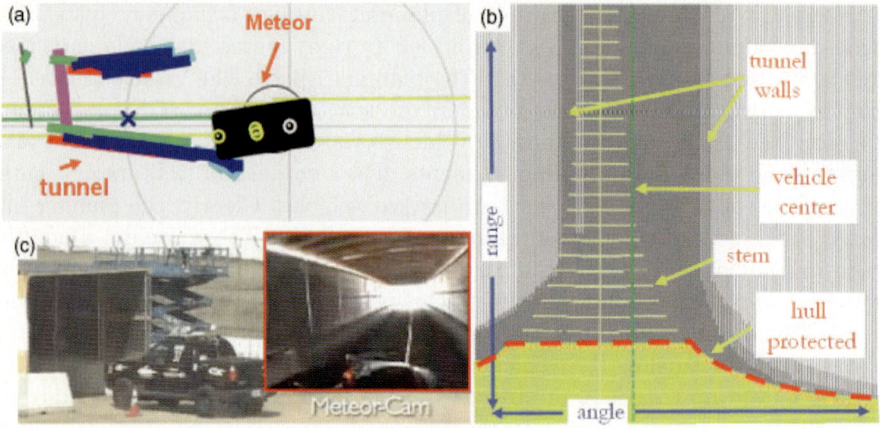

Fig. 15.13. Reactive. (a) Laser line obstacles generated as the vehicle passes through the tunnel. A mismatch in heading shows the vehicle thinking it is about to collide on the left. (b) Instantaneous lasers represented in polar coordinates with respect to vehicle centerline. The vehicle finds a steering correction angle that will guide the vehicle reactively and avoid obstacles. This representation shows the vehicle is closer to the right wall. (c) The vehicle as it enters the tunnel at the NQE. The tunnel is lined with metal sheets to block all GPS signals.

heading but instead creates its information directly from raw sensor data. Since the sensors themselves move with the vehicle, they provide a natural, vehicle centric-view of the world.

Meteor's reactive system, named Pelorus, receives inputs from the horizontal lasers and the down-lookers. The results are combined to produce a composite polar map of the open space in front of the vehicle. Pelorus is shown as an overlay of laser measurements plotted as angle vs. range in Figure 15.13b. The evenly-spaced, horizontal lines represent the projection of the vehicle in polar coordinates with respect to the front of the vehicle. This projection looks like the stem of an inverted wine glass. The shaded area represents the region Pelorus is protecting and moves up and down the stem as the speed of the vehicle changes. When the vehicle moves faster, the shaded region extends farther up the stem protecting a larger region in front of the vehicle.

If the space immediately in front of the vehicle is open, Pelorus is inactive and the planner controls. If an obstacle violates this space, Pelorus takes over and maneuvers the vehicle to avoid collision. When activated, a command is also passed to the speed controller to slow the vehicle. This shrinks the area that Pelorus protects and gives the vehicle more space to maneuver. If no viable openings can be found, the reactive system stops the vehicle.

The purpose of the reactive system is to prevent damage to the hull from obstacles that were not detected until too close to the vehicle. However, this reactive mode facilitates passing through a tunnel where GPS signals are lost. Figure 15.13 shows results from the reactive system as it passes through a tunnel at the NQE during the third run. Figure 15.13a shows the vehicle about to collide with the tunnel wall on the left. Actually, the vehicle is traveling down the center of the tunnel (Figure 15.13c). The error is the result of a poor heading estimate as GPS degraded during entry into the tunnel. Figure 15.13b shows the instantaneous laser readings being reported from the reactive system in polar coordinates with respect to the center of the vehicle. These sensor readings do not rely on reported position and heading and show the tunnel is actually closer on the right and that the vehicle is well aligned. If the vehicle followed the plan, it would collide as it steered it farther to the right. Instead, the reactive system engaged to protect the vehicle and guided it the rest of the way through the tunnel. When the vehicle emerged, the GPS signal was recovered and the vehicle relied on planning again.

Competent navigation is a balance between directed motion based on a planning and dynamic steering correction based on a reactive protection system. Based on the terrain from last year's race, Meteor placed more weight on protecting the vehicle.

15.8 Developmental Approach and Field Testing

MITRE decided to enter the Grand Challenge in October of 2004. That left eleven months to create a vehicle that could operate autonomously in desert terrain. To achieve this goal required a solid plan, rigorous development and testing, and considerable discipline.

The team laid out a series of goals in one to two month intervals. Each of the goals involved a series of tests culminating in a demonstration. Engineers from outside the team were routinely invited to observe these demonstrations and assist with ideas and suggestions. Most testing was conducted in an abandoned parking lot (Figure 15.14a).

- November. Have a vehicle that could be driven under computer control. This was achieved by purchasing goods and services from Electronic Mobility Corporation.
- December. Drive autonomously from waypoint to waypoint. That was achieved with initial versions of a simulator, vehicle throttle and brake controllers, data loggers, data replayers, map display tools, and other software architectural components.
- January. Follow multiple waypoints with horizontal obstacle avoidance. By starting with a simple configuration [one GPS, one laser measurement sensor (LMS), and a laptop], end-to-end capability was demonstrated.
- March. Drive autonomously for five miles, using sensor data from multiple sources for obstacle avoidance while operating on varied terrain. Early speed and distance tests were performed at a farm in northern Virginia (Figure 15.14b). These long-distance tests forced resolution of sensor fusion, maneuvering near obstacles, advanced vehicle performance, and speed control.
- May. Prepare for a DARPA site visit. Testing had moved from obstacle avoidance to finding a balance between speed, planning, and reaction time. At this

point, sensing strategies were unresolved with stereo vision, radar, and machine vision for road detection under consideration. The site visit required completing a 200m course while avoiding placed obstacles (trashcans). The primary focus during this period was to minimize the number of trashcans we crushed. The site visit was successful (missing five of six trashcans).

- July. Take the complete system to the Nevada desert and test on the 2004 Grand Challenge route (Figure 15.14c). This was critical preparation for the race. Finding adequate, available test environments for regression testing was difficult. Emulating a desert in urban Virginia was a challenge. Out of ten months of testing, ten days were outside a parking lot. These ten days were extremely valuable and played a large part in getting to the finals.

  The desert provided a larger and longer testing space, permitting speed tuning in the flats of a wide-open lakebed and navigating trails bounded by large desert shrubs. It also provided an opportunity to test navigation through and around obstacles like cattle guards, switchbacks, and tunnels (Figure 15.14c). It demonstrated the hazards of the desert terrain including how radically and quickly the terrain could drop off and the size of the boulders (Figure 15.14d).

  July was also a self-imposed deadline for integrating new systems. While several months remained until the race, the systems needed to be rigorously tested and tuned to be successful. The sensor suite was finalized based on existing performance. Systems that were not sufficiently trusted were dropped.

  July saw evaluation of the effect of environment on sensors and processors. Heat was a concern but testing in 130-degree temperatures showed that the vehicle's air conditioning system was adequate to maintain cabin temperatures. Sensor fouling was also a concern. Desert dust is dry and fine-grained. It builds a small charge that causes it to stick to almost everything. Repeated experiments showed that the sensors worked without degradation, leading to the conclusion that this issue could be ignored.

The team traveled to the Yuma Proving Grounds in Arizona as a graduation exercise to test the entire system on an off-road test course (Figure 15.14e).

15.8.1 The National Qualifying Event

The National Qualifying Event (NQE) was the final DARPA evaluation before the big race. It was designed to test vehicle capabilities over a 10-day period at the California Speedway and to determine which vehicles should participate in the Challenge race. The course contained two large gates, several hills, hay bales, a tunnel, a debris field, a mock switchback, several parked cars, rumble strips, and a tank trap. Runs were evaluated based on the number of obstacles avoided and the time to complete the course. Each team had up to six attempts to complete the course — three complete runs were necessary to pass.

Meteor was one of the first teams to compete on day 1. It left the starting gate and headed for the first obstacle, two large, shiny gates positioned on either side of the route. Meteor made the turn, started to point between the gates, and

Fig. 15.14. Testing. a) Most vehicle testing occurred in an empty parking lot in McLean, Virginia. Here Meteor passes through a mock tunnel constructed from wood and weather tarps. b) Testing at local sites including a farm and an off-road facility. Note deep ruts that proved hazardous to low lying sensors. c) Six days of testing in 129 degree heat of Primm Nevada in July. Here Meteor successfully exits a tunnel and passes through cattle guards. d) Testing along the 2004 route included driving in rough mountain terrain. Note the size of the boulders left of Meteor. e) Testing on an off-road autonomous vehicle test course at the Yuma Proving Grounds a week before the race.

then veered away (Figure 15.15a) when DARPA stopped it. Figure 15.15b shows the Meteor view. Even though the gates appeared to lie slightly inside the route (probably due to GPS error), the vehicle should have been able to pass through them. After reviewing data logs, it was apparent that Pelorus did not think there was enough room for the vehicle to fit through the gate. The safety margin (1m) on either side of the vehicle was discovered to be too large. After folding the rear-view mirrors to reduce vehicle width, the safety margin was reduced to a few inches.

Run 2 did not go well. Meteor left the starting gate, traveled about 40 meters, and veered to the right. It continued its slow loop backwards and as it approached the outer barriers, DARPA officials stopped it. The system generated empty log files, which prevented analysis of the anomaly. Despite the lack of log data, the GPS multipath issue was discovered (see section 15.5.1) between runs 2 and 3.

Meteor entered run 3 with the same parameter set as Run 2. During this run, it passed through the first gates, traversed a grassy area including a slight incline and decline down to an asphalt road and into the hay maze. Figure 15.15e shows it navigating through the hay bales leading to the tunnel. Meteor entered the tunnel where it veered left and stopped after nudging the left tunnel wall. Log data showed that upon entering the tunnel the transition from GPS to motion model was not quick enough. Just before the motion model engaged, the GPS error produced a large heading discrepancy and Meteor veered left too quickly for Pelorus to react. Figure 15.15f shows the final resting place of Meteor. Though the angle is shallow, Pelorus did not think there was enough room to maneuver out of it. Replay of the data exposed the failure mode. Over the next few days, handoff between the GPS and motion model was tested. The team built a mock-up of a large tunnel. Figure 15.16g shows Meteor passing into the mock tunnel. GPS loss was induced by placing a ball cap covered with tin foil over the GPS unit, effectively killing its signal (the other GPS was turned off).

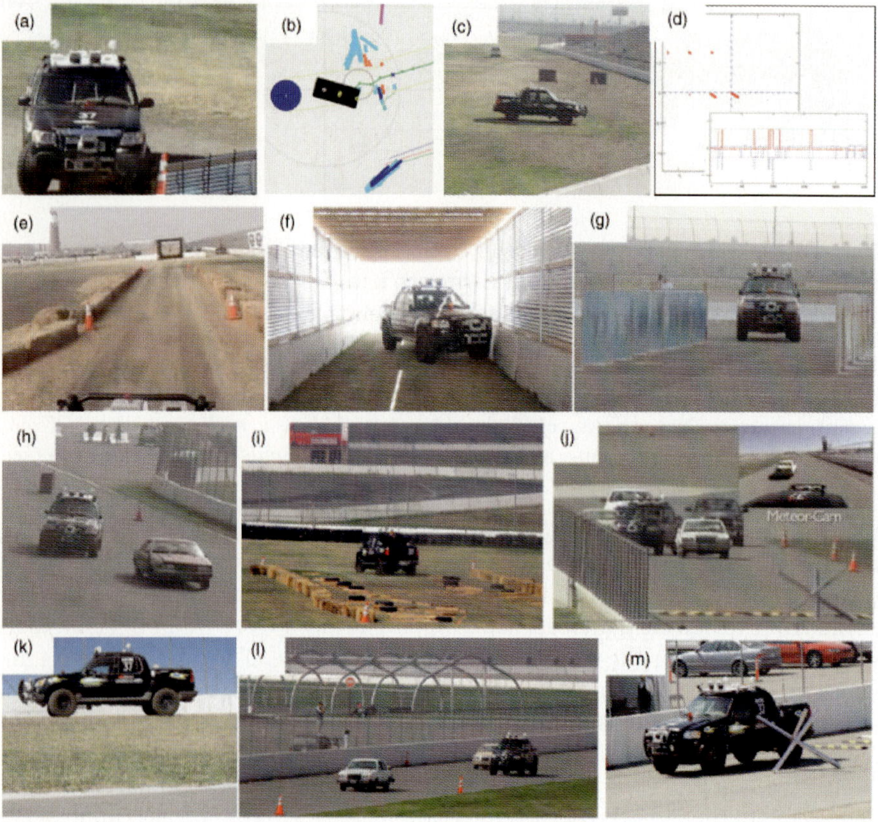

Fig. 15.15. NQE. Each team had six chances to complete the course. a,b) Meteor's first run ends as it fails to pass through the first gates. c,d) Meteor veers off course before reaching the gates. e,f,g) During the third run, Meteor clears the gate, rounds a steep hill and traverses a hay bale maze, but stops in the tunnel. Testing corrected this discrepancy. h-m) Meteor completes runs 4, 5 and 6 without incident, including avoiding several parked cars, a debris field, a mock mountain pass and a tank trap. Completing three runs qualified Meteor for the finals.

Testing showed the problem was the hand-off from planning and to reactive modes. Parameters were tweaked that make the GPS receiver threshold more sensitive causing transition to occur sooner and more often. The new parameters required a higher GPS confidence to stay in planning mode.

Once these issues were addressed, Meteor completed the next three runs with slow but repeatable times of 27 minutes each. Figures 15.15h-15.15m shows Meteor successfully avoiding obstacles at different parts of the 2.7-mile course. Figure 15.15h and 15.15j show Meteor driving around obstacles. Figure 15.15i shows Meteor driving over several small tires and 4x4s with no trouble. Figure 15.15k shows Meteor posed near the mock mountain pass. Barriers on the right and a large drop off on the left simulated terrain found on a switch back. As with the hay bales, the down-lookers kept Meteor on the path driving between positive and negative obstacles. Figure 15.16m shows the last obstacle, rumble strips a few feet before a tank trap to see how vehicles sensed in rough terrain. In all runs, Meteor avoided the tank trap and crossed the finish line.

15.8.2 The Finals

Of the 152 teams that participated in the site visit and the 43 teams that competed in the NQE, only 23 qualified to race in the finals. The Meteor team was extremely pleased to compete in such a challenging event.

Because Meteor was slowest to complete the NQE, it was last to start the finals. As the day wore on, the winds grew stronger. By the time Meteor started, there was a steady breeze with periodic gusts. Meteor left the gate, traveled by the crowd and headed out to the desert. About thirty meters before entering the dry lakebed, it encountered a large dust cloud and stopped in a field of tall weeds and bushes.

During testing in the desert before the race, occasional dust clouds appeared as winds raised fine layers of desert dust. Laser scanners detected the translucent cloud and treated it as a transient obstacle. After the dust blew beyond the range of the lasers, the offending obstacle cleared and Meteor continued. Figure 15.16b shows a dust clouds as it passed Meteor from back to front in the race. Clouds were also generated by Meteor and the chase vehicle. Figure 15.16a is an overhead reconstruction of Meteor reacting to one of these clouds. The wide line running

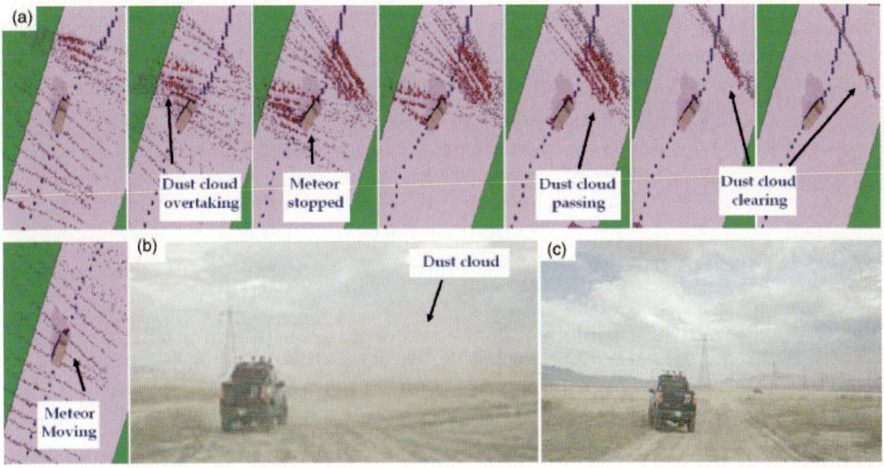

Fig. 15.16. The race. (a) Meteor approaches, slows, and attempts to avoid passing a dust cloud that periodically plagued it during its brief run. This sequence shows one of the many dust clouds as seen by the laser scanners. The thin lines are laser scans that hit the ground. The thicker lines are cause by dust. (b) A dust cloud passing in front of and away from the Meteor.

Fig. 15.17. The 2005 MITRE Meteor team. (Left to right) Frank Carr, Bob Bolling, Bob Grabowski, Richard Weatherly, Dave Smith, Ann Jones, Tiffani Horne, Mark Heslep, Keven Ring, Kevin Forbes, Mike Shadid, Laurel Riek, Alan Christiansen, and Sarah O'Donnell (inset).

through each segment is the lane defined by the route definition. Meteor is in the middle. The thin lines are laser scans that terminate at the range gate (hit the ground). The darker points are readings from perceived obstacles (Figure 15.16a middle).

The pictures start on the left where Meteor is moving on the course (as seen by the separation between laser scans). As the cloud reaches Meteor, it triggers the reactive system that forces it to veer and then stop. In most cases, as the dust clouds passed, Meteor would slow or stop and then continue along its path when the clouds cleared. This occurred 10 to 20 times during the run. The road started to turn left, so these clouds appeared to be coming from the right. The new angle pushed Meteor toward the outside of the road. Figure 15.16c shows one of these diversions. This occurred several times and ultimately pushed Meteor far enough off course that larger weeds and shrubs made the route look impassable. Since the lasers could not differentiate between weeds and large rocks, Meteor stopped just 1.1 miles from the starting gates and less than thirty meters from the wide-open expanses of the dry lakebed.

15.9 Conclusion

Although the MITRE Meteor team did not win the Grand Challenge, it did achieve its goals. MITRE's Grand Challenge experience validated three philosophies: Reliance on COTS, software location and employment transparency, and the model-build-test cycle. Purchasing COTS equipment and services focused project energy on autonomous robot control. Good examples are the installation of a COTS steering and propulsion servo control system and the vehicle suspension modification. Many teams did poorly at NQE when their homemade solutions failed.

A disciplined approach to software construction let good ideas accumulate without the system becoming fragile. One way to increase confidence in code is to employ it for as many purposes as appropriate. For the Grand Challenge, the same body of code was used to operate the robot, run non-real time laboratory simulations, and study logs recorded in the field. This transparency of employment exposed problems in the lab before they could waste field-testing time.

The model-build-test cycle accumulated experience in a tangible and persistent way. When a problem was found or a new phenomenon identified, it was first modeled in the simulation environment. With a simulation of the problem or new phenomenon in hand, the body of operational code was adjusted to deal with it. Once proven in simulation, the robot was fieldtested to evaluate the changes and improvements were fed back to the model. A result of the model-build-test approach was that the model grew in fidelity and became a lasting repository of project experience. More importantly, this experience proved that testing in real environments is the key to success.

The experience gained by taking a vehicle to the Grand Challenge, though basic in nature, is a valuable asset to many Department of Defense and civilian missions. These lessons continue to equip MITRE to help our clients succeed in their robotic endeavors.

References

[1] Behringer, R., Sundareswaran, S., Gregory, B., Elsley, R., Addison, B., Guthmiller, W. (2004, May). The DARPA Grand Challenge - Development of an Autonomous Vehicle. Paper presented at the IEEE Intelligent Vehicles Symposium, Parma Italy.

[2] Cheok, K.C., Smid, G.E., Kobayashi, K., Overholt, J.L., Lescoe, P. (1997, June). A Fuzzy Logic Intelligent Control System Architecture for an Autonomous Leader-Following Vehicle. Paper presented at the American Control Conference, Albuquerque, New Mexico.

[3] Crane, C. III, Armstrong, D. Jr., Torrie, M., Gray, S. (2005). Autonomous Ground Vehicle Technologies Applied to the DARPA Grand Challenge. Paper presented at the International Conference on Control, Automation and Systems, Bangkok, Thailand.

[4] Fulton, J., Pransky, J. (2004) DARPA Grand Challenge A Pioneering Event for Autonomous Robotic Ground Vehicles. *Industrial Robot: An International Journal* Vol. 31, No. 5, 414–422.

[5] Gillula, J. (2005). Data Fusion from Multiple Sensors: Real Time Mapping on an Unmanned Ground Vehicle. Summer Undergraduate Research Fellowship Final SURF report.

[6] Kuhl, F., Weatherly, R., Dahmann J. (2000). *Creating Computer Simulation Systems: An Introduction to the High Level Architecture.* Prentice Hall PTR.

[7] Manduchi, R., Castano, A., Talukder, A., Matthies, L (2003, September). Obstacle Detection and Terrain Classification for Autonomous Off-road Navigation. *Autonomous Robot*, Volume 18.

[8] Minsky M. (1985). *The Society of Mind.* Simon and Schuster.

[9] Ozguner, U., Redmill K., Broggi A. (2004, June). Team TerraMax and the DARPA Grand Challenge: A General Overview. Paper presented at the IEEE Intelligent Vehicles Symposium, Parma Italy.

[10] Saptharishi, M., Bhat, K.S., Diehl, C.P., Dolan, J.M., Khosla, P.K. (2000, September). CyberScout: Distributed Agents for Autonomous Reconnaissance and Surveillance. Paper presented at the Conference on Mechatronics and Machine Vision in Practice.

[11] Sato, N., Matsuno, F., Shiroma, N., Yamaski, T., Kamegawa, T., Igarashi, H (2004, September). Cooperative Task Execution by a Multiple Robot Team and Its Operators in Search and Rescue Operations. Paper presented at the IEEE/RSJ International Conference on Intelligent Robots and Systems, Sendai, Japan.

[12] Sotelo, M.A., Rodriguez, F.J., Magdalen L. (2004). VIRTUOUS: Vision-Based Road Transportation for Unmanned Operation on Urban-Like Scenarios. *IEEE Transactions on Intelligent Transportation Systems*, Vol. 5, No. 2.

[13] Urmson, C., Anhalt, J., Clark, M., Galatali, T., Gonzalez, J., Gowdy, J., Gutierrez, A., Harbaugh, S., Johnson, M., Roberson, Kato, Y., Koon, P., Peterson, K., Smith, B., Spiker, S., Tryzelaar, E., Whittaker, W. (2004). High Speed Navigation of Unrehearsed Terrain: Red Team Technology for Grand Challenge 2004. (Tech. Rep. CMU-RI-TR-04-37). Pittsburg PA: Carnegie Mellon University, Robotics Institute.

[14] Woolridge, M., Jennings, N., Kinny, D. (2000). .The Gaia Methodology for Agent-Oriented Analysis and Design. *Autonomous Agents and Multi-Agent Systems*, Volume 3, Issue 3, 285–312.

Author Index

Acknowledgements of Reviewers

The editors would like to thank the individuals who served as reviewers for the 2005 DARPA Grand Challenge Special Issues of the *Journal of Field Robotics*.

Victor Aitken	Carleton University
Robert Ambrose	NASA Johnson Space Center
Jamie Andersen	Draper Laboratory
Paraq Batavia	Applied Perception, Inc.
Reinhold Behringer	Leeds Metropolitan University
Faiz Ben Amar	University of Paris
David Bevly	Auburn University
Alberto Broggi	University of Parma
Gary Carr	ENSCO Inc.
Raja Chatila	Laboratory for Analysis and Architecture of Systems—National Center for Scientific Research (LAAS-CNRS)
Henrik Christensen	Georgia Institute of Technology
Carl Crane	University of Florida
Bruce Digney	Defense Research and Development Canada (DRDC)
Chris Dima	Carnegie Mellon University
Steven Dubowsky	Massachusetts Institute of Technology
Hugh Durrant-Whyte	University of Sydney
Scott Fish	Science Applications International Corporation (SAIC)
Alex Foessel	NavCom Technology, Inc.
Andrew Howard	NASA Jet Propulsion Laboratory
Ayanna Howard	Georgia Institute of Technology
Terry Huntsberger	Jet Propulsion Laboratory
Randy Jones	U.S. Army Engineer Research and Development Center
Steven Kalik	Toyota Technical Center
William Klarquist	PercepTek Inc.
Simon LaCroix	Laboratory for Analysis and Architecture of Systems—National Center for Scientific Research (LAAS-CNRS)
Arun Lakhotia	University of Louisiana at Lafayette
Amy Larson	University of Minnesota
Daniel Lee	University of Pennsylvania
Brett Leedy	Virginia Polytechnic Institute and State University

Anthony Levandowski	University of California, Berkeley
Matt Lichter	Massachusetts Institute of Technology
Raj Madhavan	National Institute of Standards and Technology
Arnis Mangolds	Draper Laboratory
Larry Matthies	NASA Jet Propulsion Laboratory
Isaac Miller	Cornell University
Kevin Moore	Colorado School of Mines
Stewart Moorehead	NavCom Technology, Inc.
Paul Muench	U.S. Army Tank Automotive Research, Development and Engineering Center
Richard Murray	California Institute of Technology
Eduardo Nebot	University of Sydney
Lauro Ojeda	University of Michigan
Mark Ollis	Applied Perception, Inc.
Jim Overholt	U.S. Army Tank Automotive Research, Development and Engineering Center
Ümit Özgüner	The Ohio State University
Alice Parker	University of Southern California
Liam Pedersen	NASA Ames / Carnegie Mellon University
Kevin Peterson	Carnegie Mellon University
Benoit Ricard	Defense Research and Development Canada (DRDC)
Jonathan Roberts	Commonwealth Scientific and Industrial Research Organization (CSIRO)
Aaron Saunders	Boston Dynamics
Steve Scheding	University of Sydney
Zvi Shiller	College of Judea and Samaria
Matthew Spenko	Stanford University
Gaurav Sukhatme	University of Southern California
Camillo Taylor	University of Pennsylvania
Paul Trepagnier	Gray & Company, Inc.
Chris Urmson	Carnegie Mellon University
Nicolas Vandapel	Carnegie Mellon University
Steven Velinsky	The University of California, Davis
David Wettergreen	Carnegie Mellon University
Brian Yamauchi	iRobot Corporation
Cang Ye	University of Arkansas Little Rock

Printing: Mercedes-Druck, Berlin
Binding: Stein+Lehmann, Berlin

Springer Tracts in Advanced Robotics

Edited by B. Siciliano, O. Khatib and F. Groen

Vol. 36: Buehler, M.; Iagnemma, K.; Singh, S. (Eds.)
The 2005 DARPA Grand Challenge – The Great Robot Race
520 p. 2007 [978-3-540-73428-4]

Vol. 35: Laugier, C.; Chatila, R. (Eds.)
Autonomous Navigation in Dynamic Environments
169 p. 2007 [978-3-540-73421-5]

Vol. 34: Wisse, M.; van der Linde, R.Q.
Delft Pneumatic Bipeds
136 p. 2007 [978-3-540-72807-8]

Vol. 33: Kong, X.; Gosselin, C.
Type Synthesis of Parallel Mechanisms
272 p. 2007 [978-3-540-71989-2]

Vol. 32: Milutinović, D.; Lima, P.
Cells and Robots – Modeling and Control of Large-Size Agent Populations
130 p. 2007 [978-3-540-71981-6]

Vol. 31: Ferre, M.; Buss, M.; Aracil, R.; Melchiorri, C.; Balaguer C. (Eds.)
Advances in Telerobotics
500 p. 2007 [978-3-540-71363-0]

Vol. 30: Brugali, D. (Ed.)
Software Engineering for Experimental Robotics
490 p. 2007 [978-3-540-68949-2]

Vol. 29: Secchi, C.; Stramigioli, S.; Fantuzzi, C.
Control of Interactive Robotic Interfaces – A Port-Hamiltonian Approach
225 p. 2007 [978-3-540-49712-7]

Vol. 28: Thrun, S.; Brooks, R.; Durrant-Whyte, H. (Eds.)
Robotics Research – Results of the 12th International Symposium ISRR
602 p. 2007 [978-3-540-48110-2]

Vol. 27: Montemerlo, M.; Thrun, S.
FastSLAM – A Scalable Method for the Simultaneous Localization and Mapping Problem in Robotics
120 p. 2007 [978-3-540-46399-3]

Vol. 26: Taylor, G.; Kleeman, L.
Visual Perception and Robotic Manipulation – 3D Object Recognition, Tracking and Hand-Eye Coordination
218 p. 2007 [978-3-540-33454-5]

Vol. 25: Corke, P.; Sukkarieh, S. (Eds.)
Field and Service Robotics – Results of the 5th International Conference
580 p. 2006 [978-3-540-33452-1]

Vol. 24: Yuta, S.; Asama, H.; Thrun, S.; Prassler, E.; Tsubouchi, T. (Eds.)
Field and Service Robotics – Recent Advances in Research and Applications
550 p. 2006 [978-3-540-32801-8]

Vol. 23: Andrade-Cetto, J,; Sanfeliu, A.
Environment Learning for Indoor Mobile Robots – A Stochastic State Estimation Approach to Simultaneous Localization and Map Building
130 p. 2006 [978-3-540-32795-0]

Vol. 22: Christensen, H.I. (Ed.)
European Robotics Symposium 2006
209 p. 2006 [978-3-540-32688-5]

Vol. 21: Ang Jr., H.; Khatib, O. (Eds.)
Experimental Robotics IX – The 9th International Symposium on Experimental Robotics
618 p. 2006 [978-3-540-28816-9]

Vol. 20: Xu, Y.; Ou, Y.
Control of Single Wheel Robots
188 p. 2005 [978-3-540-28184-9]

Vol. 19: Lefebvre, T.; Bruyninckx, H.; De Schutter, J. Nonlinear Kalman Filtering for Force-Controlled Robot Tasks
280 p. 2005 [978-3-540-28023-1]

Vol. 18: Barbagli, F.; Prattichizzo, D.; Salisbury, K. (Eds.)
Multi-point Interaction with Real and Virtual Objects
281 p. 2005 [978-3-540-26036-3]

Vol. 17: Erdmann, M.; Hsu, D.; Overmars, M.; van der Stappen, F.A (Eds.)
Algorithmic Foundations of Robotics VI
472 p. 2005 [978-3-540-25728-8]

Vol. 16: Cuesta, F.; Ollero, A.
Intelligent Mobile Robot Navigation
224 p. 2005 [978-3-540-23956-7]

Vol. 15: Dario, P.; Chatila R. (Eds.)
Robotics Research – The Eleventh International Symposium
595 p. 2005 [978-3-540-23214-8]

Vol. 14: Prassler, E.; Lawitzky, G.; Stopp, A.; Grunwald, G.; Hägele, M.; Dillmann, R.; Iossifidis. I. (Eds.)
Advances in Human-Robot Interaction
414 p. 2005 [978-3-540-23211-7]

Vol. 13: Chung, W.
Nonholonomic Manipulators
115 p. 2004 [978-3-540-22108-1]

Vol. 12: Iagnemma K.; Dubowsky, S.
Mobile Robots in Rough Terrain – Estimation, Motion Planning, and Control with Application to Planetary Rovers
123 p. 2004 [978-3-540-21968-2]

Vol. 11: Kim, J.-H.; Kim, D.-H.; Kim, Y.-J.; Seow, K.-T.
Soccer Robotics
353 p. 2004 [978-3-540-21859-3]

Vol. 10: Siciliano, B.; De Luca, A.; Melchiorri, C.; Casalino, G. (Eds.)
Advances in Control of Articulated and Mobile Robots
259 p. 2004 [978-3-540-20783-2]

Vol. 9: Yamane, K.
Simulating and Generating Motions of Human Figures
176 p. 2004 [978-3-540-20317-9]

Vol. 8: Baeten, J.; De Schutter, J.
Integrated Visual Servoing and Force Control – The Task Frame Approach
198 p. 2004 [978-3-540-40475-0]

Vol. 7: Boissonnat, J.-D.; Burdick, J.; Goldberg, K.; Hutchinson, S. (Eds.)
Algorithmic Foundations of Robotics V
577 p. 2004 [978-3-540-40476-7]

Vol. 6: Jarvis, R.A.; Zelinsky, A. (Eds.)
Robotics Research – The Tenth International Symposium
580 p. 2003 [978-3-540-00550-6]

Vol. 5: Siciliano, B.; Dario, P. (Eds.)
Experimental Robotics VIII – Proceedings of the 8th International Symposium ISER02
685 p. 2003 [978-3-540-00305-2]

Vol. 4: Bicchi, A.; Christensen, H.I.; Prattichizzo, D. (Eds.)
Control Problems in Robotics
296 p. 2003 [978-3-540-00251-2]

Vol. 3: Natale, C.
Interaction Control of Robot Manipulators – Six-degrees-of-freedom Tasks
120 p. 2003 [978-3-540-00159-1]

Vol. 2: Antonelli, G.
Underwater Robots – Motion and Force Control of Vehicle-Manipulator Systems
268 p. 2006 [978-3-540-31752-4]

Vol. 1: Caccavale, F.; Villani, L. (Eds.)
Fault Diagnosis and Fault Tolerance for Mechatronic Systems – Recent Advances
191 p. 2003 [978-3-540-44159-5]

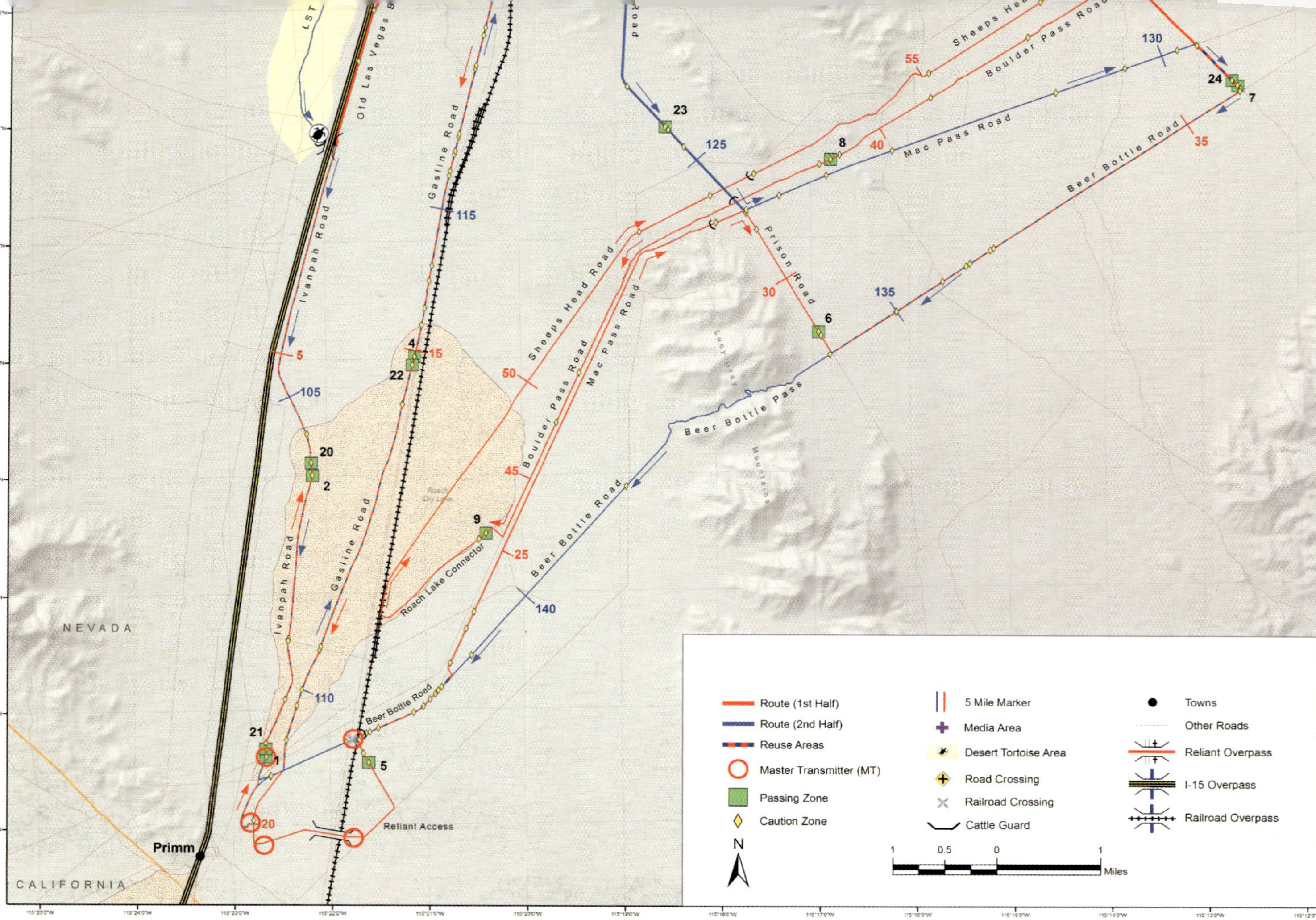

NEVADA
CALIFORNIA
Primm
Ivanpah Road
Old Las Vegas B
Gasline Road
Sheeps Head Road
Boulder Pass Road
Mac Pass Road
Beer Bottle Road
Beer Bottle Pass
Roach Lake Connector
Prison Road
Reliant Access
Lucy Gray
Mountains
Roach Dry Lake
LST
Route (1st Half)
Route (2nd Half)
Reuse Areas
Master Transmitter (MT)
Passing Zone
Caution Zone
5 Mile Marker
Media Area
Desert Tortoise Area
Road Crossing
Railroad Crossing
Cattle Guard
Towns
Other Roads
Reliant Overpass
I-15 Overpass
Railroad Overpass
N
Miles